Molecular Aspects of Oxidative Drug Metabolizing Enzymes:

Their Significance in Environmental Toxicology, Chemical Carcinogenesis and Health

NATO ASI Series

Advanced Science Institutes Series

A series presenting the results of activities sponsored by the NATO Science Committee, which aims at the dissemination of advanced scientific and technological knowledge, with a view to strengthening links between scientific communities.

The Series is published by an international board of publishers in conjunction with the NATO Scientific Affairs Division

A Life Sciences	Plenum Publishing Corporation
B Physics	London and New York
C Mathematical and Physical Sciences	Kluwer Academic Publishers Dordrecht, Boston and London
D Behavioural and Social Sciences	
E Applied Sciences	
F Computer and Systems Sciences	Springer-Verlag Berlin Heidelberg New York
G Ecological Sciences	London Paris Tokyo Hong Kong
H Cell Biology	Barcelona Budapest
I Global Environmental Change	

NATO-PCO DATABASE

The electronic index to the NATO ASI Series provides full bibliographical references (with keywords and/or abstracts) to more than 30000 contributions from international scientists published in all sections of the NATO ASI Series. Access to the NATO-PCO DATABASE compiled by the NATO Publication Coordination Office is possible in two ways:

- via online FILE 128 (NATO-PCO DATABASE) hosted by ESRIN, Via Galileo Galilei, I-00044 Frascati, Italy.

- via CD-ROM "NATO Science & Technology Disk" with user-friendly retrieval software in English, French and German (© WTV GmbH and DATAWARE Technologies Inc. 1992).

The CD-ROM can be ordered through any member of the Board of Publishers or through NATO-PCO, Overijse, Belgium.

Series H: Cell Biology, Vol. 90

Molecular Aspects of Oxidative Drug Metabolizing Enzymes:

Their Significance in Environmental Toxicology, Chemical Carcinogenesis and Health

Edited by

Emel Arınç

Middle East Technical University, Department of Biology
Inönü Bulvarı, Ankara 06531, Turkey

John B. Schenkman

University of Connecticut Health Center, Department of Toxicology
Farmington, Connecticut 06032, USA

Ernest Hodgson

North Carolina State University, Department of Toxicology
Raleigh, North Carolina 27695, USA

Springer-Verlag
Berlin Heidelberg New York London Paris Tokyo
Hong Kong Barcelona Budapest
Published in cooperation with NATO Scientific Affairs Division

Proceedings of the NATO Advanced Study Institute on molecular aspects of drug metabolizing enzymes, held in Kuşadasi, Aydın-Turkey, June 20–July 2, 1993

QP
671
.C83
M65
1995

ISBN 3-540-58856-6 Springer-Verlag Berlin Heidelberg New York

CIP-data applied for

© Springer-Verlag Berlin Heidelberg 1995
Printed in Germany

Typesetting: Camera ready by authors
SPIN 10120307   31/3130 - 5 4 3 2 1 0 - Printed on acid-free paper

Preface

The second NATO Advanced Study Institute on molecular aspects of drug metabolizing enzymes, entitled Molecular Aspects of Oxidative Drug Metabolizing Enzymes: Their Significance in Environmental Toxicology, Chemical Carcinogenesis and Health was held in Kuşadası, Aydın-Turkey, June 20-July 2, 1993. Like the first NATO Institute, Molecular Aspects of Monooxygenases and Bioactivation of Toxic Compounds, held in Çeşme, Turkey, August 27-September 7, 1989, the present workshop provided background and current concepts necessary for the understanding of the microsomal xenobiotic-metabolizing enzymes responsible for the detoxification and activation of endogenous compounds that find their way into the body. While the first workshop focused on endogenous substrates of the drug metabolizing enzymes and external toxicants such as benzene and estrogenic pesticides, the second workshop extended this theme to xenobiotic-metabolizing enzymes in aquatic species, as well as to coverage of genotoxicity and chemical carcinogenicity.

Progress in the study of xenobiotic metabolizing enzymes in the four years since the last NATO Institute on the topic has been rapid. As a result the lecturers spent time updating the participants on the latest developments and concepts. The rapid expansion of molecular biology into the fields of study of the xenobiotic metabolizing enzymes has given insight into the regulation of these enzymes. The development of heterologous expression systems has provided a basis whereby functions of the enzyme systems can be probed and its impact studied in the living cell. The growing recognition of the serious consequences of environmental pollution has spawned a number of studies on the impact of environmental contaminants on aquatic species, as well as the role of xenobiotic enzymes in chemical carcinogenesis. This book comprises the proceedings of the meeting. The chapters express the material covered by the lecturers, and contain sufficient background information for the beginning student. At the same time, the material presented is up-to-date, and we believe it will provide the latest information and citation to be of benefit to person working in these areas.

E. Arınç

J. B. Schenkman

E. Hodgson

Contents

Introduction to Cytochrome P450

John B. Schenkman and Ingela Jansson
Department of Pharmacology
University of Connecticut
Health Center
Farmington,CT06030 USA

Introduction

Cytochrome P450 is the name first given provisionally to describe a carbon monoxide-binding hemoprotein (Omura and Sato 1962), and for what has become recognized as a superfamily of monooxygenases found in just about every phylum in which the hemoprotein has been sought (Nelson et al. 1993). To date, more than 220 different cytochrome P450 genes have been identified, and, based upon determined or deduced primary structures of the proteins, fit into families of related and subfamilies of closely related enzymes, many with quite esoteric actions (see Schenkman and Greim 1993). Thirty six gene families have so far been described, 12 of which exist in mammals and comprise 22 subfamilies (Nelson et al. 1993). Most of the proteins when isolated were given common names. However, as investigators often isolated and characterized forms that turned out to be the same as forms isolated and named by other groups, common name nomenclature became extremely complex, as some forms of cytochrome P450 have as many as six common names. To obviate confusion, today most people use the genetic nomenclature. This nomenclature designates the gene family as an arabic number,e.g.,2, then a letter to designate the subfamily, if more than one exists, e.g., B, then the individual gene number, e.g.,4. Thus, if rabbit liver isozyme 2 (other common names include,LM2, B0, and PB-1) was being designated, it would be 2B4, with the prefix CYP designating CYtochrome P450, e.g., CYP2B4. Also used is the designation P450 2B4.

The provisional name cytochrome P450 was designated to indicate the hemoprotein was the microsomal pigment reported to

NATO ASI Series, Vol. H 90
Molecular Aspects of Oxidative Drug Metabolizing Enzymes
Edited by E. Arınç, J. B. Schenkman and E. Hodgson
© Springer-Verlag Berlin Heidelberg 1995

absorb at 450nm when reduced and in complex with carbon monoxide (Omura and Sato 1962). This absorption peak was unusual, because hemoproteins usually absorb at around 400nm-425nm when reduced or in complex with carbon monoxide. Cytochrome P450, even in the reduced state, has an anomalous absorption spectrum; it lacks the usual α- and β- bands. Further, in reduced minus oxidized difference spectrum a broad anomalous absorption peak was seen at around 440nm instead of a more usual Soret peak around 420nm (Omura and Sato 1964). These reports from Sato's laboratory were rapidly followed by a paper showing cytochrome P450 to be the terminal oxidase of a monooxygenase system (Estabrook et al. 1963). Since these seminal papers thousands of reports have been published, demonstrating a role for cytochrome P450 forms in steroid hormone synthesis, drug detoxification, xenobiotic metabolism, and in chemical and carcinogen activation.

Structure and multiple forms

Although the primary structure of more than 220 different cytochrome P450 forms is now known, the three dimensional structure of only one form has been determined, that of the bacterial Pseudomonas putida, cytochrome P450$_{cam}$ (Poulos et al. 1987). Another bacterial cytochrome P450 has also just been crystallized, cytochrome P450$_{terp}$, and preliminary data suggests it differs from cytochrome P450$_{cam}$ (Boddupalli et al. 1992). A third bacterial cytochrome P450 has also been crystallized, the hemoprotein domain of the fusion protein cytochrome P450$_{BM3}$. This latter form of cytochrome P450 is more like the eukaryotic microsomal forms of cytochrome P450 with respect to redox partner use, and differs enough from the bacterial forms to suggest it will turn out to be a better model than cytochrome P450$_{cam}$ when its three dimensional structure is completely worked out (Ravichandran et al. 1993). Nevertheless, studies in which 35 mammalian forms of cytochrome P450 were aligned to each other and with respect to the primary sequence of cytochrome P450$_{cam}$ revealed striking areas of similarity (Nelson and Strobel 1988; Gotoh and Fujii-Kuriyama 1989). Examples of alignment similarity are shown in Table 1. The actual residue number (#) precedes the sequence; alignment numbers are bold.

Table 1. Alignment of 37 mammalian forms of cytochrome P450 with cytochrome P450$_{cam}$ [a]

species	form	gene	#	**490**	**500**	**510**	**520**
				*	* * *		*
rat	c	1A1	453	LFGLGKRKCIGETIGRLEVFLFLAILLQQM			
mouse	P1	1a1	453	LFGLGKRKCIGETIGRSEVFLFLAILLQQI			
human	c	1A1	441	IFGMGKRKCIGETIARWEVFLFLAILLQRV			
rabbit	6	1A1	446	LFGLGKRKCIAETIGRLEVFLFLATLLQQV			
rat	d	1A2	440	LFGLGKRRCIGEIPAKWEVFLFLAILLHQL			
mouse	p3	1a2	440	LFGLGKRRCIGEIPAKWEVFLFLAILLQHL			
human	4	1A2	442	LFGMGKRRCIGEVLAKWEIFLFLAILLQQL			
rabbit	4	1A2	442	LFGLGKRRCIGEILARWEVFLFLAILLQRL			
rat	b	2B1	421	PFSTGKRICLGEGIARNELFLFFTTILQNF			
rat	e	2B2	421	PFSTGKRICLGEGIARNELFLFFTTILQNF			
rabbit	2	2B4	421	PFSLGKRICLGEGIARTELFLFFTTILQNF			
rat	a	2A1	422	PFSTGKRFCLGDGLAKMELFLLLTTILQNF			
rabbit	3b	2C3	420	PFSTGKRACVGEGLARMELFLLLTTILQHF			
rat	f	2C7	420	PFSAGKRACVGEGLARMQLFLFLTTILQNF			
rat	pb1	2C6	420	PFSAGKRMCAGEGLARMELFLFLTTILQNF			
rabbit	pc1	2C1	420	PFSTGKRVCVGEVLARMELFLFLTAILQNF			
rabbit	pc2	2C2	420	PFSTGKRVCVGEALARMELFLFLTAILQNF			
rabbit	m1	?	420	PFSAGKRICAGEALARTELFLFFTTILQNF			
rabbit	1	2C5	417	PFSAGKRVCVGEGLARMELFLFLTSILQNF			
rat	j	2E1	422	AFSAGKRVCVGEGLARMELFLLLSAILQHF			
human	j	2E1	422	PFSTGKRVCAGEGLARMELFLLLCAILQHF			
rabbit	3a	2E1	422	PFSAGKRVCVGEGLARMELFLLLSAILQHF			
chicken	pb1	2H1	421	PFSAGKRICAGEGLARMEIFLFLTSILQNF			
mouse	c21	21a-1	405	SFGCGARVCLGEPLARLELFVVLARLLQAF			
human	c21	21A2	413	AFGCGARVCLGEPLARLELFVVLTRLLQAF			
bovine	c21	21A1	412	AFGCGARVCLGESLARLELFVVLLRLLQAF			
human	17α	17	427	PFGAGPRSCIGEILARQELFLIMAWLLQRF			
bovine	17α	17	427	PFGAGPRSCVGEMLARQELFLFMSRLLQRF			
rat	pcn1	3A1	427	PFGNGPRNCIGMRFALMNMKLALTKVLQNF			
rat	pcn2	3A2	427	PFGNGPRNCIDMRFALMNMKLALTKVLQNF			
human	1p	?	427	PFGSGPRNCIGMRFALMNMKLALIRVLQNF			
human	nf	3A3	426	PFGSGPRNCIGMRFALMNMKLALIRVLQNF			
rat	1aω	4A1	440	PFSGGARNCIGKQFAMSEMKVIVALTLLRF			
rabbit	pgω	4A4	437	PFSGGARNCIGKQFAMRELKVAVALTLVRF			
human	scc	11A	446	GFGWGVRQCLGRRIAELEMTIFLINMLENF			
bovine	scc	11A	445	GFGWGVRQCVGRRIAELEMTLFLIHILENF			
bovine	11β	11B	434	AFGFGVRQCLGRRVAEVEMLLLLHHVLKNF			
P.putida	cam	101	345	TFGHGSHLCLGQHLARREIIVTLKEWLTRI			
				*	* * *		*

a. Modified from Gotoh and Fujii-Kuriyama (1989).

The asterisks at the top and bottom of the table mark residues that are 100% conserved, i.e., are present in all mammalian forms of cytochrome P450 that were aligned it the study (Gotoh and Fujii-Kuriyama 1989). In addition to these 100% conserved residues are a large number of positions which contain conservative substitutions, like a leucine-valine,or aspartate-glutamate, or a lysine-arginine substitution. The sequence identity between the different forms of cytochrome P450 is greater between members of the same subfamily (>55% identical) than between members of different subfamilies, and members of the same gene family share a ≥40% sequence identity. In some instances, different cytochrome P450 proteins have been found with >95% sequence identity, as seen with cytochrome P450 2B1 and 2B2 (Fujii-Kuriyama et al. 1982).

Instances of microheterogeneity have been suggested for rabbit cytochrome P450 2B4, and three cDNA clones coding for the isozyme 2 (B0) and two other forms,B1 and B2, were identified. Cytochrome P450 2B4 (B0) was found in the phenobarbital-induced rabbit, but not as early as 12 hours after a single phenobarbital injection of rabbits (Gasser, et al. 1988). The cDNA found after 12 hours, called B1, differed from B0 by only 6 amino acids out of 491 residues, as compared with the third clone (B2), also found after

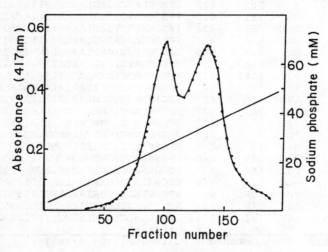

Figure 1A-Separation of two forms of P450 2B4 from phenobarbital-induced rabbit liver microsomes on carboxymethyl Sepharose.

12 hours, which differed in 11 positions, or 2% (Gasser, et al. 1988). While purifying cytochrome P450 2B4 from phenobarbital induced rabbit liver microsomes, we noted that when the rabbits are not fully induced by phenobarbital, instead of a single protein band being isolated on carboxymethyl Sepharose CM-6B the cytochrome P450 separated into two peaks (**Fig.1A**). We named the peaks LM2A and LM2B. When the animals were fully induced (microsomal cytochrome P450 >3nmol/mg) only LM2A was obtained. We isolated LM2 from untreated rats, and to our surprise the purified cytochrome obtained had a different elution pattern of on CM-Sepharose that was indicative of still another form of cytochrome P450 being present (**Fig. 1B**). The new form of cytochrome P450 was named LM2C. The amino-terminal amino acid sequences of LM2A, LM2B, and LM2C were identical for the first 24 residues, but on metabolism of different substrates the ratios of activities were strikingly different. From such data it is clear that gene duplication has occurred. Even more interesting, the different genes appear to be under different regulatory control.

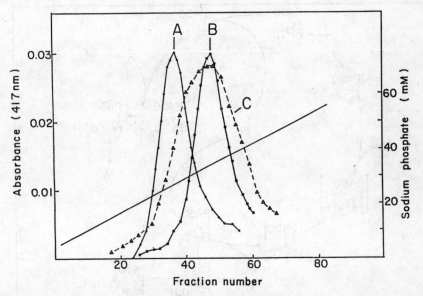

Figure 1B- Comparison of elution of three purified forms of P450 2B4 from phenobarbital induced (A and B) and untreated (C) rabbit liver microsomes on carboxymethyl Sepharose.

Reactions of cytochrome P450

A review of the types of reactions has recently appeared (Guengerich 1993) in which the thousands of reactions have been collectively grouped into six oxidative and one reductive types of reaction mechanisms. The mechanism of action of cytochrome P450 is not completely clear. Basically, the mechanisms can be viewed as an attack by atomic oxygen on carbon or a heteroatom, with or without rearrangement. In fact, cytochrome P450 has also been shown to catalyze peroxidative-like reactions, e.g., dual hydrogen atom abstraction as in conversion of testosterone to androstenedione (Cheng and Schenkman 1983) or formation of Δ^6-testosterone from testosterone (Korzekwa et al. 1990)

Cytochrome P450 cycle

The activity of the cytochrome P450 monooxygenase system can be described by the cytochrome P450 cycle. The cycle is shown in **Fig.**

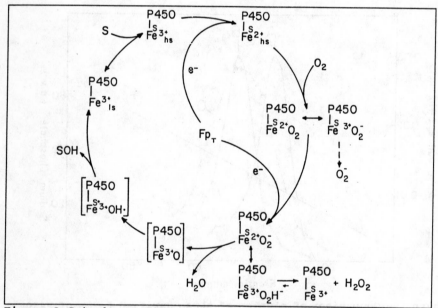

Figure 2-Cytochrome P450 cycle and its uncoupling.

2. The cycle is initiated by substrate binding to the oxidized, or ferric form of cytochrome P450. An electron from NADPH-cytochrome P450 reductase is transferred via the intramolecular flavins (FAD→FMN) to the heme iron of cytochrome P450. Oxygen then binds to the reduced hemoprotein, forming oxyP450, followed by input of a second electron from NADPH-cytochrome P450 reductase to cytochrome P450 for oxygen activation. Following input of this second electron the oxygen picks up protons and disproportionates into water (H_2O) and the perferryl $(Fe=O)^V$ heme. This is essentially atomic oxygen stabilized by the ferric heme iron. It is believed that this atomic oxygen then abstracts a hydrogen atom from the substrate in juxtaposition $[Fe=O + HS \rightarrow (Fe-OH^{\cdot})^{IV} + {\cdot}S]$, and this is followed by reaction between the radicals, $[Fe-OH^{\cdot} + {\cdot}S \rightarrow Fe^{III} + SOH]$ forming the product.

As shown in **Fig. 2**, when substrate binds to cytochrome P450 it perturbs the equilibrium between the low and high spin

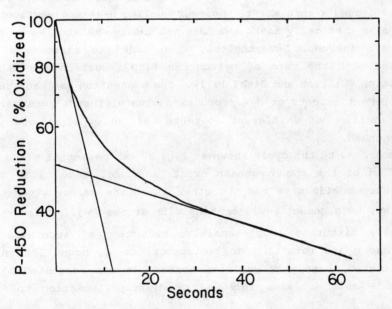

Figure 3– Reduction of cytochrome P450 in liver microsomes under carbon monoxide.

configuration of the cytochrome. This spin equilibrium perturbation has been shown to influence the midpoint potential of the cytochrome P450 and its rate of reduction (see review, Schenkman et al. 1982). Reduction of cytochrome P450 is not monophasic, but in a semilogarithmic plot appears to resemble biphasic kinetics (**Fig. 3**). Studies on the reduction kinetics indicate that, at least with liver microsomal forms of cytochrome P450, the fast phase of reduction correlates with a faster disappearance of the ferric high spin form of cytochrome P450 (Backes et al. 1985).

In addition to catalyzing the monooxygenation reaction, cytochrome P450 also releases hydrogen peroxide both in the presence and absence of substrates. The ability of liver microsomes to oxidize NADPH has been known since 1957, when this H_2O_2-producing reaction was termed NADPH oxidase (Gillette et al. 1957). Substrates of the microsomal monooxygenase system differentially stimulate the production of H_2O_2 in the microsomes (Hildebrandt et al. 1973). Perfluoro-n-hexane, like a substrate of the cytochrome P450 monooxygenase n-hexane, bound to cytochrome P450 and caused a spin shift. However, unlike n-hexane perfluoro-n-hexane is chemically inert and does not undergo metabolism by the microsomal enzymes. Nevertheless, when added to microsomes it increased both the rate of cytochrome P450 reduction and NADPH consumption (Ullrich and Diehl 1971). The suggestion was made that the different extents of H_2O_2 production with different substrates is indicative of different extents of uncoupling of the monooxygenase.

According to the cycle shown in **Fig. 2**, two potential sites of uncoupling of the monooxygenase exist. One of these is at the oxycytochrome P450 site and the other is at the peroxycytochrome P450 step, both shown equilibrating with arrows ($\leftrightarrow$). Based upon superoxide dismutase (SOD)-sensitive reduction of succinylated cytochrome c and formation of lactoperoxidase compound III in a reconstituted, partially purified system, it was suggested that superoxide anion release followed by disproportionation to H_2O_2

occurs (Kuthan et al. 1978). However, neither we (Schenkman et al. 1978) nor others (Debey and Balny 1973) were able to detect O_2^- production by lactoperoxidase complex formation when microsomes were incubated with NADPH. Instead, only the H_2O_2-lactoperoxidase complex was seen to form. Our data showed the NADPH-dependent oxidation of epinephrine to adrenochrome by liver microsomes was SOD-sensitive, indicating a role for superoxide anion in that reaction. However, calculations based upon the rate of adrenochrome formation and superoxide dismutation indicated the steady-state levels of superoxide anion would not exceed $0.3 \mu M$, a value too low to be a precursor of H_2O_2. Also, no SOD-inhibitable nor anaerobiosis inhibitable reduction of succinylated cytochrome c was seen (Jansson and Schenkman 1982), nor was SOD able to influence the apparent rate of O_2 consumption by liver microsomes in an oxygen electrode (Schenkman et al. 1979). With cytochrome P450$_{cam}$, the oxycytochrome P450 complex is stable, and has been shown to break down slowly in the absence of additional reducing equivalents, presumably releasing O_2^- (Peterson et al. 1972).

From such considerations it is considered more likely that the source of H_2O_2 is disassociation of the peroxycytochrome P450 complex. The oxygen at this redox state is a strong oxidant, and after protonation would be expected to pull the equilibrium to the right as in the equation below:

$$Fe^{II}O_2^- \leftrightharpoons Fe^{III}O_2H^- \Rightarrow Fe^{III} + H_2O_2$$

In support of this suggestion the K_m for H_2O_2 in the hydrogen peroxide-supported N-demethylation of benzphetamine by purified cytochrome P450 was 0.25M (Nordblom et al. 1976). This is more than three orders of magnitude higher than the concentrations of NADPH, O_2 or substrate needed to catalyze the monooxygenase reaction.

Stoichiometry of P450 turnover

Studies on the stoichiometry of the cytochrome P450 driven substrate metabolism indicated that the enzyme acts both as a monooxygenase and an oxidase. In the latter reaction hydrogen peroxide is the product, and must be taken into account in any stoichiometry calculations (Nordblom and Coon 1977; Jansson and Schenkman 1981). In subsequent studies using the reconstituted monooxygenase system with purified cytochrome P450 it was reported that excess NADPH and O_2 were consumed, in a ratio suggestive of a 4-electron oxidase activity yielding water as the product (Gorsky et al. 1984; Jansson and Schenkman 1987). The proportion of the P450 catalytic activity going via the monooxygenase, 2-electron and 4-electron oxidase reactions varied with the different substrates and cytochrome P450 forms investigated by the two groups.

Role of cytochrome b_5

Cytochrome b_5 has been proposed to be involved in the monooxygenase reaction. The effect of the hemoprotein was shown to depend upon the form of cytochrome P450 and upon the substrate used (Kuwahara and Omura 1980; Jansson et al. 1985). The effect of cytochrome b_5 on the stoichiometry of aminopyrine demethylation was investigated, and it was found that with rat liver microsomes the NADH stimulation of the NADPH-supported reaction approximated an increased rate of cytochrome b_5 oxidation (Jansson and Schenkman 1981). Stoichiometry experiments were carried out with the reconstituted monooxygenase system subsequently, and it was noted that with rabbit CYP2B4 cytochrome b_5 addition to the assay medium caused a metabolic switching, increasing the proportion of the monooxygenase reaction and decreasing the 2-electron oxidase reaction (Gorsky and Coon 1986). This was confirmed (Jansson and Schenkman 1987), and in addition it was shown that with a number of other forms of cytochrome P450 no metabolic switching occurs on addition of cytochrome b_5. In agreement with our earlier observations with rat liver microsomal aminopyrine demethylation (Jansson and Schenkman 1981), the stoichiometries with the reconstituted systems were satisfied by the sum of 2-electron oxidase plus monooxygenase reactions, *i.e.*, there was no excess NADPH and O_2 indicating no 4- electron oxidation had occurred (Jansson and Schenkman 1987).

Cytochrome b_5 is stimulatory of 7-ethoxycoumarin deethylation by CYP2B4, and this stimulation was shown spectrophotometrically to be due to complex formation between the two hemoproteins (Chiang 1981). Stimulation was also seen of the O-demethylation of p-nitroanisole by CYP2B4 and CYP2C11 (Tamburini et al. 1985), and binding of cytochrome b_5 to these forms of cytochrome P450 also caused spectrally observable spin shifts. In both studies, binding of cytochrome b_5 caused increased affinity of the cytochrome P450 for its substrate. An enhanced binding was also seen of cytochrome b_5 to P450 in the presence of substrate (Bonfils et al. 1981; Tamburini et al. 1985). The complexation of the cytochrome P450 by cytochrome b_5 was shown to require intact carboxyl groups on the cytochrome b_5, but modification of the lysyl amino groups was without effect on either the P450 spectrum or on substrate metabolism (Tamburini et al. 1985). The addition of cytochrome b_5 was shown not to affect the K_m for substrate, but only the V_{max} (Jansson, et al. 1985); In contrast, an earlier report (Chiang 1981) indicated that with CYP2B4 both the K_m and the V_{max} values were decreased and increased respectively with p-nitrophenetole or 7-ethoxycoumarin as substrates. The ability of cytochrome b_5 to alter the spin equilibrium of cytochrome P450 varied with the P450 form. Further, not all of the forms of cytochrome P450 were affected by cytochrome b_5 when substrate turnover was examined (Jansson et al. 1985), and with one form, rabbit CYP1A2 (LM4), cytochrome b_5 was reported to be inhibitory (Gorsky and Coon 1986).

The interaction between cytochrome b_5 and cytochrome P450 appears to be by a complementary charge-pairing mechanism involving carboxyl groups of the acidic residues on cytochrome b_5 and amino groups on cytochrome P450. Increasing the ionic content of a medium containing the two hemoproteins between 6 and 190mM increased the dissociation constant for the interaction two orders of magnitude, from $0.063 \mu M$ to $2.32 \mu M$ (Tamburini and Schenkman 1986), indicating an electrostatic interaction. The crosslinking reagent 1-ethyl-3(3-dimethylaminopropyl) carbodiimide (EDC) is a water-soluble reagent frequently used to selectively form amide bonds between complementary ion pairs; it has been called a zero-distance crosslinking reagent because when it reversibly activates a carboxyl residue, if a suitable nucleophile, like an amino group, is close enough it will react and replace the carbodiimide. When EDC was added to a mixture of cytochrome b_5 and CYP2B4 it catalyzed the formation of heterodimeric molecules composed of one cytochrome P450 and one cytochrome b_5 in covalent linkage (Tamburini et al. 1986). The covalent complex of the two hemoproteins was purified and shown to be functionally active when combined with NADPH-cytochrome P450 reductase and phospholipid (Tamburini and Schenkman 1987). Further, since the K_m for NADPH-cytochrome P450 reductase was unaltered, it was concluded that the binding site for the reductase and for cytochrome b_5 on cytochrome P450 was different.

Mechanism of action of cytochrome b_5

There have been a number of suggestions for the mechanism by which cytochrome b_5 stimulates substrate turnover. These range from providing input of the second electron necessary for the monooxygenation reaction, derived from NADH-cytochrome b_5 reductase or NADPH-cytochrome P450 reductase (Hildebrandt and Estabrook 1971), to removal of active oxygen generated during uncoupling (Staudt et al. 1974; Schenkman et al. 1976). In the complex system of liver microsomes, a decline in the absorption of a 440nm peak thought to be oxycytochrome P450 on addition of NADH was suggested to be indicative of input of an electron from cytochrome b_5 to the oxyP450 (Hildebrandt and Estabrook 1971). Electron transfer from reduced cytochrome b_5 to oxycytochrome P450 was subsequently shown using purified, photoreduced cytochromes and stopped flow spectroscopy (Bonfils et al. 1981), and it was suggested (Gorsky and Coon 1986) that the cytochrome b_5 functions to enhance the monooxygenase pathway over the autooxidative pathway by reducing the oxycytochrome P450.

In order for cytochrome b_5 to exert its stimulatory influence on cytochrome P450-dependent monooxygenation the cytochrome b_5 must

be capable of undergoing redox cycling.   By replacing the heme
moiety of cytochrome b_5 with manganese protoporphyrin IX, which is
very poorly reducible (Morgan and Coon 1984), it was found
(Tamburini and Schenkman 1987) that there was no change in the
binding affinity to cytochrome P450 2B4, but the stimulation of p-
nitroanisole oxidation to p-nitrophenol was lost.   The question
then arose as to the source of electrons transferred by cytochrome
b_5 to the oxycytochrome P450. Although it was suggested earlier,
and is generally accepted, that cytochrome b_5 provides the second
electron for the monooxygenase reaction, obtaining the necessary
electron from either NADPH-cytochrome P450 reductase of NADH-
cytochrome b_5 reductase, in fact no direct evidence existed to
support or disprove this hypothesis.   After isolation of the
cytochrome b_5-cytochrome P450 covalent complex, Tamburini and
Schenkman (1987) examined the nature of the interaction and route
of electron transfer.   Cytochrome b_5 has a number of anionic
carboxyl residues around its heme cleft that are utilized in
interaction with such redox partners as cytochrome c and NADH-
cytochrome b_5 reductase (Salemme 1976; Dailey and Strittmatter
1979).   Neutralization of these residues, as by methylamidation,
prevented such interaction.   Three carboxyl residues of acidic
amino acids and a heme propionic acid residue were shown to
interact with NADH-cytochrome b_5 reductase (Dailey and Strittmatter
1979), as well as with NADPH-cytochrome P450 reductase (Dailey and
Strittmatter 1980).   When crosslinked to NADPH-cytochrome P450
reductase cytochrome b_5 did get reduced, but was unable to transfer
its reducing equivalents to cytochrome c, indicating it uses the
same acidic residues for orienting its redox center to its redox
partners (Nisimoto and Lambeth 1985).   The same residues may be
involved in cytochrome b_5 binding to cytochrome P450; In covalent
complex with CYP2B4 the cytochrome b_5 was not reducible by
NADH-cytochrome b_5 reductase, but was reduced by NADPH-cytochrome
P450 reductase at a rate considerably faster than possible for the
reductase to reduce cytochrome b_5 in the absence of the cytochrome
P450.   The results were taken to indicate that when bound to
cytochrome P450, cytochrome b_5 obtains an electron by oxidizing the

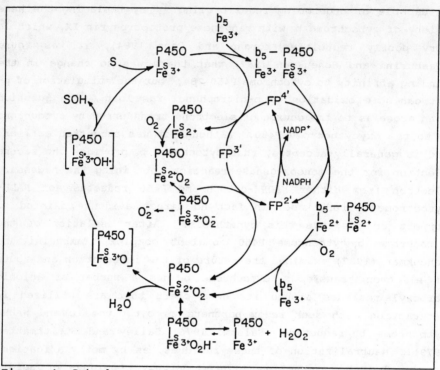

Figure 4– Cytochrome P450-cytochrome b_5 monooxygenation cycle.

reduced cytochrome P450 (see **Fig.4**). We postulate that the cytochrome P450 cycle with forms of P450 that interact with cytochrome b_5 has the two hemoprotein complex serving as a 2 electron acceptor from the NADPH-cytochrome P450 reductase: Input of an electron to cytochrome P450 is immediately followed by oxidation of the cytochrome P450 by cytochrome b_5, and re-reduction of the cytochrome P450 by the NADPH-cytochrome P450 reductase. Oxygen then binds to the reduced cytochrome P450 (since cytochrome b_5 can only accept one electron) and the oxycytochrome P450 (redox potential about +50mV (Guengerich 1983)) now oxidizes the ferrous cytochrome b_5 (redox potential +5mV) to form the peroxyP450 state (**Fig. 4**). In the absence of cytochrome b_5 P450 requires two interactions with the reductase to obtain the two needed electrons (see center of **Fig.4**).

The cytochrome b_5 binding site on cytochrome P450 is at present unknown. However, circumstantial evidence would place it somewhere near or overlapping residue S128 of CYP2B4. A number of papers have appeared in the past few years indicating that several forms of cytochrome P450 are phosphorylatable by the cAMP-dependent protein kinase (Pyerin et al. 1983; Jansson et al. 1987), which has the substrate recognition sequence RRXS. This sequence is found on a number of forms of cytochrome P450. It is found intact in many proteins of P450 family 2, 2A1, 2A3, 2B1, 2B2, 2B4, 2C11, 2D1, 2D2, 2E1 and 2G1 as well as 2C1, 2C2, 2C3 and 2C5, and is modified somewhat in other forms of cytochrome P450. Cytochrome b_5 binding to cytochrome P450 interfered with phosphorylation by the cAMP-dependent protein kinase (Jansson et al. 1987), and the phosphorylation of forms of cytochrome P450 were competitively inhibited by cytochrome b_5 binding (Epstein et al. 1989). From these studies it was inferred that the cytochrome b_5 binding site on cytochrome P450 overlapped the region containing the sequence RRFS, where the serine was residue 116 to 130 in most forms. In agreement with this suggestion, arginine 129 adjacent to serine 131 in mouse Cyp2a-5 was shown to be necessary for the binding of cytochrome b_5 to the cytochrome P450 (Juvonen et al. 1992). Although cytochrome P450$_{cam}$ also contains the sequence RRFS, the serine is at residue 293, and this is buried within the molecule at the end of the α-helix K. While P450$_{cam}$ can bind cytochrome b_5, it does so at a different site, the binding site for putidaredoxin, an acidic electron input protein for this P450 (Stayton et al. 1989).

References

Backes WL, Tamburini PP, Jansson I, Gibson GG, Sligar SG and Schenkman JB (1985) Kinetics of cytochrome P450 reduction: Evidence for faster reduction of the high spin ferric state. Biochemistry 24:5130-5136.

Boddupalli S, Hasemann C, Ravichandan K, Lu J, Goldsmith E, Deisenhofer J and Peterson J (1992) Crystallization and preliminary X-ray diffraction analyses of P450$_{terp}$ and the hemoprotein domain of P450$_{BM3}$, enzymes belonging to two distinct classes of the cytochrome P450 superfamily. Proc Natl Acad Sci USA 89:5567-5571.

Bonfils C, Balny C and Maurel P (1981) Direct evidence for electron transfer from ferrous cytochrome b_5 to the oxyferrous intermediate of liver microsomal cytochrome P450 LM2. J Biol Chem 256:9457-9465.

Cheng K-C and Schenkman JB (1983) Testosterone metabolism by cytochrome P450 isozymes RLM3 and RLM5 and by microsomes: metabolite identification. J Biol Chem 258:11738-11744.

Chiang JYL (1981) Interaction of purified microsomal cytochrome P450 with cytochrome b_5. Arch Biochem Biophys 211:662-673.

Dailey H and Strittmatter P (1979) Modification and identification of cytochrome b_5 carboxyl groups involved in protein-protein interaction with NADH cytochrome b_5 reductase. J Biol Chem 254:5388-5396.

Dailey H and Strittmatter P (1980) Characterization of the interaction of amphipathic cytochrome b_5 with the stearyl coenzyme A desaturase and NADPH-cytochrome P450 reductase. J Biol Chem 255:5184-5189.

Debey P and Balny C (1973) Production of superoxide ions in rat liver microsomes. Biochimie 55:329-332.

Epstein PM, Curti M, Jansson I, Huang C-K and Schenkman JB (1989) Phosphorylation of cytochrome P450: regulation by cytochrome b_5. Arch Biochem Biophys 271:424-432.

Estabrook RW, Cooper DY, and Rosenthal O (1963) The light-reversible carbon monoxide inhibition of the steroid C21-hydroxylase system of the adrenal cortex. Biochem Z 338:741-755.

Fujii-Kuriyama Y, Mizukami Y, Kawajiri K, Sogawa K and Muramatsu M (1982) Primary structure of cytochrome P450: Coding nucleotide sequence of phenobarbital inducible cytochrome P450 from rat liver. Proc Natl Acad Sci USA 79:2793-2797.

Gasser R, Negishi M and Philpot RM (1988) Primary structure of multiple forms of cytochrome P450 isozyme 2 derived from rabbit pulmonary and liver cDNAs. Mol Pharmacol 32:22-30.

Gillette JR, Brodie BB and LaDu BN (1957) The oxidation of drugs by liver microsomes: on the role of TPNH and oxygen. J Pharm Exp Therap 119:532-540.

Gorsky LD, Koop DR and Coon MJ (1984) On the stoichiometry of the oxidase and monooxygenase reactions catalyzed by liver microsomal cytochrome P450. J Biol Chem 259:6812-6817.

Gorsky LD and Coon MJ (1986) Effects of conditions for reconstitution with cytochrome b_5 on the formation of products in cytochrome P450-catalyzed reactions. Drug Metab Dispos 14:89-96.

Gotoh O and Fujii-Kuriyama Y (1989) Evolution, structure and gene regulation of cytochrome P450. In: Ruckpaul K and Rein H (eds) Frontiers in Biotransformation. Basis and Mechanisms of Regulation of cytochrome P450. Taylor and Francis, London, pp195-243.

Guengerich FP (1983) Oxidation reduction properties of rat liver cytochrome P450 and NADPH-cytochrome P450 reductase related to catalysis in reconstituted systems. Biochemistry 22:2811-2820.

Guengerich FP (1993) Metabolic reactions: Types of reactions of cytochrome P450 enzymes. In: Schenkman JB and Greim H (eds) Cytochrome P450. Handbook of Experimental Pharmacology vol

105, Springer-Verlag, Heidelberg,pp89-103.

Hildebrandt AG and Estabrook RW (1971) Evidence for the participation of cytochrome b_5 in hepatic microsomal mixed-function oxidation reactions. Arch Biochem Biophys 143:66-79.

Hildebrandt AG, Speck M and Roots I (1973) Possible control of hydrogen peroxide production and degradation in microsomes during mixed function oxidations. Biochem Biophys Res Commun 54:968-975.

Jansson I, Epstein PM, Bains S and Schenkman JB (1987) Inverse relationship between cytochrome P450 phosphorylation and complexation with cytochrome b_5. Arch Biochem Biophys 259:441-448.

Jansson I, Tamburini PP, Favreau LV and Schenkman JB (1985) The interaction of cytochrome b_5 with four cytochrome P450 enzymes from the untreated rat. Drug Metab Dispos 13:453-458.

Jansson I and Schenkman JB (1981) Stoichiometry of aminopyrine demethylation with and without NADH synergism. Drug Metab Dispos 9:461-465.

Jansson I and Schenkman JB (1982) Possible mechanism of coupled NADPH oxidase and P450 monooxygenase action. In: Snyder R, Parke DV, Kocsis JJ, Jollow DJ, Gibson GG and Witmer CM (eds), Biological Reactive Intermediates II, Part A, Plenum Press, NY,pp145-163.

Jansson I and Schenkman JB (1987) Influence of cytochrome b_5 on the stoichiometry of the different oxidative reactions catalyzed by liver microsomal cytochrome P450. Drug Metab Dispos 15:344-348.

Juvonen RO, Iwasaki M and Negishi M (1992) Roles of residues 129 and 209 in the alteration by cytochrome b_5 of hydroxylase activities in mouse 2A P450s. Biochemistry 31:11519-11523.

Korzekwa KR, Trager WF, Nagata K, Parkinson A and Gillette JR (1990)Isotope effect studies on the mechanism of the cytochrome P450 2A1-catalyzed formation of Δ^6-testosterone from testosterone. Drug Metab Dispos 18:974-979.

Kuthan H, Tsuji H, Graf H, Ullrich V, Werringloer J and Estabrook RW (1978) Generation of superoxide anion as a source of hydrogen peroxide in a reconstituted monooxygenase system. FEBS Lett 91:343-345.

Morgan ET and Coon MJ (1984) Effects of cytochrome b_5 on cytochrome P450-catalyzed reactions. Studies with manganese-substituted cytochrome b_5. Drug Metab Dispos 12:358-364.

Nelson DR, Kamataki T, Waxman DJ, Guengerich FP, Estabrook RW, Feyereisen R, Gonzalez F, Coon MJ, Gunsalus IC, Gotoh O, Okuda K and Nelson DW (1993) The P450 superfamily: Update on new sequences, gene mapping, accession numbers, early trivial names of enzymes, and nomenclature. DNA Cell Biol 12:1-51.

Nelson DR and Strobel HW (1988) On the membrane topology of vertebrate cytochrome P450 proteins. J Biol Chem 263:6038-6050.

Nisimoto Y and Lambeth JD (1985) NADPH-cytochrome P450 reductase-cytochrome b_5 interactions: crosslinking of the phospholipid vesicle-associated proteins by a water-soluble carbodiimide. Arch Biochem Biophys 241:386-396.

Nordblom GD and Coon MJ (1977) Hydrogen peroxide formation and stoichiometry of hydroxylation reactions catalyzed by highly purified liver microsomal cytochrome P450. Arch Biochem Biophys 180:343-347.

Nordblom GD, White RE and Coon MJ (1976) Studies on hydroperoxide-dependent substrate hydroxylation by purified liver microsomal cytochrome P450. Arch Biochem Biophys 175:524-533.

Omura T. and Sato R (1962) A new cytochrome in liver microsomes. J Biol Chem 237:1375-1376.

Omura T and Sato R (1964) The carbon monoxide binding pigment of liver microsomes. I. Evidence for its hemoprotein nature. J Biol Chem 239:2370-2378.

Peterson JA, Ishimura Y and Griffin BW (1972) Pseudomonas putida cytochrome P450: characterization of an oxygenated form of the hemoprotein. Arch Biochem Biophys 149:197-208.

Poulos TL, Finzel BC and Howard AJ (1987) High resolution crystal structure of cytochrome P450$_{cam}$. J Mol Biol 195:687-700.

Pyerin W, Wolf CR, Kinzel V and Oesch F (1983) Phosphorylation of cytochrome P450-dependent monooxygenase components. Carcinogenesis 4:573-576.

Ravichandran K, Boddupalli S, Hasemann C, Peterson JA, and Deisenhofer J (1993) Crystal structure of hemoprotein domain of P450$_{BM3}$, a prototype for microsomal P450s. Science 261:731-736.

Schenkman JB and Greim H (eds) (1993) Cytochrome P450. Handbook of Experimental Pharmacology, vol 105, Springer-Verlag, Heidelberg.

Schenkman JB, Jansson I, Powis G and Kappus H (1979) Active oxygen in liver microsomes: mechanism of epinephrine oxidation. Mol Pharmacol 15:428-438.

Schenkman JB, Jansson I and Robie-Suh KM (1976) The many roles of cytochrome b$_5$ in hepatic microsomes. Life Sci 19:611-624.

Schenkman JB, Sligar SG and Cinti DL (1982) Substrate interaction with cytochrome P450. In: Schenkman JB and Kupfer D (eds) Hepatic Cytochrome P450 Monooxygenase System. International Encyclopedia of Pharmacology and Therapeutics, Section 108, Pergamon Press, NY, pp587-615.

Staudt H, Lichtenberger F and Ullrich V (1974) The role of NADH in uncoupled microsomal monooxygenations. Eur J Biochem 46:99-106.

Stayton P, Poulos T and Sligar SG (1989) Putidaredoxin competitively inhibits cytochrome b$_5$-cytochrome P450$_{cam}$ association: a proposed model for cytochrome P450$_{cam}$ electron transfer complex. Biochemistry 28:8201-8205.

Tamburini PP, MacFarquhar S and Schenkman JB (1986) Evidence of binary complex formation between cytochrome P450, cytochrome b$_5$ and NADPH-cytochrome P450 reductase of hepatic microsomes. Biochem Biophys Res Commun 134:519-526.

Tamburini and Schenkman (1986) Mechanism of interaction between cytochromes P450 RLM5 and b$_5$ Evidence for an electrostatic mechanism involving cytochrome b$_5$ heme propionate groups. Arch Biochem Biophys 245:512-522.

Tamburini PP and Schenkman JB (1987) Purification to homogeneity and enzymological characterization of a functional covalent complex composed of cytochromes P450 isozyme 2 and b_5 from rabbit liver. Proc Natl Acad Sci USA 84:11-15.

Tamburini PP, White RE and Schenkman JB (1985) Chemical characterization of protein-protein interaction between cytochrome P450 and cytochrome b_5. J Biol Chem 260:4007-4015.

Ullrich V and Diehl H (1971) Uncoupling of monooxygenation and electron transport by fluorocarbons in liver microsomes. Eur J Biochem 20:509-512.

General Aspects of NADPH Cytochrome P450 Reductase and Cytochrome b5

Emel Arınç
Joint Graduate Program in Biochemistry
Middle East Technical University
06531 Ankara, Turkey

Introduction

In 1968, Lu and Coon demonstrated, for the first time, that the oxidative hydroxylase system of endoplasmic reticulum for liver cells is composed of three essential components: cytochrome P450 (EC 1.14.14.1), NADPH cytochrome P450 reductase (EC 1.6.2.4), and a heat stable factor subsequently shown to be phosphatidylcholine (Strobel et al. 1970), which can be reconstituted to an enzymatically active enzyme system. This study paved the way to innumerable biochemical, toxicological, and pharmacological studies.

It is well established that cytochrome P450 dependent monooxygenase system is present both in mitochondria and in endoplasmic reticulum. In the microsomal systems, FAD and FMN containing flavoprotein, NADPH dependent cytochrome P450 reductase directly catalyzes electron transfer from NADPH to cytochrome P450, whereas in the mitochondrial system, a ferredoxin type iron-sulfur protein, redoxin, acts as an electron carrier between the FAD containing redoxin reductase and cytochrome P450 (Fig. 1).

In addition, in endoplasmic reticulum, participation of cytochrome b5 in cytochrome P450 dependent oxidative hydroxylation system has been shown. Cytochrome b5 can accept electrons from cytochrome P450 reductase and transfer these electrons to some specific isozymes of cytochrome P450 (Imai, 1981; Morgan and Coon, 1984; Jansson et al. 1985) (Fig. 1).

Similar to microsomal monooxygenase system, substrate specificity of mitochondrial monooxygenase system is determined by the cytochrome P450

NATO ASI Series, Vol. H 90
Molecular Aspects of Oxidative Drug Metabolizing Enzymes
Edited by E. Arınç, J. B. Schenkman and E. Hodgson
© Springer-Verlag Berlin Heidelberg 1995

component. Although redoxin and redoxin reductase of the systems from different tissues show certain biochemical and immunological similarities, there is evidence that these may also differ from one tissue to another (Hanukoğlu and Gutfinger, 1989). Thus, redoxin from the adrenal cortex is called adrenodoxin and that from the liver as hepatoredoxin. Adrenodoxin and adrenodoxin reductase of bovine adrenal mitochondria are peripheral, soluble, single-subunit proteins having Mr of 12000 and 52000, respectively, while cytochrome P450s are integral membrane proteins (Jefcoate, 1986).

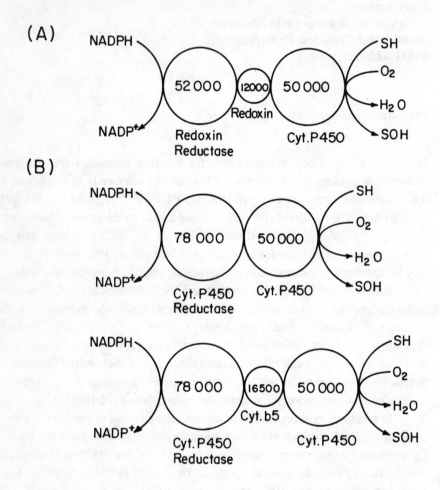

Fig. 1. Cytochrome P450 dependent monooxygenase systems of mitochondria (A) and endoplasmic reticulum (B)

NADPH Cytochrome P450 Reductase

Microsomal NADPH cytochrome P450 reductase contains one molecule each of FAD and FMN and has Mr of 76000-79500. These flavins participates in electron transport from NADPH to cytochrome P450 in a phospholipid environment. In contrast to cytochrome P450, the reductase does not exhibit substrate specificity, multiplicity of enzymatic form or significant inducibility by drugs except phenobarbital. It is encoded by a single gene in all species examined

Reductase is bound to the endoplasmic reticulum. More than three fourths of the liver microsomal reductase is located free of the lipid bilayer, whereas cytochrome P450 protein clusters appear to be more deeply embedded in the membrane (Nebert and Gonzalez, 1987).

In 1980s, structural studies with rat or rabbit liver cytochrome P450 reductase have revealed that reductase is an amphipathic protein containing a large hydrophilic catalytic domain and a small hydrophobic amino terminal domain (Gum and Strobel, 1981; Black and Coon, 1982). Like rat and rabbit liver enzyme, limited proteolysis of microsomal cytochrome P450 reductase from pig and sheep liver (Haniu et al. 1986; İşcan and Arınç, 1988) and from rabbit and sheep lung (Serabjit-Singh et al. 1979; İşcan and Arınç, 1986) yields two major peptide fragments; flavin containing hydrophilic domain having Mr of 70000-72000 that can reduce non-physiological, artificial electron acceptors such as cytochrome *c* (but not cytochrome P450) and a small N-terminal hydrophobic membrane domain (Mr 4800-6000) which is required for cytochrome P450 reduction. Hydrophilic peptide is monomeric in aqueous solution while hydrophobic peptide forms aggregates. On the other hand, the native detergent solubilized amphipathic reductase exists in aqueous solution largely as a hexamer. Black and Coon (1982) reported that the amino terminal of rabbit reductase is blocked and identified the blocking group as an acetyl group. Recently, Haniu et al. (1986) showed that the acetyl group is linked to glycyl residue of pig liver reductase. Although, the function of the acetyl group is not known, the small hydrophobic amino terminal tail has been shown to be essential for anchoring of reductase to the microsomal membrane to provide an orientation that may allow the optimal interaction for the reduction of cytochrome P450 (Black and Coon, 1982; Miwa and Lu, 1984; Strobel et al. 1992).

The presence of variable amounts of proteolytically cleaved smaller molecular weight inactive reductases has been reported in the purified rabbit liver (French and Coon, 1979), rabbit lung (Serabjit-Singh et al. 1979), pig liver (Yasukochi et al. 1980), guinea pig liver (Kobayashi and Rikans, 1984) and sheep liver and lung (İşcan and Arınç, 1986, 1988) amphipathic reductase preparations. Recently, an unusual proteolytic cleavage at the Asn-Gly (residues 502-503) linkage of the 676 residue of pig liver reductase has been reported (Haniu et al. 1986).

The complete amino acid sequence of liver cytochrome P450 reductase has been determined for the rat (Porter and Kasper, 1985), rabbit (Katagiri et al. 1986), pig (Haniu et al. 1986; Vogel and Lumper, 1986), human (Haniu et al. 1989; Yamano et al. 1989) and trout (Urenjak et al. 1987). The mammalian reductases are about 90% homologous while trout reductase shows 79% sequence homology with the rat enzyme. A recently identified nitric oxide synthase enzyme has also been shown as an FMN- and FAD-containing protein (Bredt et al. 1991) and has regions homologous to NADPH cytochrome P450 reductase.

Cytochrome P450 reductase, like cytochrome b5, is a significantly acidic protein. Rat enzyme contains 102 carboxyl residues out of total 678 amino acid residues. Sequence homology to other flavoproteins suggests that in rat, the FMN-binding site begins about 10000 Daltons from the NH_2-terminus, whereas the FAD-binding site is about 20000 Daltons form the COOH-terminus. There is only one NADP(H)-binding site and its placement is adjacent to the putative FAD site (Porter and Kasper, 1985). A detailed comparison of the amino acid sequences of cytochrome P450 reductases with flavoproteins of known three-dimensional structure has permitted the identification of the important structural and functional domains and amino acid residues of the cytochrome P450 reductase. Cytochrome P450 reductase has at least five functional domains.

The NH_2-terminal membrane binding domain: It anchors the enzyme in the membrane and is required for the reduction of cytochrome P450 (Gum and Strobel, 1981; Black and Coon, 1982). Trypsin treatment of the rat liver microsomes selectively cleaves the enzyme between the sensitive Lys 56/Ile 57 bond. The first 17 residues are variable between the rat, pig, human and rabbit liver reductases (Haniu et al. 1989). Residues from 20 to 44 is less variable and contains Leu 33 and Tyr 40 as two invariant residues. Hydropathy plot analysis of the pig liver reductase shows that two regions from residues

523-544 exhibit significant hydrophobicity suggesting to be involved in binding to the membrane and/or cytochrome P450 (Haniu et al. 1986). Residues from 45 to 48 contains a cluster of basic residues (45-Arg-Lys-Lys-Lys-48) which is characteristic of a simple transmembrane segment (von Heijne, 1985).

Flavin mononucleotide binding domain: Residues from 90-210 (Fig. 2) of rat liver reductase shows sequence similarity between the FMN binding regions of bacterial FMN-containing flavodoxins (Porter and Kasper, 1986). Tyr 178 is involved in flavin binding of rat reductase and flavin group is situated in a pocked between Tyr 140 and Tyr 178.

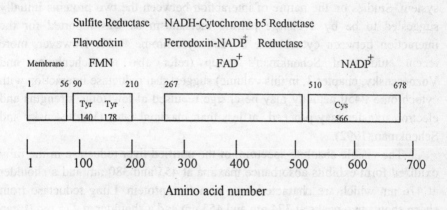

Fig. 2. The localization of membrane, FMN, FAD, NADP$^+$ binding sites for rat NADPH cytochrome P450 reductase

Flavin adenine dinucleotide binding domain: Cytochrome P450 reductase which does not interact with a ferrodoxin, shares sequence homology between its carboxyl portion and FAD containing plant ferrodoxin-NADP$^+$ reductase (FNR) (Porter and Kasper, 1986; Porter, 1991). The low but significant similarity between these enzymes is evidenced by the diagonal line in homology matrix plot, while no such line is observed in a sequence comparison of ferrodoxin reductase versus adrenodoxin reductase (Hanukoğlu, 1992).

NAPDH-binding domain: The region from residue 510 to the carboxyl end of the cytochrome P450 reductase is strongly conserved in sulfite reductase flavoprotein, spinach ferrodoxin-NADP$^+$ reductase, yeast (*S. cerevisiae, C. tropicalis*) and bacterial (P450-BM3) reductases (Shen and Kasper, 1993). There is an essential sulfhydryl group at or near the NADPH binding site. A NADPH-protected cysteine residue is identified at residue 565 of the pig and human cytochrome P450 reductases (Haniu et al. 1986, 1989).

Substrate binding domain: The manner of interaction of cytochrome P450 reductase with cytochrome P450 has been of considerable interest for a number of years. Optimal hydroxylation activity is reached at 1:1 stoichiometry between cytochrome P450 and its reductase in the reconstituted system. Studies on the nature of interaction between the two proteins initially suggested to be by a charge pairing mechanism as was reported for the interaction between cytochrome b5 and cytochrome P450. However, more recent studies of Schenkman's group (refer also to Schenkman and Voznesensky, chapter 3, in this volume) suggest that reductase interaction with cytochrome P450 actually may be charge repulsed at low ionic strengths and electron transfer may proceed at less than maximal rates (Voznesensky and Schenkman, 1992).

The visible absolute spectrum of the purified liver reductase in the fully oxidized form exhibits absorbance maxima at 455 and 380 nm and a shoulder at 476 nm which are characteristics of a flavoprotein. Lung reductase from sheep shows two peaks at 374 nm and 455 nm and a shoulder at 475 nm (İşcan and Arınç, 1986). When reductase is reduced with NADPH under aerobic conditions, absorption at 455 nm decreases and absorption in the 550 to 650 nm region increases with a peak at 585 nm due to air-stable semiquinone. Air-stable semiquinone has a single electron per reductase molecule and is not completely reduced by excess NADPH.

Both flavins, FMN and FAD are not covalently bound to reductase. Therefore, the flavins can be selectively removed from reductase in the presence of 2 M KBr at different pH values, i.e., FMN at pH 8.5 (Vermillion and Coon, 1978) and FAD at pH 6.5 (Kurzban and Strobel, 1986). The FAD depleted reductase virtually loses its ability to transfer electrons to a number of electron acceptors including cytochrome *c* and ferricyanide (Kurzban and Strobel, 1986), whereas FMN-depleted enzyme has lost most of its activity toward cytochrome c, menadione, 2,6-dichlorophenol indophenol but retained the ability to reduce ferricyanide (Vermillion and Coon, 1978). These results show that FMN is not a component electron acceptor from NADPH and that the FAD site accepts the electrons from NADPH. FAD-free reductase is capable of photoreduction when exposed to oxygen, produces an air-stable semiquinone (Kurzban and Strobel, 1986).

The above results indicate that electron transfer from NADPH to FAD to FMN to cytochrome P450 or to another electron acceptor protein. NADPH cytochrome reductase catalyzes reactions both with 1-electron acceptors such as cytochrome P450, cytochrome c and ferricyanide as well as with 2-electron acceptors like menadione and dichlorophenol indophenol. The apparent turnover numbers for artificial electron acceptors, i.e., cytochrome c ranges from 2633 min^{-1} to 5165 min^{-1} (Table 1).

Table 1. Catalytic Activities of the Purified NADPH-Cytochrome P450 Reductases from Different Sources with Various Electron Acceptors

Acceptor	Turnover Number Reductase				
	Sheep Lung[a]	Liver[a]	Rat Liver[b,c]	Guinea Pig Liver[c,d]	House Fly[a,e]
Cytochrome c	5165	5165	4010	3870	2633
Ferricyanide	6070	5606	5310	14980	4933
Menadione	N. D.	97.1	2370	2780	566
NBT	N. D.	47.1	216	1130	N. D.
Aniline (P450 liver)	6.7[f]	6.0[f]			
Ethylmorphine (P450 liver)	6.1[f]	5.3[f]			

[a]Turnover numbers for artificial acceptors are expressed as moles of acceptor reduced/min/mole enzyme. Reactions were carried out in 0.3 M phosphate buffer pH 7.8 at 37 °C (İşcan and Arınç, 1988)

[b]Reactions were carried out in 0.1 M phosphate buffer pH 7.7 at 30 °C (Vermillion and Coon, 1978)

[c]Turnover numbers for artificial acceptor are expressed as moles of acceptor reduced/min/mole of FMN

[d]Reactions were carried out in 0.3 M phosphate buffer pH 7.7 at 30 °C (Kobayashi and Rikans, 1984)

[e]Reactions were carried out in 0.05 M phosphate buffer pH 7.7 at 25 °C (Mayer and Durrant, 1979)

[f]Turnover numbers are expressed as moles of product formed/min/mole of reductase for one mole of P450

N. D. : Not Determined

The apparent K_m value for NADPH of sheep liver reductase (14.3 ± 1.23 μM) was found to be similar to K_m of sheep lung reductase (11.1 ± 0.70 μM) (İşcan and Arınç, 1988). The apparent K_m values for cytochrome c of sheep liver and lung reductases were also found to be similar, 22.2 ± 2.78 μM and 20.02 ± 2.15 μM, respectively (İşcan and Arınç, 1988). These values were in the range of K_m values reported for purified amphipathic pig liver (K_m cytochrome c : 28 μM, K_m NADPH : 14 μM) and guinea pig liver (K_m cytochrome c : 15.7 μM, K_m NADPH : 10.7 μM) reductase determined in 0.3 M phosphate buffer by Yasukochi et al. (1980) and Kobayashi and Rikans, (1984), respectively.

Kinetic studies with liver reductase by varying NADPH (Fig. 3) and cytochrome c (Fig. 4) concentrations at different fixed cytochrome c (Fig. 3) or NADPH (Fig. 4) concentrations, showed a series of parallel lines indicating a ping-pong type kinetic mechanism for the interaction of substrates NADPH and cytochrome c, with liver reductase (İşcan and Arınç, 1988). A series of parallel lines obtained with sheep lung reductase also suggested a ping-pong kinetic mechanism for NADPH dependent cytochrome c reduction (İşcan and Arınç, 1986).

However, as seen in Fig. 3, when the fixed cytochrome c concentration was increased, the slope of the lines also increased and lines converged on $1/v$ axis. In addition, in Fig. 4, it was observed that at low fixed concentrations of NADPH, the parallel lines were curved upward at high concentrations of cytochrome c. Similar results were obtained with sheep lung reductase. The converging double reciprocal lines obtained with sheep reductase at high fixed cytochrome c concentrations when NADPH was the varying substrate (Fig. 3) and the upward curvatures were observed when cytochrome c was the varying substrate (Fig. 4) suggested a competitive substrate (cytochrome c) inhibition. This phenomenon in the literature was treated as being characteristics of the ping-pong systems by Segel (1975). The data suggests that the oxidized flavoprotein reductase can be reduced by NADPH in the presence of electron acceptor generating a reduced enzyme intermediate which can react with the electron acceptor (cytochrome c, Fe^{3+}) and can form the oxidized enzyme and reduced substrate (cytochrome c, Fe^{2+}). Oxidized form of the cytochrome c (Fe^{3+}) at higher concentrations can also bind the oxidized form of the enzyme resulting the competitive substrate inhibition (Fig. 5). On the other hand, the evidence obtained by Phillips and Langdon (1962) and Baggott and Langdon (1970) suggested that a random mechanism operates for the reaction catalyzed

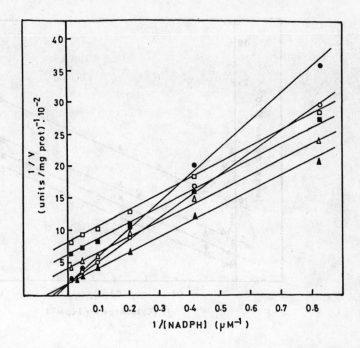

Fig. 3. Double reciprocal plots of the initial velocity for the reduction of cytochrome *c* by the purified sheep liver reductase as a function of NADPH at different concentrations of cytochrome *c*. The concentrations of NADPH were 1.2, 2.4, 4.8, 9.6, 18.5, 95.0 μM, and of cytochrome *c* were 6.7 μM (□—□), 8.5 μM (■—■), 17.8 μM (△—△), 85.4 μM (▲—▲), 127 μM (○—○) and 171 μM (●—●). The reaction medium contained 0.027 units of the purified sheep liver reductase. The points are the means of two different sets of data and each point is the mean of duplicate determinations. The lines are derived using linear regression. Reproduced from İşcan and Arınç (1988), with permission.

by the purified pig liver reductase. Subsequent to these proposals, a sequential ordered mechanism reported by Dignam and Strobel (1977) and Kobayashi and Rikans (1984) for the interaction of co-substrates with the rat and guinea pig liver microsomal cytochrome P450 reductases, respectively. In a view of the apparently contradictory data in the literature surrounding cytochrome P450 reductase kinetic mechanism with NADPH and cytochrome *c*, the ping-pong kinetic mechanism proposed for sheep liver and lung cytochrome P450 reductases (İşcan and Arınç, 1986, 1988) is in good agreement with the conclusions obtained for purified pig liver microsomal

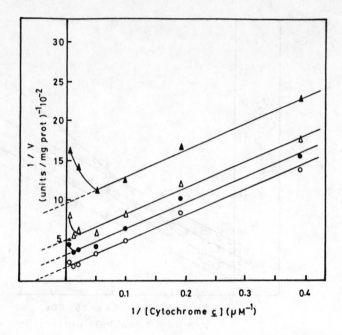

Fig. 4. Double reciprocal plots of the initial velocity for the reduction of cytochrome c by the purified sheep liver reductase as a function of cytochrome c at different fixed concentrations of NADPH. The concentrations of cytochrome c were 2.6, 5.3, 10.6, 21.1, 42.2, 84.5 and 165 μM and of NADPH were 2.4 μM (▲‒‒‒▲), 7.2 μM (△‒‒‒△), 14.1 μM (●‒‒‒●) and 120 μM (○‒‒○). The reaction medium contained 0.027 units of the purified sheep liver reductase. The points are the means of two different sets of data and each point is the mean of duplicate determinations. The lines are derived using linear regression. Reproduced from İşcan and Arınç (1988), with permission.

reductase by Masters et al. (1965), pig kidney microsomal reductase by Fan and Masters (1974), for chick kidney microsomal and mitochondrial reductase by Kulkoski et al. (1979), for house fly microsomal reductase by Mayer and Durrant (1979).

It is clear that in the view of divergent conclusions reached by several investigators, the kinetic mechanism for the NADPH dependent reduction of cytochrome c by the reductase enzyme has not been clarified yet. Further work is necessary to solve this problem.

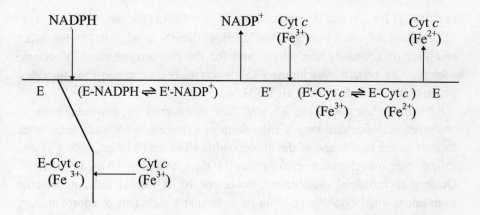

Fig. 5. A scheme for the mechanism of the NADPH dependent lung cytochrome *c* reductase. One substrate (NADPH) reacts with the enzyme (E: oxidized flavoprotein) to form product (NADP$^+$) and the modified enzyme (E': reduced flavoprotein). Then, the reduced form of enzyme, E', reacts with a second substrate (Cytochrome *c*, Fe^{3+}) and forms oxidized enzyme and reduced substrate (cytochrome *c* Fe^{2+}). Oxidized form of cytochrome *c* (Fe^{3+}), at higher concentrations (above 160 μM) also binds the oxidized form of the enzyme.

Cytochrome b5

Cytochrome b5 is a low-spin electron-transfer hemoprotein that participates in a variety of important physiological functions. It exists in three forms of which two forms occur in membrane-bound states and the third one occurs in a soluble form. The membrane-bound form has an additional small hydrophobic domain having Mr of approximately 5000 that anchors the protein to the natural or artificial membranes. One of the membrane-bound forms is located in the endoplasmic reticulum of the liver (Strittmatter and Ball, 1952; Strittmatter and Velick, 1956), lung (Güray and Arınç, 1990, 1991) and other tissues (D' Arrigo et al. 1993) while the other is present in the mitochondrial membrane (Ito, 1980; Ito et al. 1981; Lederer et al. 1983). The soluble form of the cytochrome b5 occurs naturally in erythrocytes and exists as a cytosolic protein (Passon et al. 1972; Hultquist et al. 1974; Abe and Sugita, 1979). In endoplasmic reticulum, two flavoproteins, NADH cytochrome b5 reductase and NADPH cytochrome P450 reductase reduce cytochrome b5 rapidly. Reduced cytochrome b5 then provides the electrons for Δ9-desaturation of stearyl-CoA to oleyl-CoA (Holloway and Katz, 1972; Holloway, 1971; Oshino et al. 1971) for Δ5- and Δ6 desaturation of acyl-CoAs (Oshino, 1980) for Δ12-desaturation of oleoyl-phosphatidylcholine (Smith et al. 1992) (for other reactions of b5 see Arınç, 1991) and for the monooxygenation of certain substrates by certain cytochrome P450 isozymes (Peterson and Prough, 1986; Schenkman, 1991; Bonfils et al. 1991).

Although cytochrome b5 was first discovered in the membrane of endoplasmic reticulum, very similar form of cytochrome b5 was found to be located on the membrane of the mitochondria (Raw and Mahler, 1959). It was called outer mitochondrial cytochrome b (OM cytochrome b) by Ito in 1980. Outer mitochrondrial membrane cytochrome b5 is almost indistinguishable from microsomal cytochrome b5 in its molecular weight (Mr of approximately 16000), spectral properties at room temperature and electrophoretic behavior. However, it is clearly different from microsomal cytochrome b5 in spectral properties at low temperature as well as in the amino acid composition and immunological properties (Ito, 1980). Purified tryptic fragment of outer mitochondrial membrane cytochrome b5 containing 92 amino acid residues is shown to have 60% sequence identity to the hydrophilic domain of cytochrome b5 purified from microsomes (Lederer et al. 1983). Studies with monospecific peptide antibodies for each cytochrome b5 form have demonstrated that outer mitochondrial membrane cytochrome b5 is not present on endoplasmic reticulum membranes, whereas microsomal cytochrome b5 is present on outer mitochondrial membranes in extremely low concentrations, at a level <5% of that on endoplasmic reticulum membranes (D' Arrigo et al. 1993). The results indicate that microsomal and mitochondrial cytochromes b5 have nonoverlapping distribution in the cell and suggest the existence of novel posttranslational targeting mechanisms, which could result in the restricted localization of each cytochrome b5 isoform within the cell (D' Arrigo et al. 1993).

The biological function of outer mitochondrial membrane cytochrome b5 was not known until 1981. In 1981, participation of mitochondrial b5 in NADH-semidehydroascorbic acid reductase activity of rat liver was reported and microsomal cytochrome b5 appeared to have no contribution to this activity (Ito et al. 1981).

An electron transport system, similar to cytochrome b5 system present in endoplasmic reticulum and mitochondria, also exists in erythrocytes (Hultquist and Passon, 1971; Passon et al. 1972; Hultquist et al. 1974; Abe and Sugita, 1979). Soluble cytochrome b5 participates in the reduction of methemoglobin in erythrocytes and functions to maintain the oxygen-carrying capacity of hemoglobin by keeping it in the reduced state in the red blood cells (Hegesh et al. 1986). Furthermore, recently, the involvement of erythrocyte cytochrome b5 in the hydroxylation of CMP-N-acetylneuraminic acid in the liver and other tissues has been reported (Kozutsumi et al. 1990, 1991).

Catalytic and spectral properties of soluble cytochrome b5 are found to be very similar to those of microsomal and mitochondrial membrane-bound cytochrome b5. Cytochrome b5 isolated from pig, human, rabbit and bovine erythtrocytes consists of 97 amino acid residues. The essential differences between the soluble erythrocyte and liver microsomal membrane-bound cytochrome b5 are: a) the extra 36 amino acid residues in the liver are located at the C-terminal end and serve as a membrane binding domain, and b) the amino acids 1-96 are always the same in the two forms (Kimura et al. 1984; Abe et al. 1985; Giordano and Steggles, 1991). The common amino acid change between two forms is demonstrated at residue 97, which is the C-terminus of the soluble cytochrome b5. Thr at residue 97 in the microsomal b5 is replaced by Pro in rabbit and man, and by Ser in pig and horse and no replacement is found in bovine (Kimura et al. 1984; Kozutsumi et al. 1991). Kimura et al. (1984) suggested that erythrocyte cytochrome b5 is not derived from the liver protein by proteolysis but a translational product from a distinct mRNA of cytochrome b5. Recently, Giordano and Steggles (1991) has isolated a soluble form specific cDNA from a human reticulocyte cDNA library. Their results have suggested that the human liver and reticulocyte cytochrome b5 mRNAs are products of a single gene (Giordano and Steggles, 1991). Very recently, the rabbit genomic segments for the soluble cytochrome b5 and microsomal cytochrome b5 have been amplified and isolated by means of the polymerase chain reaction using primers corresponding to various portions of the open frame of microsomal cytochrome b5 cDNA (Takematsu et al. 1992). In that study, two genes are isolated from rabbit genomic DNA, which correspond to the soluble erythrocyte form and membrane-bound form mRNAs, respectively. Therefore, they (Takematsu et al. 1992) have proposed that the soluble form and microsomal form of cytochrome b5 are encoded by separate genes.

Bovine liver microsomal b5 is a single polypeptide chain consisting of 133 amino acid residue having Mr of 16500. N-terminal amino group is blocked with an acetyl group. N-acetylalanine is identified in the amino terminal of bovine, rabbit, horse, pig, human and rat liver microsomal cytochrome b5 preparations (Ozols and Heinemann, 1982; Abe et al. 1985; Ozols, 1989; Gibson et al. 1990). Furthermore, amino acid sequence of liver microsomal cytochrome b5 from those species has been determined (for a recent review, see Arınç, 1991). Limited hydrolysis of liver microsomal cytochrome b5 yields two domains. Hydrophilic domain also called heme peptide has a Mr of 11000, contains 82-97 amino acid residues depending somewhat on the species and solubilizing conditions, occupies amino terminal of protein and resides entirely in the aqueous medium. Although it catalyzes oxidation of cytochrome b5 reductase, it is not functional in the reduction of acyl-CoA desaturases and cytochrome P450 isozymes. Hydrophobic domain containing about 40-46 amino acid residues has a Mr of approximately 5000, occupies carboxyl-terminal of the protein and is responsible for anchoring b5 tightly to the biological and synthetic membranes (Strittmatter and Dailey, 1982). In the absence of detergents, hydrophilic peptide exists as a monomer, whereas native cytochrome b5 forms octomer and the hydrophobic peptide occurs in highly aggregated state in aqueous solutions.

In many species a striking sequence homology is found among the hydrophobic peptides. Mammalian liver hydrophilic cytochrome b5 peptides show an average of 94% identity when pairs of sequences are compared, whereas hydrophobic peptides of cytochrome b5 appear to have an average sequence identity of 77% when the pairs of mammalian liver cytochrome b5 sequence are compared.

By chemical modification studies Strittmatter (1960) has shown that at least one histidine residue is involved in heme binding which is also confirmed by X-ray analysis (Mathews et al. 1971). Carboxyl residues are found to be essential in the catalytic action of liver microsomal cytochrome b5. Glu 47, Glu 48, Glu 52, Asp 64 and a heme propionate group are modified covalently with methylamine and kinetics of cytochrome b5 reduction with cytochrome b5 reductase demonstrated that Km has increased from 9 μM to 91 μM (Dailey and Strittmatter, 1979).

Further studies revealed that the interaction between the heme peptide of microsomal cytochrome b5 and cytochrome b5 reductase is dominated by complementary charge pairing involving lysyl groups on reductase with carboxyl groups on cytochrome b5. In the presence of a water soluble catalyst carbodiimide, amide bonds were formed between the carboxyl residues on liposome-bound cytochrome b5 and the lysyl residues on liposome-bound cytochrome b5 reductase and the resulted covalently bound cytochrome b5 reductase-cytochrome b5 complex has retained its catalytic activity (Hackett and Strittmatter, 1984).

Studies of Strittmatter and coworkers (for earlier reviews, see Strittmatter, 1963; Oshino, 1980; Strittmatter and Dailey, 1982) have presented a great deal of information on the hydrophilic heme-peptide of calf liver microsomal cytochrome b5, including its X-ray structure (Mathews et al. 1971; Mathews et al. 1979). Furthermore, hydrophilic peptide of cytochrome b5 in solution has recently been the subject of high-resolution, two-dimensional 500 MHz NMR studies, which have resulted in the nearly complete assignment of all the protein resonances in the protein and that the solution structure of cytochrome b5 is a more dynamic entity than that found in solid state (Guiles et al. 1990).

Recent experiments with liposome-bound bovine liver microsomal cytochrome b5 indicated that hydrophobic peptide in the membrane is in the "cis" configuration, that is, it penetrates to the middle of bilayer, loops back in a hair-pin configuration to the outer surface to place both amino and carboxyl termini on the same external surface of the phospholipid bilayer. The five carboxyl terminal residues of hydrophobic peptide including Glu 132 and Asn 133 may contribute both overall tertiary structure and ion pairing with the phospholipid head groups (Arınç et al. 1987). The subsequent results of Ozols (1989) with rabbit liver microsomal cytochrome b5 confirmed the "cis" configuration of cytochrome b5.

Cytochrome b5 was first expressed in *E. coli* by Sligar's group (Beck von Bodman et al. 1986), who constructed two synthetic double-stranded DNA molecules based on the published amino acid sequence for rat liver cytochrome b5 by Ozols and Heinemann (1982). One coded for 134 amino acid peptide, whereas the second one coded for 99 amino acid peptide. Upon expression and removal of the initiator codon, these two produce the whole rat b5 (133 residue) and the erythrocyte (98 residue) forms of the protein in *E. coli*. They also noticed that the hydrophilic peptide was expressed at a much higher level. Subsequently, the gene coding for lipase-solubilized bovine liver microsomal cytochrome b5 has been synthesized and expressed in *E. coli* by Funk et al. (1990). Recently Holloway and coworkers (Ladokhin et al. 1991) isolated cDNA for rabbit liver microsomal cytochrome b5 from a λgt11 library. This cDNA was then used to generate a mutated mRNA where the codons for Trp 108 and 112 were replaced by codons for Leu. The sequence was expressed in *E. coli* and the single Trp 109 containing mutant was isolated and sequenced. They (Ladokhin et al. 1991) showed that the mutant's ability to participate in desaturation of long-chain fatty acids is the same as that of rabbit microsomal cytochrome b5. These studies have demonstrated that mutant forms of microsomal cytochrome b5 can be isolated from *E. coli* in sufficient quantities for detailed biophysical characterization and for further understanding of structure-activity relationship of microsomal cytochrome b5.

Similar Properties of Electron Transfer Proteins and Some Problems

From a comparison of structural and functional properties of microsomal reductases and cytochrome b5, it can be concluded that cytochrome b5, NADH cytochrome b5 reductase and NADPH cytochrome P450 reductase, in spite of differences in their prosthetic groups and molecular weights, share similar properties.

- All three proteins are amphipathic in nature, consist of a larger hydrophilic domain which contains the prosthetic groups and a smaller hydrophobic domain which anchors in microsomal membrane.
- The N-terminal amino acid residues are blocked. N-acetylglycine and N-acetylalanine are detected in the amino terminal of cytochrome P450 reductase and cytochrome b5, respectively, while cytochrome b5 reductase has N-myristoylglycine in its amino terminal.
- NADH cytochrome b5 reductase and NADPH cytochrome P450 reductase exhibit a ping-pong type kinetic mechanism for NADH and NADPH mediated cytochrome b5 and cytochrome *c* reduction, respectively.
- In addition to their membrane anchoring hydrophobic domains, presence of phospholipids are required for their physiological catalytic functions.

In oxidative hydroxylation reactions catalyzed with cytochrome P450 as terminal oxidase, the results so far obtained indicate that cytochrome b5 stimulates cytochrome P450 dependent monooxygenase reactions. In recent studies with reconstituted systems, the influence of cytochrome b5 was shown to depend on cytochrome P450 isozyme used, the choice of substrate and on the stoichiometry of the constituting components (for reviews, see Bonfils et al. 1991; Schenkman, 1991). In addition, experiments carried out in our laboratory have shown that different fractions of purified cytochrome b5 obtained from the same tissue also influence the degree of stimulation in a different manner in the reconstituted systems (Pasha, 1992).

Two cytochrome b5 fractions were obtained in pure form from sheep lung microsomes. The reconstituted systems contained the same amount of purified lung cytochrome P450LgM2 (Adalı and Arınç, 1990, 1992) (0.48 nmol), 150 units (System 1) or 300 units (System 2) of lung cytochrome P450 reductase (İşcan and Arınç, 1986) and increasing concentrations of lung cytochrome b5 Fraction 1 (Fig. 6) or lung b5 Fraction 2 (Fig. 6). The addition of cytochrome b5 Fraction 1 enhanced N-demethylation of benzphetamine 2.48 fold and 1.98 fold when it was present 0.5:1 ratio with respect to lung cytochrome P450LgM2 in System 1 and System 2, respectively. By the increasing concentrations of cytochrome b5 Fraction 1, enhancement of the rate of N-demethylation continued and reached to 3.60 fold when cytochrome b5:cytochrome P450LgM2 ratio was 3:1 in System 1. With cytochrome b5 fraction 2, the highest stimulation of N-demethylation activity occurred at a 0.25:1 and 0.5:1 molar ratio of cytochrome b5 to cytochrome P450LgM2. Besides, the greater fold increase in benzphetamine N-demethylation activity due to addition of either cytochrome b5 was observed in the system with the lower concentration of cytochrome P450 reductase, most resembling the conditions prevailing in the intact microsomes. However, unlike the effect of cytochrome b5 Fraction 1, Fraction 2 cytochrome b5 exerted inhibitory effect when cytochrome b5:cytochrome P450 ratio was 2:1 or more in both System 1 and System 2 (Pasha, 1992).

38

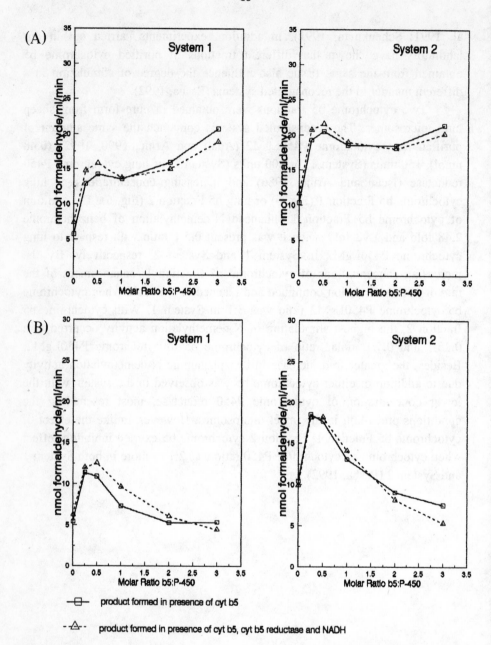

- ▭ product formed in presence of cyt b5

- △ product formed in presence of cyt b5, cyt b5 reductase and NADH

Fig. 6. Influence of lung cytochrome b5 fraction 1 (A) and lung cytochrome b5 fraction 2 (B) concentrations on cytochrome P450LgM2 dependent benzphetamine N-demethylation. The reconstituted medium contained 0.48 nmol of lung cytochrome P450LgM2, 0.08 mg phosphatidylcholine dilauroyl, 150 units (System 1) or 300 units (System 2) of lung P450 reductase and increasing concentrations of lung cytochrome b5 fraction 1 (A) or cytochrome b5 fraction 2 (B) in a final volume 1 ml.

References

Abe K, Sugita Y (1979) Properties of cytochrome b_5 and methemoglobin reduction in human erythrocytes. Eur J Biochem 101:423-428

Abe K, Kimura S, Kizawa R, Anan FK, Sugita Y (1985) Amino acid sequences of cytochrome b5 from human, porcine, and bovine erythrocytes and comparison with liver cytochrome P450. J Biochem (Tokyo) 97:1659-1668

Adalı O, Arınç E (1990) Electrophoretic, spectral, catalytic and immunochemical properties of highly purified cytochrome P450 from sheep lung. Int J Biochem 22:1433-1444

Adalı O, Arınç E (1992) Further characterization of sheep lung cytochrome P450LgM2. In: Archakov AI, Bachmanova GI (eds) Cytochrome P450: biochemistry and biophysics. INCO-TNC, Moscow, pp 51-53

Arınç E (1991) Essential features of NADH dependent cytochrome b5 reductase and cytochrome b5 of liver and lung microsomes. In: Arınç E, Schenkman JB, Hodgson E (eds) Molecular aspects of monooxygenases and bioactivation of toxic compounds. Plenum Press, New York, pp 149-170

Arınç E, Rzepecki LM, Strittmatter P (1987) Topography of the C terminus of cytochrome b5 tightly bound to dimyristoylphosphatidylcholine vesicles. J Biol Chem 262:15563-15567

Baggot JP, Langdon RG (1970) The relation of reduced triphosphopyridine nucleotide cytochrome c reductase structure to its interaction with cofactors. J Biol Chem 245:5888-5896

Beck von Bodman S, Schuler MA, Jollie DR, Sligar ST (1992) Synthesis, bacterial expression and mutagenesis of the gene coding for mammalian cytochrome b_5. Proc Natl Acad Sci USA 83:9443-9447

Black S, Coon MJ (1982) Structural features of liver microsomal NADPH-cytochrome P450 reductase. Hydrophobic domain, hydrophilic domain, and connecting region. J Biol Chem 257:5929-5938

Bonfils C, Saldana J-L, Balny C, Maurel P (1991) Electron transfer from cytochrome b5 to cytochrome P450. In: Arınç E, Schenkman JB, Hodgson E (eds) Molecular aspects of monooxygenases and bioactivation of toxic compounds. Plenum Press, New York, pp 171-183

Bredt DS, Hwang PM, Glatt CE, Lowenstein C, Reed RR, Snyder SH (1991) Cloned and expressed nitric oxide synthase structurally resembles cytochrome P450 reductase. Nature 351:714-718

Dailey HA, Strittmatter P (1979) Modification and identification of cytochrome b_5 carboxyl groups involved in protein-protein interactions with cytochrome b_5 reductase. J Biol Chem 254:5388-5396

D' Arrigo A, Manera E, Longhi R, Borgese N (1993) The specific subcellular localization of two isoforms of cytochrome b_5 suggests novel targeting pathways. J Biol Chem 268:2802-2808

Dignam JD, Strobel HW (1977) NADPH-cytochrome *P*-450 reductase form rat liver: purification by affinity chromatography and characterization. Biochemistry 16:1116-1123

Fan LL, Masters BSS (1974) Properties of purified kidney microsomal NADPH-cytochrome *c* reductase. Arch Biochem Biophys 165:665-671

French JS, Coon MJ (1979) Properties of NADPH-cytochrome *P*-450 reductase purified from rabbit liver microsomes. Arch Biochem Biophys 195:565-577

Funk WD, Lo TP, Mauk MR, Brayer GD, MacGillioray RTA, Mauk AG (1990) Mutagenic, electrochemical and crystallographic investigation of the cytochrome b_5 oxidation-reduction equilibrium: Involvement of asparigine-57, serine-64, and heme propionate-7. Biochemistry 29:5500-5508

Gibson BW, Falick AM, Lipka JJ, Waskel LA (1990) Mass spectrometric analysis of rabbit and bovine trypsin-solubilized cytochrome b_5. J Prot Chem 9:695-703

Giordano SJ, Steggles AW (1991) The human liver and reticulocyte cytochrome b_5 mRNAs are products from a single gene. Biochem Biophys Res Commun 178:38-44

Guiles RD, Altman J, Kuntz ID, Waskel L, (1990) Structural studies of cytochrome b_5: Complete sequence-specific resonance assignments for the trypsin-solubilized microsomal ferrocytochrome b_5 obtained from pig and calf. Biochemistry 29:1276-1289

Gum JR, Strobel HW (1981) Isolation of the membrane-binding peptide of NADPH-cytochrome P-450 reductase. J Biol Chem 256:7478-7486

Güray T, Arınç E (1990) Purification of NADH-cytochrome b_5 reductase from sheep lung and its electrophoretic, spectral and some other properties. Int J Biochem 22:1029-1037

Güray T, Arınç E (1991) Kinetic properties of purified sheep lung microsomal NADH-cytochrome b5 reductase. Int J Biochem 23:1315-1320

Hackett CS, Strittmatter P (1984) Covalent cross linking of the active site of vesicles-bound cytochrome b_5 and NADH-cytochrome b_5 reductase. J Biol Chem 259:3275-3282

Haniu M, Iyanagi T, Miller P, Lee TD, Shively JE (1986) Complete amino acid sequence of NADPH-cytochrome P-450 reductase from porcine hepatic microsomes. Biochemistry 25:7906-7911

Haniu M, McManus ME, Birkett DJ, Lee TD, Shively JE (1989) Structural and functional analysis of NADPH-cytochrome P-450 reductase from human liver: complete sequence of human enzyme and NADPH-binding sites. Biochemistry 28:8639-8645

Hanukoğlu I (1992) Reductases of the bacterial, mitochondrial and microsomal cytochrome P450 systems: Structural comparisons. In: Archakov AI, Bachmanova GI (eds) Cytochrome P450: biochemistry and biophysics. INCO-TNC, Moscow, pp 339-344

Hanukoğlu I, Gutfinger T (1989) cDNA sequence of adrenodoxin reductase-identification of NADP-binding sites in oxidoreductases. Eur J Biochem 180:479-484

Hegesh E, Hegesh J, Kaftory A (1986) Congenital methemoglobinemia with a deficiency of cytochrome b5. N Eng J Med 314:757-761

Holloway PW (1971) A requirement for three protein components in microsomal stearyl coenzyme A desaturation. Biochemistry 10:1556-1560

Holloway PW, Katz JT (1972) A requirement for cytochrome b_5 in microsomal stearyl coenzyme A. Biochemistry 11:3689-3695

Hultquist DE, Passon PG (1971) Catalysis of methemoglobin reduction by erythrocyte cytochrome b_5 and cytochrome b_5 reductase. Nature 229:252-254

Hultquist DE, Dean RT, Douglas RH (1974) Homogenous cytochrome b_5 from human erythrocytes. Biochem Biophys Res Commun 60:28-34

Imai Y (1981) The roles of cytochrome b_5 in reconstituted monooxygenase systems containing various forms of hepatic microsomal cytochrome P-450. J Biochem (Tokyo) 89:351-362

Ito A (1980) Cytochrome b_5-like hemoprotein of outer mitochondrial membrane; OM cytochrome b. J Biochem (Tokyo) 87:73-80

Ito A, Hayashi S, Yoshida T (1981) Participation of a cytochrome b_5-like hemoprotein of outer mitochondrial membrane (OM cytochrome b) in NADH-semidehydroascorbic acid reductase activity of rat liver. Biochem Biophys Res Commun 101:591-598

İşcan MY, Arınç E (1986) Kinetic and structural properties of biocatalytically active sheep lung microsomal NADPH-cytochrome c reductase. Int J Biochem 18:731-741

İşcan MY, Arınç E (1988) Comparison of highly purified sheep liver and lung NADPH-cytochrome P-450 reductases by the analysis of kinetic and catalytic properties. Int J Biochem 20:1189-1196

Jansson I, Tamburini PP, Favreau LV, Schenkman JB (1985) The interaction of cytochrome b_5 with four cytochrome P-450 enzymes from untreated rat. Drug Metab Dispos 13:453-458

Jefcoate CR (1986) Cytochrome P-450 enzymes in sterol biosynthesis and metabolism. In: Ortiz de Montellano PR (ed) Cytochrome P450, structure, mechanism and biochemistry. Plenum, New York, pp 387-428

Katagiri M, Murakami H, Yabusaki Y, Sugiyama T, Okamoto M, Yamano T, Ohkawa H (1986) Molecular cloning and sequence analysis of full length cDNA for rabbit liver NADPH-cytochrome P450 reductase mRNA. J Biochem (Tokyo) 100:945-954

Kimura S, Abe K, Sugita Y (1984) Differences between erythrocyte and liver cytochrome b_5 isolated from pig and human. FEBS Letters 169:143-146

Kobayashi S, Rikans LE (1984) Kinetic properties of guinea-pig liver microsomal NADPH cytochrome P-450 reductase. Comp Biochem Physiol 77B:313-318

Kozutsumi Y, Kawano T, Yamakawa T, Suzuki A (1990) Participation of cytochrome b_5 in CMP-N-acetylneuraminic acid hydroxylation in mouse liver cytosol. J Biochem (Tokyo) 108:704-706

Kozutsumi Y, Kawano T, Kawasaki H, Suzuki K, Yamakawa T, Suzuki A (1991) Reconstitution of CMP-N-acetylneuraminic acid hydroxylation activity using mouse liver cytosol fraction and soluble cytochrome b_5 purified horse erythrocytes. J Biochem (Tokyo) 110:429-435

Kulkoski JA, Weber JL, Ghazarian JG (1979) NADPH-cytochrome c reductase in outer membrane of kidney mitochondria. Arch Biochem Biophys 192:539-547

Kurzban GP, Strobel HW (1986) Preparation and characterization of FAD-dependent NADPH-cytochrome P-450 reductase. J Biol Chem 261:7824-7830

Ladokhin AS, Wang L, Steggles AW, Holloway PW (1991) Flourescence study of a mutant cytochrome b_5 with a single tryptophan in the membrane-binding domain. Biochemistry 30:10200-10206

Lederer F, Ghrir R, Guiard B, Cortial S, Ito A (1983) Two homologous cytochromes b_5 in a single cell. Eur J Biochem 132:95-102

Lu AYH, Coon MJ (1968) Role of hemoprotein P-450 in fatty acid ω-hydroxylation in a soluble enzyme system from liver microsomes. J Biol Chem 243:1331-1332

Masters BSS, Kamin H, Gibson QH, Williams CH (1965) Studies on the mechanism of microsomal triphosphopyridine nucleotide-cytochrome c reductase. J Biol Chem 240:921-931

Mathews FS, Argos P, Levine P (1971) The structure of cytochrome b_5 at 2.0 Å resolution. Cold Spring Harbor Symp Quant Biol 36:387-395

Mathews FS, Czerwinski EW, Argos P (1979) The X-ray crystallographic structure of calf liver cytochrome b_5. In: Dolphin D (ed) The porphyrins vol VII. Academic Press, New York, pp 107-147

Mayer RT, Durrant JL (1979) Preparation of homogeneous NADPH cytochrome c (P-450) reductase from house flies using affinity chromatography techniques. J Biol Chem 254:756-761

Miwa FT, Lu AYH (1984) The association of and NADPH-cytochrome P450 reductase in phospholipid membranes. Arch Biochem Biophys 234:161-166

Morgan ET, Coon MJ (1984) Effects of cytochrome b_5 on cytochrome P-450 catalyzed reactions. Drug Metab Dispos 12:358-364

Nebert DW, Gonzalez FJ (1987) P450 genes: structure, evolution and regulation. Ann Rev Biochem 56:945-993

Oshino N (1980) Cytochrome b_5 and its physiological significance. In Schenkman JB, Kupfer D (eds) Hepatic cytochrome P-450 monooxygenase system. Pergamon Press, New York, pp 407-447

Oshino N, Imai Y, Sato R (1971) A function of cytochrome b_5 in fatty acid desaturation by rat liver microsomes. J Biochem (Tokyo) 69:155-167

Ozols J (1989) Structure of cytochrome b_5 and its topology in the microsomal membrane. Biochim Biophys Acta 997:121-130

Ozols J, Heinemann FS (1982) Chemical structure of cytochrome b5. Isolation of peptides by high pressure liquid chromatography. Biochim Biophys Acta 704:163-173

Pasha RPK (1992) Effects of liver and lung cytochrome b5 on lung cytochrome P-450LgM2 catalyzed benzphetamine N-demethylase reactions. MSc Thesis, Middle East Technical University, Ankara, Turkey, pp 114

Passon PG, Reed DW, Hultquist DE (1972) Soluble cytochrome b_5 from human erythrocytes. Biochim Biophys Acta 275:51-61

Peterson JA, Prough RA (1986) Cytochrome *P*-450 reductase and cytochrome b_5 in cytochrome *P*-450 catalysis. In: Ortiz de Montellano PR (ed) Cytochrome P450, structure, mechanism and biochemistry. Plenum, New York, pp 89-117

Phillips AH, Langdon RG (1962) Hepatic triphosphopyridine nucleotide-cytochrome *c* reductase: isolation, characterization and kinetic studies. J Biol Chem 237:2652-2660

Porter TD (1991) An unusual yet strongly conserved flavoprotein reductase in bacterial and mammals. Trends Biochem Sci 16:154-158

Porter TD, Kasper CB (1985) Coding nucleotide sequence of rat NADPH-cytochrome P-450 oxidoreductase cDNA and identification of flavin-binding domains. Proc Natl Acad Sci USA 82:973-977

Porter TD, Kasper CB (1986) NADPH-cytochrome P-450 oxidoreductase: flavin mononucleotide and flavin adenine dinucleotide domains evolved from different flavoproteins. Biochemistry 25:1682-1687

Raw I, Mahler HR (1959) Studies on electron transfer enzymes. III Cytochrome b_5 of pig liver mitochondria. J Biol Chem 234:1867-1873

Schenkman JB (1991) Cytochrome P450-dependent monooxygenase: An overview. In: Arınç E, Schenkman JB, Hodgson E (eds) Molecular aspects of monooxygenases and bioactivation of toxic compounds. Plenum Press, New York, pp 1-10

Shen AL, Kasper CB (1993) Protein and gene structure and regulation of NADPH-cytochrome P450 oxidoreductase. In Schenkman JB, Greim H (eds) Cytochrome P450. Handbook of experimental pharmacology vol 105 Springer-Verlag, Heildelberg, pp 35-59

Segel IH (1975) Enzyme kinetics. Wiley, New York, pp 826-829

Serabjit-Singh CJ, Wolf RJ, Philpot RM (1979) The rabbit pulmonary monooxygenase system. Immunological and biochemical characterization of enzyme components. J Biol Chem 254:9901-9907

Smith MA, Jonsson L, Stymne S, Stobart M (1992) Evidence for cytochrome b_5 as an electron donor in ricinoleic acid biosynthesis in microsomal preparations from developing castor bean (*Ricinus communis* L.). Biochem J 287:141-144

Strittmatter P (1960) The nature of heme binding in microsomal cytochrome b_5. J Biol Chem 235:2492-2497

Strittmatter P (1963) Microsomal cytochrome b5 and cytochrome b5 reductase. The Enzymes 8:113-145

Strittmatter CP, Ball EG (1952) A hemochromogen component of liver microsomes. Proc Natl Acad Sci USA 38:19-25

Strittmatter P, Velick SF (1956) The isolation and properties of microsomal cytochrome. J Biol Chem 221:253-264

Strittmatter P, Dailey HA (1982) Essential structural features and orientation of cytochrome b_5. In: Mortonosi N (ed) Membranes and transport, vol 1. Plenium Press, New York, pp 71-82

Strobel HW, Lu AYH, Heidema J, Coon MJ (1970) Phoshatidylcholine requirement in the enzymatic reduction of hemoprotein P-450 and in fatty acid, hydrocarbon, and drug hydroxylation. J Biol Chem 245:4851-4854

Strobel HW, Shen S, Nadler SG (1992) Electrostatic components in cytochrome P450 reductase: Cytochrome P450 interaction. In: Archakov AI, Bachmanova GI (eds) Cytochrome P450: biochemistry and biophysics. INCO-TNC, Moscow, pp 254-259

Takematsu H, Kozutsumi Y, Suzuki A, Kawasaki T (1992) Molecular cloning of rabbit cytochrome b5 genes: Evidence for the occurrence of two separate genes encoding the soluble and microsomal forms. Biochem Biophys Res Commun 185:845-851

Urenjak J, Linder D, Lumper L (1987) Structural comparison between the trout and mammalian hydrophilic domain of NADPH-cytochrome P-450 reductase. J Chromatogr 397:123-136

Vermillion JL, Coon MJ (1978) Identification of the high and low potential flavins of liver microsomal NADPH-cytochrome P-450 reductase. J Biol Chem 253:8812-8819

Vogel F, Lumper L (1986) Complete structure of the hydrophobic domain in the porcine NADPH-cytochrome P-450 reductase. Biochem J 236:871-878

von Heijne G (1985) Structure and thermodynamic aspects of the transfer of proteins into and across membranes. Curr Top Membr Trans 24:151-

Yamano S, Aoyama T, McBride OW, Hardwick JP, Gelboin HV, Gonzalez FJ (1989) Human NADPH-P450 oxidoreductase: complementary DNA cloning, sequence and vaccinia virus-mediated expression and localization of the CYPOR gene to chromosome 7. Mol Pharmacol 36:83-88

Yasukochi Y, Okita RT, Masters BSS (1980) Comparison of properties of detergent solubilized NADPH-cytochrome P-450 reductases from pig liver and kidney: Immunological, kinetic and reconstitutive properties. Arch Biochem Biophys 202:491-498

Voznesensky AI, Schenkman JB (1992) The cytochrome P450 2B4-NADPH cytochrome P450 reductase complex is not formed by charge-pairing. J Biol Chem 267:14669-14676

Interaction Between Cytochrome P450 and Reductase

John B. Schenkman and Andrei I. Voznesensky
Department of Pharmacology
University of Connecticut
Health Center
Farmington, CT066030 USA

Introduction

The cytochrome P450 monooxygenase systems consist of two membrane-bound proteins, a hemoprotein terminal oxidase (cytochrome P450) and a pyridine nucleotide-utilizing reductase as a source of reducing equivalents. Mitochondrial forms of the monooxygenase make use of a nonheme iron sulfur protein as an electron carrier between the reductase and terminal oxidase, while in the endoplasmic reticulum NADPH-cytochrome P450 reductase serves as the direct source of reducing equivalents for the monooxygenase. This latter reductase is a flavoprotein, containing two flavin molecules per molecule of enzyme, one of which is flavin mononucleotide (FMN) and the other is flavin adenine dinucleotide (FAD) (Iyanagi et al.,1974). The two flavin molecules have different functions in the reductase, one bringing electrons into the flavoprotein, and the other serving to transfer electrons out to a redox partner, which acts as an acceptor molecule. FAD was shown to receive reducing equivalents from NADPH, while FMN was shown to function at the donor side to such acceptors as cytochrome c and cytochrome P450 (Vermilion and Coon,1978).

Although the two components of the microsomal (endoplasmic reticulum) P450 monooxygenase system are membrane-bound, the incubation of the microsomes with dilute trypsin solution will solubilize a catalytically active fragment of the reductase (Nishibayashi and Sato, 1967), leaving a hydrophobic NH2-terminal 56 amino acid oligopeptide anchored in the membrane bilayer (Black et al. 1979; Enoch and Strittmatter,1979). In contrast, cytochrome P450 was reportedly not solubilized by protease, but was instead

NATO ASI Series, Vol. H 90
Molecular Aspects of Oxidative Drug Metabolizing Enzymes
Edited by E. Arınç, J. B. Schenkman and E. Hodgson
© Springer-Verlag Berlin Heidelberg 1995

converted to a denatured form, cytochrome P420 (Nishibayashi and Sato, 1967) when the protease concentration was raised too high. The catalyltic subunit of NADPH-cytochrome P450 reductase is able to transfer electrons to a number of redox acceptors, but not to cytochrome P450 (Black and Coon, 1982) . With these thoughts in mind, a study was begun into the nature of the interaction between cytochrome P450 and its reductase.

Stoichiometry

The amount of NADPH-cytochrome P450 reductase present in liver microsomes has been estimated as approximately 5% of the molar amount of cytochrome P450 in rat (Peterson et al.,1976). Thus, it is clear that one molecule of reductase must somehow service many molecules of cytochrome P450, as well as provide reducing equivalents for other microsomal electron transfer pathways, such as fatty acid desaturases. While suggestions were made that the reductase exists in the membrane among a cluster of cytochromes P450, it is probable that the proteins are freely diffusable. Nisimoto and coworkers (1983) studied the intraction between cytochrome P450 and NADPH-cytochrome P450 reductase in reconstituted phospholipid vesicles. They concluded, based upon fluorescence intensity decreases in labelled reductase that significant molecular interactions occur between these proteins. Studies on the stoichiometry and kinetics of interactions between NADPH-cytochrome P450 reductase and cytochrome P450 indicated the formation of a 1:1 transient complex between the proteins with a calculated dissociation constant of $0.1\mu M$ (Miwa and Lu, 1984). A K_m for reductase in cytochrome P450-dependent substrate oxidation was found to be $0.38\mu M$ (Tamburini and Schenkman, 1986c). The binding site of NADPH-cytochrome P450 reductase on cytochrome cytochrome P450 was suggested to differ from the binding site of cytochrome b_5, since the K_m for reductase was the same in the presence as in the absence of cytochrome b_5 (Tamburini and Schenkman, 1987).

Reductase interactions

Studies on the interaction between NADPH-cytochrome P450 reductase and cytochrome b_5 in phospholipid vesicles revealed that the membrane-binding tail was required (Enoch and Strittmatter, 1979). The hydrophobic carboxyl-terminal tail was suggested to be necessary for binding of the cytochrome b_5 to the membrane (Dailey and Strittmatter, 1978). The interaction between native (amphipathic) cytochrome b_5 and NADPH-cytochrome P450 reductase was found to involve complementary charge pairing (Dailey and Strittmatter, 1980) utilizing carboxyl residues on the hemoprotein around the exposed heme crevice. The interaction was demonstrated to be electrostatic by an increase in K_m for cytochrome b_5 on neutralization of anionic charges on cytochrome b_5. Further, using a water-soluble carbodiimide (EDC) cytochrome b_5 could be crosslinked to NADPH-cytochrome P450 reductase in 1:1 heterodimers (Nisimoto and Lambeth, 1985). The crosslinked cytochrome b_5 was reducible, but cytochrome c could not be reduced subsequently, indicating a common electron output site on the reductase for both cytochromes.

Although both cytochrome c and cytochrome b_5 appear to compete for electrons from NADPH-cytochrome P450 reductase, the latter has been shown to make use of anionic residues around its heme cleft for interaction with electron transfer proteins (Dailey and Strittmatter, 1980). In contrast, cytochrome c has been shown to make use of cationic residues around its heme cleft for interaction with its redox partners (Ng et al., 1977). Tamburini et al (1985) demonstrated that the reduction of native cytochrome b_5 by NADPH-cytochrome P450 reductase was actually greatly accelerated when a few carboxyl groups on cytochrome b_5 were neutralized by methylamidation, in contrast to the observations of Dailey and Strittmatter (1980). Similarly, methylamidation of carboxyl residues on NADPH-cytochrome P450 reductase caused a stimulation of the rate of cytochrome b_5 reduction (Tamburini and Schenkman, 1986c). Such observations are suggestive of electrostatic interactions, but not necessarily of complementary charge-pairing.

While the native reductase readily reduces cytochrome b_5 the neutraliztion of the cytochrome b_5 heme propionate residues decreased the K_m for cytochrome b_5. Depending upon the ionic strength, neutralization of the heme propionate affected the V_{max} as well; at low ionic strength V_{max} was increased, while at higher ionic strength it was unchanged (Tamburini and Schenkman, 1986d). Early studies on the kinetics of NADPH-cytochrome P450 reductase revealed that electron transfer by the proteolytically prepared catalytic fragment to cytochrome c was strongly sensitive to ionic strength (Phillips and Langdon, 1962), increasing greatly with increasing salt concentration. The proteolytically generated NADPH-cytochrome P450 reductase is devoid of ability to reduce cytochrome b5 , as is lipase-prepared reductase (Bilimoria and Kamin, 1973); activity could, however be demonstrated in the presence of high salt concentrations.

Influence of protein modifiers

Ferric cytochrome P450 and oxidized NADPH-cytochrome P450 reductase interact in a reconstituted system and such interaction can often be monitored spectrophotometrically as a shift in the spin equilibrium of the cytochrome P450 form (Tamburini et al. 1986a). The dissociation constants for the interactions ranged from 0.28-0.80 μM with 6 rat forms of cytochrome P450 and was 3.0μM with CYP2B4. The magnitudes of the spectral changes varied widely.

Modifiers of amino acid residues of the reductase were used to examine the nature of its interaction with a number of redox partners. When, for example, the lysine residues on the reductase were amidinated (reacted with ethyl acetimidate), the resultant amidines were still cationic, but contained a bulky ethyl amidine group. Such treatment of the reductase did not appear to affect the absorption spectrum of the reductase, but did stimulate electron transfer from the reductase to potassium ferricyanide to cytochrome c and cytochrome b_5, and decreased the rate of reduction of two forms of cytochrome P450 (Tamburini and Schenkman, 1986c). On the other hand, product formation by these forms of cytochrome

P450 was stimulated more than three-fold. The meaning of such contrasting observations is unclear, but may be an indication that caution must be used in interpretation of effects of structural alterations of this protein; alterations in the charged residues of the reductase could cause structural modifications and confuse the understanding of the nature of the protein-protein interactions. The reaction of cytochrome P450 2B4 with fluorescein isothiocyanate (FITC) resulted in a single covalently bound FITC at the α-amino group; a second FITC binds at higher FITC concentrations to lysine residue 382 (Bernhardt et al., 1984). The effect of these modifications was a stepwise diminution of benzphetamine N-demethylase activity and the cytochrome P450 reduction rate. From such data the authors concluded that interaction between NADPH-cytochrome P450 reductase and cytochrome P4502B4 was electrostatic in nature, involving a charge-pairing mechanism between the NH_2-terminal methionine of the cytochrome P450 with the reductase.

While complete amidination of amino residues on NADPH-cytochrome P450 reductase did not impair substrate turnover, and only decreased the rate of cytochrome P450 reduction by about 50%, a much stronger influence was observed when the carboxyl residues of the reductase were charge-neutralized by methylamidation. The neutralization of as few as 3 of the 100 carboxyl residues of NADPH-cytochrome P450 reductase were enough to diminish its ability to reduce cytochrome c by 50% (Tamburini and Schenkman, 1986c). As few as 7 carboxyl residues neutralized by methylamidation was sufficient to remove more than 90% of the cytochrome c reductase activity. Cytochrome P450 reduction was also sensitive to neutralization of NADPH-cytochrome P450 reductase carboxyl residues. Reduction of cytochrome P450 is biphasic; on semilogarithmic paper the plot appears to resemble two concomitant first order reactions, a fast phase and a slow phase. It is the fast phase which appears to relate to substrate turnover (Schenkman et al., 1982). Neutralization of carboxyl residues on the reductase caused strong inhibition of the fast phase, but was without effect on the slow phase of cytochrome P450 reduction (Tamburini et al., 1986c). Substrate turnover by three different forms of rat cytochrome P450 were similarly inhibited by neutralization of carboxyl groups on NADPH-cytochrome P450 reductase. Of special interest, while methylamidation inhibited substrate turnover and the fast phase of cytochrome P450 reduction, it was without influence on the spectrophotometrically observed interaction between CYP2C11 and NADPH-cytochrome P450 reductase (Table 1). The calculated dissociation constant was the

Table 1
Effect of chemical neutralization of 7.2 carboxyl residues on NADPH-cytochrome P450 reductase

CYP2C11 binding	Kd	Vm
control	213 ± 55nM	47
amidinated	350 ± 80nM	38
amidinated,neutralized	329 ± 103nM	34

CYP2B4 Ethoxycoumarin dealk.	Km	Vm
control	378nM	2.9
amidinated	84	9.1
amidinated,neutralized	108	3.4

Data from Tamburini and Schenkman (1986c)

same before and after neutralization of 7 residues on NADPH-cytochrome P450 reductase. The K_m for ethoxycoumarin in the O-dealkylation by amidinated reductase and CYP2B4 was decreased, and carboxyl neutralization was virtually without influence on the K_m. However, amidination, a structural change, increased V_{max} 3-fold, and this was lost by carboxyl neutralization (Table 1).

Based upon our prior observations it was concluded that charged group modification on the reductase does not influence the affinity, but does affect the functionality of the complex of the cytochrome P450 and NADPH-cytochrome P450 reductase (Tamburini and Schenkman, 1986c). Using the water-soluble carbodiimide 1-ethyl-3(3-dimethylaminopropyl) carbodiimide (EDC) we were able to show that in the presence of NADPH-cytochrome P450 reductase and cytochrome b_5, with or without phospholipid, heterodimeric covalent complexes could be seen on denaturing gel electrophoresis (Tamburini et al., 1986b). Similarly, a high proportion of cytochrome b_5 and cytochrome P450 2B4 formed covalent heterodimer complexes which migrated at the appropriate molecular mass on denaturing electrophoresis conditions. However, only what appeared to be trace amounts of cytochrome P4502B4/NADPH-cytochrome P450 reductase heterodimers were seen when these two proteins were incubated with EDC (Tamburini et al., 1986b). Subsequent studies revealed these to only be artifacts, and that, in fact, no heterodimer complexes were produced (Voznesensky and Schenkman, 1992a). Similar observations were observed when CYP2B1 and rat NADPH-cytochrome P450 reductase were incubated with EDC; no evidence for covalently linked heterodimers was obtained (Nisimoto and Otsuka-Murakami, 1988).

Examination of electrostatic interactions

In our earlier studies on the interaction between cytochrome b_5 and cytochrome P4502C11 we observed (Tamburini and Schenkman, 1986d) that increases in the ionic strength of the medium impaired the interaction between these hemoproteins; changes in the sodium phosphate buffer concentration from 6 to 290mM raised the dissociation constant from 63nM to 2322nM. According to the Debye-Huckel relationship, the electrical potential of a medium will increase with increasing ionic content (Koppenol, 1980). Since the force of attraction varies inversely with the distance between the ions, the greater the salt concentration the closer the ions will be to each other, and the stronger the attraction. Charged residues on a protein will thus tend to become surrounded by ions in solution as the salt concentration is raised, thereby diminishing the interaction between charged residues on one protein with those on another protein in the solution. If the redox interaction between two such proteins is by complementary charge pairing, then increased salt levels would interfere with the interactions. In fact, when the effect of ionic strength (the conductance) of the medium containing a reconstituted monooxygenase system consisting of NADPH-cytochrome P450 reductase, CYP2B4 and dilaurylphosphatidylcholine (molar ratio 1:10:880) was examined, it was observed that increasing the ionic strength actually stimulated electron transfer to the hemoprotein (Figure 1). The stimulation was seen in both the apparent fast phase rate constant

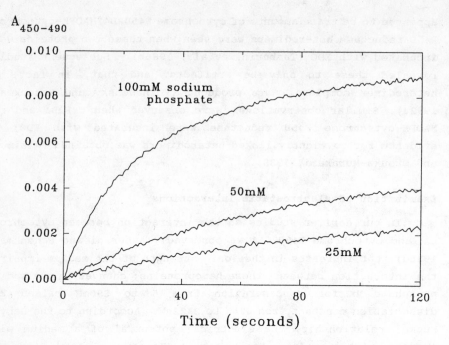

Figure 1.Effect of ionic strength on the reduction of CYP2B4 in the reconstituted system.

of reduction and in the slow phase rate constant. The effect was not limited to the reconstituted system, and was observed in microsomes as well (Voznesensky and Schenkman, 1992a). Further, substrate (benzphetamine) metabolism was stimulated by increasing ionic strength in a manner similar to cytochrome P450 reduction. Studies on the reductase dependence of cytochrome P450 reduction at low and high salt concentration revealed the K_m for reductase to be 26 fold higher at low ionic strength.A similar stimulation in the reduction of cytochrome c by increasing ionic strength was observed, even as reported by Phillips and Langdon (1962) and by Bilimoria and Kamin(1973).

Nisimoto and Edmondson (1992) also examined the effect of

ionic strength on reduction of cytochrome c, cytochrome b_5, and cytochrome P4502B4 by NADPH-cytochrome P450 reductase. However, when they examined turnover of NADPH in the reconstituted cytochrome P450 monooxygenase system, only marginal stimulation was observed with increasing ionic strength. Their conclusion was that the stimulatory influence of increased ionic strength was due to increased hydrophobic-hydrophobic interactions bringing the flavin and heme prosthetic groups closer. Our data, while in agreement with those of Nisimoto and Edmondson (1992), would suggest, rather, that charge repulsion between the highly cationic and anionic reductase and charged redox acceptors limits access to electron transfer interactions. An alternative possibility might be that complementary charge pairing traps the redox partners in a nonproductive orientation. This latter possibility, which was shown with yeast cytochrome c peroxidase and horse heart cytochrome c (Hazzard et al., 1988), probably does not apply, since it was not possible to covalently link complementary charged pairs of proteins with EDC.

As shown in Figure 1, with cytochrome P450 and reductase from microsomes of the phenobarbital-treated rabbit, the NADPH-dependent reduction of cytochrome P450 increased with increasing conductance of the medium. In addition, the reduction of cytochrome c and potassium ferricyanide also increased with ionic strength (Voznesensky and Schenkman, 1992a). What these three redox partners have in common is their ionic nature. The two cytochromes have both anionic and cationic charged residues and ferricyanide is an anionic molecule. If charge repulsion is the cause of the lower activity at lower ionic strength, then an uncharged redox acceptor should not be affected by altered ionic strength of the medium. In fact, that is what was seen (Fig. 2). When 1,4-benzoquinone was used as the electron acceptor, the rate of electron transfer from NADPH-cytochrome P450 reductase was uninfluenced by the conductivity of the medium.

The picture we are presenting is one in which charges on the reductase repulse interactions by other charged redox acccceptors, and surrounding these charged amino acid residues by oppositely

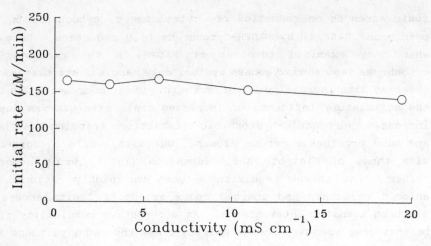

Figure 2.The lack of effect of ionic strength on electron transfer from P450 reductase to 1,4-benzoquinone.

charged ions in the medium blocks the repulsive effects thereby allowing other forces of attraction to come to bear. Because in the ionic medium it is not possible to determine whether the repulsion is between the anions or cations in solution, we turned to a more specific method of charge neutralization. The polyionic oligopeptide polylysine was added to the medium at a concentration of 1μM. In a medium containing 5mM sodium phosphate buffer, this low conentration of polylysine would scarcely influence the ionic strength of the medium. However, polylysine will bind to anion-dense regions on the proteins (Cheddar and Tollin, 1990), and thereby influence those protein-protein interactions affected by arionic centers. It was observed that in the presence of 1μM polylysine the fast phase rate constant of CYP2B4 reduction was increased by 74%. In addition the rate of ferricyanide reduction was increased by 40%, and almost no change in cytochrome c reduction was seen (Voznesensky and Schenkman, 1992a).

Effects of Glycerol

During the purification of cytochrome P450 forms,the hemoprotein is carried through solubilization and fractionation stages in the presence of glycerol. This solvent is generally added as a protein stabilizer (Gekko and Timasheff, 1981), as it was shown earlier that glycerol could convert cytochrome P420, a breakdown product of cytochrome P450, back to cytochrome P450 when added to the medium at levels of about 20% (Ichikawa and Yamano, 1967). The protein stabilizing effect of glycerol is probably by virtue of ordering water molecules around the protein, making protein unfolding (denaturation) thermodynamically less favorable (Gekko and Timasheff, 1981). Other effects of glycerol on the aqueous medium include a decrease in the dielectric constant;the dielectric constant of glycerol (42 at 25°) is about half of that of water. Although the change in aqueous solutions is small, it causes a pronounced effect on the conductivity of the solution and on the protein-protein interactions.

The influence of glycerol in the system is by virtue of its effect on the dielectric constant, and since the force of electrostatic interaction is inversely related to the dielectric constant, the force of interaction between NADPH-cytochrome P450 reductase and its charged redox partners is increased. If, as we suggest, the interaction is a repulsive one, then as the force of electrostatic interaction is increased, and activity will be decreased.

In Figure 3 is shown the reduction of phenobarbital-treated rabbit liver cytochrome P450 in phosphate buffer from 5mM to 200mM concentration (□). As described above, the increase in ionic strength resulted in an increased rate of electron transfer to the cytochrome. At 200mM buffer (20mS) activity was about maximal, and the stepwise addition of glycerol (Δ) or ethylene glycol (O) from 0% to 40% caused a stepwise decrease in the conductivity from 20 mS to about 7mS, with a concommittant decrease in the rate. Similar effects were observed with cytochrome c as the redox acceptor and with ferricyanide (Voznesensky and Schenkman, 1992b). With 1,4-benzoquinone as the redox partner, electrostatic interaction does not occur, since the acceptor is uncharged; increasing content of glycerol in the 100mM buffer only lowered the conductance, but did

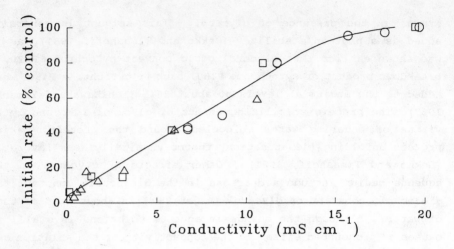

Figure 3. Effects of ionic strength and glycerol concentration on phenobarbital rabbit liver microsomal P450 reduction.

not affect the rate of reduction (Voznesensky and Schenkman, 1992b).

The interesting observation that glycerol and ethylene glycol interfered with electron transfer to charged redox acceptors from NADPH-cytochrome P450 reductase was further demonstrated to be the result of unfavorable electrostatic interactions by examining the NADPH-cytochrome P450 reductase dependence of CYP2B4 reduction in the presence of glycerol. The reduction of the cytochrome P450 was compared in the presence of 5% and 25% glycerol and 100mM sodium phosphate buffer. Iterative fit of the data to the Michaelis-Menten equation indicated the K_m for NADPH-cytochrome P450 reductase was 9 times larger at the higher glycerol level (Voznesensky and Schenkman, 1992b). From such observations it is clear that the nature of redox interaction between CYP2B4 and NADPH-cytochrome P450 reductase is not by complementary charge pairing. The effect is not restricted to this rabbit cytochrome P450, but was also observed with several forms of rat liver cytochrome P450 (Voznesensky and Schenkman submitted).

Role of hydrophobic interactions

Nisimoto (1986) has examined the interaction between cytochrome c and NADPH-cytochrome P450 reductase . This interaction was suggested to be electrostatic, since components could be trapped in covalent linkage with EDC. A 1:1 covalent complex was isolated and shown to block electron transfer to exogenous cytochrome c or cytochrome P450. However, reduction of the covalent complex proceeded very slowly, suggesting that the trapped proteins were not in a favorable configuration for electron transfer. Nevertheless, separation of proteolytically cleaved fragments of the heterodimer indicated the region of the NADPH-cytochrome P450 reductase that was involved in the complementary charge pairing with the cytochrome c was an oligopeptide encompassing residues 207 to 215, a highly anionic region of the reductase (Katagiri et al., 1986). The inability of the reductase to transfer reducing equivalents to exogenous cytochrome c or cytochrome P450 suggested that the covalently bound cytochrome c was blocking the docking site for electron transfer. In view of our studies suggesting a charge repulsion between charged redox partners and NADPH-cytochrome P450 reductase, it is probable that the covalent linking of charge-paired proteins reflected a non-specific binding, rather than a redox pairing of the two proteins. If this is the case, how does NADPH-cytochrome P450 reductase interact with its redox acceptors, and in particular, with cytochrome P450? There is some evidence at present which would point to hydrophobic interactions.

Earlier studies on the nature of cytochrome P450 interaction with NADPH-cytochrome P450 reductase demonstrated the need for the membrane-binding NH_2-terminal segment of the protein. In agreement with Black and Coon (1982), we observed that NADPH-cytochrome P450 reductase could not reduce cytochrome P450 when the hydrophobic NH_2-terminal segment was removed by trypsin (Voznesensky and Schenkman, 1992a). The proteolytically cleaved reductase was, however, able to reduce cytochrome c. The catalytic fragment of NADPH-cytochrome P450 reductase was unable to reduce cytochrome P450 even at elevated ionic strength. A possible role of the hydrophobic peptide region was suggested by the observation (Black et al.,

1979) that the hydrophobic peptide, cleaved from the reductase, was able to interfere with substrate turnover by CYP2B4 when added to the reconstituted system. Finally, hydropathy plots of a large number of cytochrome P450 forms, aligned relative to the cytochrome P450$_{cam}$ amino acid sequence, revealed some eleven hydrophobic regions on the cytochromes (Nelson and Strobel, 1988). While the topology of the eukaryotic cytochrome P450 forms is still not known, the possibility exists that one (or more) of these hydrophobic regions is involved in the redox docking between NADPH-cytochrome P450 reductase and cytochrome P450. Eukaryotic cytochrome P450 forms all have an NH$_2$-terminal region of high hydrophobicity. While the need for this region in interaction with the reductase, or with the endoplasmic reticulum membrane has not been shown, recent studies would suggest its noninvolvement in the former, and to possibly be unnecessary in the latter role. The recent observation (Larson et al., 1991) that cytochrome P450 cDNA truncated at the 5'region could be expressed in E.coli was a relevant finding. Further, since the expressed protein was nevertheless membrane-bound, it would appear that the NH$_2$terminal region of the protein is not critical for this function. The truncated protein could also function in the reconstituted system containing phospholipid and native NADPH-cytochrome P450 reductase, further indicating that the NH$_2$-terminal region on cytochrome P450 is not required for protein-protein interation. Clearly more studies are needed to further elucidate the manner of functional interaction between NADPH-cytochrome P450 reductase and cytochrome P450 .

References

Bernhardt R, Makower A, Janig G-R and Ruckpaul K (1984) Selective chemical modification of a functionally linked lysine in cytochrome P450 LM$_2$ Biochim Biophys Acta 785:186-190
Bilimoria MH, and Kamin H (1973) The effect of high salt concentrations upon cytochrome c,cytochrome b$_5$,and iron-EDTA reductase activities of liver microsomal NADPH-cytochrome c reductase. Ann NY Acad Sci 212:428-446
Black S, and Coon M (1982) Structural features of liver microsomal NADPH-cytochrome P450 reductase. Hydrophobic

domain, hydrophilic domain, and connecting region. J Biol Chem 257:5929-5938.

Black S, French J, Willians C, Jr, and Coon M (1979) Role of a hydrophobic polypeptide in the N-terminal region of NADPH-cytochrome P450 reductase in complex formation with P450LM. Biochem Biophys Research Commun 91:1528-1535.

Cheddar G and Tollin G (1990) Electrostatic effects on the spectral properties of Clostridium pasteurianum flavodoxin:Effects of salt concentration and polylysine. Arch Biochem Biophys 278:25-28

Dailey HA and Strittmatter P (1978) Structural and functional properties of the membrane binding segment of cytochrome b_5. J Biol Chem 253:8203-8209

Dailey HA and Strittmatter P (1980) Characterization of the interaction of amphipathic cytochrome b_5 with stearyl coenzyme A desaturase and NADPH-cytochrome P450 reductase.J Biol Chem 255:5184-5189

Enoch HG and Strittmatter P (1979) Cytochrome b_5 reduction by cytochrome P450. J Biol Chem 254:8976-8981

Gekko K and Timasheff SN (1981 Mechanism of protein stabilization by glycerol:preferential hydration in glycerol-water mixtures. Biochemistry 20:467-4676

Hazzard JT,McLendon G, Cusanovich MA and Tollin G (1988) Formation of electrostatically stabilized complex at low ionic strength inhibits interprotein electron transfer between yeast cytochrome c and cytochrome c peroxidase. Biochem Biophys Res Commun 151:429-434

Iyanagi T, Makino N, and Mason H (1974) Redox properties of the reduced nicotinamide adenine dinucleotide phosphate-cytochrome P-450 and reduced nicotinamide adenine dinucleotide-cytochrome b5 reductases. Biochemistry 13:1701-1710.

Ichikawa Y and Yamano T (1967) Reconversion of detergent-and sulfhydryl reagent-produced P-420 to P-450 by polyols and glutathione. Biochim Biophys Acta 131:490-497

Katagiri M, Murakami H, Yabusaki Y, Sugiyama T, Okamoto M, Yamano T, and Ohkawa H (1986) Molecular cloning and sequence analysis of full length cDNA for rabbit liver NADPH-cytochrome P450 reductase mRNA. J Biochem 100:945-954.

Koppenol W (1980) Effect of a molecular dipole on the ionic strength dependence of a bimolecular rate constant. Identification of the site of reaction. Biophys J 29:493-507.

Larson J, Coon M, and Porter T (1991) Purification and Properties of a Shortened Form of Cytochrome-P-450 2E1 - Deletion of the NH2-terminal membrane-insertion signal peptide does not alter the catalytic activities. Proc Natl Acad Sci USA 88:9141-9145.

Miwa FT and Lu AYH (1984) The association of and NADPH-cytochrome P450 reductase in phospholipid membranes. Arch Biochem Biophys 234:11-166

Nelson D, and Strobel H (1988) On the membrane topology of vertebrate cytochrome P-450 proteins. J Biol Chem 263:6038-6050.

Ng D, Smith MB, Smith HT and Millet F (1977) Effect of modification of individual cytochrome c lysines on the reaction with cytochrome b_5. Biochemistry 166:4975-4978

Nishibayashi H and Sato R (1967) On the reduced minus oxidized difference spectrum of P-450 in liver microsomes. J Biochem 61:491-496

Nisimoto Y (1986) Localization of cytochrome c-binding domain on NADPH-cytochrome P450 reductase. J Biol Chem 21:14232-14239

Nisimoto Y and Edmondson DE (1992) Effect of KCl on the interaction between NADPH-cytochrome P450 reductase and either cytochrome c,cytochrome b_5 or cytochrome P450 in octyl glucoside micelles. Eur J Biochem 204:1075-1082

Nisimoto Y,Kinosita K, Ikegami A, Kawai N, Ichihara I and Shibata Y (1983) Possible association of NADPH-cytochrome P450 reductase and cytochrome P450 in reconstituted phospholipid vesicles. Biochemistry 22:3586-3594

Nisimoto Y and Lambeth JD (1985) NADPH-cytochrome P450 reductase - cytochrome b_5 interactions:crosslinking of the phospholipid vesicle-associated proteins by a water-soluble carbodiimide. Arch Biochem Biophys 241:38-396

Nisimoto Y and Otsuka-Murakami H (1988) Cytochrome c and cytochrome P450 interactions with NADPH-cytochrome P450 reductase in phospholipid vesicles. Biochemistry 27:5869-5876

Peterson JA, Ebel RE, O'Keefe DH, Matsubara D, and Estabrook RW (1976) Temperature dependence of cytochrome P450 reduction. J Biol Chem 251:4010-4016.

Phillips AH and Langdon RG (1962) Hepatic triphosphopyridine nucleotide-cytochrome c reductase: isolation, characterization, and kinetic studies.J Biol Chem 237:2652-2660

Schenkman JB, Sligar SG and Cinti DL (1982)Substrate interaction with cytochrome P450. In:Schenkman JB and Kupfer D (eds)Hepatic cytochrome P-450 monooxygenase system. Pergamon Press Ltd.,Great Britain,587-615

Tamburini PP, Jansson I, Favreau LV,Backes WL and Schenkman JB (1986a) Differences in the spectral interactions between NADPH-cytochrome P450 reductase and its redox partners. Biochem Biophys Res Commun 137:437-442

Tamburini PP, MacFarquhar S and Schenkman JB (1986b) Evidence of binary complex formations between cytochrome P450,cytochrome b_5 , and NADPH-cytochrome P450 reductase of rat hepatic microsomes. Biochem Biophys Res Commun 134:519-526

Tamburini PP and Schenkman JB (1986c)Differences in the mechanism of functional interaction between NADPH-cytochrome P450 reductase and its redox partners. Mol Pharmacol 30:178-185

Tamburini PP and Schenkman JB (1986d) Mechanism of interaction between cytochromes P450, RLM5 and b_5: evidence for an electrostatic mechanism involving heme propionate groups. Arch Biochem Biophys 245:512-522

Tamburini PP and Schenkman JB (1987) Purification to homogeneity and enzymological characterization of a functionally active complex composed of cytochrome P450 isozyme 2 and b_5 from rabbit liver. Proc Natl Acad Sci USA 84:11-15

Tamburini PP, White RE and Schenkman JB (1985) Chemical characterization of protein-protein interactions between cytochrome P450 and cytochrome b_5 . J Biol Chem 260:4007-4015

Vermilion J, and Coon M (1978) Identification of the high and low potential flavins of liver microsomal NADPH-cytochrome P-450 reductase. J Biol Chem 253:8812-8819.

Voznesensky, AI, and Schenkman JB (1992a) The cytochrome P450 2B4-NADPH cytochrome P450 reductase electron transfer commplex is not formed by charge-pairing.J Biol Chem 267:14669-14676.

Voznesensky AI, and Schenkman JB (1992b) Inhibition of cytochrome P450 reductase by polyols has an electrostatic nature. Eur J Biochem 210:741-746.

Molecular Biology of Cytochrome P-450: Evolution, Structure and Regulation

Yoshiaki Fujii-Kuriyama, Osamu Gotoh[1]
Department of Chemistry
Faculty of Science
Tohoku University
Aoba-ku, Sendai 980
Japan

Introduction

Cytochrome P450 is a class of hemoproteins that function as the terminal monooxygenases in microsomal, mitochondrial, or cytosolic NADPH-dependent electron transport pathways. P450 genes are widely distributed in nature from bacteria to higher animals and plants. Each mammalian genome possesses at least 50 P450 genes located on various chromosomes. Mammalian P450s as a whole metabolize an uncountable number of endogenous and exogenous substrates, such as steroids, fatty acids, vitamin D3, drugs, toxicants, environmental pollutants, and so on (Gonzalez 1990). The P450 enzymes involved in biosynthesis of physiologically active compounds (steroid hormones, prostaglandins, bile acids, vitamins, thromboxane, etc) play critical roles in development and maintenance of normal animal life. Other P450s mainly work in detoxication systems to convert lipophilic xenobiotics to more water-soluble forms. Some of the products of these reactions are chemically unstable and often produce DNA adduct which may trigger chemical carcinogenesis.

With a single exception, all eukaryotic P450s known to date are bound to either microsomal or mitochondrial membranes. (Nuclear outer membrane is considered to be a kind of microsomal membranes.) On the other hand, all known bacterial P450s are cytosolic soluble proteins. Microsomal P450s receive electrons from a membrane-bound flavoprotein, NADPH-cytochrome P450 reductase, whereas mitochondrial P450s as well as most bacterial P450s receive electrons from a soluble iron-sulfur

[1]Department of Biochemistry
Saitama Cancer Center Research Institute
Inamachi, Saitama 362
Japan

NATO ASI Series, Vol. H 90
Molecular Aspects of Oxidative Drug Metabolizing Enzymes
Edited by E. Arınç, J. B. Schenkman and E. Hodgson
© Springer-Verlag Berlin Heidelberg 1995

protein related to ferredoxin. Because of the soluble nature, it is easier to purify bacterial P450 proteins than eukaryotic proteins. P450cam isolated from soil bacterium Pseudomonas putida is the P450 protein best characterized in terms of physico-chemical properties, and is the sole P450 molecule whose three-dimensional (3-D) structure has been determined (Poulos 1987). Animal tissues such as liver contain multiple forms of P450s which have similar physico-chemical properties and overlapping substrate specificities. This molecular multiplicity made purification of individual forms of membranous P450 proteins even more difficult. However, several research groups succeeded in purifying some hepatic P450s in the late 1970s, partly taking advantage of the fact that only a specific molecular species of P450 is induced by treatment of animals with an inducer such as phenobarbital (PB) or 3-methylcholanthrene (3MC). The elevation in P450 contents is mainly controlled at the transcriptional level. Elucidation of the molecular mechanisms of the transcriptional regulation is currently one of the major fields of P450 research.

Classification of genes and proteins of P450 superfamily

In the last decade, a large number of P450 genes and cDNAs have been cloned from various sources. Consequently, more than 200 P450 protein sequences are now available. In accordance with the functional versatility, these sequences are divergently related to each other (Table I). However, there is strong evidence that all the existing P450 genes were derived from a common ancestor. As shown in Table I, the Z-scores for any comparisons suggest statistically significant relatedness in protein sequences, indicating the true homology among all P450s. Similar tests with other hemoproteins (Table II) confirm that P450 comprises a unique protein superfamily. This table also indicates that chloroperoxidase, H450 (a hepatic soluble protein), and the 'P450 domain' of nitric oxide synthase are neither related to the 'genuine' cytochrome P450 nor to one another, in spite of the fact that all these proteins show very similar spectral properties clearly distinct from those of other hemoproteins. Most probably, cytochrome P450, chloroperoxidase, H450, and nitric oxide synthase, each contains a cysteine residue that provides the thiolate ligand to the heme, but these proteins belong to different (super)families and fold into different tertiary structures.

Cloning, sequencing, and heterologous expression of cDNAs for individual forms of P450 have provided us with profound information about their intrinsic functions and mutual relationships. In particular, it becomes possible to identify or discriminate various P450 species prepared by different research groups. In 1987, Nebert et al. proposed a system of nomenclature of all characterized P450 genes and proteins based

Table 1. Relatedness of protein sequences in P450 families   (Z \ D)

P450	Family	1	2	21	17	71	3	6	4	19	7	11A	11B	27	51	52	101	102
Rat	1A1	524	1.59	1.76	1.70	2.04	2.21	2.22	2.72	2.45	2.63	2.52	2.55	2.28	2.74	2.33	3.19	2.14
Rat	2B1	28.0	491	1.69	1.79	2.01	1.90	2.47	2.31	2.17	2.58	2.43	2.60	2.35	2.59	2.44	2.78	2.22
Human	21A2	24.2	21.4	494	1.45	1.81	2.14	2.40	1.82	2.83	2.58	2.13	2.40	1.95	2.37	2.32	2.40	2.35
Bovine	17A	26.5	24.7	32.7	509	2.08	1.94	2.56	2.45	2.27	2.69	2.47	2.59	2.34	2.69	2.19	2.99	2.50
Avocado	71A	21.9	21.8	16.2	23.5	471	2.17	2.11	2.18	2.24	2.78	2.13	2.41	2.19	2.34	2.23	3.56	2.07
Rat	3A2	15.7	15.5	16.9	17.0	16.9	504	1.48	2.07	2.39	3.03	2.29	2.42	2.13	2.29	1.90	2.73	1.80
House fly	6A	12.7	13.1	12.7	9.0	13.4	29.5	509	1.96	2.43	2.82	2.28	2.24	2.23	2.31	2.33	2.92	2.11
Rabbit	4A4	9.1	13.1	13.0	10.8	13.9	20.0	17.1	506	2.55	2.75	2.38	2.66	2.31	2.34	2.19	2.86	1.87
Human	19A	10.4	9.9	10.7	11.0	8.7	8.7	6.9	8.5	503	2.45	2.39	2.52	2.65	3.12	3.13	2.86	2.55
Rat	7A	4.4	6.8	6.4	7.1	5.6	5.3	7.2	5.5	5.1	503	2.26	3.10	2.66	2.16	2.51	2.97	2.55
Bovine	11A	9.2	10.0	11.4	8.2	11.8	13.2	10.4	10.2	11.9	6.8	520	1.26	1.58	2.50	2.57	3.10	2.77
Bovine	11B1	10.7	10.9	9.1	7.4	7.4	11.2	9.1	9.4	8.1	4.0	40.5	503	1.69	2.39	2.67	2.87	2.67
Rabbit	27A	10.6	14.0	13.3	12.1	15.1	19.0	9.8	14.4	8.5	8.2	26.5	26.3	535	2.65	2.38	2.57	2.34
Yeast	51A	8.2	6.8	6.9	8.8	9.5	10.6	8.9	10.5	9.5	12.7	9.0	6.6	6.4	530	2.73	3.22	2.70
Yeast	52A1	5.2	8.4	7.6	7.2	6.8	14.7	11.4	12.5	4.7	5.2	9.0	8.0	10.1	7.8	543	3.04	1.81
P.putida	101A	2.7	4.1	4.7	3.4	2.4	4.1	2.7	3.3	2.7	2.0	4.3	3.0	4.1	6.8	2.7	415	2.71
B.mega.	102A	11.7	12.0	13.5	14.2	12.6	19.5	14.3	20.6	10.5	7.4	9.9	8.7	9.1	11.0	14.2	8.1	460

Values below the diagonal indicate significance of sequence relatedness measured in standard deviation unit (Z-score) each of which was estimated by a shuffle test with 250 pairs of randomized sequences. Values above the diagonal are estimated numbers of mutations per 100 residues after correction for multiple hits (D-score). The lengths of protein sequences are shown on the diagonal.

Table II. Sequece Relatedness between Hemoproteins

	101	2B1	CPO	NOS	H450	Catl	POX	Hbα
Rat CYP2B1	4.97							
C. fumago CPO	−1.43	0.66						
Human endothelial NOS	−0.16	−0.77	−0.62					
Rat H450	0.73	−0.59	1.20	0.89				
Rat catalase	1.16	1.53	−1.19	2.47	1.41			
Horseradish peroxydase	2.14	−0.96	−0.58	−0.46	1.40	1.35		
Human hemoglobin	1.35	−0.29	−1.12	−0.25	0.80	0.13	0.75	
Rat cyt. b5	0.06	−0.94	1.79	0.73	0.36	−0.86	−0.52	1.36

The significance of sequence relatedness was examined in the same way as described in the footnote to Table 1.

on relatedness in their protein sequences (Nebert et al. 1987). This system has been updated every other year, and is now widely accepted because it has helped to reduce former terminological confusions. The recommended naming of a gene and cDNA (Italic) or protein and mRNA (Roman) starts with the root symbol 'CYP' followed by an Arabic number designating a family, an optional alphabetic character indicating a subfamily, and then an optional serial number identifying individual gene within a subfamily. For example, CYP2B1 denotes the major PB-inducible P450 in rat liver, and CYP2B1 indicates its gene or cDNA. According to the latest version (Nelson et al. 1993), the P450 genes are classified into 36 families, each of which is composed of one to ten subfamilies. Table III lists representative P450 families together with some of their characteristics. You will find the wide spectrum of P450-mediated chemical processes.

Phylogenetic relationships of P450 genes

Fig. 1 shows a phylogenetic tree of P450 genes. To construct this tree, single representative protein sequences were chosen from individual subfamilies, their multiple sequence alignment was obtained, the degree of sequence divergence was calculated for each pair of sequences to generate a distance matrix, and then the neighbor-join method (Saitou and Nei 1987) was applied to the distance matrix. From the tree it is obvious that P450 proteins are divided into two major classes, B-class and E-class. All but one eukaryotic P450s belong to E-class whereas the majority of bacterial P450s belong to B-class. *Fusarium oxisporum* P450NOR (CYP55) is the sole known eukaryotic (fungal) P450 of B-class, and the protein is the unique eukaryotic

Table III Enzymatic Characteristics of Representative Mammalian P450s

P450	Trivial names	Inducer/Regulator[a]	Representative catalytic function
1A1	c,LM6,P_1,MC–1...	TCDD,MC,β–NF...	Hydroxylation of aryl hydrocarbons
1A2	d,LM4,P_3,MC–2...	TCDD,MC,ISF...	N–hydroxylation of aromatic amines
2A5,2A7	coh		Coumarin hydroxylation
2B1,2B4	b,LM2,PB–4...	PB	7–Ethoxycoumarin O–deethylation
2C11	h,male,M–1	Male specific	Testosterone 2α,16α–hydroxylation
2D1,2D6	db1		Debrisoquine 4–hydroxylation
2E1	j,LM3a	Alcohol	Aniline p–oxidation
3A2	PCN2	PCN	Aflatoxine B_1 metabolic activation
4A1	LAω	Clofibrate	Fatty acid ω and ω–1 hydroxylation
5A	TXA	$1,25(OH)_2D_3$?	Thromboxane synthesis
7A	7α,C7	Cholestyramine	Cholesterol 7α–hydroxylation
11A[b]	SCC	Gonadotropin/cAMP	Cholesterol side chain cleavage
11B1[b]	11β,C11	ACTH/cAMP	Steroid 11β–hydroxylation
17A	17α,C17	ACTH/cAMP	Steroid 17α–hydroxylation/17,20 lyase
19A	aromatase	cAMP,TPA,Dex...	Estrogen synthesis
21A	C21,21–OHase	ACTH/	Steroid 21–hydroxylation
24A[b]	cc24	$1,25(OH)_2D_3$	25–OH vitamin D_3 24–hydroxylation
27A[b]	C27/25		Vitamin D_3 25–, cholesterol 27–hydroxylation

[a] Abbreviations used: TCDD, 2, 3, 7, 8,-tetrachlorodibenzo-*p*-dioxin; MC, 3-methylcholanthrene; β-NF, β-naphthoflavone; ISF, isosafrole; PB, phenobarbital; PCN, pregnenolone 16α-carbonitrile; ACTH, adrenocorticotropic hormone; TPA, 12-O-tetradecanoylphorbol-13-acetate; Dex, dexamethasone; 1, 25(OH)2D3, 1α, 25-dihydroxy vitamin D3.
[b] Mitochondrial P450.

soluble P450. The similarity between P450NOR and *Streptomyces* CYP105 sequences is higher than those between most bacterial P450s, and is interpreted as a result of a recent (in paleontological time scale) horizontal gene transfer event (Kizawa et al. 1991). On the other hand, two prokaryotic E-class P450s have been identified to date. *Bacillus megaterium* P450BM3 (CYP102) is particularly interesting because it consists

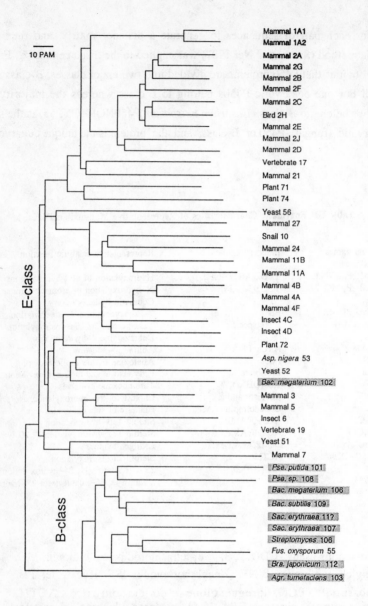

Fig. 1. A phylogenetic tree of P450 proteins. Forty six representative P450 protein sequences were aligned and a phylogenetic tree was constructed by the neighbor-join method (Saitou and Nei 1987) from their mutual sequence divergences. Bacterial P450s are indicated by shaded boxes. One PAM (percent accepted mutations) roughly corresponds to ten million years of divergence time from the common ancestor. Two major classes of P450s, E-class and B-class, are indicated.

of two distinct domains of the N-terminal P450 domain and the C-terminal NADPH-cytochrome P450 reductase domain (Ruettinger et al. 1989). This structural organization is very similar to that of nitric oxide synthases (Bredt et al. 1991), and the reductase domains of both of the fused structures are clearly homologous to the membranous P450 reductases and some bacterial flavoproteins. Our consideration that the 'P450 domain' of nitrite oxide synthase is not homologous to the ordinary P450s suggests that the fusion of the two domains occurred independently in the two lineages.

The presence of E-class genes both in eukaryotes and prokaryotes supports the idea that the divergence that led to the contemporary B- and E-class P450s had occurred before eukaryotic cells emerged on the earth. This is consistent with the estimated divergence time longer than 1.5 billion years (Fig. 1), although this estimation might be subject to large errors because of the inhomogeneous amino-acid substitution rates along different branches. However, it seems certain that the earliest P450 gene appeared quite early in the history of life.

Structure of P450 genes

Like most genes of higher organisms, all mammalian P450 genes are interrupted by several introns. One of the most remarkable features of P450 genes is that the exon-intron organizations are specific to individual families, i.e., all genes in a family have basically the same number of introns inserted in the corresponding nucleotide positions on the mature mRNAs, whereas virtually no intron positions are shared by genes belonging to different families. For example, the locations of the six introns in CYP1 genes and eight in CYP2 genes are strictly conserved within each family, but none of them is common across the families (Sogawa et al. 1984). Although several exceptions have been found to this simple rule (Picardo-Leonard and Miller 1987; Ohyama et al. 1993), the general specificity in gene organization provides an additional adequacy to the proposed classification of P450 families. As opposed to the exon-shuffling hypothesis (Gilbert et al. 1986) dynamic gains and losses of introns were likely to be the major causes of the high-degree of discordance in the present-day P450 gene structures.

There have been found many transcripts of various P450 genes that are likely to be produced by alternative splicing. Most of these alternative transcripts do not have the capability to encode active proteins, and therefore they are thought to arise from erroneous splicing mediated by inaccurate machinery. Alternative use of distinct promoters is another mechanism of generation of multiple transcriptional products from a single gene. Tissue-specific expression of the human aromatase gene (CYP19) is proven to be controlled by this mechanism; at least four distinct promoters are differentially active in placenta, skin fibroblasts, ovary, and prostatic gland (Mahedroo et al. 1991; Harada 1992). This kind of very intriguing gene regulation mechanisms have not been found for other P450 genes.

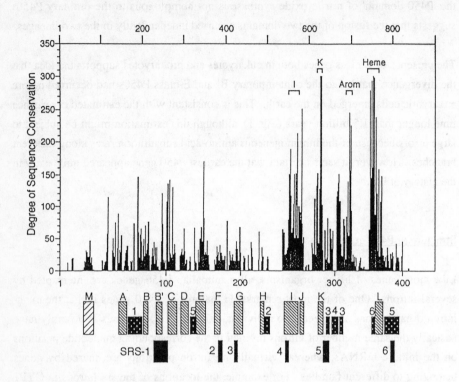

Fig. 2. Sequence conservation profile of E-class P450 proteins. 174 E-class P450 protein sequences were aligned and the degree of sequence conservation at each alignment position i, C_i, was evaluated according to the equation: $C_i = N_{Max}\ p_k\ \log(20\ p_k)$, where N_{Max} denotes the number of sequences with the most common amino acid at that position, and p_k denotes the fraction of amino acid k found at the position. Four highly conserved regions are indicated within the panel. The numbering above the profile corresponds to the alignment position, whereas that below the profile indicates equivalent P450cam site. The locations of helixes and β structures predicted from the 3D structure of P450cam and an alignment of all the P450 sequences are indicated by hatched and dark boxes, respectively. The N-terminal hydrophobic region possibly inserted into the microsomal membrane is indicated by a striped box (M). The six putative substrate recognition sites are also shown below the structural map.

Structure of P450 proteins

As noted earlier, all P450 genes are certainly derived from a common ancestor. This implies that the tertiary structures of all the P450 proteins are analogous to each other, since it is a general rule that homologous proteins have essentially the same 3-D folding pattern. Several lines of evidence confirm that this is really the case (Gotoh and Fujii-Kuriyama 1989). First, the residues identified from its 3-D structure to be important for the structure and function of P450cam are also well conserved in other P450 proteins. Second, local hydrophobicities and secondary structures predicted for various families of P450 proteins show very high correlation to each other, and are consistent with the 3-D structure of P450cam. Third, substrate recognition sites in mammalian P450s identified by experiments with site-directed mutagenesis and chimeric constructs agree nearly perfectly with those predicted from the structure of P450cam and a sequence alignment.

Based on an alignment of all P450 protein sequences, we can now infer various structural units in any P450 proteins. Fig. 2 shows such a structural map superimposed on a sequence-conservation profile obtained from all known E-class P450 sequences. Four clusters of well-conserved sites are recognized in the C-terminal half of the map. These conserved regions are called I-helix (distal helix), K-helix, aromatic, and heme-binding regions from the left (N-proximal) to the right (C-proximal).

The cysteine residue that is the fifth ligand to the heme is located near the center of the heme binding region consisting of about 20 residues. The conserved sequence pattern, **F**-[s,g]-x-**G**-x-[R,H]-x-**C**-x-**G**-x-x-@-a-x-x-e-@, is the most characteristic 'fingerprint' of P450 superfamily proteins, where the bold-face, capital, and lower-case letters indicate strictly, tightly, and intermediately conserved amino acids, respectively, @ indicates a site occupied by large aliphatic amino acids (Ile, Leu, Met, Val, and Phe), and x indicates a variable site.

The I helix is located on the opposite side of the heme plane to the heme-chelating cysteine residue. This helix has a distorted helical pattern, plays essential roles in binding of the molecular oxygen, internal water molecules, and substrates, and thus is the major constituent of the catalytic site of the enzymes. The helix is also close to the sixth coordinate position of the heme, but no amino acid residue directly participates in the coordination; in the substrate-free low-spin ferric state of P450cam, the sixth ligand position is occupied by a water molecule or hydroxyl anion, while the position is empty in the substrate-bound form (Poulos et al. 1986). There have been many site-directed mutagenesis studies supporting that the I helix is concerned with not only catalytic activities but also substrate specificities. We will discuss later this and other substrate recognition sites in more detail.

The K helix region contains two invariant residues E287 and R290 in the P450cam numbering system. Besides a few sites in the heme binding region, these two sites are the only sites where no amino acid replacement is found in any P450 sequences. The aromatic region (Gotoh and Fujii-Kuriyama 1989) is characterized by a peculiar amino acid sequence pattern, A1-[G]-x-x-p1-x-x-A2-x-p2-x-b-A3, where Ai represents an aromatic amino acid (Phe, Tyr, or Trp), and pi, b, x, and [G] indicate a Pro, a basic residue (Arg or His), a variable site, and an optional Gly found in some P450s, respectively. This pattern is found in only E-class P450s, and so is a fingerprint that discriminate E-class P450s from B-class P450s. The K helix and possibly the aromatic regions are located in the proximal side of the heme plane. Site-directed mutagenesis studies have suggested that these regions are important for retention of the heme (Shimizu et al. 1991), although neither region makes direct contact with the heme. Chemical modification and mutagenesis studies have also suggested involvement of the K helix region in the interaction with the electron donor protein (Tsubaki et al. 1989; Tuls et al. 1989; Wada and Waterman 1992). However, the functional roles of these two conserved regions remain to be elucidated in more detail.

Substrate Recognition Sites and Adaptive Evolution of Drug-Metabolizing P450s

Identification of the substrate binding/recognition sites in mammalian P450s has been difficult since the high variabilities in sequence around the candidate sites hampered accurate alignment with the P450cam sequence. However, recent progress in sequence analysis methods and accumulation of experimental data have been successfully combined to predict six separate putative substrate recognition sites, SRS-1 to SRS-6, in CYP2 family proteins (Gotoh 1992). These sites are also indicated in Fig. 2. As shown in Fig. 3, all experimentally identified sites that are responsible for substrate specificity are mapped within these SRSs. Moreover, most of the SRSs coincide with the regions where nucleotide substitutions accompanying amino acid replacements occur more frequently than those in the rest of sequences. These observations are reasonably interpreted as a result of adaptive evolution of the CYP2 genes, because high variability in substrate recognition sites is beneficial to metabolize a wider range of xenobiotics by cooperation of various P450 species. P450 genes seemingly have adopted the same strategy from the very beginning of their evolution. Thus the basic protein structure has remained invariant while the substrate recognition sites have undergone large changes to accommodate various types of substrate molecules. This strategy will be useful to design new engineered P450 molecules with a desired catalytic activity.

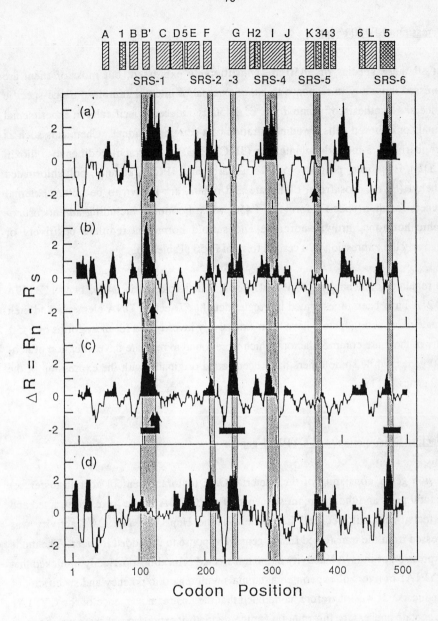

Fig. 3. Local variations in difference between rates of amino-acid replacing nucleotide substitutions and silent substitutions (adapted from Gotoh 1992). Positive R means that nucleotide substitutions replacing amino acids within a window (nine codons) have occurred more frequently than the average rates. Calculations were made for four CYP2 subfamilies, CYP2A (a), CYP2B (b), CYP2C (c), and CYP2D (d). Shaded areas indicate the SRSs. The locations of mutations that alter substrate specificities are indicated by arrows, while the chimeric fragments responsible for substrate specificities are indicated by thick bars in (c). The potential secondary structures are shown above the panels by boxes.

Expression of P450 genes

Not all these species of P450 are constitutively expressed, but most of them are expressed inducibly in response to either external or internal stimuli in tissue-specific manners. Synthesis of some form of P450 is induced specifically in experimental animals or cultured cells by administration of a certain chemical. Chemicals such as phenobarbital, 3-methylcholanthrene (3MC), 2, 3, 7, 8-tetrachlorodibenzo-p-dioxin (TCDD), isosafrole, pregnenolone 16 a- carbonitril (PCN), ethanol, polychlorinated biphenyl (PCB), isosafrole, clofibrate and others, are known to be potent external inducers for their specific forms of P450, while hormones including adrenocortico-trophic hormone, growth hormone and steroid hormones regulate positively or negatively the expression of a certain form of P450 (Table III).

The regulatory mechanisms of P4501A1, 4A and steroidogenic P450, 11A, 11B, 19A and 21A have been investigated in greater detail. *Cis*-acting DNA elements and their cognate *trans*-acting factors have been extensively investigated for these forms of P450. Many of them are common factors which were found to regulate the expression of non-P450 genes, while some others have been found originally with the expression of the P450 genes.

Cis-acting DNA elements of CYP1A1 gene

A fusion genes consisting of the bacterial CAT (chloramphenicol acetyltransferase) structural gene and the upstream sequence of the CYP1A1 gene were constructed and transfected into cultured cell lines, Hepa-1 and Hep G2. The CAT activity was expressed from the transfected fusion gene in response to the added inducer. The mode of expression of the CAT activity in the transfected cells almost perfectly minicked that of CYP1A1 in livers of experimental animals with regard to potency and specificity of the inducers. It was, therefore, concluded that the upstream sequence of the CYP1A1 gene contains necessary information for the regulatory expression of the gene (Sogawa et al. 1986). External and internal deletion in the upstream sequence of the fusion gene and subsequent transfection of the fusion genes into the cultured cell lines allowed us to determine the expressed CAT activity driven by the upstream sequence of the CYP1A1 gene and thereby, to localize the regulatory DNA elements in this region. This transient transfection system was able to define two kinds of *cis*-acting regulatory sequences in the upstream sequence of the CYP 1A1 gene. As shown in Fig 4, one is designated BTE, basic transcription element, at the -44 nucleotide position relative to the transcription initiation site and is involved in the constitutive expression of the gene

(Yanagida et al. 1990), while the other is XRE or xenobiotic responsive element which is distributed at least 5 times in the 5' upstream region of the gene and enhances the expression of the gene in response to the inducers such as TCDD and 3MC (Fujisawa-Sehara et al. 1987). In addtion to the TATA sequence, the two kinds of *cis*-acting DNA elements are necessary for a high level of the inducible expression of the gene. This arrangement of the reglutatory DNA elements is found to be conserved in mouse, rat, and human CYP1A1 genes (Fig. 4).

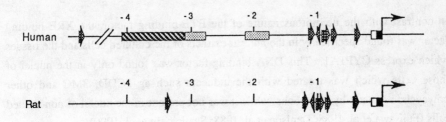

Fig. 4 Locations of regulatory DNA elements upstream of CYP1A1 genes. Solid arrows, transcription start site; shadowed arrows, XRE; open boxes, BTE; closed boxes, TATA sequences; doted boxes, Alu sequences; hatched boxes, SINE-R elements. Minus numbers indicate the distance in kb from the transcription initiation site.

The BTE (AGAAGGAGGCGTGGCC) or its analogous sequences were found in the upstream sequence of various P450 genes such as CYP2B, 21A, 11A and 2E genes (Yanagida et al. 1990). *In vitro* transcription assays also demonstrated that the BTE sequence has an enhancer effect on the transcription of the fusion gene. Systematic introduction of mutations in the BTE sequence defined an essential core sequence (AGGCGT) for the enhancer activity (Yasumoto et al. Unpublished observation) and clearly showed that the core sequence matches the GC box consensus sequence. The XRE seuqnece contains an invariant sequence of pentanucleotide, CACGC and was found in other genes such as glutathione S-transferase Ya, quinone reduatase, and CYP1A2 genes (Rushmore and Pickett 1993).

Trans-acting regulatory factors

Trans-acting factors on these regulatory DNA elements were detected in the nuclear extracts from the cells which express CYP1A1 by gel mobility shift assay (Fujisawa-

Sehara, 1988) and DNA methylation interference analysis (Neuhold et al. 1989; Shen and Whitlock 1989; Cuthill et al. 1991). DNA-binding factor(s) to the BTE sequence was found ubiquitously in the nuclear extracts from various cell lines. Judging from the sequence of the DNA element and some biochemical properties, one of the BTE-binding factors was considered to be Sp1 (Yanagida et al. 1990). This factor was a known factor which acts *in trans* on the GC-box sequence. cDNA cloning by using the BTE sequence as a labeling probe isolated two other BTE-binding factors, BTEB and BTEB2, zinc finger proteins.

In contrast with the ubiquitous nature of the BTE-binding factor, the XRE-binding factor was found specifically in the nuclear extracts of the cultured cells and the tissues which express CYP1A1. This DNA binding factor was found only in the nuclei of those cells which was treated with the inducers such as TCDD, 3MC and other polycyclic aromatic hydrocarbons, while it was not present in the nuclei of non-treated cells (Fujisawa et al. 1988; Denison et al. 1988; Saaticioglu et al. 1990).

In consistence with this observation, recently, DNA cross-linking experiments demonstrated that two proteins with molecular weights of 100 and 110 KDa were labeled with the ^{32}P-labeled XRE probe, when the ^{32}P-labeled XRE and the nuclear extracts from the cells treated with the inducer were irradiated by UV-light. (Elferink et al. 1990)

Structure and function of the two protein factors.

Elaborate experiments of rescuing the C4 mutant of Hepa-1 cells which is defective in the nuclear translocation of Ah receptor (AhR) with normal human genomic DNA isolated the gene for the Arnt protein or AhR nuclear translocator and subsequently, its cDNA clone. Transfection of the cDNA in the expression vector clearly restored the Arnt function of the C4 mutant, resulting in the induction of CYP1A1 in response to the inducer. Sequence analycis of the cDNA revealed that the deduced primary structure of Arnt consists of 789 amino acids and the helix-loop-helix domain characteristic of the transcriptional regulatory protein (Hoffman et al. 1991).

The XRE-binding factor was found in a latent form in the cytoplasm of non-treated cells (Fujisawa-Sehara et al. 1988). The cytoplamic form of the binding factor was found in association with HSP90 and showed no XRE-binding activity (Perdew 1988; Denis et al. 1988; Wilhemsson et al. 1990). When treated in vitro with the inducer, the cytoplasmic fraction prepared from the non-treated cells exhibited the XRE binding activity. Upon treatment of the cells with the inducer, the XRE-binding factor in the cytoplasm was translocated to the nuclei in association with the inducer (Fujisawa-Sehara et al. 1988) and bound to the XRE sequence in the major groove as clarified by the methylation interference experiment (Shen and Whitlock 1989). The regulatory factor bound with the XRE sequence was found associated with the inducer. The behavior of the *trans*-acting factor in response to the inducer suggested that the factor is the AhR or contains at least the AhR as a constituent. The AhR has been known for about 20 years and suggested to be closely associated with the induction of the drug-metabolizing enzymes by aromatic compounds. Mutation analysis in the induction process of CYP1A1 in Hepa-1 cells suggested that at least two genes are necessary for the inducible expression of CYP1A1 gene, in addition to CYP1A1 gene itself. One encodes AhR protein which binds the inducer and activate the expression CYP1A1 gene and the other is the gene for a protein which is involved in translocation of AhR to the nuclei (Hankinson 1983; Legaraverend et al. 1982).

Recently, cDNA clones for AhR protein have been isolated from a Hepa-1 cDNA library by using oligonucleotide probes synthesized from the reported N-terminal amino acid of the purified AhR (Ema et al. 1993; Burback et al. 1993). Determination of the cDNA sequence shows that the AhR protein is composed of 805 amino acids and, interestingly, contains a helix-loop-helix (HLH) domain at the N-terminus, similar to the case with the Arnt protein. The sequences of this part of the AhR and Arnt proteins significantly resemble each other and the sequence similarity is found to extend to those of other proteins, such as mouse TFEB and Drosophila Sim. Another part of this protein just C-terminal to the HLH domain shows some sequence similarity to those of Arnt, Sim and Drosophila Per proteins (Fig. 5). This conserved region is designated PAS domain, an acronym of Per, Arnt or AhR and Sim. The function of this domain will be described.

To investigate the roles of the two proteins, AhR and Arnt, in the XRE-binding factor, the antibody against AhR or Arnt was prepared in rabbits by using oligopeptides synthesized according to the amino acid sequences deduced from the cDNAs. Gel mobility shift assay using these antibodies demonstrated that the XRE-binding factor is composed of at least the two proteins, AhR and Arnt (Reyes et al. 1992; Whitelaw et al. 1993; Matsushita et al. 1993). The two proteins are separate entities in the cytoplasm of the non-treated cells. When the inducer is taken up in the cells, the two proteins form a complex with the XRE-binding activity and translocate to the nuclei.

The isolated cDNA of AhR was inserted into a expression vector of pSRα and transfected into CV-1 cells which do not show any AhR activity. The cytoplasmic

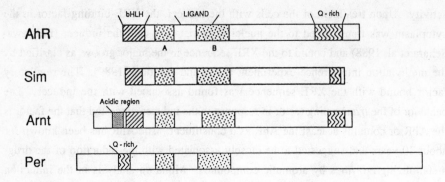

Fig. 5. Locations of various domains in the AhR, Arnt and the related proteins. bHLH, basic-helix-loop helix domains; spotted boxes, PAS domains; Q-rich, Gln-rich domain; acidic region, region rich in acidic amino acids; LIGAND, ligand-binding domain; AhR, Arylhydrocarbon receptor; Arnt, AhR nuclear translocator;Sim, Drosophila single minded protein; Per, Drosophila period protein.

fractions were prepared from the transfected cells and examined for the ligand-binding activity. The cytoplasmic fraction was incubated with ^{3}H-TCDD and then subjected to glycerol gradient centrifugation. A ^{3}H-TCDD-binding component peaked at the 9S position, the same position as observed with Hepa-1 cells and this peak was competed out with an excess amount of TCDF, a competitor. This ^{3}H-TCDD-binding peak was observed in only CV-1 cells transfected with the expression plasmid (Ema et al. Unpublished observation). Formation of the XRE-binding complex with AhR and Arnt was observed by reconstitution systems using C4 or C12 mutants of Hepa-1 cells. C4 mutant of Hepa-1 cells is know to be defective in the Arnt function, while C12 mutant lacks the normal AhR function (Hankinson 1983). Cytoplasmic fractions were prepared from the C4 mutant cells and incubated with the *in vitro* synthesized Arnt. The XRE-binding activity was induced by addition of TCDD in the reaction mixture. This binding activity was inhibited by anti AhR or anti-Arnt antibody.

On the other hand, the *in vitro* synthesized AhR was found to restore the XRE-binding activity in the cytoplasmic fraction prepared from the C12 mutant in the presence of the inducer such as TCDD and 3MC. Since either *in vitro* synthesized AhR or Arnt protein alone does not show any XRE-binding activity even in the presence of the inducer, both proteins are necessary for the activity. Formation of the complex between these two factors was demonstrated by the coimunoprecipitation with either anti Arnt or anti AhR antibody (Matsushita et al. 1993). Although the mechanism of the complex formation between Arnt and AhR remain to be seen, it has recently been suggested that the PAS domain is involved in the heterodimer formation of Drosophila Sim and Per (Huang et al. 1993).

Transcriptional activity of AhR and Arnt proteins was cleary demonstrated by the transient expression system using CV-1 cell. We constructed expression plasmids of the AhR and Arnt proteins under the control of the cytomegalovirus promoter and enhancer and transfected them either separately or in combination into CV-1 cells with a reporter plasmid pMC6.3k consisting of the CYP1A1 regulatory upstream sequence and the CAT structural gene. The CAT activity in the cells transfected with both of the effector plasmids was synergistically increased as compared with those expressed in the cells transfected with either alone. Although the synergistic effect was very distinct and the inducibility was reproducible, the inducibility of the CAT expression seemed lower than that observed with Hepa-1 cells which express AhR and Arnt endogenously. This low inducibility is not due to a lowered level of the induced expression of the CAT activity, but rather due to elevated constitutive expression without treatment of the inducer, 3MC. The reason for this high constitutive expression remains to be seen (Matsushita et al. 1993). In any case, this result strongly suggest that the two proteins function cooperatively as a transcriptional activator in response to the inducer.

Summary

More than 200 primary structures of cytochrome P450 have been reported and are divided structurally into two main groups, E and B groups by the presence and absence of the aromatic region, respectively, which contains three conserved aromatic amino acids. Structural relation among these P450s are summarized in a phylogenetic tree. Functional domains of these P450s are discussed with special reference to the conserved and the variable regions.

Current studies on the regulatory mechanisms of drug-metabolizing CYP1A1 are summarized. Two kinds of *cis*-acting DNA elements are defined in the 5' upstream flanking sequence. Their cognate *trans*-acting factors were idenfied and elucidated structurally from cDNA cloning. The mode of gene regulation of CYP1A1 has begun to be understood more in detail by molecular biological technology using the cloned cDNAs.

This work was supported in part by a Grant-in-Aid for Scientific Research from the Ministry of Education, Science, and Culture of Japan and by a grant from Nissan Science Foundation.

Reference

Adesnik M, Atchison M (1985) Gene for cytochrome P-450 and their regulation. CRC Crit Rev Biochem 19:247-305

Bredt D S, Hwang P M, Glatt C C, Lowenstein C, Reed R R, Snyder S H (1991) Cloned and expressed nitric oxide synthase structurally resembles cytochrome P-450 reductase. Nature 351:714-718

Burback K M, Poland A, Bradfield C A (1993) Cloning of the Ah-receptor cDNA reveals a distinctive ligand-activated transcription factor. Proc Natl Acad Sci USA 89:8185-8189

Cuthill S, Wilhelmsson A, Poellinger L (1991) Role of the ligand in intracellular receptor function: receptor affinity determines activation *in vitro* of the latent dioxin receptor to a DNA-binding form. Mol Cell Biol 11:401-411

Denis M, Cuthill S, Wikström A-C, Poellinger L, Gustaffsson J-Å (1988) Association of the dioxin receptor with the Mr 90,000 heat shock protein: a structural kinship with the glucocorticoid receptor. Biochem Biophys Res Commun 155:801-807

Denison M S, Fisher J M, Whitlock J P Jr (1988) The DNA recognition site for the dioxin-Ah receptor complex. J Biol. Chem 263:17221-17224

Elferink C J, Gasiewicz T A, Whitlock J P Jr (1990) Protein-DNA interactions at a dioxin-responsive enhancer. J Biol Chem 265:20708-20712

Ema M, Sogawa K, Watanabe N, Chujoh Y, Matsushita N, Gotoh O, Funae Y, Fujii-Kuriyama Y (1993) cDNA cloning and structure of mouse putative Ah receptor. Biochem Biophys Res Commun 184:246-253

Ema M, Sogawa K, Fujii-Kuriyama Y Unpublished observation.

Fujisawa-Sehara A, Sogawa K, Yamane M, Fujii-Kuriyama Y (1987) Characterization of xenobiotic responsive elements of the drug-metabolizing cytochrome P-450c gene: a similarity to glucocorticoid regulatory elements. Nucl Acids Res 15:4179-4191

Fujisawa-Sehara A, Yamane M, Fujii-Kuriyama Y (1988) A DNA-binding factor specific for xenobiotic response elements of P450c gene exists as cryptic form in cytoplasm: its possible translocation to nucleus. Proc Natl Acad Sci USA 85:5859-5863

Gilbert W, Marchionni M, McKnight G (1986) On the antiquity of introns. Cell 46:151-154

Gonzalez F J (1990) Molecular genetics of the P-450 superfamily. Pharmacol Ther 45:1-38

Gotoh O, Fujii-Kuriyama Y (1989) Evolution, structure and gene regulation of cytochrome p-450. in Frontiers in Biotransformation (Ruckpaul K, Hein H eds.) Vol. 1, pp. 195-243 Akademie-Verlag, Berlin

Gotoh O (1992) Substrate recognition sites in cytochrome P-450 family 2 (CYP2) proteins inferred from comparative analyses of amino acid and coding nucleotide sequences. J Biol Chem 267:83-90

Hankinson O (1983) Dominant and recessive aryl hydrocarbon hydroxylase-deficient mutants of mouse hepatoma line, Hepa-1, and assignment of recessive mutants to three complementation groups. Somat Cell Genet 9:497-514

Harada N (1992) A unique aromatase (P-450AROM) mRNA fomed by alternative use of tissue-specific exons 1 in human skin fibroblasts. Biochem Biophys Res Commun 189:1001-1006

Hoffman E C, Reyes H, Chu F-F, Sander F, Conley L H, Brooks B A, Hankinson O (1991) Cloning of a factor required for activity of the Ah (dioxin) receptor. Science 252:954-958

Huang Z J, Edery I, Rosbash M (1993) PAS is a dimerization domain common to Drosophila period and several transcription factors. Nature 364:259-262

Kizawa H, Tomura D, Oda M, Fukamizu A, Hoshino T, Gotoh O, Yasui T, Shoun H (1991) Nucleotide sequence of the unique nitrate/nitrite-inducible cytochrome P-450 cDNA from *Fusarum oxysporum*. J Biol Chem 266:10632-10637

Legraverend C, Hannch R R, Eisen H J, Owens I S, Nebert D W, Hankinson O (1982) Regulatory gene product of the Ah locus. J Biol Chem 257:6402-6407

Mahendroo M S, Means G D, Mendelson C R, Simmpson E R (1991) Tissue-specific expression of human P-450arom. The promoter responsible for expression in adipose tissue is different from that utilized in placenta. J Biol Chem 266:11276-11281

Matsushita N, Sogawa K, Ema M, Yoshida A, Fujii-Kuriyama Y (1993) A factor binding to the xenobiotic responsive element (XRE) of P-4501A1 gene consisits of at least two helix-loop-helix proteins, Ah receptor and Arnt. J Biol Chem in print

Nebert D W, Adesnik M, Coon M J, Estabrook R W, Gonzalez E L, Guengerich F P, Gunsalus I C, Johnson E F, Kemper B, Levin W, Phillips I R, Sato R, Waterman

M R (1987) The P450 gene superfamily. Recommended nomenclature. DNA 6:1-11

Nelson D R, Kamataki T, Waxman D J, Guengerich P, Estabrook R W, Feyereisen R, Gonzalez F J, Coon M J, Gunsalus I C, Gotoh O, Okuda K, Nebert D W (1993) The P450 superfamily: Update on new sequences, gene mapping, accession numbers, early trivial names of enzymes, and nomenclature. DNA Cell Biol 12:1-51

Neuhold L A, Shirayoshi Y, Ozato K, Jone J E, Nebert D W (1989) Regulation of mouse *CYP*1A1 gene expression by dioxin: requirement of two *cis*-acting elements during induction. Mol Cell Biol 9:2378-2386

Ohyama Y, Noshiro M, Eggertsen G, Gotoh O, Kato Y, Bjorkhem I, Okuda K (1993) Structural characterization of the gene encoding rat 25-hydroxyvitamin D3 24-hydroxylase. Biochemistry 32:76-82

Perdew G H (1988) Association of the Ah receptor with the 90 kDa heat shock protein. J Biol Chem 263:13802-13805

Picado-Leonard J, Miller W. L (1987) Cloning and sequence of the human gene for P450c17 (steroid 17α-hydroxylase/17, 20, lyase): similarity with the gene for P450c21. DNA 6:439-449

Poulos T L, Finzel B C, Howard A J (1986) Crystal structure of substrate-free pseudomonas putida cytochrome P-450. Biochemistry 25:5314-5322

Poulos T L, Finzel B C, Howard A J (1987) High-resolution crystal structure of cytochrome P450cam. J Mol Biol 195:687-700

Reyes H, Reisz-Porozasz S, Hankinson O (1992) Identification of the Ah receptor nuclear translocator protein (Arnt) as a component of the DNA binding form of the Ah receptor. Science 256:1193-19

Ruettinger R T, Wen L-P, Fulco A J (1989) Coding nucleotide, 5' regulatory and deduced amino acid sequences of P-450BM-3 , a single peptide cytochrome P-450: NADPH-P-450 reductase from *Bacillus megaterium*. J Biol Chem 246:10987-100995

Rushmore T H, Pickett C B (1993) Glutathione S-transferases, structure, regulation and therapeutic implications. J Biol Chem 268:11475-11478

Saaticioglu F, Perry D J, Pasco D S, Fagan J B (1990) Aryl hydrocarbon (Ah) receptor DNA-binding activity. J Biol Chem 265:9251-9258

Saitou N, Nei M (1987) The neighbor-joining method: A new method for reconstructing phylogenctic trees. Mol Biol Evol 4:406-425

Shen E S, Whitlock J P Jr (1989) The potential role of DNA methylation in the response to 2, 3, 7, 8-tetrachlorodibenzo-p-dioxin. J Biol Chem 264:17754-17758

Shimizu T, Tateishi T, Hatano M, Fujii-Kuriyama Y (1991)Probing the role of lysines and arginines in the catalytic function of cytochrome P450d by site-directed mutagenesis. Interaction with NADPH-cytochrome P450 reductase. J Biol Chem 266:3372-3375

Sogawa K, Gotoh O, Kawajiri K, Harada T, Fujii-Kuriyama Y (1984) Distinct organization of methylcholanthrene- and phenobarbital-inducible cytochrome P-450 genes in the rat. Proc Natl Acad Sci USA 81:5066-5070

Sogawa K, Fujisawa-Sehara A, Yamane M, Fujii-Kuriyama Y (1986) Location of regulatory elements resposible for drug induction in rat cytochrome P-450c gene. Proc Natl Acad Sci USA 83:8044-8048

Tuls J, Geren L, Millett F (1989) Fluorescein isothiocyanate specifically modifies lysine 338 of cytochrome P-450scc and inhibits adrenodoxin binding. J Biol Chem 264:16421-16425

Tsubaki M, Iwamoto Y, Hiwatashi A, Ichikawa Y (1989) Inhibition of electron transfer from adrenodoxin to cytochrome P-450scc by chemical modification with pyridoxal 5'-phosphate: identification of adrenodoxin-binding site of cytochrome P-450scc. Biochemistry 28:68899-6907

Wada A, Waterman M R (1992) Identification by site-directed mutagenesis of two lysine residues in cholesterol side chain cleavage cytochrome P450 that are essential for adrenodoxin binding. J Biol Chem 267:22877-22882

Whitelaw M, Pongratz I, Wilhelmsson A, Gustafsson J.-Å, Poellinger L (1993) Ligand-dependent recruitment of the Arnt coregulator determines DNA recognition by the dioxin receptor. Mol Cell Biol 13:2504-2514

Wilhelmsson A, Cuthill S, Denis M, Wikström A-C, Gustafsson J-Å, Poellinger L (1990) The specific DNA binding activity of the dioxin receptor is modulated by the 90 kd heat shock protein. EMBO J 9:69-76

Yanagida A, Sogawa K, Yasumoto K, Fujii-Kuriyama Y (1990) A novel cis-acting DNA element required for a high level of inducible expression of the rat P-450c gene. Mol Cell Biol 10:1470-1475

Yasumoto K, Imataka H, Sogawa K, Fujii-Kuriyama Y Unpublished observation

The Role of Nuclear Receptors in the Regulation of P450s and Other Genes by Peroxisome Proliferators

Eric F. Johnson, Colin N. A. Palmer, Mei-H. Hsu, and Keith J. Griffin
The Scripps Research Institute
Division of Biochemistry, NX-4
10666 North Torrey Pines Road
La Jolla, CA 92037 USA

Introduction

Peroxisome proliferators are a large and varied group of chemicals that can induce P450s as well as other drug metabolizing enzymes. These chemicals can cause the number of peroxisomes in the livers of some sensitive species to increase with subsequent enlargement of the liver. This may ultimately lead to the development of liver tumors, but peroxisome proliferators or their metabolites do not appear to be mutagens (Rao, Reddy, 1991). Thus, they are considered to be non-genotoxic carcinogens. The increase in peroxisomes may in turn reflect the induction of several peroxisomal enzymes that metabolize fatty acids by ß-oxidation. In contrast to mitochondrial ß-oxidation, the peroxisomal pathway appears to function as pathway for the elimination of fatty acids and the production of heat rather than conservation of energy through the formation of ATP. In addition, long-chain, branched chain and dicarboxylic fatty acids are preferentially metabolized by the peroxisomal pathway. The latter can arise via the ω-hydroxylation of monocarboxylic acids, a reaction that is catalyzed by microsomal P450s of the *CYP4A* family and subsequent oxidation of the alcohol to a second carboxyl

NATO ASI Series, Vol. H 90
Molecular Aspects of Oxidative Drug Metabolizing Enzymes
Edited by E. Arınç, J. B. Schenkman and E. Hodgson
© Springer-Verlag Berlin Heidelberg 1995

moiety by other enzymes. The P450 4A enzymes are the P450s most prominently induced by peroxisome proliferators (Gibson, 1992; Lock et al. 1989; Kaikaus et al. 1993).

The induction of these pathways of fatty acid metabolism may actually occur in response to an increase of free cellular fatty acids elicited by peroxisome proliferators rather than from their direct action (Lock et al. 1989). This would be a case of substrate induction for these enzymes. Kaikaus et al. (1993) have presented evidence that the induction of the P450 4A enzymes by peroxisome proliferators may contribute to the induction of the peroxisomal ß-oxidation pathway as well as of the liver fatty acid binding protein. When these investigators inactivated the P450 4A enzymes as well as most other P450s by treatment with a suicide substrate, the induction by peroxisome proliferators of the mRNAs encoding the fatty acyl-CoA oxidase (FACO), the rate-limiting and initial enzyme of the peroxisomal ß-oxidation pathway and of the fatty acid binding protein (FABP) are greatly diminished in cultured hepatocytes. In contrast, the induction of mRNAs encoding the P450 4A enzymes are not greatly affected. Kaikaus et al. (1993) hypothesized that dicarboxylic acids might be the proximate inducers of the FACO and FABP. Inactivation of the P450 4A enzymes would block ω-hydroxylation of monocarboxylic acids, the first step in the conversion of monocarboxylic to dicarboxylic fatty acids. In support of this hypothesis, they demonstrated that mRNAs encoding FACO and FABP could be induced by dicarboxylic acids following the destruction of the P450 4A enzymes. In contrast, the P450 4A mRNAs did not appear to be induced by the dicarboxylic acids (Kaikaus et al. 1993).

The results of Kaikaus et al. (1993) suggest that the expression of the CYP4A genes and those encoding the FACO and FABP are under differential control. Additional evidence supports this possibility. Several investigators have demonstrated that the induction of P450 4A mRNAs precedes that of FACO mRNAs and that the latter largely blocked by cyclohexamide, an inhibitor protein synthesis (Bell, Elcombe, 1991).

RABBIT P450 4A ENZYMES

Four P450 4A enzymes are known in rabbits, designated P450 4A4 (Yamamoto *et al*. 1984; Williams *et al*. 1984; Matsubara *et al*. 1987), 4A5 (Johnson *et al*. 1990; Yokotani *et al*. 1991), 4A6 (Johnson *et al*. 1990; Yokotani *et al*. 1989), and 4A7 (Johnson *et al*. 1990; Yokotani *et al*. 1989). Of these, the latter three are each induced by peroxisome proliferators with P450 4A6 being the most highly induced (12-fold in liver and 6-fold in kidney). This largely reflects a much lower level of expression for P450 4A6 mRNAs when compared to P450 4A5 and P450 4A7 in untreated rabbits. Following induction, similar mRNA abundances are seen for all three enzymes. In contrast, P450 4A4 mRNAs are normally not detected in liver, kidney or lung, or following treatment with peroxisome proliferators. However, P450 4A4 mRNAs are expressed at levels similar to that of the other three P450 4A mRNAs in the livers of pregnant rabbits and at higher abundance in the lung where the other three P450 4A mRNAs are not detected in either pregnant or non-pregnant rabbits. P450 4A4 can also be induced by dexamethasone and progesterone (Muerhoff *et al*. 1987; Matsubara *et al*. 1987; Palmer *et al*. 1993a).

The four P450 4A enzymes exhibit similar but distinct patterns of substrate specificity (Masters *et al*. 1992). All four catalyze the ω-hydroxylation of palmitic acid, and all but *CYP4A4* also catalyze the ω-hydroxylation of lauric acid. Of these *CYP4A5* exhibits the most restricted specificity preferring lauric acid as a substrate and exhibiting low relative rates toward palmitic and arachidonic acid. The latter is metabolized well by the remaining three enzymes. The ω-hydroxylation of arachidonic acid is of particular interest as it produces a precursor for a potent vasoconstrictor in kidney (Schwartzman *et al*. 1989). The pregnancy related enzyme, *CYP4A4*, is distinguished from the other three by its preference for prostaglandins relative to fatty acids as substrates (Masters *et al*. 1992).

THE PEROXISOME PROLIFERATOR ACTIVATED RECEPTOR

By analogy to the induction of P450 1A1 by dioxin, it has long been suspected that the induction of other P450s by xenobiotics is mediated by a cellular receptor. Work in our laboratory indicates that the transcriptional activation of the *CYP4A6* gene by peroxisome proliferators can be mediated by the murine peroxisome proliferator activated receptor (PPAR) (Muerhoff *et al.* 1992). The latter is a member of a subfamily of the nuclear receptor superfamily which includes the steroid hormone receptors and receptors for thyroid hormone, vitamin D_3, and retinoic acid (Issemann, Green, 1990). These receptors, which can be found in either cytosol or nuclei, are transcription factors that recognize specific sequences in the gene. In the case of the hormone receptors, the effects of the receptor on gene transcription are modulated by hormone binding to the receptor. However, a number of proteins in this family have been described for which ligands have not been identified, and these are often termed orphan receptors. The PPAR can be considered as an orphan receptor because although it is activated by peroxisome proliferators, they have not been demonstrated to be ligands that bind to the protein (Issemann, Green, 1990). Fatty acids have also been shown to activate PPAR (Göttlicher *et al.* 1992), and these might be ligands for this receptor. However, the ED_{50} values for fatty acids and for peroxisome proliferators range from 10^{-6} to 10^{-3} M, and if these reflect the K_D for the binding of these putative ligands to PPAR, specific binding could be difficult to detect when the concentration of the receptor is low. Although only a single mouse (Issemann, Green, 1990) and rat PPAR (Göttlicher *et al.* 1992) have been described, a human orphan receptor, NUC1, has been described that is activated by fatty acids and the peroxisome proliferator, Wy-14,643 (Schmidt *et al.* 1992). This receptor is less closely related to the murine PPAR than the human PPAR (Sher *et al.* 1993), 62% and 85%, respectively. In addition three members of the nuclear receptor family have been described in *Xenopus laevis*

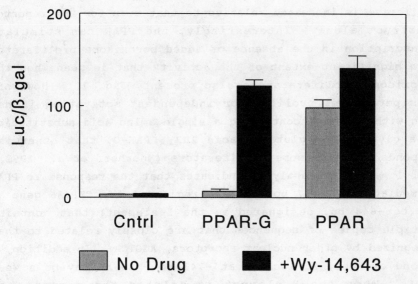

Fig. 1. Transactivation of reporter gene transcription by PPAR-G or PPAR. RK13 cells were transfected with a luciferase reporter gene construct under control of *CYP4A6* gene promoter, -880 to +18bp, and expression plasmids for PPAR-G or PPAR in the pCMV expression vector as well as for ß-galactosidase. The results determined in the presence or absence of 50 μM Wy-14,643 were normalized to that for ß-galactosidase.

that differ by roughly 30% amino acid sequence identity and each can be activated by peroxisome proliferators (Dreyer *et al*. 1992).

PPAR MEDIATES THE TRANSCRIPTIONAL ACTIVATION OF *CYP4A6* GENE BY PEROXISOME PROLIFERATORS

Gene transfer experiments have been used to link the PPAR to the transcriptional activation of the *CYP4A6* gene by peroxisome proliferators. In these experiments, the luciferase reporter gene under control of the promoter and a portion of the 5' flanking sequence of the *CYP4A6* gene (-880 to +17 bp) are transfected into one of several cells lines together with expression vectors for PPAR (Muerhoff *et al*. 1992). Our results, Fig. 1, indicate that when the PPAR and a peroxisome proliferator are present, the expression of luciferase is increased relative to that seen for the reporter construct alone. Interestingly, the PPAR can stimulate transcription in the absence of added peroxisome proliferator to a significant extent of the activity that is seen when the peroxisome proliferator is also present, Fig. 1. However, this peroxisome proliferator independent activation is not seen with a mutant containing a single amino acid substitution of a glycine for glutamic acid 282, PPAR-G, that otherwise responds to peroxisome proliferators (Muerhoff et al. 1992), Fig. 1. Deletion analysis indicates that the response to PPAR is mediated by a 37 bp upstream region of the CYP4A6 gene (-677 to -644 bp, designated as the Z-element) that contains multiple copies of sequences that are closely related to that recognized by other nuclear receptors, AGGTCA. In addition, a second element, X, located at -743 to -725 bp may be a weak PPRE. When the X element is deleted the response to peroxisome proliferators drops 2 to 3-fold. However, a single copy of the X element does not confer a response when incorporated into the -155 reporter construct. These results indicate that PPAR can mediate the transcriptional activation of the CYP4A6 gene by peroxisome proliferators.

The PPAR has also been shown mediate the transcriptional activation of the FACO (Tugwood et al. 1992; Dreyer et al. 1992) and FABP (Issemann et al. 1992) genes as well as the gene encoding the peroxisomal bifunctional enzyme (Zhang et al. 1992; Bardot et al. 1993) by peroxisome proliferators. The PPRE of the FACO was first defined by deletion analysis employing the rat hepatoma cell line, H4IIEC3 (Osumi et al. 1991). This study defined a segment of the FACO gene, -578 to -576, that contained two regions, A and B, protected by nuclear proteins in DNase I footprinting experiments. The B element alone did not mediate a response to peroxisome proliferators, whereas the A element exhibited a weak response, 02-fold, that was increased to 04-fold when the B element was incorporated together with the A element in the reporter construct. Tugwood et al. (1992) subsequently demonstrated that the A element could regulate the response of

```
Z:       ACGCAAACACTGAACT AGGGCA A AGTTGA GGGCAGTG
FACO:            GGGACC AGGACA A AGGTCA CGTTCGGG
FABP:            AAATAT AGGCCA T AGGTCA GTGATTG
BIF:  CTCAAATGTAGGTAATA GTTCAA T AGGTCA AAGGA
```

Fig. 2. Response Elements for PPAR. Regions of the genes for P450 4A6 (Z element), FACO (A-element), FABP, and the bifunctional enzyme (BIF) that contain PPREs. The imperfect, direct repeat separated by a single nucleotide is set off by spaces.

a reporter gene to peroxisome proliferators mediated by the murine PPAR in cotransfection experiments. The sequence of the FACO A element exhibits extensive sequence identity with a portion of the *CYP4A6* Z and X elements as shown in Fig. 2. Deletion analysis has also defined similar PPREs for the FABP gene (Issemann *et al.* 1992) as well as for the gene encoding the peroxisomal bifunctional enzyme (Zhang *et al.* 1992; Bardot *et al.* 1993), Fig. 2. An examination of these sequences reveals that each PPRE exhibits an imperfect direct repeat of a sequence related to AGGTCA, a sequence recognized by the zinc finger motif of the nuclear hormone receptors.

Zinc finger motifs are found in a variety of DNA binding proteins, and the nuclear receptors exhibit a characteristic pair of zinc finger motifs. The domain containing this motif is sufficient for DNA binding to hormone response elements, and the structure of the complex formed between the glucocorticoid response element and the zinc finger of the glucocorticoid receptor has been determined experimentally (Luisi *et al.* 1991). The principal DNA contacts are found in the N-terminal zinc finger, and this region is highly conserved among the subfamily of the nuclear receptors that includes the glucocorticoid, androgen, progesterone and mineralocorticoid receptors. These receptors recognize a palindromic repeat of the sequence, AGAACA, separated by three nucleotides, and the receptors bind to the palindromic response element as a dimer. The structure of the complex between the two DNA binding domains and the palindromic response element indicates that the interface between the two DNA binding domains of the dimer is encoded in the C-terminal zinc fingers, and that the three nucleotides separating the inverted repeat allows each of the N-terminal zinc fingers to contact the major groove of the DNA at the recognition sequence. As the hormone responsive elements are similar or identical for each of these receptors, the differential regulation by hormones of genes bearing these response elements is thought to involve several factors such as (1) differential expression of the receptors, (2) differential metabolism of the hormone signals in tissues, or (3) the differential capacity of each receptor to alter the actions of other transcription factors that bind at or near the hormone response element (Funder, 1993).

The N-terminal zinc finger of the PPARs are highly similar to that of the thyroid hormone, estrogen, retinoic acid, retinoid X, vitamin D_3 receptors as well as of a number of orphan receptors such as COUP-TF, HNF4, and ARP-1, but is distinct from that of glucocorticoid family. These receptors recognize both palindromic and direct repeats of sequences related to AGGTCA. Although the spacing of these repeats can underlie selective responses, there is a considerable degree of overlap in recognition by these transcription factors. Moreover, with the exception of the estrogen receptor, these receptors have been shown to bind *in vitro* to responsive elements as heterodimers with the retinoid X receptor, RXRα, another member of this family (Bugge *et al.* 1992; Hallenbeck *et al.* 1992; Carlberg *et al.* 1993; Husmann *et al.* 1992; Durand *et al.* 1992).

PPAR/PPRE BINDING IS FACILITATED BY RXR

Highly purified PPAR does not appear to bind to the Z-element of the *CYP4A6* gene, but it does so .when supplemented with the retinoid X receptor, RXRα, another member of the nuclear receptor family. This is similar to the behavior of other members of subfamily II of the nuclear receptor family that appear to require RXR proteins for high affinity binding

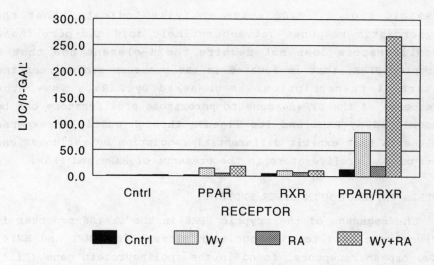

Fig. 3. Synergism between 50μM Wy-14,643 (Wy) and 1μM retinoic acid (RA) following heterologous expression of both PPAR-G and RXRα. Results are shown for the luciferase reporter containing -880 to +18 of the *CYP4A6* gene in RK13 cells normalized to the values obtained for the ß-galactosidase control. Results obtained following transfection of expression plasmids for the individual receptors or a vector control are also shown.

to their recognition sequences. Moreover, there is an RXR requirement for the binding of PPAR to the PPREs of the other genes that have been investigated.

The formation of heterodimers between PPAR and other receptor proteins provides a means for cross talk to occur through interaction of the other receptor with its ligand (Kliewer *et al*. 1992). RXRα can be activated by inclusion of retinoic acid in the cell culture medium. In the case of the Z-element coupled to a heterologous promoter, retinoic acid and RXRα do not increase the response in reporter gene expression to peroxisome proliferators above that seen with PPAR alone. However, under the same conditions, transcription of the reporter construct under control of the *CYP4A6* gene is greatly enhanced in a synergistic fashion (Palmer *et al*.

1993b), Fig. 3. Deletion analysis indicates that the synergistic response between retinoic acid and peroxisome proliferators does not require the Z-element and that a cryptic PPRE, that is located in the promoter region near the start of transcription, is unmasked by RXRα. Thus, the response of the *CYP4A6* gene to peroxisome proliferators can be modulated by RXRα and its ligands through multiple response elements that exhibit differential regulation by retinoids and peroxisome proliferators in the presence of RXRα and PPAR.

PPREs INTERACT WITH OTHER PROTEINS

The sequence of the cryptic PPRE in the *CYP4A6* promoter is closely related to a recognition element for ARP1 and HNF4, two orphan receptors, found in the apolipoprotein gene (CIII) (Ladias *et al.* 1992). Electrophoretic mobility shift analysis indicates that ARP1 binds to both the cryptic PPRE in the *CYP4A6* promoter as well as the upstream, Z-element. In contrast to PPAR, this binding does not require RXRα.

When increasing amounts of an expression vector for ARP1 are transfected together with those for RXRα and PPAR, the response of the *CYP4A6* reporter constructs to peroxisome proliferators and retinoic acid are diminished. ARP1 was originally shown to be a negative regulator of apolipoprotein A1 gene expression (Ladias, Karathanasis, 1991). These results suggest that *CYP4A6* as well as other genes that encode enzymes or proteins that function in lipid homeostasis are likely to be regulated by the same group of transcription factors. This also suggests that competition may exist among the various members of the nuclear receptor family for binding to PPREs. Moreover, the specific sequence of the PPRE, the relative cellular concentration of the nuclear receptors, and the modulation of their activity by ligands and/or phosphorylation could determine the overall expression of the *CYP4A6* gene as well as of the FACO and FABP genes.

ENDOGENOUS ACTIVATORS OF PPAR

Peroxisome proliferators can activate the PPAR, but it has not yet been shown that they can activate the receptor directly by ligand binding. As we indicated earlier, the murine PPAR can activate the transcription of *CYP4A6* reporter constructs in the absence of added peroxisome proliferators and that this can be further augmented by the inclusion of peroxisome proliferators. A single mutation, a glycine substitution for glutamic acid residue 282, abolishes significant activation of the reporter in the absence of peroxisome proliferators, but the mutant fully activates *CYP4A6* reporter expression in their presence.

This difference may reflect the effect of the mutation on the affinity of the murine PPAR for an endogenous ligand. Dose-response studies indicate that the apparent ED_{50} of the PPAR-G is shifted to higher concentrations for Wy-14,643 by roughly 10-fold. The effect of the mutation on the dose-response curves for PPAR is highly suggestive that the dose-response curve reflects the activation of the receptor by a ligand. If peroxisome proliferators cannot bind directly with PPAR, they must effect proportionate increases in the concentrations of the endogenous activating agent. The concentration of the latter may be sufficiently high to partially populate the PPAR leading to significant reporter gene activation in the absence of added peroxisome proliferators. In contrast, the concentration of the endogenous activator may be too low to populate the mutant receptor.

As we indicated earlier, it has been proposed that peroxisome proliferators can lead to elevations of free monocarboxylic fatty acids and that these may mediate the effects of peroxisome proliferators on *CYP4A* gene expression. The P450 4A enzymes can in turn catalyze the formation ω-hydroxylated monocarboxylic fatty acids, and the ω-hydroxyl group can be further oxidized to form dicarboxylic fatty acids that can contribute to the induction of the FACO and FABP in response to peroxisome proliferators. This possibility was supported by the initial observation that monocarboxylic fatty acids can activate a chimeric receptor constructed from the DNA binding domain of the glucocorticoid receptor and the putative ligand binding domain of the rat PPAR (Göttlicher *et al*. 1992). Issemann *et al*. (1992) have also reported that palmitic acid activates the mPPAR. In addition, the *xenopus laevis* PPARα (Keller *et al*. 1993) mediates the transcriptional activation of a reporter gene containing the PPRE from the FACO gene by both saturated and unsaturated monocarboxylic fatty acids including arachidonic acid. Interestingly, the synthetic eicosinoid, ETYA (5,8,11,14-eicosatetraynoic acid), proved to be a more potent activator of the *xenopus laevis* PPARα than Wy-14,643. In contrast, our studies on the activation of the murine PPAR or PPAR-G reveal similar potencies for these two compounds. Although the effects of ETYA and arachidonic acid on PPAR suggest that eicosinoids could play a role in the response to peroxisome proliferators, a variety of inhibitors of cyclooxygenases and of lipoxygenases did not affect the activation of the *xenopus laevis* PPARα (Keller *et al*. 1993) indicating that this activation is unlikely to involve signal transduction pathways utilizing arachidonic acid.

We have compared the transcriptional activation of the *CYP4A6* reporter constructs by fatty acids mediated by PPAR and PPAR-G. These results indicate that the mutation affects the dose-response to fatty acids in the same manner as seen for peroxisome proliferators. Interestingly, the dicarboxylic fatty acid hexadecanedioic acid activates PPAR-G, Fig. 4, and this response is exhibited by both reporter constructs containing the PPRE from the *CYP4A6* gene (Z-element) or the FACO gene (AB-element). In general, the activation of PPAR by either monocarboxylic or dicarboxylic fatty acids requires their presence in the culture medium at concentrations that exceed those required for the peroxisome proliferator Wy-14,643. However, potential differences in their uptake and metabolism of these compounds by the host cells confound direct comparisons. The response to peroxisome proliferators is seen in cell lines, such as COS-1 cells, that exhibit undetectable levels of fatty acid ω-hydroxylation. Thus, it

99

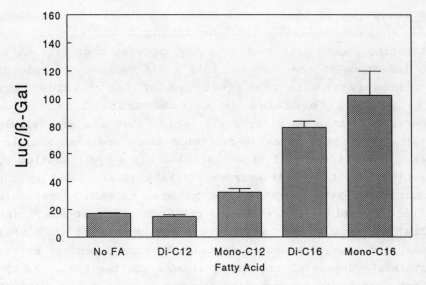

Fig. 4. Activation of *CYP4A6* driven luciferase expression by the lauric (mono-C12), dodecanedioic (di-C12), thapsic (di-C16), or palmitic (mono-C16) acids (50 μM) mediated by PPAR-G in RK13 cells. The reporter construct contained the -880 to +18 bp region of the *CYP4A6* gene, and the results were normalized to the ß-galactosidase control.

is unlikely that expression of cytochrome P450 4A enzymes is essential for the activation of PPAR by peroxisome proliferators.

Summary

PPAR can mediate the transcriptional activation of both the *CYP4A6* gene and of the genes encoding peroxisomal enzymes that participate in the ß-oxidation of fatty acids. In addition to peroxisome proliferators, both dicarboxylic and monocarboxylic fatty acids can activate PPARs. The latter represent endogenous compounds that could regulate the normal functions of PPARs, and peroxisome proliferators could elicit their effects indirectly through elevations of cellular fatty acids that in turn activate PPARs. Alternatively, the elevation of cellular fatty acids could augment a direct response to peroxisome proliferators by increasing the overall concentration of ligands for PPARs. In a similar fashion, P450 4A participation in ω-oxidation would convert one activating ligand into another. The importance of P450 4A in the induction of the FACO and FABP could reflect the role of ω-hydroxylation in the diversion of fatty acids from triglyceride formation or mitochondrial ß-oxidation sequestering them as dicarboxylic acids that are more slowly eliminated by peroxisomal ß-oxidation alone and that activate PPAR. The differential effect of blocking ω-hydroxylation by inactivation of P450 4A enzymes (Kaikaus *et al*. 1993) on the induction of P450 4A mRNAs when compared to FACO mRNAs could reflect kinetic differences in the induction process. The latter is slow when compared to the response of the *CYP4A* genes, and thus, its induction could be more dependent on the sustained presence of activating ligands for the PPAR. As the latter process is supported by the presence of active P450 4A enzymes, inhibition of these enzymes could blunt the induction of the FACO gene by peroxisome proliferators before it can fully develop. On the other hand, several members of the nuclear receptor family can recognize PPREs, and differences in the sequences of the PPREs and their contexts between PPAR responsive genes could lead to differential expression. The details of this regulatory network are unclear and represent an active area for future investigation.

Acknowledgment

The work of the authors' laboratory is supported by USPHS Grant HD04445. Colin N. A. Palmer is the recipient of a research Fellowship from the American Heart Association, California Affiliate.

Reference

Bardot O, Aldridge TC, Latruffe N, Green S (1993) PPAR-RXR heterodimer activates a peroxisome proliferator response element upstream of the bifunctional enzyme gene. Biochem Biophys Res Commun 192:37-45

Bell DR, Elcombe CR (1991) Induction of acyl-CoA oxidase and cytochrome P450IVA1 RNA in rat primary hepatocyte culture by peroxisome proliferators. Biochem J 280:249-253

Bugge TH, Pohl J, Lonnoy O, Stunnenberg HG (1992) RXRα, a promiscuous partner of retinoic acid and thyroid hormone receptors. EMBO J 11:1409-1418

Carlberg C, Bendik I, Wyss A, Meier E, Sturzenbecker LJ, Grippo JF, Hunziker W (1993) Two nuclear signalling pathways for vitamin D. Nature 361:657-660

Dreyer C, Krey G, Keller H, Givel F, Helftenbein G, Wahli W (1992) Control of the peroxisomal beta-oxidation pathway by a novel family of nuclear hormone receptors. Cell 68:879-887

Durand B, Saunders M, Leroy P, Leid M, Chambon P (1992) All-trans and 9-cis retinoic acid induction of CRABPII transcription is mediated by RAR-RXR heterodimers bound to DR1 and DR2 repeated motifs. Cell 71:73-85

Funder JW (1993) Mineralocorticoids, glucocorticoids, receptors and response elements. Science 259:1132-1133

Gibson GG (1992) Co-induction of cytochrome P4504A1 and peroxisome proliferation: A causal or casual relationship. Xenobiotica 22:1101-1109

Göttlicher M, Widmark E, Li Q, Gustafsson J-Å (1992) Fatty acids activate a chimera of the clofibric acid-activated receptor and the glucocorticoid receptor. Proc Natl Acad Sci USA 89:4653-4657

Hallenbeck PL, Marks MS, Lippoldt RE, Ozato K, Nikodem VM (1992) Heterodimerization of thyroid hormone (TH) receptor with H-2RIIBP (RXRß) enhances DNA binding and TH-dependent transcriptional activation. Proc Natl Acad Sci USA 89:5572-5576

Husmann M, Hoffmann B, Stump DG, Chytil F, Pfahl M (1992) A retinoic acid response element from the rat CRBPI promoter is activated by an RAR/RXR heterodimer. Biochem Biophys Res Commun 187:1558-1564

Issemann I, Prince R, Tugwood J, Green S (1992) A role for fatty acids and liver fatty acid binding protein in peroxisome proliferation. Biochem Soc Trans 20:824-827

Issemann I, Green S (1990) Activation of a member of the steroid hormone receptor superfamily by peroxisome proliferators. Nature 347:645-650

Johnson EF, Walker DW, Griffin KJ, Clark JE, Okita RT, Muerhoff AS, Masters BS (1990) Cloning and expression of three rabbit kidney cDNAs encoding lauric acid omega-hydroxylases. Biochemistry 29:873-879

Kaikaus RM, Chan WK, Lysenko N, Ray R, Ortiz de Montellano PR, Bass NM (1993) Induction of peroxisomal fatty acid ß-oxidation and liver fatty acid-binding protein by peroxisome proliferators. Mediation via the cytochrome P-450IVA1 omega-hydroxylase pathway. J Biol Chem 268:9593-9603

Keller H, Dreyer C, Medin J, Mahfoudi A, Ozato K, Wahli W (1993) Fatty acids and retinoids control lipid metabolism through activation of peroxisome proliferator-activated receptor-retinoid X receptor heterodimers. Proc Natl Acad Sci USA 90:2160-2164

Kliewer SA, Umesono K, Noonan DJ, Heyman RA, Evans RM (1992) Convergence of 9-cis retinoic acid and peroxisome proliferator signalling pathways through heterodimer formation of their receptors. Nature 358:771-774

Ladias JAA, Hadzopoulou-Cladaras M, Kardassis D, Cardot P, Cheng J, Zannis V, Cladaras C (1992) Transcriptional regulation of human apolipoprotein genes ApoB, ApoCIII, and ApoAII by members of the steroid hormone receptor superfamily HNF-4, ARP-1, EAR-2, and EAR-3. J Biol Chem 267:15849-15860

Ladias JAA, Karathanasis SK (1991) Regulation of the apolipoprotein AI gene by ARP-1, a novel member of the steroid receptor superfamily. Science 251:561-565

Lock EA, Mitchell AM, Elcombe CR (1989) Biochemical mechanisms of induction of hepatic peroxisome proliferation. Annu Rev Pharmacol Toxicol 29:145-163

Luisi BF, Xu WX, Otwinowski Z, Freedman LP, Yamamoto KR, Sigler PB (1991) Crystallographic analysis of the interaction of the glucocorticoid receptor with DNA. Nature 352:497-505

Masters BSS, Clark JE, Roman LJ, McCabe TJ, Helm CB, Johnson EF, Ma Y-H, Kauser K, Harder DR, Roman RJ (1992) Structure-Function studies and physiological roles of eicosanoids metabolized by cytochrome P450 omega-hydroxylases. In: Bailey JM (ed) Prostaglandins, Leukotrienes, Lipoxins and PAF. Plenum Publishing Co. New York pp 59-66

Matsubara S, Yamamoto S, Sogawa K, Yokotani N, Fujii-Kuriyama Y, Haniu M, Shively JE, Gotoh O, Kusunose E, Kusunose M (1987) cDNA cloning and inducible expression during pregnancy of the mRNA for rabbit pulmonary prostaglandin omega-hydroxylase (cytochrome $P-450_{p-2}$). J Biol Chem 262:13366-13371

Muerhoff AS, Williams DE, Leithauser MT, Jackson VE, Waterman MR, Masters BSS (1987) Regulation of the induction of a cytochrome P-450 prostaglandin omega-hydroxylase by pregnancy in rabbit lung. Proc Natl Acad Sci 84:7911-7914

Muerhoff AS, Griffin KJ, Johnson EF (1992) The peroxisome proliferator activated receptor mediates the induction of CYP4A6, a cytochrome P450 fatty acid omega-hydroxylase, by clofibric acid. J Biol Chem 267:19051-19053

Osumi T, Wen J-K, Hashimoto T (1991) Two cis-acting regulatory sequences in the peroxisome proliferator-responsive enhancer region of rat acyl-CoA oxidase gene. Biochem Biophys Res Commun 175:866-871

Palmer CNA, Griffin KJ, Johnson EF (1993a) Rabbit prostaglandin omega-hydroxylase (CYP4A4): Gene structure and expression. Arch Biochem Biophys 300:670-676

Palmer CNA, Hsu M-H, Muerhoff AS, Griffin KJ, Johnson EF (1993b) RXRα reveals a cryptic PPAR responsive element in the CYP4A6 promoter. FASEB J 7:A1166

Rao MS, Reddy JK (1991) An overview of peroxisome proliferator-induced hepatocarcinogenesis. Environ Health Perspect 93:205-209

Schmidt A, Endo N, Rutledge SJ, Vogel R, Shinar D, Rodan GA (1992) Identification of a new member of the steroid hormone receptor superfamily that is activated by a peroxisome proliferator and fatty acids. Mol Endocrinol 6:1634-1641

Schwartzman ML, Falck JR, Yadagiri P, Escalante B (1989) Metabolism of 20-hydroxyeicosatetraenoic acid by cyclooxygenase. J Biol Chem 264:11658-11662

Sher T, Yi H-F, McBride OW, Gonzalez FJ (1993) cDNA cloning, chromosomal mapping, and functional characterization of the human peroxisome proliferator activated receptor. Biochemistry 32:5598-5604

Tugwood JD, Issemann I, Anderson RG, Bundell KR, McPheat WL, Green S (1992) The mouse peroxisome proliferator activated receptor recognizes a response element in the 5' flanking sequence of the rat acyl CoA oxidase gene. EMBO J 11:433-439

Williams DE, Hale SE, Okita RT, Masters BSS (1984) A prostaglandin omega-hydroxylase cytochrome P-450 (P-450 PG-omega) purified from lungs of pregnant rabbits. J Biol Chem 259:14600-14608

Yamamoto S, Kusunose E, Ogita K, Kaku M, Ichihara K, Kusunose M (1984) Isolation of cytochrome P-450 highly active in prostaglandin omega-hydroxylation from lung microsomes of rabbits treated with progesterone. J Biochem 96:593-603

Yokotani N, Bernhardt R, Sogawa K, Kusunose E, Gotoh O, Kusunose M, Fujii-Kuriyama Y (1989) Two forms of omega-hydroxylase toward prostaglandin A and laurate. J Biol Chem 264:21665-21669

Yokotani N, Kusunose E, Sogawa K, Kawashima H, Kinosaki M, Kusunose M, Fujii-Kuriyama Y (1991) cDNA cloning and expression of the mRNA for cytochrome P-450$_{kd}$ which shows a fatty acid omega-hydroxylating activity. Eur J Biochem 196:531-536

Zhang B, Marcus SL, Sajjadi FG, Alvares K, Reddy JK, Subramani S, Rachubinski RA, Capone JP (1992) Identification of a peroxisome proliferator-responsive element upstream of the gene encoding rat peroxisomal enoyl-CoA hydratase/3-hydroxyacyl-CoA dehydrogenase. Proc Natl Acad Sci USA 89:7541-7545

Mapping Determinants of the Substrate Specificities of P450s by Site-Directed Mutagenesis

Toby H. Richardson, Mei-H. Hsu, and Eric F. Johnson
The Scripps Research Institute
Division of Biochemistry, NX-4
10666 North Torrey Pines Road
La Jolla, CA 92037 USA

Introduction

Differences between individuals as well as between species in the disposition of toxic, foreign chemicals can reflect differences in the expression and properties of cytochrome P450 enzymes. More than 35 genes encoding P450 enzymes have been described in the rat (Nelson et al. 1993), and similar or greater numbers are likely to exist in other mammals. The individual enzymes that constitute this family of proteins often exhibit unique but overlapping substrate selectivities. Together these enzymes provide an almost limitless capacity to metabolize foreign, lipophilic compounds.

Comparisons of the amino acid sequences of these proteins suggests that there has been extensive duplication and divergence of the genes encoding P450s in mammalian species since they last shared a common ancestor (Nebert et al. 1987; Nelson, Strobel, 1987; Gotoh, Fujii-Kuriyama, 1989). As a result, each mammalian species exhibits a distinct spectrum of P450 enzymes. This duplication and divergence has led to pairs of highly similar proteins that, perhaps surprisingly, exhibit very distinct substrate selectivities. This suggests that genetic changes can readily alter substrate selectivity without disrupting the basic catalytic features exhibited by

NATO ASI Series, Vol. H 90
Molecular Aspects of Oxidative Drug Metabolizing Enzymes
Edited by E. Arınç, J. B. Schenkman and E. Hodgson
© Springer-Verlag Berlin Heidelberg 1995

all P450s, and that selective pressure during evolution leads to acquisition of new functions which enhance the overall capacity of the system to metabolize foreign compounds.

Site-directed mutagenesis has demonstrated that single amino acid substitutions can dramatically alter the substrate selectivity profile of individual P450s. We will review those studies which have defined elements of the primary sequences of these enzymes that determine differences in substrate selectivity among P450 enzymes. These will be discussed in terms of current models for the structures of eukaryotic P450s based on the structure of the bacterial enzyme, P450 cam.

DIFFERENCES BETWEEN CLOSELY RELATED FORMS OF P450 THAT DETERMINE DISTINCT CATALYTIC PROPERTIES.

The functional differences exhibited by closely related pairs of enzymes reflect one or a few differences in their amino acid sequences. Several distinct pairs of P450 enzymes have been examined, and these studies have identified critical residues that determine differences in substrate specificity. This approach has been applied in detail to several pairs of enzymes in the 2A, 2B, 2C and 2D subfamilies. This in turn has lead to more detailed mutagenesis studies and formed a basis for engineering P450s to catalyze additional reactions.

Substrate Switching between P450 2A4 and 2A5. These two mouse P450s differ by only 11 of 494 amino acids (Lindberg et al. 1989), yet the two enzymes catalyze distinctly different reactions, the 15α-hydroxylation of testosterone (2A4 or P45015α) and the 7-hydroxylation of coumarin (2A5 or P450coh) respectively with each exhibiting no detectable capacity to catalyze the other reaction (Negishi et al. 1989). Lindberg and Negishi (1989) introduced single mutations into 2A5, the coumarin hydroxylase, corresponding to the amino acid found at the same position for the steroid hydroxylase, and the 11 mutant enzymes were characterized following their expression in yeast. Three mutations in 2A5 affected the specific activity of cell homogenates when they were assayed for coumarin 7-hydroxylase activity, A117V, F209L, and M365L. The concentration of each enzyme was similar to that of the wild type as judged by Western blotting suggesting that the catalytic differences resulted from the mutation rather than differences in the level of expression of the mutants.

Only one of the mutations, F209L,[1] conferred steroid 15α-hydroxylase activity to the mutant P450coh. Thus, in contrast to either parent enzyme, the P450coh-F209L mutant enzyme exhibits both activities although the rate of each was roughly 35-50% and 70%, respectively, of the rates exhibited by the parental enzymes. A mutant in which all three residues of the coumarin hydroxylase were changed to that of the steroid hydroxylase, A117V, F209L, and M365L, produced an enzyme that catalyzed the steroid 15α-hydroxylase activity but that did not catalyze coumarin hydroxylation.

Interestingly, the reciprocal mutation, L209F, in the steroid hydroxylase, 2A4, conferred only low coumarin hydroxylase activity to the mutant although it greatly diminished the capacity of the enzyme to perform the steroid hydroxylation. In addition, the A117V and L365M mutations were each able to confer coumarin hydroxylase activity to the mutant, and in contrast to the L209F change, the mutants exhibited higher steroid hydroxylase activity than the parent (Lindberg, Negishi, 1989). Introduction of all three reciprocal mutations produced a fully active coumarin hydroxylase with little steroid hydroxylase activity (Lindberg, Negishi, 1991).

These results demonstrate that a single mutation can change the substrate profile of a P450. In this case, the F209L mutation conferred steroid 15α-hydroxylase activity to the coumarin hydroxylase, and the mutant exhibited both activities. However, all three changes were necessary to switch the coumarin hydroxylase to an exclusive steroid

---

[1]Mutations will be described using the one-letter abbreviation of the amino acid that is altered, its position in the sequence, and the one-letter abbreviation for the new amino acid in the indicated order. The same convention will be used to designate differences between two sequences.

hydroxylase. In contrast, the coumarin hydroxylase activity was affected by cumulative changes at all three positions, and each reciprocal mutant exhibited some coumarin hydroxylase activity.

The effects of other amino acid substitutions at position 209 of P450coh were also examined. In general, Km and Vmax increased as the size of the amino acid side chain decreased for the hydrophobic amino acids leucine, valine, alanine, and glycine (Juvonen et al. 1991). The Km also increased when either asparagine or serine were substituted for the native phenylalanine, but Vmax was not greatly affected. Lysine was found to greatly diminish activity, and no activity was detected for an aspartic acid substitution although the enzyme exhibited a normal P450 CO-difference spectrum (Juvonen et al. 1991). With the exception of the F209A mutant that exhibited a reverse type I binding spectrum, most of the mutants exhibited type I binding spectra in the presence of coumarin. Dissociation constants determined from the dependence of the spectral changes on the concentration of coumarin were found to parallel the changes in Km effected by the mutations indicating that the changes in Km are likely to reflect changes in substrate binding (Iwasaki et al. 1991).

The mutations also affected the relative proportion of high and low spin character of the ferric cytochrome P450 (Iwasaki et al. 1991). The F209L mutant exhibited predominately high spin spectral characteristics that slightly exceeded that of the wild type. As the size of the hydrophobic side chain diminished or its polarity increased, the low spin character increased. Occupancy of the sixth coordination site of the heme by water is thought to result in a low spin cytochrome, whereas the high spin state is thought to reflect an unoccupied axial coordination site. Thus, the increase in low spin character is suggestive of an increase in the accessibility of water to the active site (Iwasaki et al. 1991). This did not generally lead to an increase in uncoupling as judged by hydrogen peroxide formation during substrate metabolism (Juvonen et al. 1992). Although the formation of hydrogen peroxide was elevated for some mutants, there was no clear dependence on the size or charge of the amino acid for this effect. It was also found that mutation of the adjacent glycine 207 to a proline produced a low spin enzyme. This mutation also increased the Km for coumarin greatly, whereas an alanine substitution had little effect on either property. The authors suggest that the proline substitution resulted in a larger substrate binding site (Juvonen et al. 1993).

In addition to the F209L substitution, two amino acid substitutions at residue 209, valine and asparagine, were also found to confer significant steroid 15α-hydroxylase activity to P450coh (approximately 20% of wild type P450 15α)(Iwasaki et al. 1991). Moreover, the F209N mutant exhibits a capacity to hydroxylate corticosterone that is not normally a substrate for P450 15α (Iwasaki et al. 1993a). The observation that 11ß-deoxycorticosterone is not 15α-hydroxylated by a wide range of mutants led to the suggestion that asparagine 209 might interact directly with the 11ß-hydroxyl group of corticosterone. The mutations at residue 209 did not, however, appear to alter the regiospecificity of steroid hydroxylation exhibited by P450 15α as is seen for mutants at other positions in the polypeptide chain that are described later for subfamilies 2B and 2C.

Interestingly, an asparagine is found at this alignment position for a third mouse P450 2A that exhibits only 70% amino acid identity with the P450 15α and that hydroxylates steroids in the 7α position (Iwasaki et al. 1993b). Substitution of leucine that is found in P450 15α for the asparagine of the P450 7α diminished the apparent Km and Vmax for testosterone 7α-hydroxylation as well as the apparent binding constant for testosterone determined spectrally, although the ratio of Vmax/Km was slightly increased. This mutation did not confer 15α hydroxylation to the enzyme. The 7α-hydroxylation of testosterone catalyzed by P450 7α but not that catalyzed by the leucine mutant was inhibited by corticosterone consistent with the affects of these amino acids on the P450 15α. However, corticosterone was not found to be hydroxylated by P450 7α.

Rat P450 2A1 also catalyzes the 7α-hydroxylation of steroids, whereas rat P450 2A2 catalyzes predominantly 15α- and 12α- hydroxylation with a broader range of hydroxylations evident including 7α- and 6ß-hydroxylation (Hanioka et al. 1990; Hanioka et al. 1992a; Hanioka et al. 1992b). The two proteins exhibit 88% amino acid identity. Chimeric constructs generated from the 2A1 and 2A2 cDNAs were analyzed in order to identify regions of the primary structures of the two enzymes that might determine the distinct regiospecificity of the two enzymes (Hanioka et al. 1990; Hanioka et al. 1992b). Most of the chimeras appeared to mimic the broad regiospecificity of P450 2A2 with a selective loss of 15α-hydroxylase activity. Only one chimera exhibited 15α-hydroxylase activity leading the authors to examine differences between the two proteins in the region between amino acids 275 to 356 for possible determinants of the 15α-hydroxylase activity. Individual substitutions in P4502A1 were made that corresponded to the residue found in P450 2A2 (Hanioka et al. 1992a; Hanioka et al. 1992b). However, none of these substitutions conferred 15α-hydroxylation to P450 2A1, whereas several of the reverse substitutions in P450 2A2 were found to diminish 15α-hydroxylation to a greater extent than 12α-hydroxylation to yield a pattern of regiospecificity similar to that of the chimeric enzyme. These correspond to G302E, S303T and H310Y. The reciprocal changes to P450 2A1 did, however, affect Vmax and Km values for 7α-hydroxylation although with the exception of the T303S change they did not alter regiospecificity. The specificity of the T303S mutant appeared to be relaxed as several additional unidentified metabolites were apparent (Hanioka et al. 1992a). Thus, these studies did not clearly identify a key amino acid residue that would confer the regiospecificity of one enzyme to the other, although substitutions that altered the regiospecificity of the enzymes were identified.

<u>Mutations that selectively delete one of several reactions</u>
<u>catalyzed by variant forms of P450 2C3.</u> Earlier work in our
laboratory described preparations of P450 2C3 isolated from
different inbred strains of rabbits (Schwab, Johnson, 1985) as
well as from outbred rabbits (Dieter, Johnson, 1982) that
exhibited selective differences in their catalytic properties.
One of the two forms of the enzyme which is expressed
exclusively in some inbred strains (Schwab, Johnson, 1985;
Dieter, Johnson, 1982) does not exhibit significant
progesterone 6ß-hydroxylase activity. Both forms of the
enzyme exhibit progesterone 16α-hydroxylase activity, but the
two forms differ significantly in their kinetic parameters.
The enzyme which catalyzes progesterone 6ß-hydroxylase
activity catalyzes 16α-hydroxylation with a high efficiency,
whereas the other enzyme exhibits a much lower efficiency.
The *CYP2C3* gene characterized by Chan and Kemper (1990)
encodes the latter enzyme as judged by the expression of the
enzyme from its cDNA in COS-1 cells or in *E. coli* (Richardson
et al. 1993). A cDNA encoding the other form of the enzyme
was generated by the use of specific primers based on the
sequence of 2C3 that flanks the coding region and
amplification by polymerase chain reaction of cDNAs generated
from mRNAs encoding both enzymes (Hsu et al. 1993). COS-1
cells were transfected with the cloned cDNAs to assess the
enzymic phenotypes of the cognate proteins. Both phenotypes
were observed among the clones. Sequence analysis indicated
that the variant that exhibits 6ß-hydroxylase activity
(*CYP2C3v*) differs at 5 amino acids from the other form:
I178M, S256L, S364T, E472D, and V476L (Hsu et al. 1993).
Analysis of hybrid enzymes expressed from chimeric cDNAs as
well as site-specific mutants revealed that the mutation S364T
conferred the progesterone 6ß-hydroxylase activity of P450
2C3v to P450 2C3. These enzymes were each expressed in *E.
coli*, purified and reconstituted with P450 reductase to
characterize their enzymic properties. The single mutant,
S364T, was found to exhibit a Km that was roughly 4-fold

greater than P450 2C3v. Characterization of additional mutants indicated that the Ile/Met difference at position 178 was responsible for this difference of Km (Hsu et al. 1993).

Variants of human P450 2C9. Four variant cDNAs encoding human P450 2C9 exhibit one of two codons at each of four alignment positions: 144, 358, 359 and 417. Examination of the four variants and 2 additional single mutants revealed that all of the enzymes catalyzed tolbutamide hydroxylation but that one of the variants exhibited a relatively low capacity to hydroxylate phenytoin when compared to the others. A higher ratio of phenytoin to tolbutamide metabolism was restored by exchanging residue 359 (Veronese et al. 1993). In addition, an independent study observed that the same isoleucine/leucine difference at residue 359 of P450 2C9 altered the regio- and stereospecificity of warfarin metabolism from a preference for the 7-hydroxylation of (S)-warfarin to the 4'-hydroxylation of (R) warfarin (Kaminsky et al. 1993).

Allelic variants of rat P450 2D1. Matsunaga *et al.* (1990) have described a variant of rat P450 2D1 that catalyzes the 4-hydroxylation of debrisoquine like the wild type enzyme but which exhibits a greatly diminished ability to hydroxylate bufuralol. The wild type and variant enzymes exhibit 4 differences of amino acid sequence: V122I, F124L, R173N, and L380F. Analysis of hybrid enzymes expressed from chimeric cDNAs in COS-1 cells indicated that the single amino acid substitution F380L in the variant enzyme was sufficient to confer bufuralol activity to the mutant.

Determinants of Regioselectivity for rat P450 2B1. Studies of closely related P450 2B enzymes have revealed additional key amino acids. Rat 2B1 and 2B2 differ at 14 of 501 amino acids, and they generally display the same substrate and product profiles although 2B1 usually exhibits roughly 10-fold higher activities. A variant cDNA encoding 2B2 was isolated that differed at 3 amino acid residues from wild type 2B2 (Aoyama et al. 1989). The enzyme expressed from this cDNA in HepG2 cells using a recombinant vaccinia virus vector did not catalyze the 16ß-hydroxylation of testosterone, a metabolite produced by both wild type 2B1 and 2B2. The variant enzyme did, however, produce the 16α-hydroxy and 17-keto products seen for the wild type 2B1 and 2B2. The 3 amino acid differences are L58F, I114F, and E282V. The first two differences occur in a region where 2B1 and 2B2 are identical. When these two changes were incorporated into 2B1, a selective loss in the capacity of the mutant 2B1 to catalyze the 16ß-hydroxylation was seen. The capacity to produce the 16α-hydroxyl and 17-keto products was retained although the rate was diminished to different extents for the two products. Similar changes in the metabolism of androstenedione were also observed in the formation of these two metabolites, whereas an increase in the production of the 15α-hydroxyl product was seen for this substrate. In contrast, the rate of formation of products that eluted with 16α-hydroxyprogesterone was greater for the double mutant when compared to the wild type 2B1 using progesterone as a substrate. Thus, the effect of the double mutation was to alter the regiospecific metabolism of these steroid substrates in distinctly different ways.

It is not clear from this study whether both mutations are required to produce these effects or only the L58F mutation. The L58F mutant was not expressed at appreciable concentrations, and its properties could not be evaluated independently of the I114F mutation. The I114F mutant catalyzed the formation of 16ß-hydroxytestosterone, but the rate and distribution of products were altered to some extent when compared to wild type 2B1. The metabolism of androstenedione and progesterone by the I114F mutant was not reported. Thus, the I114F mutation alone is not sufficient to delete the 16ß-hydroxylase activity, but it may be required together with the L58F mutation to alter regiospecificity and possibly to compensate for the effect of the L58F mutation on the stability of 2B1.

Although the activity exhibited by 2B1 toward most substrates generally exceeds that of 2B2, the rate for the 12-hydroxylation of 7,12-dimethylbenz[a]anthracene (DMBA) is higher for 2B2 (Christou et al. 1992). The double mutation, L58F, I114F, diminished the turnover number for DMBA when introduced into 2B2 but increased the rate exhibited by the corresponding mutants of 2B1. A similar increase was also seen for an E282V mutant of 2B1. However, a triple mutant that included all three changes in 2B1 did not exhibit any gain in activity.

Interestingly, an allozymic form of P450 2B1 also displays a reduced 16ß-hydroxylase activity toward androstenedione and testosterone relative to 16α-hydroxylase capacity which is relatively unaffected (Kedzie et al. 1991). This *CYP2B1* allele is expressed in Wistar Munich rats, and it exhibits a single mutation, A478G, when compared to the wild-type *CYP2B1*. A number of additional amino acids where introduced at position 478 of P450 2B1 and evaluated for their effects on steroid metabolism (He et al. 1992). Increasing the size of the hydrophobic residue at 478 generally increased the 16ß- to 16α- hydroxylation ratio, and a serine mutant was found to exhibit the highest ratio of 15α- to 16-hydroxylated products. In addition, the susceptibility of the enzyme to irreversible inhibition by chloramphenicol was found to diminish as the size of the hydrophobic residue increased (He et al. 1992). Thus, alignment position 478 is also a key residue.

These results demonstrated that a single mutation at position 478 produced a change similar to that described for the L58F,I114F double mutant of P450 2B1 (Aoyama et al. 1989). Halpert and He (1993) examined a number of additional mutations at position 114 of P450 2B1. They noted that the ratio of androgen 16ß:16α-hydroxylation decreased and the ratio of 15α:16-hydroxylation increased when the less bulky hydrophobic residues valine or alanine were substituted at position 114. In addition, the trends seen earlier for substitutions at 478 were also seen for either mutant at 114. As a result, the double mutant I114A,G478S exhibited a 15α:16-hydroxylation ratio 1000-fold greater than that of wild-type 2B1 and closely approximated the regiospecificity of P450 2A4 discussed earlier (Halpert, He, 1993).

<u>Determinants of Differences in Catalytic Efficiency between</u>
<u>P450 2C4 and P450 2C5</u>. Rabbit P450 2C5 is rather unusual
because it catalyzes the 21-hydroxylation of progesterone to
produce the mineralocorticoid, deoxycorticosterone. This
metabolic pathway is generally confined to the adrenal gland
where it is catalyzed by P450 21A and where
deoxycorticosterone serves as a precursor for the formation of
adrenal steroids. In contrast, 2C5 is expressed in the liver
where, like other class 2C P450s, it exhibits a wide-ranging
substrate profile that includes foreign compounds such as the
carcinogens, benzo(a)pyrene (Raucy, Johnson, 1985) and 2-
acetylaminofluorene (McManus et al. 1984). P450 2C4 exhibits
24 differences among 487 amino acids when compared to 2C5.
Like 2C5, it also catalyzes steroid 21-hydroxylation, but it
is roughly 20-fold less efficient due largely to a difference
in apparent Km for progesterone (Kronbach et al. 1989).

In order to determine which of these 24 differences
contribute to this difference in apparent Km, a series of
chimeric cDNAs were constructed from the cDNAs for 2C4 and
2C5, and expressed in COS cells. All of the chimeras that
contained a small segment of 2C5 that codes for three
differences of amino acid sequence from 2C4 exhibit the high
affinity activity. The chimera, G, constructed largely from
2C4 and which contains only the three differences, 2C4-V113A,
-S115T, and -N118K, exhibits a Km of 3 μM which is only
slightly higher than the 2 μM Km exhibited by 2C5 (Kronbach et
al. 1989).

Single mutations were then introduced into 2C4. The V113A
2C4-mutant was the only one of three that exhibited a marked
decrease of Km, whereas the other two mutants where similar to
2C4 in their kinetic properties (Kronbach, Johnson, 1991).
The Km for the 2C4-V113A mutant is 7μM which is intermediate
between that of 2C4, 25μm and 2C5, 2μM. Thus, although the
S115T and N118K mutations do not individually alter the
catalytic properties of 2C4, one or both appear to contribute
together with the V113A mutation to confer the low Km of 2C5
to 2C4.

A Critical Residue. It is interesting to note that when the amino acid sequences of 2A5, 2B1 and 2C4 are aligned, Fig. 1, the V113A difference effecting the difference in Km between 2C4 and 2C5 corresponds directly with the positions of the A117V difference for the mouse P450s 2A that alters coumarin hydroxylase activity, as well as the I114F difference that contributes to differences for the regiospecific metabolism of steroids by allelic forms of rat 2B2. The occurrence of amino acid differences at the same residue among functionally distinct but closely related class 2 P450s suggests that this might be a residue that is under positive selection for P450 diversity.

ENGINEERING NEW SUBSTRATE SELECTIVITIES

The studies above were extended to address whether the 21-hydroxylase activity characteristic of 2C5 could be transferred by site-specific mutagenesis to class 2C P450s that are more distantly related to 2C5 than 2C4. We therefore asked whether alterations of the amino acid at alignment position 113 could confer progesterone 21-hydroxylation to P450 2C1 which exhibits only 75% amino acid sequence identity with 2C5 and that exhibits no detectable progesterone hydroxylase activity when expressed in COS-1 cells.

Initial studies indicated that a chimera which fused the N-terminal 128 residues of 2C5 with the C-terminal 129-490 residues of 2C1 produced an active 21-hydroxylase with a similar Km for progesterone, 3 μM, as seen for 2C5, 2 μM, when expressed in COS cells (Kronbach et al. 1990). These experiments suggested that differences between 2C5 and 2C1 in this N-terminal region were sufficient to confer 21-hydroxylase activity to 2C1 even though the C5/C1 chimera exhibits 99 differences of amino acid sequence when compared to 2C5.

The introduction of the A113V mutation into 2C1 was sufficient to confer a low but readily detectable progesterone 21-hydroxylase activity to 2C1 (Kronbach et al. 1991). As the activity was low, specific kinetic parameters were not

117

```
2D1v MELLNGTGLWSMAIFTVIFILLVDLMHRRHRWTSRYPPGPVPWPVLGNLLQVDLSNMPYSLYKLQHRYGDVFSLQKGWKP  80
2B1  ...MEPTILLL..LALLVGFLLLLVRGHPKSR.GNFPPGPRPLPLLGNLLQLDRGGLLNSFMQLREKYGDVFTVHLGPRP  74
2A5  ..MLTSGLLLVAAVAFLSVLVLMSVWKQRKLS.GKLPPGPTPLPFIGNFLQLNTEQMYNSLMKISQRYGPVFTIYLGPRR  77
2A2  ..MLDTGLLLVVILASLSVMFLVSLW.QQKIR.ERLPPGPTPLPFIGNYLQLNMKDVYSSITQLSERYGPVFTIHLGPRR  76
2C3  ...MDLLIILGIC...LSCVVLLSLWKKTHGK.GKLPPGPTPLPVVGNLLQLETKDINKSLSMLAKEYGSIFTLYFGMKP  73
2C9  ...MDSLVVLVLC...LSCLLLLSLWRQSSGR.GKLPPGPTPLPVIGNILQIGIKDISKSLTNLSKVYGPVFTLYFGLKP  73
2C4  ...MDPVAGLVLG...LCCLLLLSLWKQNSGR.GKLPPGPTPFPIIGNILQIDVKDISKSLTKFSERYGPVFTVYLGMKP  73

                                            SRS1
2D1v MVIVNRLKAVQEVLVTHGEDTADRPPVPIFKCLGVKPRSQGVILASYGPEWREQRRFSVSTLRTFGMGKKSLEEWVTKEA 160
2B1  VVMLCGTDTIKEALVGQAEDFSGRGTIAVIEPIF...KEYGVIFAN.GERWKALRRFSLATMRDFGMGKRSVEERIQEEA 150
2A5  IVVLCGQEAVKEALVDQAEEFSGRGEQATFDWLF...KGYGVVFSS.GERAKQLRRFSIATLRDFGVGKRGIEERIQEEA 153
2A2  IVVLYGYDAVKEALVDQAEEFSGRGELPTFNILF...KGYGFSLSN.VEQAKRIRRFTIATLRDFGVGKRDVQECILEEA 152
2C3  AVVLYGYEGVIEALIYRGEEFSGRGIFPVFDRVT...KGLGIVFSS.GEKWKETRRFSLTVLRNLGMGKKTIEERIQEEA 149
2C9  IVVLHGYEAVKEALIDLGEEFSGRGIFPLAERAN...RGFGIVFSN.GKKWKEIRRFSLMTLRNFGMGKRSIEDCVQEEA 149
2C4  TVVLHGYKAVKEALVDLGEEFAGRGHFPIAEKVN...KGLGIYFTN.ANTWKEMRRFSLMTLRNFGMGKRSIEDRVQEEA 149

                                              SRS2
2D1v GHLCDAFTAQAGQSINPKAMLNKALCNVIASLIFARRFEYEDPYLIRMVKLVEESLTEVSGFIPEVLNTFPALLR.IPGL 239
2B1  QCLVEELRKSQGAPLDPTFLFQCITANIICSIVFGERFDYTDRQFLRLLELFYRTFSLLSSFSSQVFEFFSGFLKYFPGA 230
2A5  GFLIDSFRKTNGAFIDPTFYLSRTVSNVISSIVFGDRFDYEDKEFLSLLRMMLGSFQFTATSMGQLYEMFSSVMKHLPGP 233
2A2  GYLIKTLQGTCGAPIDPSIYLSRTVSNVINSIVFGNRFDYEDKEFLSLLEMIDEMNIFAASATGQLYDMFHSVMKYLPGP 232
2C3  LCLIQALRKTNASPCDPTFLLFCVPCNVICSVIFQNRFDYDDEKFKTLIKYFHENFELLGTPWIQLYNIFPILGHYLPGS 229
2C9  RCLVEELRKTKASPCDPTFILGCAPCNVICSIIFHKRFDYKDQQFLNLMEKLNENIKILSSPWIQICNNFSPIIDYFPGT 229
2C4  RCLVEELRKTNALPCDPTFILGCAPCNVICSVILHNRFDYKDEEFLKLMERLNENIRILSSPWLQVYNNFPALLDYFPGI 229

      SRS3                                              SRS4
2D1v ADKVFQGQKTFMALLDNLLAENRTTWDPAQPPRNLTDAFLAEVEKAKGNPESSFNDENLRMVVVDLFTAGMVTTATTLTW 319
2B1  HRQISKNLQEILDYIGHIVEKHRATLDP.SAPRDFIDTYLLRMEKEKSNHHTEFHHENLMISLLSLFFAGTETSSTLRY 309
2A5  QQQAFKELQGLGLEDFITKKVEHNQRTLDP.NSPRDFIDSFLIRMLEEKKNPNTEFYMKNLVLTTLNLFFAGTETVSTLRY 312
2A2  QQQIIKVTQKLEDFMIEKVRQNHSTLDP.NSPRNFIDSFLIRMQEE.KYVNSEFHMNNLVMSSLGLLFAGTGSVSSTLYH 310
2C3  HRQLFKNIDGQIKFILEKVQEHQESLDS.NNPRDFVDHFLIKMEKEKHKQSEFTMDNLITTIWDVFSAGTDTTSNTLKF 308
2C9  HNKLLKNVAPMKSYILEKVKEHQESMDM.NNPQDFIDCFLMKMEKEKHNQPSEFTIESLENTAVDLFGAGTETTSTTLRY 308
2C4  HKTLLKNADYTKNFIMEKVKEHQKLLDV.NNPRDFIDCFLIKMEKENN...LEFTLGSLVIAVFDLFGAGTETTSTTLRY 305

                                                    SRS5
2D1v ALLLMILYPDVQRRVQQEIDEVIGQVRCPEMTDQAHMPYTNAVIHEVQRFGDIAPLNLPRFTSCDIEVQDFVIPKGTTLI 399
2B1  GFLLMLKYPHVAEKVQKEIDQVIGSHRLPTLDDRSKMPYTDAVIHEIQRFSDLVPIGVPHRVTKDTMFRGYLLPKNTEVY 389
2A5  GFLLLMKHPDIEAKVHEEIDRVIGRNRQPKYEDRMKMPYTEAVIHEIQRFADMIPMGLARRVTKDTKFRDFLLPKGTEVF 392
2A2  GFLLLMKHPDVEAKVHEEIERVIGRNRQPQYEDHMKMPYTQAVINEIQRFSNLAPLGIPRRIIKNTTFRGFFLPKGTDVF 390
2C3  ALLLLLKHPEITAKVQEEIEHVIGRHRSPCSQDRSRMPYTDAVMHEIQRYVDLVPTSLPHAVTQDIEFNGYLIPKGTDII 388
2C9  ALLLLLKHPEVTAKVQEEIERVIGRNRSPCMQDRSHMPYTDAVVHEVQRYIDLLPTSLPHAVTCDIKFRNYLIPKGTTIL 388
2C4  SLLLLLKHPEVAARVQEEIERVIGRHRSPCMQDRSHMPYTDAVIHEIQRFIDLLPTNLPHAVTRDVKFRNYFIPKGTDII 385


2D1v INLSSVLKDETVWEKPHRFHPEHFLDAQGNFVKHEAFMPFSAGRRACLGEPLARMELFLFFTCLLQRFSFSVPVG.QPRP 479
2B1  PILSSALHDPQYFDHPDSFNPEHFLDANGALKKSEAFMPFSTGKRICLGEGIARNELFLFFTTILQNFSVSSHLAPKDID 469
2A5  PMLGSVLKDPKFFSNPKDFNPKHFLDDKGQFKKNDAFVPFSIGKRYCFGEGLARMELFLFLTNIMQNFHFKSTQAPQDID 472
2A2  PIIGSLMTEPKFFPNHKDFNPQHFLDDKGQLKKNAAFLPFSIGKRFCLGDSLAKMELFLLLTTILQNFRFKFPMNLEDIN 470
2C3  PSLTSVLYDDKEFPNPEKFDPGHFLDESGNFKKSDYFMPFSTGKRACVGEGLARMELFLLLTTILQHFTLKPLVDPKDID 468
2C9  ISLTSVLHDNKEFPNPEMFDPHHFLDEGGNFKKSKYFMPFSAGKRICVGEALAGMELFLFLTSILQNFNLKSLVDPKNLD 468
2C4  TSLTSVLHDEKAFPNPKVFDPGHFLDESGNFKKSDYFMPFSAGKRMCVGEGLARMELFLFLTSILQNFKLQSLVEPKDLD 465

      SRS6
2D1v STHGFFAFPVAPLPYQLCAVVREQGL                                                      505
2B1  LTPKESGIGKIPPTYQICFSAR....                                                      491
2A5  VSPRLVGFATIPPTYTMSFLSR....                                                      494
2A2  EYPSPIGFTRIIPNYTMSFMPI....                                                      492
2C3  PTPVENGFVSVPPSYELCFVPV....                                                      490
2C9  TTPVVNGFASVPPFYQLCFIPV....                                                      490
2C4  ITAVVNGFVSVPPSYQLCFIPI....                                                      487
```

Fig. 1. Location of substrate recognition sites (shaded) as defined by Gotoh (1992) and critical amino acids identified by site-directed mutagenesis (reversed letters) as discussed in the text.

determined for the 2C1-A113V mutant. This result extends previous observations that this is a critical residue for determining substrate selectivity. In addition, it suggests that context is likely to be of importance. The results described earlier for the A113V mutant of 2C4 are also suggestive of this possibility. Although this mutant was the only one which dramatically altered the 21-hydroxylase activity of 2C4, the apparent Km was greater than that of the minimal chimera which contained the cluster of three altered amino acids in 2C4 (A113V, S115T, N118K) or 2C5, 7 μM as compared to 2 μM for the latter two enzymes.

Alanine 113 of P450 2C5 occurs in a region where other residues in P450 2C proteins are highly variable. The sequences flanking this region vary much less and are predicted to form helical elements as described by Edwards *et al.* (1989). These regions are generally predicted to be helices in a large number of mammalian P450s, and by alignment with P450cam, they are thought to correspond to helices B and C (Hudecek, Anzenbacher, 1988; Edwards et al. 1989; Nelson, Strobel, 1989; Gotoh, 1992; Ouzounis, Melvin, 1991) which are found in the latter enzyme by x-ray diffraction studies (Poulos, 1991). The region between these two predicted-helices, which exhibits the high variation, corresponds to a substrate contacting, surface loop in P450cam that contains a short stretch of alpha-helical structure, the B' helix (Poulos, 1991). Similar structural features are seen for P450 102 (BM-3) as determined by x-ray diffraction analysis (Ravichandran et al. 1993).

This loop is also on the surface of the protein, and it, together with surface loops at distal segments of the protein, may form a flexible access channel for entry of the substrate to the enzyme (Poulos, 1991). Studies employing antibodies elicited to peptides derived from this region of P450 2B1 (De Lemos-Chiarandini et al. 1987) and the mapping of the epitope recognized by a monoclonal antibody to P450 2C5 (Kronbach, Johnson, 1991) indicate that residues in this region are on the surface of these two mammalian P450s as well. Tyr 96 of

P450cam is found in this loop, and the side-chain of this residue forms a hydrogen bond with the substrate camphor that contributes to the orientation of the substrate with respect to the site of oxygen binding and reduction. Phenylalanines 87 and 98 that are found in this segment also form close contacts with the substrate. If this region has corresponding functions in mammalian P450s as it does in P450cam, it might easily accommodate genetic variation without disrupting the overall structure of P450s because of its surface localization and loop structure. On the other hand, if it also forms a substrate contact loop as suggested by studies of mutants and chimeric enzymes, genetic variation in this region could lead to differences in substrate selectivity and kinetic properties between closely related enzymes (Kronbach et al. 1989).

With this model in mind, a chimera was constructed to incorporate the hypervariable region of 2C5, residues 95-123 into 2C1, Fig. 2 (Kronbach et al. 1991). The resulting chimera exhibits 21-hydroxylase activity with an apparent Km that is close to that of wild type 2C5, even though, the chimera shares only 77% sequence identity with 2C1.

A similar chimera was also generated by placing the hypervariable region of 2C5 into 2C3v. The latter exhibits 68% amino acid identity with 2C5, and in contrast to 2C5, 2C3v catalyzes progesterone 6ß- and 16α-hydroxylations rather than 21-hydroxylation. The resulting chimera was expressed in *E. coli*, and following purification and reconstitution it exhibited 21-hydroxylase activity rather than 16α-hydroxylase activity, but retained 6ß-hydroxylase activity. Introduction of a single V113A substitution into 2C3v also conferred 21-hydroxylase activity to 2C3v at the expense of 16α-hydroxylase activity while leaving the 6ß-hydroxylase activity largely unaffected (Richardson, Johnson, 1993). This differential effect may reflect two different orientations for the binding of progesterone at the active site of 2C3v, one for 6ß-hydroxylation and a second orientation for 16α-hydroxylation. The V113A mutation may not perturb the binding of progesterone for 6ß-hydroxylation but may affect the binding of

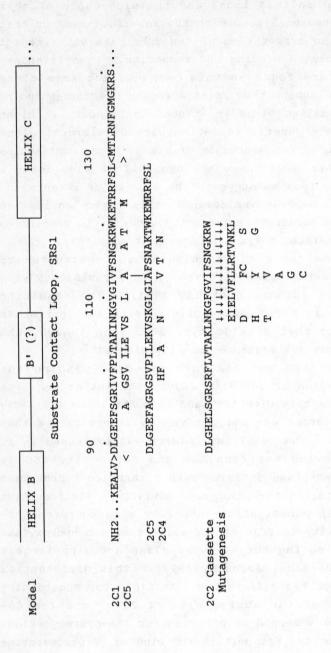

Fig. 2. A comparison of the SRS 1 regions of P450s 2C1, 2C2, 2C4, and 2C5. The predicted topology based on alignments with P450 cam is shown above the sequence of 2C1. The changes introduced into P450 2C1 when the segment between the angle brackets, ><, was replaced with the corresponding segment of 2C5 are shown below the sequence of 2C1. The corresponding segment of 2C4 is compared to that of 2C5, and the valine/alanine difference that determines the Km for progesterone 21-hydroxylation is indicated. In addition, various mutations introduced into this region of P450 2C2 are shown.

progesterone for 16α-hydroxylation to a greater extent shifting the orientation of the substrate to favor 21- versus 16α-hydroxylation.

The S364T difference that determines the difference in 6β-hydroxylation exhibited by 2C3v and 2C3 (Hsu et al. 1993) may largely affect the other binding orientation. We therefore examined whether substitution of an asparagine found in 2C5 for the threonine found at position 364 in 2C3v would diminish 6β-hydroxylase activity and increase 21-hydroxylase activity of the mutant. As expected, the 2C3v:V113A,T364N mutant did not exhibit 6β-hydroxylase activity. Moreover, both the 21- and 16α-hydroxylase activities were increased when compared to the V113A single mutant of 2C3V. In addition, a new metabolite tentatively identified as 17α-hydroxyprogesterone was apparent. These results indicate that the mutations in 2C3v have relaxed the regiospecificity of the enzyme to permit hydroxylation of 3 sites, 16α, 17α and 21, that are in close proximity on the progesterone molecule.

Thus, the studies described here indicate that residues corresponding to position 113 of P4502C5, when class 2 P450s are aligned, may have a dramatic affect on the catalytic properties of these enzymes. Additional amino acids in this region are also likely to contribute to the active site, and these studies indicate that adjacent residues contribute to the affects of substitutions at residues corresponding to position 113 of P450 2C5. The region corresponding to the hypervariable region of 2C5 (Kronbach et al. 1991) has been designated as substrate recognition site 1 (SRS1) by Gotoh (1992). However, it is unlikely that this entire region serves as a substrate recognition site particularly as only some residues are likely to extend inward and others are likely to be on the surface. In P450 cam, only the two short segments of this region that flank the B' helix form part of the substrate binding site.

A series of mutations in the SRS1 region have been examined by cassette mutagenesis (Straub et al. 1993a; Straub et al. 1993c; Straub et al. 1993b) in P450 2C2, a lauric acid ω-1 hydroxylase, and in a chimera, C2MstC1. The C2MstC1 chimera comprises the 184 N-terminal amino acids of P450 2C2 and the 306 C-terminal amino acids of P450 2C1, and it exhibits progesterone 21-hydroxylase activity in addition to lauric acid ω-1 hydroxylase activity (Straub et al. 1993a). A number of mutants bearing alternative hydrophobic amino acid residues at residue 113 were characterized. Incorporation of radiolabeled amino acid precursors and specific immunoprecipitation indicated that similar steady-state levels of each mutant were expressed in COS-1 cells. The laurate ω-1 hydroxylase activity of 2C2 was not greatly affected by the substitution of A, V, L or F for I, whereas, the activity of the G, Y, or C mutants were very low (Straub et al. 1993a). The C2MstC1 chimera appeared to be more sensitive to the substitution of A, V, L or F for I, and this effect was greater for progesterone hydroxylation than for laurate hydroxylation. However, these effects were small. Interestingly, the alanine substitution did not enhance progesterone 21-hydroxylation of the C2MstC1 chimera as was seen for 2C4, 2C1 and 2C3v. However, the apparent Km was not assessed in these studies, and under the conditions of the assays the apparent rate might not be very sensitive to changes in Km of <10-fold. These results indicate that a number of hydrophobic residues of different bulk could be accommodated in the two enzymes at this alignment position. In addition, the hydroxylation of the more flexible and less bulky substrate, lauric acid, was affected to a lesser extent than that of the rigid, planar progesterone molecule by amino acid side-chains of increasing size.

A degenerate cassette was synthesized, and at least one mutant for each residue between 107-120 was obtained for 2C2 (Straub et al. 1993c) and the 2C2MstC1 chimera (Straub et al. 1993b), Fig. 2. In some cases, the resulting mutants exhibited relatively conservative changes whereas in others they were non-conservative. For the mutants obtained, the alteration of amino acids from 107 to 110 and from 116 to 119 did not greatly diminish lauric acid ω-1 hydroxylase activity of 2C2. A relatively non-conservative G109E mutation had little affect on activity, but the G111V and G117V mutations diminished the activity of 2C2 (Straub et al. 1993c) by 50-fold and 7-fold less activity, respectively, and affected the activity of the C2MstC1 chimera similarly (Straub et al. 1993b). In like manner, the substitution of Leu for the highly conserved Trp 120 greatly reduced activity for both but did not eliminate it. The S115R mutants of both enzymes also exhibited low activity. In addition, as was seen for the substitutions at 113, the V112F and F114L mutations reduced activity and in the case of the C2MstC1, the progesterone 21-hydroxylase activity appeared to be more sensitive to the substitution at 112. These results are consistent with the likelihood that the short stretch of hydrophobic amino acids flanked by the two highly conserved glycine residues forms a portion of the substrate recognition site. Glycine 111 may serve to terminate the putative B' helix in the SRS1 region of mammalian P450s (Straub et al. 1993c) as it does in P450 BM3 (Ravichandran et al. 1993).

MOST MUTATIONS AFFECTING P450 SPECIFICITY MAP TO SRS REGIONS

A number of sites that determine the catalytic properties of mammalian P450s map to a second region, Fig. 1, predicted to be a substrate recognition site (Johnson, 1992; Gotoh, 1992). These are the L365M mutation that confers coumarin hydroxylase activity to P450 2A4, and the catalytically distinct I359L, L380F, and T364S variants of 2C9, 2D1 and 2C3, respectively. As shown in the alignment of these sequences, Fig. 1, these residues are clustered in a short segment that maps to the region of P450 cam between helix K that appears to be highly conserved for mammalian P450s (Hudecek, Anzenbacher, 1988; Edwards et al. 1989; Nelson, Strobel, 1989; Gotoh, 1992; Ouzounis, Melvin, 1991) and the beta-sheet 3. This region of P450cam includes Val295 that forms a contact residue with the substrate camphor (Poulos et al. 1987) and that has been designated as SRS5 by Gotoh (1992). The topology of this region is also conserved in P450 BM3 (Ravichandran et al. 1993).

Thus, a pattern has emerged from these studies that has identified regions of mammalian P450s where critical amino acid residues that lead to enzymic variation among closely related proteins reside. Moreover, the location of these residues is in good agreement with regions predicted to form substrate recognition sites based on the structure of P450cam. In this regard, the A478G difference that underlies the allelic variation in catalytic properties of P450 2B1 (Kedzie et al. 1991) also maps to such a region, SRS6, Fig. 1. In addition, the mutations that appeared to alter the regiospecificity of 2A2 map to SRS4, Fig. 2.

Several alignments appear to map residue 209 of the mouse 2A enzymes to helix E of P450cam as pointed out by Poulos (1989). Helix E is well-removed from the substrate binding site of P450cam. When the structures of P450 cam and P450 BM-3 are compared, these regions show some of the largest differences in the length and spatial orientations of the helices and intervening loops that form this portion of the structure (Ravichandran et al. 1993). Thus it may be difficult to align these regions unambiguously. An alternative alignment by Gotoh (1992) predicts that residue 209 of P450 2A4 maps to the region between helices F and G of P450cam that forms SRS2, Fig. 1.

Residue 58 of the 2B enzymes maps to a region of P450cam, between helices A and B, that is also remote from the substrate binding site of P450 cam. However, this segment may contribute to the proposed substrate binding pocket of P450 BM-3 (Ravichandran et al. 1993). Alternatively, this and other mutations could effect a change in the structure of the protein that affects the active site although they are not in close proximity to it.

TARGETED MUTAGENESIS OF PUTATIVE HELIX I AND SRS4

The identification of segments of mammalian P450s that correspond to helix I and the observation that several residues along this helix contact the substrate in P450cam led investigators to target residues associated with this segment for mutagenesis. Furuya *et al*. (1989b; 1989a) mutated residues in the region of rat P450 1A2 corresponding to the SRS4 region of helix I to residues found in either rat or mouse P450 1A1 or in rabbit or mouse P450 1A2. Point mutants in which other residues of this region were altered were also generated. These mutants were then expressed in *S. cerevisiae*. Microsomes were isolated and supplemented with NADPH-cytochrome P-450 reductase and assayed for enzymic activity using benzphetamine (Furuya et al. 1989b), ethoxycoumarin (Furuya et al. 1989b) and acetanilide (Furuya et al. 1989a). Almost all mutants with the exception of one, G316E, exhibited greatly diminished activity with acetanilide and a loss of regioselectivity when compared to the wild-type rat 1A2 expressed in this manner. This includes mutations which mimic the corresponding sequences in the mouse and rabbit 1A2 enzymes which also metabolize acetanilide. The activity of the mutants with benzphetamine as substrate was also diminished when compared to the wild type enzyme. In contrast, the activities determined with ethoxycoumarin as the substrate were not affected as greatly with several of the mutants exhibiting values of kcat or kcat/Km that exceeded that of the wild-type. However, these values remained low when compared to the kcat and kcat/Km values reported for rat or mouse P450 1A1 suggesting that none of the mutations in this region, including those that introduced all of amino acids corresponding to the 1A1 enzymes, confer the high activity of 1A1 for ethoxycoumarin O-deethylation to 1A2. Characterization of these mutant enzymes by ESR at low temperatures (Sotokawa et al. 1990) also indicated that the high-spin character of 1A2 was not altered by any of these mutations to exhibit the low-spin state characteristic of the 1A1 enzymes.

The effects of mutating threonine 319 in the SRS 4 region of helix I of rat 1A2 is of particular interest because threonine is highly conserved at this alignment position for many P450s (Gotoh, Fujii-Kuriyama, 1989). The T319A mutant exhibits no detectable activity toward benzphetamine (Furuya et al. 1989b) and very low but detectable activity with acetanilide (Furuya et al. 1989a), whereas the activity exhibited with ethoxycoumarin is higher than that of the wild-type enzyme (Furuya et al. 1989b). Moreover, this mutation opens up the active site to permit phenyldiazene to attack the B-ring of the iron proporphyrin that is normally protected by helix I (Tuck et al. 1993). It should be noted, however, that none of these substrates is efficiently metabolized by P450 1A2 which exhibits kcat/Km values for these substrates that are roughly 100-fold or more lower than that exhibited by 2C5 or 2C3 for progesterone hydroxylation.

The presence of threonine at this alignment position is thought to contribute to a deformation of helix I in P450 cam that accommodates the binding of molecular oxygen to the iron of the heme (Poulos et al. 1985; Poulos et al. 1986). The conservation of this threonine among most P450s suggests that this structural feature might be conserved in them as well. This would not be surprising given that oxygen binding and reduction are shared functions of these enzymes. However, other amino acids are found at this position for some P450 enzymes, and when other amino acids were substituted for this threonine in rat 1A2 the metabolism of some substrates was not greatly affected whereas that of other substrates was. These differential affects are likely to reflect the function of this amino acid as a component of SRS4 rather than an essential role in oxygen activation. Similar observations were made by Imai and Nakamura (1988; 1989) in earlier studies where they mutated this highly conserved threonine in 2C2 and 2C14. The initial T301H mutants were devoid of laurate 11-hydroxylase or testosterone 16α-hydroxylase activity, respectively. Moreover, lauric acid did not effect a conversion of the 2C2-T301H mutant from a predominantly low-spin to high-spin form as it does for the wild type 2C2 (Imai, Nakamura, 1988). Interestingly, the 2C2-T301K mutant exhibits a visible absorption spectrum indicating that the ϵ-amino group of the lysine coordinates with the heme iron of the enzyme (Imai, Nakamura, 1991). This suggests that this portion of the protein is in close proximity to the catalytic site as predicted. The 2C2-T301S mutant exhibits roughly 40% of the lauric acid 11-hydroxylase activity of the wild-type, but interestingly it exhibits a slightly higher (110%) activity with caproic acid as a substrate. In contrast, the 2C2-T301N and 2C2-T301V mutants exhibit 40% and 20% of the lauric acid 11-hydroxylase activity of the wild-type, respectively, but neither exhibits detectable activity with caproic acid. The T301S mutant of 2C14 also displayed differential effects on the 16α-hydroxylation of testosterone and progesterone. The activity toward testostosterone was diminished by 50% whereas little or no effect on the metabolism of progesterone was seen (Imai, Nakamura, 1989). Differential effects were also observed for the binding of 1-methylimidazole to 2C2 as judged by the conversion of the ferric protein from low-spin to high-spin forms. In contrast, the mutants did not differ greatly in their apparent affinity for ethyl isocyanide (Imai, Nakamura, 1989).

These results indicate that modifications of residues in the SRS4 region can lead to differences in the metabolic properties of P450 enzymes. Furya *et al*. (1989b) noted that the affect of alterations of the conserved threonine on the relative activity for different substrates of these enzymes is consistent with the location of the threonine in P450 cam being in close proximity to the substrate (Poulos et al. 1985). The effect of a given mutation is clearly dependent on the substrate. This is not surprising when one considers that the contributions of individual residues to binding are likely to differ for structurally distinct substrates which may interact differently with the protein.

The observation that the highly conserved threonine of helix I is not essential among the mammalian P450s examined in these studies is somewhat more surprising given its proposed role in the binding of oxygen in a favorable orientation relative to the heme. However, mutagenesis of P450 cam also indicates that this residue is not essential. The Ser, Val and Ala mutants each metabolize camphor although at reduced rates compared to the wild type. However, the rate of formation of the hydroxylated product is uncoupled from oxygen reduction in the mutants which produced H_2O_2 as 15%, 45% and 83% of the oxygen consumed (Imai et al. 1989). The authors of the latter study suggest that this might reflect the loss of the threonine's function as a hydrogen ion donor and/or an acid-base catalyst in oxygen scission. Alternatively, the changes in oxygen and/or camphor binding might lead to a greater access by water molecules to the site of oxygen reduction thereby facilitating the oxidase activity. This might also contribute to the loss of activity seen with some but not all of the substrates of the mammalian enzymes when the corresponding threonine was substituted with another amino acid. Hatano and his colleagues have examined the effects of mutations of Thr319 of rat 1A2 as well as of adjacent residues on coupling in the metabolism of ethoxycoumarin (Ishigooka et al. 1992) and methanol (Hiroya et al. 1992). In general, the efficiency of oxygen incorporation into these substrates is low, 1.7% and 9.4%, respectively. Surprisingly, the T319A mutation increased the efficiency of oxygen incorporation into both substrates as did a T319D mutation. The latter also increased the turnover number for ethoxycoumarin by several fold, and it is interesting to note that an aspartic acid is found for 1A2 enzymes from several other species. These effects may, therefore, reflect a close proximity to the site of oxygen reduction and substrate binding rather than a specific role for the side-chain of this threonine in catalysis.

TARGETED MUTAGENESIS OF SRS3

Krainev et al. (1992) have mutated the region of rat 1A2 corresponding to SRS3. This region contains three basic residues on the surface of the N-terminus of helix G. The basic residues were each mutated individually to leucine residues, and the three mutants, K250L, R251L and K253L were each characterized for their capacity to catalyze methoxyresorufin and ethoxyresorufin O-dealkylations. In general, each of the leucine substitutions increased the apparent turnover number of the enzyme for these substrates. However, the alteration of these charges did not diminish the capacity of a series of substituted pyridine inhibitors bearing an ionic side chain to inhibit the enzyme. Thus, it is unlikely that the charged residues in the wild-type enzyme interact with the ionic side-chain as had been proposed (Krainev et al. 1991).

Summary

Perhaps the most remarkable aspect of these studies is the capacity of the P450 enzymes to accommodate genetic change. The examples shown here indicate that many distinct alterations of amino acid sequence can impart changes in specificity to a P450 without a loss of the basic capacity of the enzyme to function as a monooxygenase. In some cases, a single amino acid change can alter the substrate selectivity of a P450 enzyme. This is not surprising given that P450 enzymes have evolved to catalyze so many different monooxygenations in nature.

The studies reviewed here encompass two diametrically different approaches. One, examines which of a limited number of differences between two P450s contribute to their distinct catalytic properties. The second selects residues for mutation based on the structure of the very distantly related enzyme, P450 cam. Each has resulted in a demonstration of how the substrate selectivity of a mammalian P450 can be altered. As we discussed earlier, some but not all functionally important differences that were defined by switching residues between closely related pairs fit easily into models based on the structure of P450 cam. In this regard, it is particularly interesting that some of the same residues are seen to vary between different pairs of functionally distinct but highly similar (>95% identical) pairs of P450s. These results are suggestive that hot spots for the generation of functional diversity may occur.

Acknowledgment

The authors acknowledge Keith J. Griffin for critical comments during the preparation of this manuscript as well as the support of USPHS Grant GM31001.

Reference

Aoyama T, Korzekwa K, Nagata K, Adesnik M, Reiss A, Lapenson DP, Gillette J, Gelboin HV, Waxman DJ, Gonzalez FJ (1989) Sequence requirements for cytochrome P-450IIB1 catalytic activity. Alteration of the stereospecificity and regioselectivity of steroid hydroxylation by a simultaneous change of two hydrophobic amino acid residues to phenylalanine. J Biol Chem 264:21327-21333

Chan G, Kemper B (1990) Structure of the rabbit cytochrome P450IIC3 gene, a constitutive member of the P450IIC subfamily. Biochemistry 29:3743-3750

Christou M, Mitchell MJ, Aoyama T, Gelboin HV, Gonzalez FJ, Jefcoate CR (1992) Selective suppression of the catalytic activity of cDNA-expressed cytochrome P4502B1 toward polycyclic hydrocarbons in the microsomal membrane: Modification of this effect by specific amino acid substitutions. Biochemistry 31:2835-2841

De Lemos-Chiarandini C, Frey AB, Sabatini DD, Kreibich G (1987) Determination of the membrane topology of the phenobarbital-inducible rat liver cytochrome P-450 isoenzyme PB-4 using site-specific antibodies. J Cell Biol 104:209-219

Dieter HH, Johnson EF (1982) Functional and structural polymorphism of rabbit microsomal cytochrome P-450 form 3b. J Biol Chem 257:9315-9323

Edwards RJ, Murray BP, Boobis AR, Davies DS (1989) Identification and location of α-helices in mammalian cytochrome P450. Biochemistry 28:3762-3770

Furuya H, Shimizu T, Hatano M, Fujii-Kuriyama Y (1989a) Mutations at the distal and proximal sites of cytochrome P-450$_d$ changed regio-specificity of acetanilide hydroxylations. Biochem Biophys Res Commun 160:669-676

Furuya H, Shimizu T, Hirano K, Hatano M, Fujii-Kuriyama Y, Raag R, Poulos TL (1989b) Site-directed Mutageneses of rat liver Cytochrome P-450d: Catalytic activities toward benzphetamine and 7-ethoxycoumarin. Biochemistry 28:6848-6857

Gotoh O (1992) Substrate recognition sites in cytochrome P450 family 2 (CYP2) proteins inferred from comparative analyses of amino acid and coding nucleotide sequences. J Biol Chem 267:83-90

Gotoh O, Fujii-Kuriyama Y (1989) Evolution, structure, and gene regulation of cytochrome P-450. In: Ruckpaul K, Rein H (eds) Frontiers in biotransformation. Basis and mechanisms of regulation of cytochrome P-450. Taylor & Francis New York pp 195-243

Halpert JR, He Y (1993) Engineering of cytochrome P450 2B1 specificity. Conversion of an androgen 16ß-hydroxylase to a 15α-hydroxylase. J Biol Chem 268:4453-4457

Hanioka N, Korzekwa K, Gonzalez FJ (1990) Sequence requirements for cytochromes P450IIA1 and P450IIA2 catalytic activity: Evidence for both specific and non-specific substrate binding interactions through use of chimeric cDNAs and cDNA expression. Protein Eng 3:571-575

Hanioka N, Gonzalez FJ, Lindberg NA, Liu G, Gelboin HV, Korzekwa KR (1992a) Site-directed mutagenesis of cytochrome P450s CYP2A1 and CYP2A2: Influence of the distal helix on the kinetics of testosterone hydroxylation. Biochemistry 31:3364-3370

Hanioka N, Gonzalez FJ, Lindberg NA, Liu G, Korzekwa KR (1992b) Chimeric cDNA expression and site directed mutagenesis studies of cytochrome P450s CYP2A1 and CYP2A2. J Steroid Biochem Mol Biol 43:1037-1043

He Y, Balfour CA, Kedzie KM, Halpert JR (1992) Role of residue 478 as a determinant of the substrate specificity of cytochrome P450 2B1. Biochemistry 31:9220-9226

Hiroya K, Ishigooka M, Shimizu T, Hatano M (1992) Role of Glu318 and Thr319 in the catalytic function of cytochrome P450$_d$ (P4501A2): Effects of mutations on the methanol hydroxylation. FASEB J 6:749-751

Hsu M-H, Griffin KJ, Wang Y, Kemper B, Johnson EF (1993) A single amino acid substitution confers progesterone 6ß-hydroxylase activity to rabbit cytochrome P450 2C3. J Biol Chem 268:6939-6944

Hudecek J, Anzenbacher P (1988) Secondary structure prediction of liver microsomal cytochrome P-450; proposed model of spatial arrangement in a membrane. Biochim Biophys Acta 955:361-370

Imai M, Shimada H, Watanabe Y, Matsushima-Hibiya Y, Makino R, Koga H, Horiuchi T, Ishimura Y (1989) Uncoupling of the cytochrome P-450cam monooxygenase reaction by a single mutation, threonine-252 to alanine or valine: A possible role of the hydroxy amino acid in oxygen activation. Proc Natl Acad Sci USA 86:7823-7827

Imai Y, Nakamura M (1988) The importance of threonine-301 from cytochromes P-450 (laurate (w-1)-hydroxylase and testosterone 16α-hydroxylase) in substrate binding as demonstrated by site-directed mutagenesis. FEBS Lett 234:313-315

Imai Y, Nakamura M (1989) Point mutations at threonine-301 modify substrate specificity of rabbit liver microsomal cytochromes P-450 (laurate (0-1)-hydroxylase and testosterone 16α-hydroxylase). Biochem Biophys Res Commun 158:717-722

Imai Y, Nakamura M (1991) Nitrogenous ligation at the sixth coordination position of the Thr-301 to Lys-mutated P450IIC2 heme iron. J Biochem (Tokyo) 110:884-888

Ishigooka M, Shimizu T, Hiroya K, Hatano M (1992) Role of Glu318 at the putative distal site in the catalytic function of cytochrome P450$_d$. Biochemistry 31:1528-1531

Iwasaki M, Juvonen R, Lindberg R, Negishi M (1991) Alteration of high and low spin equilibrium by a single mutation of amino acid 209 in mouse cytochromes P450. J Biol Chem 266:3380-3382

Iwasaki M, Darden TA, Pedersen LG, Davis DG, Juvonen RO, Sueyoshi T, Negishi M (1993a) Engineering mouse P450coh to a novel corticosterone 15α-hydroxylase and modeling steroid-binding orientation in the substrate pocket. J Biol Chem 268:759-762

Iwasaki M, Lindberg RLP, Juvonen RO, Negishi M (1993b) Site-directed mutagenesis of mouse steroid 7α-hydroxylase (cytochrome P-450$_{7\alpha}$): Role of residue-209 in determining steroid-cytochrome P-450 interaction. Biochem J 291:569-573

Johnson EF (1992) Mapping determinants of the substrate selectivities of P450 enzymes by site-directed mutagenesis. Trends Pharmacol Sci 13:122-126

Juvonen RO, Iwasaki M, Negishi M (1991) Structural function of residue-209 in coumarin 7-hydroxylase (P450coh). Enzyme-kinetic studies and site-directed mutagenesis. J Biol Chem 266:16431-16435

Juvonen RO, Iwasaki M, Negishi M (1992) Roles of residues 129 and 209 in the alteration by cytochrome b5 of hydroxylase activities in mouse 2A P450S. Biochemistry 31:11519-11523

Juvonen RO, Iwasaki M, Sueyoshi T, Negishi M (1993) Structural alteration of mouse P450coh by mutation of glycine-207 to proline: spin equilibrium, enzyme kinetics, and heat sensitivity. Biochem J 294:31-34

Kaminsky LS, De Morais SMF, Faletto MB, Dunbar DA, Goldstein JA (1993) Correlation of human cytochrome P4502C substrate specificities with primary structure: Warfarin as a probe. Mol Pharmacol 43:234-239

Kedzie KM, Balfour CA, Escobar GY, Grimm SW, He Y, Pepperl DJ, Regan JW, Stevens JC, Halpert JR (1991) Molecular basis for a functionally unique cytochrome P450IIB1 variant. J Biol Chem 266:22515-22521

Krainev AG, Weiner LM, Kondrashin SK, Kanaeva IP, Bachmanova GI (1991) Substrate access channel geometry of soluble and membrane-bound cytochromes P450 as studied by interactions with type II substrate analogues. Arch Biochem Biophys 288:17-21

Krainev AG, Shimizu T, Hiroya K, Hatano M (1992) Effect of mutations at Lys250, Arg251, and Lys253 of cytochrome P450 1A2 on the catalytic activities and the bindings of bifunctional axial ligands. Arch Biochem Biophys 298:198-203

Kronbach T, Larabee TM, Johnson EF (1989) Hybrid cytochromes P-450 identify a substrate binding domain in P-450IIC5 and P-450IIC4. Proc Natl Acad Sci USA 86:8262-8265

Kronbach T, Kemper B, Johnson EF (1990) Multiple determinants for substrate specificities in cytochrome P450 isozymes. In: Villafranca JJ (ed) Current Research in Protein Chemistry: Techniques, Structure, and Function. Academic Press Inc. San Diego pp 481-488

Kronbach T, Kemper B, Johnson EF (1991) A hypervariable region of P450IIC5 confers progesterone 21-hydroxylase activity to P450IIC1. Biochemistry 30:6097-6102

Kronbach T, Johnson EF (1991) An inhibitory monoclonal antibody binds in close proximity to a determinant for substrate binding in cytochrome P450IIC5. J Biol Chem 266:6215-6220

Lindberg R, Burkhart B, Ichikawa T, Negishi M (1989) The structure and characterization of type I P-45015α gene as a major steroid 15α-hydroxylase and its comparison with type II P-45015α gene. J Biol Chem 264:6465-6471

Lindberg RLP, Negishi M (1989) Alteration of mouse cytochrome P450coh substrate specificity by mutation of a single amino-acid residue. Nature 339:632-634

Lindberg RLP, Negishi M (1991) Modulation of specificity and activity in mammalian cytochrome P-450. Methods Enzymol 202:741-752

Matsunaga E, Zeugin T, Zanger UM, Aoyama T, Meyer UA, Gonzalez FJ (1990) Sequence requirements for cytochrome P-450IID1 catalytic activity: A single amino acid change (ILE^{380}PHE) specifically decreases V_{max} of the enzyme for bufuralol but not debrisoquine hydroxylation. J Biol Chem 265:17197-17201

McManus ME, Minchin RF, Sanderson N, Schwartz D, Johnson EF, Thorgeirsson SS (1984) Metabolic processing of 2-acetylaminofluorene by microsomes and six highly purified cytochrome P-450 forms from rabbit liver. Carcinogenesis 5:1717-1723

Nebert DW, Adesnik M, Coon MJ, Estabrook RW, Gonzalez FJ, Guengerich FP, Gunsalus IC, Johnson EF, Kemper B, Levin W, Phillips IR, Sato R, Waterman MR (1987) The P450 Gene Superfamily. Recommended Nomenclature. DNA 6:1-11

Negishi M, Lindberg R, Burkhart B, Ichikawa T, Honkakoski P, Lang M (1989) Mouse steroid 15α-hydroxylase gene family: Identification of type II P-45015α as coumarin 7-hydroxylase. Biochemistry 28:4169-4172

Nelson DR, Kamataki T, Waxman DJ, Guengerich FP, Estabrook RW, Feyereisen R, Gonzalez FJ, Coon MJ, Gunsalus IC, Gotoh O, Okuda K, Nebert DW (1993) The P450 superfamily: Update on new sequences, gene mapping, accession numbers, early trivial names of enzymes, and nomenclature. DNA Cell Biol 12:1-51

Nelson DR, Strobel HW (1987) Evolution of cytochrome P-450 proteins. Mol Biol Evol 4:572-593

Nelson DR, Strobel HW (1989) Secondary structure prediction of 52 membrane-bound cytochromes P450 shows a strong structural similarity to P450$_{cam}$. Biochemistry 28:656-660

Ouzounis CA, Melvin WT (1991) Primary and secondary structural patterns in eukaryotic cytochrome P-450 families correspond to structures of the helix-rich domain of *Pseudomonas putida* cytochrome P-450$_{cam}$--Indications for a similar overall topology. Eur J Biochem 198:307-315

Poulos TL, Finzel BC, Gunsalus IC, Wagner GC, Kraut J (1985) 2.6 Angstrom crystal structure of the Pseudomonas putida cytochrome P-450. J Biol Chem 260:16122-16130

Poulos TL, Finzel BC, Howard AJ (1986) Crystal structure of substrate-free Pseudomonas putida cytochrome P-450. Biochemistry 25:5314-5322

Poulos TL, Finzel BC, Howard AJ (1987) High-resolution crystal structure of cytochrome P450cam. J Mol Biol 195:687-700

Poulos TL (1989) Site-directed mutagenesis: Reversing enzyme specificity. Nature 339:580-581

Poulos TL (1991) Modeling of mammalian P450s on Basis of P450cam X-ray structure. Methods Enzymol 206:11-30

Raucy JL, Johnson EF (1985) Variations among untreated rabbits in benzo(a)pyrene metabolism and its modulation by 7,8-benzoflavone. Mol Pharmacol 27:296-301

Ravichandran KG, Boddupalli SS, Hasemann CA, Peterson JA, Deisenhofer J (1993) Crystal structure of hemoprotein domain of P450BM-3, a prototype for microsomal P450's. Science 261:731-736

Richardson TH, Hsu M-H, Kronbach T, Barnes HJ, Chan G, Waterman MR, Kemper B, Johnson EF (1993) Purification and characterization of recombinant-expressed cytochrome P450 2C3 from Escherichia coli: 2C3 encodes the 6ß-hydroxylase deficient form of P450 3b. Arch Biochem Biophys 300:510-516

Richardson TH, Johnson EF (1993) Alteration by site-directed mutagenesis of the regioselectivity of P450 2C3v catalyzed progesterone hydroxylation. FASEB J 7:A1166

Schwab GE, Johnson EF (1985) Two catalytically distinct subforms of P-450 3b as obtained from inbred rabbits. Biochemistry 24:7222-7226

Sotokawa H, Shimizu T, Furuya H, Sadeque AJM, Hatano M, Ohba Y, Iwaizumi M, Fujii-Kuriyama Y (1990) Electron spin resonance studies of wild-type and mutant cytochromes $P-450_d$: Effects of mutations at proximal, aromatic and distal sites on g values. Biochim Biophys Acta 1037:122-128

Straub P, Johnson EF, Kemper B (1993a) Hydrophobic side chain requirements for lauric acid and progesterone hydroxylation at amino acid 113 in cytochrome P450 2C2, a potential determinant of substrate specificity. Arch Biochem Biophys (In Press)

Straub P, Lloyd M, Johnson EF, Kemper B (1993b) Differential Effects in Mutations in SRS1 of cytochrome P450 2C2 on lauric acid and progesterone metabolism. Biochemistry (In Press)

Straub P, Lloyd M, Johnson EF, Kemper B (1993c) Cassette-mutagenesis of a potential substrate recognition region of cytochrome P450 2C2. J Biol Chem (In Press)

Tuck SF, Hiroya K, Shimizu T, Hatano M, Ortiz de Montellano PR (1993) The cytochrome P450 1A2 active site: Topology and perturbations caused by glutamic acid-318 and threonine-319 mutations. Biochemistry 32:2548-2553

Veronese ME, Doecke CJ, Mackenzie PI, McManus ME, Miners JO, Rees DLP, Gasser R, Meyer UA, Birkett DJ (1993) Site-directed mutation studies of human liver cytochrome P-450 isoenzymes in the CYP2C subfamily. Biochem J 289:533-538

Diversity and Regulation of Cytochromes P450 in Aquatic Species

John J. Stegeman
Biology Department
Woods Hole Oceanographic Institution,
Woods Hole, Massachusetts 02543 USA

Introduction

Knowledge of the multiplicity, function and regulation of cytochrome P450 forms in non-mammalian and non-traditional species continues to grow in importance. Microsomal cytochromes P450 that are involved in xenobiotic metabolism can play critical roles in determining the susceptibility or resistance of organisms to the toxic action of numerous anthropogenic chemicals and natural products. Currently, these processes are understood far better in rodent models than in wild or cultivated species that provide food and material resources, or in species that are important to the functioning of ecosystems that sustain those resource populations. Knowledge of P450 systems in these various groups is important for chemical therapeutics of species in culture, and to management decisions related to the assessment of chemical effects, and the protection of wild species. Non-traditional species are also important as potential models for novel features of P450 systems, and to discerning the molecular basis underlying species differences and similarities in responses to xenobiotics. In this paper we summarize aspects of P450 systems in selected groups of aquatic species, focusing first on recent findings regarding the diversity of P450 forms. Secondly, we summarize aspects of the use of induction of cytochrome P4501A forms as a biomarker of exposure to Ah-receptor agonists.

Research on aquatic species monooxygenase (MO) systems expanded rapidly in the mid-1970s so that by the late 1970s and early 1980s several major reviews of microsomal P450 systems in aquatic species appeared (Bend and James, 1978; Stegeman, 1981; Lech et al., 1982) Interest has continued to grow, and from 1980 to 1992, there were more than 800 papers published on P450 and monooxygenases in aquatic species. Much of the earlier work dealt

NATO ASI Series, Vol. H 90
Molecular Aspects of Oxidative Drug Metabolizing Enzymes
Edited by E. Arınç, J. B. Schenkman and E. Hodgson
© Springer-Verlag Berlin Heidelberg 1995

with characterization of hepatic microsomal MO activities, and the modulation of these in animals treated with or exposed to various chemicals. Since the early 1980s the effort has shifted to characterization of the catalysts themselves, and subsequently to characterization of the regulatory mechanisms. Research on the occurrence and regulation of P450 systems in extrahepatic organs of aquatic species also has grown substantially in recent years. Several more recent reviews summarize many of the important progress that has been made recently in understanding of P450 systems in aquatic species (James, 1989; Livingstone et al., 1989; Stegeman, 1989; Stegeman and Hahn, 1993). This paper will draw on these reviews, and consider the diversity of P450 forms in aquatic vertebrates, their relationships and their regulation, the latter focusing on CYP1A forms. Many aspects of these systems in fish have been known for many years, but are only now being recognized by the larger P450 research community.

Diversity of P450 Forms

Direct information regarding the identity of P450 forms in aquatic species is most abundant for fish. The first P450s purified from fish were the two forms, DBAI and DBAII, from the elasmobranch little skate (Bend et al., 1977). To date there have been more than 40 P450 forms reported purified, partially purified or cloned from aquatic species (Stegeman and Hahn, 1993). Thirty-five of these have been obtained from five species of teleost fish, and only four from invertebrates. The majority of the fish forms have been described in liver, although one form has been purified from kidney and three P450s that are involved in steroidogenesis have been cloned from rainbow trout ovary. Probing with antibodies to known teleost or mammalian forms has indicated the presence of related P450 forms in other aquatic species groups, including reptiles, amphibians, aquatic birds and marine mammals. The detection and probable identity of P450 forms in another important group, marine mammals, is presented in part to illustrate outstanding questions concerning these enzymes in many groups.

Cytochrome P450 (CYP) subfamilies (Nelson et al., 1993) represented among the P450 forms purified from fish are indicated in Table 1, along with the nature of the evidence indicating these assignments. In addition to these subfamilies, the specific recognition of fish liver microsomal proteins by antibodies to mammalian P450s (Stegeman et al., 1990), and the sequence comparisons of cloned cDNAs (Kaplan et al., 1991) indicate that representatives of subfamilies 2E, 2C and 4A (at minimum) occur in fish. However, until such proteins have been purified, or expressed in heterologous systems, functional relationships to presumed mammalian homologues will be uncertain.

Table 1. Apparent Subfamilies of Hepatic Cytochrome P450 Proteins Purified from Teleost Fish

Subfamily	Trivial Name	Evidence
1A	Scup P450 E	P-C, F, I, R, S
	Trout LM4b	P-C, F, I, R, S
	Cod c	P-C, F, I, R,
	Perch V	P-C, F, I, R
2B	Scup P450 B	I, S
	Trout LMC 1	I, F
2K (nov)	Trout LM 2	S
	Trout KM 2	I
3A	Scup P450 A	F, I
	Trout LMC 5	F, I
	Trout P450 con	F, I
	Cod b	I

Evidence refers to physico-chemical (P-C), functional (F), immunological (I), regulatory (R) or protein or cDNA sequence (S) similarities to known mammalian subfamily members. Data are found in Stegeman and Hahn (1993). The assignment of trout KM2 to subfamily 2K is speculative. Relationships of the various putative 3A forms is from Celander, Stegeman and Buhler (Unpubl.)

NADPH-cytochrome c reductase, the enzyme dominant in electron transfer to P450, has also been purified from liver of scup and trout, and cytochrome b5 has been purified from scup liver (Klotz et al., 1986). Reconstitution studies with various combinations of reductase and P450 derived from fish or rat confirmed that heterologous reductase is able to support activity of teleost P450 (Williams and Buhler, 1983; Williams and Buhler, 1984; Klotz et al., 1986). However, species-specific preferences for reductase have also been seen with some teleost P450s (Klotz et al., 1986). Thus, scup P450A showed a much greater activity when reconstituted with scup reductase than with rat reductase, while this preference was not seen with scup liver P450E (Klotz et al, 1986). A possible role of cytochrome b5 in electron transfer to P450 in fish has also been indicated by reconstitution studies (Klotz et al, 1986). As in mammalian systems, there appear to be some substrates (e.g. aminopyrine) whose metabolism is preferentially enhanced by the addition of b5. The molecular basis for the reductase preference or for the distinct effects of cytochrome b5 have not been established in fish.

Gene Family 1

Members of gene subfamily 1A are prominent in the activation of many environmental carcinogens. P450 forms purified from several fish species, trout P450LM4b, scup P450E and cod P450c (Williams and Buhler, 1982; Klotz et al., 1983; Goksøyr, 1985), and perch P450V (Zhang et al., 1991) all have physico-chemical, catalytic and regulatory properties in common with each other and with mammalian CYP1A forms, particularly CYP1A1. Properties of scup P450E indicating a relationship to mammalian CYP1A1 are summarized in Table 2.

Table 2. Properties of scup liver P450E (CYP1A1)

Molecular properties
    54 Kda molecular weight
    Fe^{2+}-CO absorption max @ 447nm
    Low spin heme iron
Catalytic properties
    Ethoxyresorufin O-deethylase
    aryl hydrocarbon hydroxylase
    Benzo[a]pyrene 7,8- and 9,10-epoxide formation
    Ethoxycoumarin-O-deethylase
    strongly inhibited by alpha-naphthoflavone
Regulatory properties
    Induced by planar aromatic and halogenated hydrocarbons
    Induced in liver and in extrahepatic organs

(Modified from Stegeman, 1989)

Each of the teleost CYP1A-like proteins purified to date shares these properties. Antibodies have been made to these proteins, and they are immunologically cross-reactive with each other and with CYP1A forms in mammals, particularly CYP1A1 (Stegeman, 1989). All are inducible by polynuclear aromatic hydrocarbons (PAHs) and by planar halogenated aromatic hydrocarbons, including polychlorinated biphenyls (PCBs) and 2,3,7,8-tetrachlorodibenzodioxin (TCDD) and dibenzofuran (TCDF). Fish CYP1As catalyze ethoxyresorufin O-deethylase (EROD) and aryl hydrocarbon hydroxylase (AHH), and metabolize benzo(a)pyrene to the 7,8- and 9,10-dihydrodiols. This preference for metabolism on the benzo-ring of B[a]P is significant, as it is strongly associated with the activation of such PAHs to mutagenic and carcinogenic derivatives.

P450 Sequence Considerations

The first amino acid sequence reported for a fish P450 was the N-terminal sequence of scup P450E (Klotz et al, 1983). Subsequently, a P450 cDNA was cloned from liver of methylcholanthrene-induced trout (Heilmann et al., 1988). That gene was classified in the CYP1A subfamily, and particularly as a CYP1A1, based on the sequence comparisons with mammalian CYP genes. The two members of the CYP1A subfamily, CYP1A1 and CYP1A2, occuring in a given mammalian species are paralogous, having been diverged from a common ancestor. Heilmann et al. (1988) reported that the inferred sequence of the trout CYP1A had 57-59% of the amino acid residues identical with various mammalian CYP1A1 forms and 51-53% of the residues identical with mammalian CYP1A2 forms, concluding that the trout CYP1A is a CYP1A1.

Based on the rates of molecular evolution within other protein families, it was suggested that the mammalian CYP1A1 and CYP1A2 genes diverged from one another 250 million years ago or less (Nelson and Strobel, 1987). Thus, the divergence of the mammalian CYP1A1 and CYP1A2 genes is thought to have occurred *after* divergence of the lines leading to teleosts and mammalian lines. If so, then fish would not be expected to have diverged P450 genes that would be paralogous to one another, with one being orthologous to mammalian CYP1A1 and the other orthologous to mammalian CYP1A2. If but a single CYP1A gene occurs in fish, such a CYP1A would be orthologous to both mammalian CYP1A1 and CYP1A2, and could be considered to represent a type ancestral to both CYP1A1 and CYP1A2.

There is some evidence, in addition to the speculations based on molecular clocks, that is consistent with the idea that teleost CYP1A forms are possibly orthologous to both mammalian CYP1A forms. Some of this evidence has been discussed previously (Stegeman, 1989). Some positions of the inferred trout CYP1A sequence have amino acids identical to all known mammalian CYP1A1 but not CYP1A2, and others identical to all CYP1A2 but not CYP1A1 (Heilmann et al., 1988; Stegeman, 1989). One such region is near the phe-gly-ala-gly-phe (FGAGF) sequence at about residue position 320, shown in Table 3.

Table 3. Sequence Comparison of CYP1A in fish and mammals

Group	Protein	Sequence
Teleost	CYP1A	KIVGIVNDLFGAGFDT
Mammalian	CYP1A1	KVITIV*DLFGAGFDT
Mammalian	CYP1A2	KIVNIVNDI FGAGFDT

The fish sequences include three; trout Heilmann et al; 1988; plaice (George 1992, and scup (Morrison, Sogin and Stegeman, unpublished). The mammalian sequences include rat, mouse, rabbit and human. Examples of similarities are highlighted (N and L in the fish sequences). * indicates variation at this position, but no identity between the 1A1 and 1A2 sequences.

This could represent a "hybrid" condition to be expected if the known fish CYP1A gene represents a type ancestral to both CYP1A1 and CYP1A2 in mammals. The degree to which this indicated structural relationship of fish CYP1A to both CYP1A1 and CYP1A2 proteins is reflected also in catalytic and other properties is uncertain, but evidence indicates a closer similarity to CYP1A1 than to CYP1A2. Clearly, additional sequences might negate these arguments. For the moment, however, the known teleost CYP1A proteins appear to be more like the CYP1A1 proteins in mammals, based on sequence and other properties, and thus are classified as such.

Whether there has been divergence of the CYP1A gene line in fish is important to evolutionary and environmental considerations. Evidence used by some has been detailed elsewhere (Stegeman, 1989), and involves different temporal patterns or degree of induction of AHH and EROD activities. However, the possibility of molecular differences other than the production of different proteins makes this line of argument suspect (See, Philpot, this volume). Wirgin and colleagues (Kreamer et al., 1991) found evidence for variants of CYP1A at the DNA and RNA levels in tomcod (*Microgadus tomcod*), but not two distinct genes. The possibility that CYP1A genes diverged in some fish groups since separation of mammalian and fish lines, for example in salmonids in which there has been genome duplication, had been suggested (Stegeman, 1989). Until recently, there was no substantive information concerning the presence of multiple CYP1A genes. Recently, there have been two genes cloned from rainbow trout liver, both of which are CYP1A, and which are greater than 90% identical at the amino acid level (Berghard and Chen , pers. comm.). The two genes are sufficiently similar to suggest strongly that they have diverged quite recently, perhaps within the last 20 million years. Whether these two genes are transcribed *and* translated is not certain. Given the similarity, they both could be CYP1A1 like proteins, and distinguishing between them may depend on the differences in non-coding regions (Chen, pers. comm.). The finding could indicate a new member of the teleost CYP1A subfamily.

Questions regarding the possibility of multiple CYP1A genes in fish most often ask whether there is a "CYP1A2" in fish, the arguments above notwithstanding. There is continuing speculation that a CYP1A2-like form exists, based on a number of results involving distinct patterns of response to the typical inducers of CYP1A1 and CYP1A2 in mammals, BNF and isosafrole (Leaver et al., 1988; Celander and Forlin, 1991). Celander reported striking differences not only in the degree of induction of selected alkoxycoumarin O-deethylase activities, but also saw differences in the way that the substrates could inhibit the activity of the enzymes induced by ISF as opposed to BNF. Nevertheless, the results regarding induction of multiple CYP1A proteins or mRNAs in fish by either 1A1 or 1A2 inducers (BNF or isosafrole) are ambiguous at best.

Comparison of CYP1A1 from trout with its homologue from scup (P450E) shows that about 85% of 25 residues in the N-terminal region are identical. This degree of similarity is higher than might be expected based on the phylogenetic relationship of the families containing scup and trout, which may have diverged as recently as 65 million years ago, but more likely as long as 200 million years ago. By comparison, more recently diverged mammalian groups show greater difference in CYP1A1 sequence in the N-terminal region and overall, suggesting possible slower rates of evolution in the fish P450 genes. The recent sequencing of a cDNA for the PAH-inducible P450 from plaice (George, 1992) indicates a close structural similarity between it and the other forms for which sequence is available. The greater than expected degree of sequence similarity in the N-terminus of CYP1A from distant teleost families supports the idea that the rate of evolution could be slower in the P450 genes in fish than in mammals. Whether the suggested slower rate of change in fish CYP1A merely reflects a slower rate of sequence change common to many fish proteins, or is unique to fish P450s (or even CYP1A alone, perhaps related to some physiological or ecological variable peculiar to fish) is not yet known. This issue will be resolved as additional teleost P450 full length sequences are obtained. If the accumulation of mutations within different gene families proceeds at different rates, as is believed to be the case, then the "clock" for P450 must be established by analysis of homologous P450 sequences in many species and taxa.

Gene Families 2 and 3

Immunochemical studies have indicated the presence in fish liver of homologues to mammalian CYP2B and CYP2E forms (e.g., 4). Additional evidence has been provided by use of nucleic acid probes for mammalian family 2 forms. Haasch and coworkers (Haasch et al., 1990; Kleinow et al., 1990) showed that a rat CYP2B1 cDNA probe hybridized with trout DNA on Southern blots. Kaplan et al (Kaplan et al., 1991) used an oligonucleotide (49-mer) specific for rat CYP2E1 to probe Northern blots of hepatic mRNA from *Poeciliopsis* spp.; a 3.3 kb band was detected.

The first family 2 cDNA to be cloned from fish was recently reported by Buhler (Buhler, 1992). This form, which corresponds to P450LM2 purified by his group, has been classified as a CYP2K. As discussed elsewhere (Buhler, this volume; Stegeman, 1992), this novel P450 is important in the metabolism of many substrates in trout. It also appears to be similar to a form purified for trout kidney (see Table 1).

Several lines of evidence indicate a relationship of teleost forms to members of subfamily 3A. Microsomal activities attributed to trout LMC5 include steroid 6B-hydroxylase (Table 2), catalyzed in mammals largely by 3A proteins (Waxman et al., 1991) Immunological cross-reactivity has been detected between LMC5 and both rat CYP3A1 and human CYP3A4 (Miranda et al., 1991). The structural similarity was supported by inhibition of progesterone 6B-hydroxylase in trout liver microsomes by anti-human CYP3A4. The evidence that scup P450A is a steroid 6B-hydroxylase suggests that the scup and trout proteins may be related. A relationship between P450con and rat subfamily 3 proteins was suggested by (Celander et al., 1989), who found the amount of this protein was induced about 40% by high-dose PCN. This and the other relationships suggested here are being investigated by reciprocal immunological analysis.

Steroidogenic P450s

Several genes coding P450s homologous to mammalian P450 forms have been described. A cDNA clone encoding a P450c17 (17a-hydroxylase/17,20-lyase) has been isolated from a rainbow trout ovarian follicle cDNA library (Sakai et al., 1992). The amino acid sequence of the 522-residue protein encoded by this cDNA shows a high degree of sequence identity (64%) to chicken CYP17, and lesser identity (46-48%) to P45017 sequences from human, cow, and rat. Recently a cDNA encoding a P450 aromatase (CYP19) was isolated from a rainbow trout ovarian cDNA library (Tanaka and Sakai, 1991). The sequence of this cDNA appears to encode a protein of 522 amino acids that is slightly more than 50 % identical with previously determined aromatase coding sequences from human, rat, mouse and chicken. The trout CYP19 gene expression appears to be closely linked to the production of estradiol by the ovary and consequently to the stimulation of hepatic vitellogenesis by estradiol. This consistent with a transcriptional activation, which is known to be a mechanism of CYP19 gene activation in mammalian steroidogenic organs.

P450 Diversity in Marine Mammals

The bulk of the data on aquatic P450 systems has been derived from studies of fish. Information concerning the number, identity and characteristics of P450s in many other groups is still scant. The marine mammals provide a case in point.

Studies have identified P450 or P450-associated catalytic activities in pinnipeds (Engelhardt, 1982; Addison and Brodie, 1984; Addison et al., 1986; Hahn et al., 1991; Goksøyr et al., 1992) and a few species of cetaceans (minke whales, pilot whales, killer whales, beluga, and dolphins(Goksøyr et al., 1986; Goksøyr et al., 1988; Watanabe et al., 1989; Hahn and Stegeman, 1990) (White et al., 1993)). There are also a growing number of studies that have examined the presence of proteins cross-reacting with antibodies to known P450 forms. The results of such studies in two species are provided in Table 4.

Based on the immunochemical and catalytic studies some generalizations can be made regarding P450 forms in marine mammals: 1) Pinnipeds and cetaceans possess proteins recognized by antibodies to mammalian and fish CYP1A1 forms. 2) Good correlations between the amount of such proteins and the degree of contamination by PCBs suggests that these putative CYP1A forms are inducible, as in other vertebrates. 3) Catalytic and immunochemical data obtained so far are consistent with reduced expression and/or altered structural specificity of 2B forms in aquatic mammals.

Table 4.   Immunoblot Detection of Hepatic Microsomal
           Proteins In Cetaceans

P450	Antibody type	Immunogen	#Reactive Bands
CYP1A	MAb 1-12-3	scup 1A	1
	PAb	chick 1A	1
CYP2B	PAb	scup 2B	0
	PAb	rat 2B1	0
	MAb 2-63-3	rat 2B1	0
	PAb	rabbit 2B4	1
CYP2E	PAb	rat 2E1	1
	PAb	rat 2E1	2

The results are from Goksøyr et al, 1989, and White et al., 1993, for minke and beluga whale respectively. MAb refers to monoclonal and PAb refers to polyclonal antibody.

The results of the studies in whales raise further questions concerning the evolution and significance of particular P450 forms. For example, is the relative lack of CYP2B cross-reactive proteins and activities the result of long evolution with a diet lacking in the plant products thought to have forced the diversification within the CYP2 gene family? The role of natural products as modulators of cytochrome P450 function or regulation is an area where we can expect to see growing attention in aquatic systems.

Regulation of CYP1A

Mammalian CYP1A genes are induced by polynuclear aromatic hydrocarbons, and planar halogenated aromatic hydrocarbons including non-ortho-substituted PCBs, and laterally substituted dibenzo-p-dioxins, most potent of which is TCDD. These compounds act through binding to a cytosolic receptor (the Ah-receptor) (Poland and Knutson, 1982). The mammalian receptor recently was identified as a helix-loop-helix protein (Ema et al., 1992) (Burbach et al., 1992). Fish CYP1A proteins are induced by PAH, BNF, planar PCB congeners, TCDD and TCDF suggesting a similar mechanism in fish. A similar receptor is presumed to mediate the induction of CYP1A genes in fish as well. A TCDD-binding protein thought to correspond to the Ah-receptor was found by traditional methods in trout liver (Heilmann et al., 1988).

Hahn et al. (Hahn et al., 1992) using a photo-affinity ligand detected Ah-receptor-proteins in cartilaginous and bony fish, but not in jawless fish (hagfish or lamprey). There is a marked similarity in the phylogenetic distribution of the recognizable receptor, and the inducibility of a protein recognized by antibodies specific to CYP1A forms (Hahn et al., 1993). The presence of receptor and its apparent activation imply that the complex of proteins involved in the transcriptional activation of the mammalian CYP1A1 genes operate also in "lower" vertebrates. This likewise implies the presence of xenobiotic response elements (XRE) in the 5' regulatory regions of fish CYP1A genes to which activated receptor would bind. That does not require that the inducer structure-activity relationships would be the same for fish and mammals. Evidence obtained with PCB congeners has suggested differences between fish and mammals, with some fish showing a somewhat greater requirement for planarity of the inducers (Gooch et al., 1989).

Variables Affecting Induction

Studies by many investigators have shown that the induction of CYP1A activity or protein is affected by several internal and external variables. These have been treated extensively in the literature, but the basis for the effects is still unclear. Several examples are given here.

Temperature: P450 enzymes have been studied in temperature acclimated fish by several investigators. Earlier studies in *F. heroclitus* for example (Stegeman, 1979) showed that PAH induction of CYP1A activities was suppressed in cold acclimated animals. More recent studies (Kloepper-Sams and Stegeman, 1992) addressed whether this resulted from differences in transcriptional or translational processes (CYP1A mRNA and/or protein content) or effects on the function of the enzyme itself. CYP1A content and activity rose only slowly over time after treatment of cold ($6\,^{\circ}C$) acclimated fish with a known inducer, but were strongly and quickly induced in warm ($16\,^{\circ}C$) acclimated fish. However, CYP1A mRNA content was substantially and similarly elevated in both warm and cold fish, with an apparently longer half-life in the cold fish. Whether the persistence of mRNA in cold-acclimated fish is the result of a slower turnover of mRNA, or of slower metabolism of inducer and hence retention of active levels, is not clear. Regardless, the different patterns in CYP1A mRNA and protein contents indicate that important effects of temperature acclimation on CYP1A induction in fish involve post-transcriptional events.

Hormones: There are strong sex differences in total P450 and CYP1A content in liver of spawning fish. Estradiol treatment can alter the expression of P450s in fish liver (Hansson et al., 1982; Pajor et al., 1990). In both naturally maturing fish and in fish treated with estradiol there is suppression of the induction of CYP1A mRNA as well as protein (Elskus et al., 1992) indicating a pretranslational block. However, the mechanism by which estradiol exerts this effect on CYP1A is not known. It is possible that it may be mediated by estradiol receptors, which would imply the presence of estradiol response elements in fish CYP1A genes. Steroid response elements occur in mammalian 1A1 genes; Prough and others found glucocorticoid response elements (GRE) in the first intron of rat *CYP1A1*. Devaux and Pesonen (Devaux et al., 1992) showed enhancement of CYP1A induction in trout hepatocytes by dexamethasone (which acts through glucocorticoid receptors) suggesting that GRE occur also in fish CYP1A genes. The interactions between CYP1A and hormones in aquatic species require further research.

Table 5. Modulation Of CYP1A Expression In *F. heteroclitus* Liver

Variable	Effect on CYP1A	Apparent site of action	Process Affected	Mech.
Temperature (low T acclim.)	Suppressed Induction	Post-transcription	Translation Efficiency?	?
Reproduction (estradiol in F)	Suppressed induction	Pre-translation	Transcription? Splicing?	?
Development (pre-hatching)	Less sensitive induction	?	?	?

The results are for *Fundulus heteroclitus* and are drawn from Kloepper-Sams and Stegeman (1992), Elskus (1991), and Binder et al, (1985).

Development: Induction of CYP1A catalytic activity was detected in fish embryos some years ago (Binder and Stegeman, 1983). In those studies CYP1A induction was found to be much less sensitive before hatching than after hatching. CYP1A has been localized in many organs of embryos, and studies using immunohistochemistry support the observation that induction is less sensitive in pre than in post-hatched larvae.

Organ and Cellular Distribution of CYP1A

It is now recognized that CYP1A is induced in most extrahepatic organs in fish, as well as in mammals. Microsomal CYP1A can be strongly induced in organs proximal to the environment (gill; gut) and in excretory organs (kidney). The regulation of extrahepatic CYP1A has been linked to protective mechanisms in some cases. For example, low concentrations of PAH in the diet of fish can produce a strong induction of intestinal CYP1A, without induction occurring in the liver (Van Veld et al., 1988). This could result from enhanced metabolism of inducer in the gut sufficient to prevent active concentrations from reaching the liver.

Many of the questions concerning extrahepatic induction of CYP1A require an understanding of the cellular targets for that induction. These targets can be identified by immunohistochemical approaches. The first demonstration of cellular localization of a CYP in liver of a vertebrate was in rat liver, by Baron et al in 1978 (Baron et al., 1978). Demonstration of cellular localization in fish tissues was reported at about the same time by several groups (Goksøyr et al., 1987) (Stegeman et al., 1987) (Lorenzana et al., 1988). Subsequent studies (Smolowitz et al., 1991) noted CYP1A induction in multiple extrahepatic organs. CYP1A is commonly induced in epithelial cells of many organs, including liver, kidney, gut, gill, and skin. Miller et al (Miller et al., 1989) first noted a strong induction in the endothelium of fish. Studies since have shown that this is a very common result of exposure of fish to CYP1A inducers. However, the common site of induction in extrahepatic organs is in endothelial cells, regardless of inducer or route of exposure.

The ubiquity of CYP1A induction in endothelium indicates that this is a fundamental response in this cell type in fish, and implies that Ah-receptors are operating in endothelium. Situated between the blood and underlying tissues, endothelium is a potential target site where xenobiotics could be intercepted and exert toxicity. CYP1A induction in endothelium throughout the organism could serve a protective role, but it could lead to enhanced activation of foreign compounds in the endothelium. In induced fish CYP1A amounts to nearly 25% of the ER protein in endothelial cells (Stegeman et al., 1989). This P450 could be involved in other molecular interactions in endothelium, such as with NO (nitric oxide). NO is an endothelium derived relaxing factor, which can be avidly bound by heme proteins including P450. Sequestration of NO by induced P450 could affect vascular function.

Environmental Induction Of CYP1A

Cytochrome P4501A proteins are induced in vertebrates from fish to mammals by polynuclear aromatic and halogenated aromatic hydrocarbons. CYP1A and Ah-receptors have been identified in all vertebrate groups, from elasmobranchs to mammals, implicating Ah-receptor mediated mechanism in CYP1A induction. There is great and growing interest in using CYP1A induction to indicate the exposure of organisms in the wild to inducing compounds, i.e., to Ah-receptor ligands, to evaluate the degree and possible risk of environmental contamination (Stegeman et al., 1992).

The induction of monooxygenase (AHH) activity in fish liver was first suggested as an indicator in the early 1970s (Payne, 1976). Analysis of environmental induction by immunoassay with antibodies to fish CYP1A was demonstrated in 1986 (Stegeman et al., 1986). Data indicating that mRNA detection could be used appeared in 1989 (Haasch et al., 1989). The first immunohistochemical analysis showing the multiple organ and cellular localization of *environmentally* induced CYP1A in fish appeared in 1991 (Stegeman et al., 1991). These approaches complement one another and all have value in detecting induction as a marker of exposure.

There has been an explosion in the numbers of investigators using induction as a biomarker (Stegeman et al., 1992). Each new study shows that the approach is highly promising. Induction is being repeatedly tested as a marker for contamination in coastal waters, rivers and lakes around the world, not only in fish, but in all vertebrate groups, including reptiles, amphibians, birds and mammals (Ellenton et al., 1985; Hoffman et al., 1987; Bellward et al., 1990; Stegeman, ; White et al., 1993). Present knowledge indicates less potential for using monooxygenase activity or P450 levels in many invertebrates, such as molluscs or crustaceans, to indicate their exposure to compounds such as the aromatic and chlorinated hydrocarbons. This is due to the current lack of convincing evidence for induction of known P450 forms or of monooxygenase activity, by any known mechanism.

That CYP1A induction occurs commonly in the environment is no longer in question. However, there are important questions concerning the environmental chemicals contributing to that induction, the geographic and temporal distribution and the biological significance of that induction.

Distribution and Significance of Environmental Induction

Studies around the world are showing that environmental induction of CYP1A is a common occurrence in natural populations. Many studies have shown that fish from coastal regions near population centers have induced levels of P4501A in their livers (Stegeman and Binder, 1979; Spies et al., 1982; Foureman et al., 1983; Varanasi et al., 1986; Luxon et al., 1987; Stegeman et al., 1987; Spies et al., 1988; Stegeman et al., 1988; Elskus and Stegeman, 1989;

Elskus et al., 1989; Stegeman et al., 1990; Van Veld et al., 1990; Kreamer et al., 1991; Monosson and Stegeman, 1991; Stegeman et al., 1991). A great weight of evidence by now indicates that we can reliably predict this condition to occur in fish from such areas. At greater distances from urban centers there is usually a decrease in the content of P4501A. However, disseminated sources and local sources both have been associated with strong induction signals in fish at remote sites. The apparent induction in the liver of fish in the deep ocean is the prime example (Stegeman et al., 1986). Recent studies in the Antarctic and the central Pacific show that even in these remote areas there are local sources causing this biochemical effect. Whether slight induction seen in fish at sites distant from any known local sources reflects the influence of natural products, the effect of the global chemical "background" condition or some unknown source of inducers is not known. In the case of marine mammals showing high levels of apparent CYP1A, it is likely that the signal reflects their position at the top of the food chain.

Studies have shown that the degree of CYP1A induction in fish from the wild can be closely linked to the degree of contamination by chemicals such as PCBs or PAH in the animals themselves, or in other biota or sediments at the same site (Stegeman et al., 1986; Payne et al., 1987; Spies et al., 1988; Stegeman et al., 1988; Elskus and Stegeman, 1989; Elskus et al., 1989; Van Veld et al., 1990). Studies in marine mammals are also revealing close correlations between tissue residues of candidate inducers and the amount of CYP1A in liver (White et al., 1993). The results are convincing enough so that we can safely predict that such use of CYP1A induction will become commonplace not only in aquatic species, but in terrestrial animals of all kinds. The development of biopsy or other non-destructive sampling methods will ensure a widespread use. However, there is a continuing need to establish the identity of the most active agents contributing to induction in specific sites and environments. Given our growing knowledge of induction mechanisms and the structural features of compounds capable of eliciting induction, it is increasingly possible, with knowledge of chemistry in a given area, to name suspect active agents in that area, a possibility suggested long ago. The transition from suspected to confirmed cause of environmental induction is more difficult, as illustrated by studies of induced CYP1A in fish exposed to bleached and unbleached pulp mill effluent (Lindstrom-Seppa et al., 1992; Munkittrick et al., 1992), the latter having lesser amounts of dioxins. The prospect that natural products in which the unbleached effluents are acting as inducers is clear.

The significance of environmental induction is yet to be fully understood. Indeed, the significance of this or any biochemical effect in the field is central to environmental toxicology, and cannot be answered lightly or briefly. The significance of CYP1A induction will depend on 1) the catalytic function(s) of the protein(s), 2) the relative rates of activation and detoxification of the inducer and other compounds, 3) inducer avidity for Ah-receptors and hence efficacy in eliciting CYP1A induction or other gene regulatory changes, and 4) the sites where these events occur. At minimum, we can say that the induction will aid in clearing many substrates for CYP1A from the body. How far we can interpret the induction to indicate toxic consequences, as it does indicate exposure, is less certain. Adequate answers must consider the full range of possible toxic mechanisms involving P450 function or Ah-receptor function, discussed above. The most obvious possibility involves carcinogenesis.

Activation of many pro-carcinogens requires the function of P450. A cell or organ devoid of the requisite catalyst(s) will not transform such a compound into a carcinogen. As CYP1A is a prominent catalyst through which some PAH are activated in fish, some degree of CYP1A induction is a likely prerequisite for the initial steps in environmental carcinogenesis involving those compounds. Carcinogenic PAH that are active inducers are the dominant compounds in some regions (Malins et al., 1984; Malins et al., 1985; Elskus and Stegeman, 1989). Greater P450 induction could contribute to a higher steady state level of activated carcinogens, and consequently to a higher degree of persistent and relevant DNA adduct formation or to enhanced oxidative DNA damage; there is a correlation between induction and carcinogenesis in mammals. But highly induced P450 is not *necessarily* associated with a greater risk of carcinogenesis. Formation and persistence of critical genetic lesions may be influenced as much by detoxication or repair processes as by the oxidative metabolism creating the activated carcinogenic derivative. The many questions yet to be answered include the consequences of long-term, low-level exposure to compounds producing sustained induction of CYP1A (Haasch et al., 1993) .

Though most often studied in liver, induction and attendant changes in extrahepatic sites may determine the toxicity of inducers. Immunohistochemical analyses have disclosed organ- and cell-specific expression of CYP1A in adults, embryos or larvae of numerous species of freshwater and marine fish treated with BNF, benzo(a)pyrene, 3,3',4,4'-tetrachlorobiphenyl, 2,3,7.8-tetrachlorodibenzofuran, 2,3,7,8-tetrachlorodibenzo-p-dioxin, PCB mixtures or petroleum, by injection, in the diet or in the water (Stegeman et al unpublished). The analysis of fish from the environment has shown similar patterns of induction in animals from chemically-contaminated sites (see for example scup, Stegeman, et al, 1991).

Determining the significance of binding the Ah-receptor by low doses of external ligands is particularly important to determing the significance of environmental induction. Further studies on the cellular localization and degree of induction in wildlife species will be important to our eventual understanding on the sources and significance of this evidence for exposure to Ah-receptor agonists.

Support was provided by the U.S. NIH grant ES-04220, the Donaldson Charitable Trust and the Stanley Watson Chair.

References

Addison R and Brodie P (1984) Characterization of ethoxyresorufin O-deethylase in grey seal *Halichoerus grypus*. Comp Biochem Physiol 79C: 261-263.

Addison R, Brodie P, Edwards A and Sadler M (1986) Mixed function oxidase activity in the harbour seal (*Phoca vitulina*) from Sable Is., N.S. Comp Biochem Physiol 85C: 121-124.

Bellward GD, Norstrom RJ, Whitehead PE, Elliot JE, Bandiera SM, Dworschak C, Chang T, Forbes S, Cadario B, Hart LE and Cheng KM (1990) Comparison of polychlorinated dibenzodioxin levels with hepatic mixed-function oxidase induction in great blue herons. J Toxicol Environ Health 30: 33-52.

Bend JR and James MO (1978) Xenobiotic metabolism in marine and freshwater species. In Biochemical and Biophysical Perspectives in Marine Biology. DC Malins and JR Sargent Academic Press New York. 125-188.

Bend JR, Pohl RJ, Arinc E and Philpot RM (1977) Hepatic microsomal and solubilized mixed-function oxidase systems from the little skate, *Raja erinacea*, a marine elasmobranch. In Microsomes and Drug Oxidations. V Ullrich, I Roots, A Hildebrandt, RW Estabrook and AH Conney Pergamon Oxford. 160-169.

Binder RL and Stegeman JJ (1983) Basal levels and induction of hepatic aryl hydrocarbon hydroxylase activity during the embryonic period of development in brook trout. Biochem Pharmacol 32: 1324-1327.

Buhler D (1992) Personal communication.

Burbach KM, Poland A and Bradfield CA (1992) Cloning of the Ah receptor cDNA reveals a distinctive ligand-activated transcription factor. Proc Nat Acad Sci USA 89: 8185-8189.

Celander M and Forlin L (1991) Catalytic activity and immunochemical quantification of hepatic cytochrome P-450 in beta -naphthoflavone and isosafrole treated rainbow trout (*Oncorhynchus mykiss*). Fish Physiol Biochem 9: 189-197.

Celander M, Ronis M and Forlin L (1989) Initial characterization of a constitutive cytochrome P-450 isoenzyme in rainbow trout liver. MarEnviron Res 28: 9-13.

Devaux A, Pesonen M, Monod G and Andersson T (1992) Glucocorticoid-mediated potentiation of P450 induction in primary culture of rainbow trout hepatocytes. Biochem Pharmacol 43: 898-901.

Ellenton J, Brownlee L and Hollebone B (1985) Aryl hydrocarbon hydroxylase levels in herring gull embryos from different locations on the Great Lakes. Environ Toxicol Chem 4: 615-622.

Elskus AA, Pruell R and Stegeman JJ (1992) Endogenously-mediated, pretranslational suppression of cytochrome P4501A in PCB-contaminated flounder. Mar Environ Res 34: 97-101.

Elskus AA and Stegeman JJ (1989) Induced cytochrome P-450 in *Fundulus heteroclitus* associated with environmental contamination by polychlorinated biphenyls and polynuclear aromatic hydrocarbons. Mar Environ Res 27: 31-50.

Elskus AA, Stegeman JJ, Susani LC, Black D, Pruell RJ and Fluck SJ (1989) Polychlorinated biphenyls concentration and cytochrome P-450E expression in winter flounder from contaminated environments. Mar Environ Res 28: 25-30.

Ema M, Sogawa K, Watanabe N, Chujoh Y, Matsushita N, Gotoh O, Funae Y and Fuji-Kuriyama Y (1992) cDNA cloning and structure of mouse putative Ah receptor. Biochem Biophys Res Comm 184: 246-253.

Engelhardt F (1982) Hydrocarbon metabolism and cortisol balance in oil-exposed ringed seals, *Phoca hispida*. Comp Biochem Physiol 72C: 133-136.

Foureman G, White N and Bend J (1983) Biochemical evidence that winter flounder (*Pseudopleuronectes americanus*) have induced hepatic cytochrome P-450-dependent monooxygenase activities. Can J Fish Aquat Sci 40: 854-865.

George S (1992) personal communication.

Goksøyr A (1985) Purification of hepatic microsomal cytochromes P-450 from ß-naphthoflavone-treated Atlantic cod (*Gadus morhua*), a marine teleost fish. Biochim Biophys Acta 840: 409-417.

Goksøyr A, Andersson T, Forlin L, Stenersen J, Snowberger EA, Woodin BR and Stegeman JJ (1988) Xenobiotic and steroid metabolism in adult and foetal piked (minke) whales, *Balaenoptera acutorostrata*. Mar Environ Res 24: 9-13.

Goksøyr A, Andersson T, Hansson T, Klungsoyr J, Zhang Y and Forlin L (1987) Species characteristics of the hepatic xenobiotic and steroid biotransformation systems of two teleost fish, Atlantic cod (*Gadus morhua*) and rainbow trout (*Salmo gairdneri*). Toxicol Appl Pharmacol 89: 347-360.

Goksøyr A, Beyer J, Larsen HE, Andersson T and Forlin L (1992) Cytochrome P450 in seals: monooxygenase activities, immunochemical cross-reactions and response to phenobarbital treatment. Mar Environ Res 34: 113-116.

Goksøyr A, Solbakken J, Tarlebø J and Klungsøyr J (1986) Initial characterization of the hepatic microsomal cytochrome P-450-system of the piked whale (minke) *Balaenoptera acutorostrata*. Mar Environ Res 19: 185-203.

Gooch JW, Elskus AA, Kloepper-Sams PJ, Hahn ME and Stegeman JJ (1989) Effects of *ortho* and non-*ortho* substituted polychlorinated biphenyl congeners on the hepatic monooxygenase system in scup (*Stenotomus chrysops*). Toxicol Appl Pharmacol 98: 422-433.

Haasch ML, Quardokus EM, Sutherland LA, Goodrich MS and Lech JJ (1993) Hepatic CYP1A1 induction in rainbow trout by continuous flowthrough exposure to ß-naphthoflavone. Fundam Appl Toxicol 20: 72-82.

Haasch ML, Wejksnora PJ and Lech JJ (1990) Molecular aspects of cytochrome P450 induction in rainbow trout. Toxicologist 10: 322

Haasch ML, Wejksnora PJ, Stegeman JJ and Lech JJ (1989) Cloned rainbow trout liver P1450 complementary DNA as a potential environmental monitor. Toxicol Appl Pharmacol 98: 362-368.

Hahn ME, Poland A, Glover E and Stegeman JJ (1992) The Ah receptor in marine animals: Phylogenetic distribution and relationship to P4501A inducibility. Mar Environ Res 34: 87-92.

Hahn ME, Poland A, Glover E and Stegeman JJ (1993) Photoaffinity labeling of the Ah receptor: Phylogenetic survey of diverse vertebrate and invertebrate species. Arch Biochem Biophys (submitted).

Hahn ME and Stegeman JJ (1990) The role of biotransformation in the toxicity of marine pollutants. In Pour L'Avenir du Beluga Proc of the International Forum for the Future of the Beluga, Tadoussac, Quebec, Canada, 29 September - 1 October, 1988). J Prescott and M Gauquelin Presses de l'Universite du Quebec Sillery, Quebec. 185-198.

Hahn ME, Steiger GH, Calambokidas J, Shaw SD and Stegeman JJ (1991) Immunochemical characterization of cytochrome P450 in harbor seals (*Phoca vitulina*). 9th Biennial Conf on Biology of Marine Mammals p 30.

Hansson T, Forlin L, Rafter J and Gustafsson J-A (1982) Regulation of hepatic steroid and xenobiotic metabolism in fish. In Cytochrome P-450, Biochemistry, Biophysics and Environmental Implications. E Hietanen, M Laitinen and O Hanninen Elsevier Biomedical Amsterdam. 217-224.

Heilmann LJ, Sheen Y-Y, Bigelow SW and Nebert DW (1988) Trout P450IA1: cDNA and deduced protein sequence, expression in liver, and evolutionary significance. DNA 7: 379-387.

Hoffman DJ, Rattner BA, Sileo L, Doucherty D and Kubiak TJ (1987) Embryotoxicity, teratogenicity, and aryl hydrocarbon hydroxylase activity in Forster's terns on Green Bay, Lake Michigan. Environ Res 42: 176-184.

James MO (1989) Cytochrome P450 monooxygenases in crustaceans. Xenobiotica 19: 1063-1076.

Kaplan LAE, Schultz ME, Schultz RJ and Crivello JF (1991) Nitrosodiethylamine metabolism in the viviparous fish *Poeciliopsis*: Evidence for the existence of liver P450pj activity and expression. Carcinogenesis 12: 647-652.

Kleinow KM, Haasch ML, Williams DE and Lech JJ (1990) A comparison of hepatic P450 induction in rat and trout (*Oncorhynchus mykiss*): Delineation of the site of resistance of fish to phenobarbital-type inducers. Comp Biochem Physiol C 96C: 259-270.

Kloepper-Sams PJ and Stegeman JJ (1992) The effect of temperature acclimation on the expression of cytochrome P4501A mRNA and protein in the fish *Fundulus heteroclitus*. Arch Biochem Biophys 299: 38-46.

Klotz AV, Stegeman JJ and Walsh C (1983) An aryl hydrocarbon hydroxylating hepatic cytochrome P-450 from the marine fish *Stenotomus chrysops*. Arch Biochem Biophys 226: 578-592.

Klotz AV, Stegeman JJ, Woodin BR, Snowberger EA, Thomas PE and Walsh C (1986) Cytochrome P-450 isozymes from the marine teleost *Stenotomus chrysops*: their roles in steroid hydroxylation and the influence of cytochrome b_5. Arch Biochem Biophys 249: 326-338.

Kreamer GL, Squibb K, Gioeli D, Garte SJ and Wirgin I (1991) Cytochrome P450IA mRNA expression in feral Hudson River tomcod. Environ Res 55: 64-78.

Leaver MJ, Burke MD, George SG, Davies JM and Raffaelli D (1988) Induction of cytochrome P450 monooxygenase activities in plaice by 'model' inducers and drilling muds. 24: 27-30.

Lech JJ, Vodicnik MJ and Elcombe CR (1982) Induction of monooxygenase activity in fish. In Aquatic Toxicology. Raven Press 107-148.

Lindstrom-Seppa P, Huuskonen S, Pesonen M, Muona P and Hanninen O (1992) Unbleached pulp mill effluents affect cytochrome P450 monooxygenase enzyme activities. Mar Environ Res 34: 157-161.

Livingstone DR, Kirchin MA and Wiseman A (1989) Cytochrome P450 and oxidative metabolism in molluscs. 19: 1041-1062.

Lorenzana RM, Hedstrom OR and Buhler DR (1988) Localization of cytochrome P-450 in the head and trunk kidney of rainbow trout (*Salmo gairdneri*). Toxicol Appl Pharmacol 96: 159-167.

Luxon PL, Hodson PV and Borgmann U (1987) Hepatic aryl hydrocarbon hydroxylase activity of lake trout (Salvelinus namaycush) as an indicator of organic pollution. Environ Toxicol Chem 6: 649-657.

Malins DC, Krahn MM, Brown DW, Rhodes LD, Myers MS, McCain BB and Chan S-L (1985) Toxic chemicals in marine sediment and biota from Mukilteo, WA: Relationships with hepatic neoplasms and other hepatic lesions in English sole (*Parophrys vetulus*). J Natl Cancer Inst 74: 487-494.

Malins DC, McCain BB, Brown DW, Chan SL, Myers MS, Landahl JT, Prohaska PG, Friedman AJ, Rhodes LD, Burrows DG, Gronlund WD and Hodgins HO (1984) Chemical pollutants in sediments and diseases of bottom-dwelling fish in Puget Sound, WA. Environ Sci Technol 18: 705-713.

Miller MR, Hinton DE and Stegeman JJ (1989) Cytochrome P-450E induction and localization in gill pillar (endothelial) cells of scup and rainbow trout. Aquat Toxicol 14: 307-322.

Miranda CL, Wang J-L, Henderson MC, Zhao X, Guengerich FP and Buhler DR (1991) Comparison of rainbow trout and mammalian cytochrome P450 enzymes: Evidence for structural similarity between trout P450 LMC5 and human P450IIIA4. Biochem Biophys Res Commun 176: 558-563.

Monosson E and Stegeman JJ (1991) Cytochrome P450E (P450IA) induction and inhibition in winter flounder by 3,3',4,4'-tetrachlorobiphenyl: Comparison of response in fish from Georges Bank and Narragansett Bay. Environ Toxicol Chem 10: 765-774.

Munkittrick KR, Kraak GJVD, McMaster ME and Portt CB (1992) Response of hepatic MFO activity and plasma sex steroids to secondary treatment of bleached kraft pulp mill effluent and mill shutdown. Environ Tox Chem 11: 1427-1439.

Nelson D and Strobel H (1987) Evolution of cytochrome P450 proteins. Mol Biol Evol 4: 572-593.

Nelson DR, Kamataki T, Waxman DJ, Guengerich FP, Estabrook RW, Feyereisen R, Gonzalez FJ, Coon MJ, Gunsalus IC, Gotoh O, Okuda K and Nebert DW (1993) The P450 Superfamily - Update on New Sequences, Gene Mapping, Accession Numbers, Early Trivial Names of Enzymes, and Nomenclature. DNA Cell Biol 12: 1-51.

Pajor AM, Stegeman JJ, Thomas P and Woodin BR (1990) Feminization of the hepatic microsomal cytochrome P-450 system in brook trout by estradiol, testosterone, and pituitary factors. J Exper Zool 253: 51-60.

Payne J, Fancey L, Rahimtula A and Porter E (1987) Review and perspective on the use of mixed-function oxygenase enzymes in biological monitoring. Comp Biochem Physiol 86C: 233-245.

Payne JF (1976) Field evaluation of benzopyrene hydroxylase induction as a monitor for marine pollution. Science 191: 945-946.

Poland A and Knutson JC (1982) 2,3,7,8-Tetrachlorodibenzo-p-dioxin and related halogenated aromatic hydrocarbons: examination of the mechanism of toxicity. Ann Rev Pharmacol Toxicol 22: 517-554.

Sakai N, Tanaka M, Adachi S, Miller WL and Nagahama Y (1992) Rainbow trout cytochrome-P-450c17 (17-alpha-Hydroxylase/ 17,20-Lyase) - cDNA cloning, enzymatic properties and temporal pattern of ovarian P-450c17 messenger RNA expression during oogenesis. FEBS Lett 301: 60-64.

Smolowitz RM, Hahn ME and Stegeman JJ (1991) Immunohistochemical localization of cytochrome P450IA1 induced by 3,3',4,4'-tetrachlorobiphenyl and by 2,3,7,8-tetrachlorodibenzofuran in liver and extrahepatic tissues of the teleost *Stenotomus chrysops* (scup). Drug Metab Dispos 19: 113-123.

Spies R, Felton J and Dillard L (1982) Hepatic mixed function oxidases in California flatfishes are increased in contaminated environments and by oil and PCB ingestion. Marine Biol 70: 117-127.

Spies RB, D.W. Rice J and Felton J (1988) Effects of organic contaminants on reproduction of the starry flounder Platichthys stellatus in San Francisco Bay. I. Hepatic contamination and mixed-function oxidase (MFO) activity during the reproductive season. Mar Biol 98: 181-189.

Stegeman JJ (1979) Temperature influence on basal activity and induction of mixed function oxygenase activity in *Fundulus heteroclitus*. J Fish Res Board Can 36: 1400-1405.

Stegeman JJ (1981) Polynuclear aromatic hydrocarbons and their metabolism in the marine environment. In Polycyclic Hydrocarbons and Cancer. HV Gelboin and POP Ts'o Academic Press New York. 1-60.

Stegeman JJ (1989) Cytochrome P450 forms in fish: catalytic, immunological and sequence similarities. Xenobiotica 19: 1093-1110.

Stegeman JJ (1993) unpublished results.

Stegeman JJ and Binder RL (1979) High benzo[a]pyrene hydroxylase activity in the marine fish *Stenotomus versicolor*. Biochem Pharmacol 28: 1686-1688.

Stegeman JJ, Brouwer M, DiGiulio RT, Forlin L, Fowler BM, Sanders BM and Van Veld P (1992) Molecular responses to environmental contamination: Enzyme and protein systems as indicators of contaminant exposure and effect. In Biomarkers. Biochemical, Physiological, and Histological Markers of Anthropogenic Stress. RJ Huggett, RA Kimerle, PM Mehrle and HL Bergman Lewis Boca Raton. 235 - 335.

Stegeman JJ and Hahn ME (1993) Biochemistry and molecular biology of monooxygenases: Current perspectives on the forms, function and regulation of cytochromes P450. In Aquatic Toxicology: Cellular, Molecular and Biochemical Perspectives. D Malins and G Ostrander Lewis Boca Raton Fla. 87-206.

Stegeman JJ, Kloepper-Sams PJ and Farrington JW (1986) Monooxygenase induction and chlorobiphenyls in the deep-sea fish *Coryphaenoides armatus*. Science 231: 1287-1289.

Stegeman JJ, Miller M, Singh H and Hinton D (1987) Cytochrome P450E induction and localization in liver and endothelial tissue of extrahepatic organs of scup. Fed Proc 46: 379.

Stegeman JJ, Miller MR and Hinton DE (1989) Cytochrome P450IA1 induction and localization in endothelium of vertebrate (teleost) heart. Mol Pharmacol 36: 723-729.

Stegeman JJ, Renton KW, Woodin BR, Zhang Y-S and Addison RF (1990) Experimental and environmental induction of cytochrome P450E in fish from Bermuda waters. J Exp Mar Biol Ecol 138: 49-67.

Stegeman JJ, Smolowitz RM and Hahn ME (1991) Immunohistochemical localization of environmentally induced cytochrome P450IA1 in multiple organs of the marine teleost *Stenotomus chrysops* (scup). Toxicol Appl Pharmacol 110: 486-504.

Stegeman JJ, Teng FY and Snowberger EA (1987) Induced cytochrome P-450 in winter flounder (*Pseudopleuronectes americanus*) from coastal

Massachusetts evaluated by catalytic assay and monoclonal antibody probes. Can J Fish Aquat Sci 44: 1270-1277.

Stegeman JJ, Woodin BR and Goksøyr A (1988) Apparent cytochrome P-450 induction as an indication of exposure to environmental chemicals in the flounder *Platichthys flesus*. Mar Ecol Prog Ser 46: 55-60.

Stegeman JJ, Woodin BR and Smolowitz RM (1990) Structure, function and regulation of cytochrome P-450 forms in fish. Biochem Soc Trans 18: 19-21.

Tanaka M and Sakai NN Y. (1991). cDNA encoding P450 aromatase and P450c17 lyase of medaka and rainbow trout; their expression and cloning in the ovary. 2nd International Marine Biotech Conf (IMBC) Baltimore 1991

Van Veld PA, Patton JS and Lee RF (1988) Effect of preexposure to dietary benzo[a]pyrene (BP) on the first-pass metabolism of BP by the intestine of toadfish (*Opsamus tau*): *in vivo* studies using portal vein-catheterized fish. Toxicol Appl Pharmacol 92: 255-265.

Van Veld PA, Westbrook DJ, Woodin BR, Hale RC, Smith CL, Huggett RJ and Stegeman JJ (1990) Induced cytochrome P-450 in intestine and liver of spot (*Leiostomus xanthurus*) from a polycyclic aromatic hydrocarbon contaminated environment. Aquatic Toxicol 17: 119-132.

Varanasi U, Collier TK, Williams DE and Buhler DR (1986) Hepatic cytochrome P-450 isozymes and aryl hydrocarbon hydroxylase in English sole (*Parophyrs vetulus*). Biochem Pharmacol 35: 2967-2971.

Watanabe S, Shimada T, Nakamura S, Nishiyama N, Yamashita N, Tanabe S and Tatsukawa R (1989) Specific profile of liver microsomal cytochrome P-450 in dolphin and whales. Mar Environ Res 27: 51-65.

Waxman DJ, Pampori NA, Ram PA, Agrawal AK and Shapiro BH (1991) Interpulse interval in circulating growth hormone patterns regulates sexually dimorphic expression of hepatic cytochrome P450. Proc Natl Acad Sci USA 88: 6868-6872.

White RD, Hahn ME, Lockhart WL and Stegeman JJ (1993) The hepatic microsomal cytochrome P450 system in beluga whales (*Delphinapterus leucas*). Toxicol Appl Pharmacol (in press)

Williams DE and Buhler DR (1982) Purification of cytochromes P-448 from ß-naphthoflavone-treated rainbow trout. Biochim Biophys Acta 717: 398-404.

Williams DE and Buhler DR (1983) Comparative properties of purified cytochrome P-448 from b-naphthoflavone treated rats and rainbow trout. Comp Biochem Physiol 75C: 25-32.

Williams DE and Buhler DR (1984) Benzo[a]pyrene hydroxylase catalyzed by purified isozymes of cytochrome P-450 from ß-naphthoflavone-fed rainbow trout. Biochem Pharm 33: 3743-3753.

Zhang YS, Goksoyr A, Andersson T and Forlin L (1991) Initial purification and characterization of hepatic microsomal cytochrome P-450 from BNF-treated perch (Perca fluviatilis). Comp Biochem Physiol B 98B: 97-103.

Cytochrome P450 Expression in Rainbow Trout: An Overview

Donald R. Buhler
Department of Agricultural Chemistry, Toxicology Program and
Marine/Freshwater Biomedical Research Center
Oregon State University
Corvallis, Oregon, U.S.A.

Introduction

The cytochromes P450 are the major enzymes involved in the
oxidative metabolism of both endogenous and exogenous chemicals
in fish (Buhler and Williams, 1989; Goksoyr and Förlin, 1992;
Guengerich, 1992) and other animals (Guengerich, 1990). The
membrane-bound cytochromes P450, predominately localized in the
liver but also found in lower concentrations in other tissues,
comprise a superfamily of closely related heme-containing
enzymes of similar structure and diverse functions (Nelson et
al., 1993). Initially, it was thought that fish lacked
cytochrome P450-linked monooxygenases (Brodie and Maickel,
1962) but studies carried out in the late 1960's (Buhler, 1966;
Buhler and Rasmusson, 1968; Dewaide and Henderson, 1968) showed
livers of rainbow trout and other fishes contained these
enzymes. Enzyme activities were generally lower than in
mammals and the fish P450s had temperature optima of about 25°C,
explaining why they were not detected in the earlier studies
where incubations at a higher temperature were employed.
Subsequently, as observed in mammals, multiple cytochrome P450
forms were discovered in fishes (Klotz et al., 1983; Miranda et
al., 1989; Williams and Buhler, 1982, 1983a, 1984) and several
laboratories have succeeded in purifying and characterizing
cytochrome P450 enzymes from numerous marine and freshwater
fishes (for reviews see Andersson and Förlin, 1992; Buhler and
Williams, 1989; Goksoyr and Förlin, 1992; Stegeman, 1989).

NATO ASI Series, Vol. H 90
Molecular Aspects of Oxidative Drug Metabolizing Enzymes
Edited by E. Arınç, J. B. Schenkman and E. Hodgson
© Springer-Verlag Berlin Heidelberg 1995

The rainbow trout (*Oncorhynchus mykiss*, formerly named *Salmo gairdneri*) has been used extensively as a sensitive test species for toxicity studies (Buhl and Hamilton, 1991; McKim et al., 1987) and to investigate carcinogenic mechanisms of chemicals that have been shown to also cause cancer in trout (Bailey et al., 1984; Fong et al., 1993). Perhaps because of the prominence of the rainbow trout as an experimental animal in toxicity studies and the importance of cytochrome P450s in the detoxification and activation of xenobiotics (Buhler and Williams, 1988; Stegeman and Lech, 1991; Watkins, 1990), these trout enzymes have been the most intensively studied among aquatic species.

To date, at least twelve different trout cytochrome P450s have been isolated, extensively purified and characterized to varying degrees (Andersson, 1992; Arinç and Adali, 1983; Celander et al., 1989; Miranda et al., 1989; Sakai et al., 1992; Takahashi et al., 1993; Tanaka et al., 1992; Williams and Buhler, 1982, 1983a, 1984). These investigations include the purification of five different hepatic P450s from β-naphthoflavone (BNF) pretreated rainbow trout (Williams and Buhler, 1982, 1983a, 1984), five P450s from the livers of untreated trout (Miranda et al., 1989) and two P450 forms from the trunk kidney of sexually mature male trout (Andersson, 1992). Two of these isolated hepatic P450 forms were subsequently cloned and sequenced (Buhler et al., 1994; Heilmann et al., 1988). Besides the direct purification of hepatic and renal P450s from trout and their biochemical and catalytic characterization, recent studies (Sakai et al., 1992; Takahashi et al., 1993; Tanaka et al., 1992) have made use of cDNA probes from orthologous mammalian P450s to clone and sequence three trout ovarian P450 forms involved in steroid biosynthesis. These latter three P450s then were expressed in COS-1 cells and their catalytic properties verified. The characteristics of the reported rainbow trout cytochrome P450s isoforms are summarized in Table 1.

Table 1. Rainbow Trout (*Oncorhynchus mykiss*) Cytochrome P450s

Trivial Names	Gene Family	Tissue	Relative M.W.	Residues	Actual M.W.	Inducers	Reference
LM4a	1A2?	liver	58,000			PAHs, BNF	Williams and Buhler, 1984
LM4b	1A1	liver	58,000	522	59,241	PAHs, PCBs, BNF	Williams and Buhler, 1984
LMC1	?	liver	50,000				Miranda et al., 1989
LMC2, LM2	2K1	liver	54,000	504	56,795		Williams and Buhler, 1984; Miranda et al., 1989
LMC3	?	liver	56,000				Miranda et al., 1989
LMC4	?	liver	58,000				Miranda et al., 1989
LMC5, P450con	3A?	liver	59,000				Celander et al., 1989; Miranda et al., 1989
KM1	?	kidney	54,000				Andersson, 1992
KM2	?	kidney	52,000			androgens	Andersson, 1992
P450scc	11A1	ovaries		514			Takahashi et al., 1993
P450C17	17	ovaries		514			Sakai et al., 1992
P450arom	19	ovaries		522			Tanaka et al., 1992

Hepatic Cytochrome P450s

Pretreatment of trout with polycyclic aromatic hydrocarbons
(PAHs) such as 3-methylcholanthrene (3-MC), and similar
inducers including 2,3,7,8-tetrachlorodibenzo-p-dioxin (TCDD),
BNF and coplanar polychlorinated biphenyls (PCBs), produce
significant increases in hepatic P450 levels and in microsomal
benzo[a]pyrene (BaP)-hydroxylase and ethoxyresorufin-O-
deethylase (EROD) activities (Lech et al., 1982). Initial
studies from my laboratory, therefore, were performed with BNF-
pretreated Mt. Shasta strain adult rainbow trout and resulted
in the purification of five hepatic microsomal cytochrome P450s
(Williams and Buhler, 1982, 1983a, 1984). The membrane-bound
proteins were solubilized with 3-[(3-cholamidopropyl)-
dimethylammonio]-1-propane sulfonate (CHAPS) and then isolated
by chromatography on tryptamine Sepharose 4B, DEAE-Sepharose
and hydroxylapatite columns. Only three of the purified P450s,
initially designated P450s LM4a, LM4b and LM2, had specific
contents sufficiently high (10.3-11.9 nmoles P450/mg protein)
to warrant further characterization. Later, using essentially
the same isolation procedures, five constitutive P450 forms
(designated LMC1, LMC2, LMC3, LMC4 and LMC5) also were purified
to high specific contents from the livers of untreated rainbow
trout (Miranda et al., 1989).

<u>Cytochrome P4501A (P450 LM4a and LM4b)</u>. Although
isozymes LM4a and LM4b were resolved on DEAE-Sepharose
(Williams and Buhler, 1984), they appeared to have identical M_r
values of 58,000 on SDS-PAGE electrophoresis. Both LM4a and
LM4b had the same 447.0 nm λ_{max} value and gave similar
(approximately 0.4 nmoles/min/nmole P450) turnover numbers with
BaP in reconstitution systems employing rat NADPH-cytochrome
P450 reductase and dilauroylphosphatidylcholine (DLPC).
Similar results with the two forms were obtained with
acetanilide. Both LM4a and LM4b were inhibited by α-
naphthoflavone (ANF) and by rabbit antibodies raised against

each form. However, slight differences in the amino acid composition of the two purified P450s and disparities in the HPLC metabolite profiles following incubation of the reconstituted isozymes with [^{14}C]-BaP, however, suggested that LM4a and LM4b might be different P450 isoforms (Williams and Buhler, 1984).

In a subsequent investigation Heilmann et al. (1988), using anti-P450 LM4b IgG (Williams and Buhler, 1984), isolated and sequenced cDNA clones from a liver cDNA library prepared from 3-MC-treated rainbow trout. The cDNAs hybridized to a 3-MC-inducible 2.8 kb mRNA that contained an open reading frame encoding for a 522 residue protein with a calculated M_r of 59,241. Comparisons of the predicted amino acid sequence with those of mammalian P450s suggested that the trout form was a P4501A1 (CYP1A1) ortholog. The presence of a single P4501A1 gene in trout liver was indicated by Southern blot analysis of DNA digested with restriction enzymes and hybridized with a trout P4501A1 cDNA probe.

There has been an ongoing controversy, however, regarding the possible existence of a second P4501A form in the livers of rainbow trout and other fishes (Andersson and Förlin, 1992). Two gene subfamilies, P4501A1 and P4501A2, are found in rats and other mammals (Soucek and Gut, 1992). However, based on a higher amino acid sequence homology (57-59%) with mammalian P4501A1 forms than with P4501A2 forms, Heilmann et al. (1988) concluded that there was only a single rainbow trout gene and that it belonged to the P4501A1 subfamily. Stegeman (1994), however, based on amino acid sequence similarities between the trout P4501A1 as reported by Heilmann et al. (1988) and those of various mammalian P4501A1 and P4501A2 forms, has reasoned that the trout form should be designated as P4501A. Stegeman suggested further that trout P4501A is hybrid protein, coded for by a gene ancestral to both mammalian P4501A1 and P4501A2 forms and that the trout P4501A form really contains characteristics of both mammalian forms.

Various groups have attempted to find a P4501A2 form in rainbow trout and other fishes. In earlier studies our laboratory isolated and purified two closely related BNF-inducible P450s (LM4a and LM4b) from trout liver (Williams and Buhler, 1984). P450 LM4b appears to be P4501A1 isoform (Heilmann et al., 1988) but LM4a could be the P4501A2 ortholog.

Pretreatment of trout with isosafrole (ISF), an inducer of P4501A2 in mammals (Okey, 1990), caused a selective increase in the microsomal conversion of 7,12-dimethylbenz[a]anthracene (DMBA) to DMBA-t-8,9-diol compared to that seen with microsomes from BNF induced trout (Henderson et al., 1992). ISF was a weak inducer of P4501A and the anti-P4501A1 IgG cross-reacting pattern in Western blots was similar to that seen with BNF microsomes. Differences in the catalytic and immunological properties of liver microsomes from BNF and ISF treated trout also have been reported by Celander and Förlin (1991).

Studies of trout P450s at the mRNA level further support the conclusion that two forms of P4501A are being expressed in trout. Pesonen et al. (1992) found that primary culture of trout liver cells treated with 2,3,7,8-tetrachlorodibenzo-p-dioxin (TCDD) exhibited two P4501A1-hybridizable mRNA bands upon probing with [^{32}P]-labelled pSg15 cDNA, a probe derived from the cloning of trout P4501A1 (Heilmann et al., 1988). Through use of another cDNA probe designated pfP$_1$P450 derived from the same trout P4501A1 cloning study (Heilmann et al., 1988), we also found that BNF-treated yearling trout contain two P4501A1-hybridizable m
RNA bands at 2.8 (major) and 1.9 kb (minor) (Miranda et al., in preparation). A similar pattern was seen upon exposure to ISF. Whether these two P450 mRNA forms are allelic variants of CYP1A1, the products of alternate splicing of a single gene or are CYP1A1 and CYP1A2 orthologs remains to be established.

This issue has been clarified substantially by recent studies of Berndtson and Chen (Dr. Thomas Chen, Center for Marine Biotechnology, University of Maryland at Baltimore, personal communication) who have cloned two P4501A genes, corresponding to P4501A1 and P4501A2, from the liver of 3-MC induced trout. The two genes, both induced by 3-MC treatment, code for proteins of very similar size that have a predicted

96% amino acid sequence homology with one another. Striking differences were noted, however, in the 5' flanking sequences of the two genes. Identification as the P4501A1 and P4501A2 genes, respectively, were based upon sequence analysis of the two trout genes in the 5' regulatory regions by comparisons to known mammalian P4501A gene sequences. It thus appears that rainbow trout may express both P4501A1 and P4501A2 enzymes.

The levels of the P4501A1 protein found in the livers of rainbow trout is low in fish not exposed to polycyclic aromatic hydrocarbons (PAHs) or halogenated aromatic compounds. In trout unexposed to such inducers, therefore, P4501A1 was barely detectable upon immunohistochemical examination of liver sections (Lorenzana et al., 1989) but was present in the cytoplasm of trunk kidney proximal tubules (Lorenzana et al., 1988) probed with rabbit antibodies raised against purified trout P4501A1. However, strong induction of P4501A1 was seen in trout pretreated with BNF or other inducers. Antibodies raised against the purified P4501A1 orthologs of rainbow trout, scup and Atlantic cod cross-reacted against P4501A1 forms from 7 different BNF-induced fish species and inhibited EROD in the same fish microsomal samples (Goksoyr et al., 1991) suggesting the close structural similarity of this isoform between fish species. Antibodies against trout P4501A1, therefore, have proved useful in determining environmental induction of this P450 form in other fishes (Varanasi et al., 1986).

Cytochrome P450 LMC1. A constitutive P450 designated P450 LMC1 was purified to a specific content of 11.6 nmol P450/mg protein from trout unexposed to chemicals (Miranda et al., 1989). The P450 had a λ_{max} of 450 nm and an estimated M_r of 50,000. Rabbit polyclonal antibodies raised against trout P450 LMC1 and LMC2 cross-reacted with one another and the anti-LMC1 IgG reacted strongly with rat CYP2B1 (Miranda et al., 1990a). Reconstitution experiments with LMC1 performed with rat NADPH-cytochrome P450 reductase and DLPC demonstrated that this isoform had modest activity toward 17β-estradiol, testosterone and progesterone but BaP and benzphetamine were not

metabolized.   Reconstituted LMC1, however, had appreciable
lauric acid hydroxylase activity.   In a subsequent study
(Miranda et al., 1990b), P450 LMC1 was shown to catalyze the ω-
hydroxylation of lauric acid.   While lauric acid
ω-hydroxylase activities were high in liver microsomes from
juvenile and sexually mature female trout, microsomes from
sexually mature male trout reflected only limited enzyme
activity (Miranda et al., 1990b).

Hydroxylation of lauric acid at the ω- and (ω-1) (12- and
11-positions) is catalyzed by mammalian P450s belonging to the
clofibrate-inducible CYP4A family (Aoyma et al., 1990).
Clofibrate, injected ip into 18-month old trout at 10 mg/kg or
50 mg/kg, every other day for two weeks, revealed no detectable
increases in the levels of P450 LMC1 or LMC2 (CYP2K1) shown by
Western blots (Miranda et al., to be published).   Clofibrate
also had no effect on the ω- and (ω-1)-hydroxylation of lauric
acid.   Western blots of liver microsomes from clofibrate-
treated trout probed with anti-rat CYP4A IgG (gift from Dr.
R.T. Okita, Washington State University) also exhibited no
increase in any immunoreactive proteins.   Anti-rat CYP4A IgG
cross-reacted slightly with trout P450 LMC2 and to a lesser
extent, trout P450 LMC1.   Thus, although P450 LMC1 seemed to be
a lauric acid ω-hydroxylase, the evidence obtained to date does
not support its assignment to the P4504A subfamily.

Cytochrome P4502K1 (P450 LM2/LMC2).   A major constitutive
trout P450 form called LM2 was isolated with a specific content
of 10.8 nmol P450/mg protein from the livers of BNF induced
trout (Williams and Buhler, 1984).   P450 LM2 is biochemically,
immunologically and catalytically indistinguishable from P450
LMC2, the major constitutive isozyme subsequently purified
(specific content 14.0 nmol P450/mg protein)from liver of
untreated trout (Miranda et al., 1989).   This isozyme exhibited
a M_r of 54,000 and a λ_{max} in the CO-reduced difference spectrum
of 449.5 nm and showed no detectable BaP hydroxylase activity.

P450 LMC2, however, was highly active toward aflatoxin B1 (AFB1) with strong regioselectivity in the formation of AFB1-8,9-epoxide, the ultimate carcinogenic form of AFB1. In the reconstituted P450 LMC2 system (Williams and Buhler, 1983b), production of the highly reactive AFB1-8,9-epoxide was detected by the appearance of its hydrolysis product, AFB1-8,9-dihydroxy-8,9-dihydrodiol. High covalent binding of [G-^{3}H]-AFB1 to calf thymus DNA also was observed when the DNA was added to the P450 LMC2 reconstituted incubation mixture. This constitutive P450 form is the major P450 present in the livers of rainbow trout and is, thus, likely responsible for the acute sensitivity of this species to the carcinogenic effects of AFB1 (Bailey et al., 1984).

P450 LMC2 also is a major catalyst for the (ω-1)-hydroxylation of lauric acid (Miranda et al., 1990b; Williams et al., 1984). This isoform also catalyzes the 2-hydroxylation of 17β-estradiol, the 16β-hydroxylation of testosterone, the 16α-hydroxylation of progesterone and benzphetamine N-demethylase (BND) activity (Miranda et al., 1989).

Rabbit polyclonal antibodies prepared against P450 LMC2 had a significant inhibitory effect on trout hepatic microsomal lauric acid hydroxylase and BND activities and the microsome-mediated covalent binding of AFB1 to added DNA, strongly suggesting that P450 LMC2 was primarily responsible for these microsomal activities (Miranda et al., 1989). Anti-LMC2 IgG was used in an enzyme-linked immunosorbent assay (ELISA) with alkaline phosphatase to show the existence of a cross-reactivity between these antibodies and trout P450 LMC1 (Miranda et al., 1990a). Antibodies to LMC2 also cross-reacted strongly with rat P4502B1 and with P450s in phenobarbital (PB)-induced rat liver microsomes. While in mammals PB readily induces P4502B1 isoforms having high BND activities, trout are refractory to PB-type induction (Kleinow et al., 1990) and show no increases in LMC2 levels after such pretreatment. Nevertheless, Southern blot analysis of trout genomic DNA hydrolyzed with various restriction enzymes and probed with a rat P4502B1 cDNA indicated that sequences analogous to mammalian P4502B1 are present in their genome but that this isoform is apparently not expressed in trout (Kleinow et al., 1990).

Immunohistochemical analysis established that P450 LMC2 was localized in the second portion of the proximal tubules (P_2) in the trunk kidney of trout (Lorenzana et al., 1988). More intense staining appeared in trunk kidney tissue from mature male animals as compared to the female kidney sections. By contrast, head kidney displayed only faint staining of interrenal cells by anti-P450 LMC2 IgG. Immunostaining for P450 LMC2 in liver was observed along the sinusoidal border of some parenchymal cells coupled with moderated staining within the cytoplasm of most cells (Lorenzana et al., 1989). There was intense, uniform immunostaining for P450 LMC2 within the cytoplasm of the bile duct cells, in the endothelial lining of arterioles, and along the epithelial surface of the gall bladder.

P450 LMC2 is developmentally expressed (Buhler et al., 1994; Williams et al., 1986). Through immunoblotting, levels of this P450 were shown to be almost 2-fold higher in liver and 25-fold higher in trunk kidney microsomes of sexually mature male trout compared to females. Differences in the relative amounts of the LMC2 form between mature males and females was correlated with significantly higher microsomal lauric acid hydroxylase activity and the activation of AFB1 and its covalent binding to DNA.

To learn more about its expression and relationship with mammalian P450 forms, we have cloned and sequenced a full-length cDNA encoding for P450 LMC2 (Buhler et al., 1994). The 1859 bp cDNA, isolated from a 3-year old male rainbow trout liver cDNA library, contained an open reading frame encoding a protein of 504 amino acids with a calculated M_r of 56,795. From amino acid sequence comparisons, the trout P450 LMC2 has been assigned to a new cytochrome P450 gene subfamily designated P4502K1 or CYP2K1 (Nelson et al., 1993). Investigation of the expression of this gene at the transcriptional and translational level were made in sexually mature rainbow trout liver and trunk kidney. Transcriptional expression investigated by Northern analysis of total RNA using a 440 bp 3'-terminal cDNA probe (2K1,7c) suggested sexual and organ differences. Mature male trunk kidney expressed 2K1,7c hybridizable mRNA to a much greater extent than did female trunk kidney, with multiple mRNA bands appearing at approximately the 2.8 kb and 1.9 kb region. A 1.9 kb mRNA bands were found at similar concentrations in the liver samples of both sexes. The livers of some fish also displayed a separate 2.8 kb hybridizable band but such bands were much weaker or nondetectable in female liver samples. Considering the individual biological variabilities, it was apparent that mature male trout expressed the 2.8 kb mRNA much more strongly in trunk kidney than in liver. Translational expression, analyzed by Western blotting of microsomes separated on SDS-PAGE and probed with anti-P450 LMC2 IgG, revealed corresponding sex and organ-related differences in P450 protein expression.

Cytochrome P450 LMC3 and LMC4. Isolation of P450s LMC3 and LMC4 from the livers of untreated trout yielded incompletely purified proteins with specific contents of 2.8 and 9.0 nmol P450/mg protein, respectively (Miranda et al., 1989). These two isoforms had reduced CO-absorption maxima of 450 nm and by SDS-PAGE their M_r's were estimated at 56,000 and 58,000, respectively. Using an ELISA assay with alkaline phosphatase, strong cross-reactivity was observed between rabbit anti-trout LMC3 IgG and trout LMC4 (Miranda et al., 1990a). Although insufficient enzyme was available for reconstitution experiments, trout hepatic microsomal inhibition studies using rabbit antibodies prepared against these isforms failed to show any diminution in testosterone, progesterone, 17β-estradiol or lauric acid hydroxylation. The endogenous substrates and catalytic activities toward xenobiotics for these two isozymes yet remains to be determined.

Cytochrome P450 LMC5. A highly purified (specific content 14.9 nmol P450/mg protein) hemoprotein, designated P450 LMC5, also was isolated from the livers of untreated trout (Miranda et al., 1989). This P450 form had an apparent M_r of 59,000, a λ_{max} of 448 nm and in reconstituted systems displayed significant activity for the 6β-hydroxylation of testosterone and progesterone and for the N-demethylation of benzphetamine. There was little or no cross-reactivity of anti-LMC5 IgG with other trout P450 forms.

Since steroid 6β-hydroxylase activity is associated with mammalian P4503A enzymes, further studies were carried out to determine the immunochemical homology between trout P450 LMC5 and mammalian P4503A enzymes (Miranda et al., 1991). Polyclonal antibodies generated against trout P450 LMC5 reacted strongly on immunoblots with P4503A1 in dexamethasone-induced rat liver microsomes and with P4503A4 in human liver microsomes. Reciprocal immunoblots using anti-rat P4503A1 showed that this antibody did not recognize trout P450 LMC1 or LMC5 but antihuman P4503A4 IgG was found to cross-react strongly with these trout P450s. Progesterone 6β-hydroxylase activity of trout liver microsomes, a reaction catalyzed by P450 LMC5 and characteristic of P4503A isoforms, was markedly inhibited by anti-P4503A4 and by gestodene, a mechanism-based inactivator of human P4503A4. These results provide compelling evidence for a similarity in the structure and catalytic function between trout P450 LMC5 and human P4503A4.

Cytochrome P450$_{con}$. Celander et al. (1989), have partially purified a P450 from the livers of trout pretreated with various inducers. This P450 called P450$_{con}$, had an apparent M_r of 54,000 on SDS-PAGE but its catalytic properties were not determined. Polyclonal antibodies raised against P450$_{con}$ showed in Western blots that this isoform was a major constitutive form in juvenile trout. However, hepatic concentrations of P450$_{con}$ were about twice as high in sexually mature males compared to females. While preexposure of trout to dexamethasone failed to increase levels of this P450, cortisol or pregnenolone-16α-carbonitrile (PCN) treatment resulted in a significant elevation of P450$_{con}$ concentrations 7 days after injection.

Recently, in collaboration with M. Celander and L. Förlin (University of Göteborg) we have found (Celander et al., to be published) by Western blotting that rainbow trout P450con (Celander et al., 1989) is very similar if not identical with our P450 LMC5. These results suggest that P450con may belong to the P4503A subfamily (Miranda et al., 1991).

Renal Cytochrome P450s

Anderson (1992) observed that male trout kidney contained considerably higher P450 concentrations than did females during the late reproductive stage. Trunk kidney microsomes from sexually mature male trout were solubilized in Emulgen 911 and sodium cholate and then fractionated on a CH-Sepharose 4B column coupled to p-chloramphetamine. The two resulting P450 peak fractions were pooled and then chromatographed on DEAE-Sepharose to yield two partially purified proteins, designated P450 KM1 and KM2, with estimated M_r's of 54,000 and 52,000, respectively. P450 KM1 was not further studied but this isoform may be the same as LMC2 (P4502K1) based on its estimated 54,000 molecular weight.

Rabbit antibodies against P450 KM2 were prepared after it was further purified on SDS-PAGE (Andersson, 1992). Antiserum against P450 KM2 did not cross-react with the KM1 form. Quantitative Western blotting established that P450 KM2 constituted about 66% of the total P450 in the kidney of sexually mature trout. Levels of this isoform were substantially lower (120-fold) in the kidneys of sexually mature females where P450 KM2 accounted for 15% of the total P450. Treatment of trout with 11-ketotestosterone (0.2 mg weekly for 6 weeks) significantly induced kidney P450 KM2 concentrations whereas this isoform was unaffected by testosterone treatment. In the 11-ketotestosterone dosed trout, total renal P450 levels were increased (4-fold) with similar increases in microsomal 7-ethoxycoumarin-O-deethylase and progesterone-6β-hydroxylase activities.

Ovarian Cytochrome P450s

Cytochrome P45011A1 (P450$_{scc}$). One of the more important P450s involved in mammalian steroid biosynthesis is the cholesterol side-chain cleavage P450 (P450$_{scc}$, now designated P45011A1). Takahashi et al. (1993) recently were successful in cloning and sequencing a cDNA encoding for the trout ortholog of this isoform. These investigators initially prepared a cDNA library from poly (A)$^{+}$ isolated from ovarian thecal cells of a 3-year-old female trout. Using a [^{32}P]-labeled 285 bp fragment of a human P45011A1 cDNA probe, they then screened the library to obtain a full-length trout cDNA clone encoding for a protein of 514 amino acid residues. The predicted amino acid sequence showed a 48% homology with that of P45011A1 isoforms from rat, bovine and pig. Northern blot analysis with a trout P45011A1 cDNA indicated that high concentrations of a single 1.8 kb P45011A1 mRNA in postovulatory follicles.

Verification of the identity of the cloned trout P450 as a P45011A1 ortholog was based on the expression of the cloned trout cDNA in COS-1 monkey kidney cells. After transfection the non-steroidogenic cells became able to convert 25-hydroxycholesterol to pregnenolone, a characteristic catalytic activity of P45011A1.

Cytochrome P45017 (P450$_{c17}$). Mammalian P45017 has both steroid 17α-hydroxylase and 17,20-lyase activities. Sakai et al. (1992) has employed a 630 bp fragment of a human P45017 (P450$_{c17}$) cDNA as a probe to screen a thecal cell cDNA library obtained from the ovaries of 3-year-old female trout. This yielded a full-length cDNA containing an open reading frame of

1,542 nucleotides encoding for a P450 of 514 amino acid residues. Comparisons of the predicted amino acid sequences of the trout P450 with that of other members of the P450 17 subfamily, showed a 64% homology with the chicken isoform but a lesser similarity (46-48% homology) with human, bovine and rat P450 17. Northern hybridization analysis of trout ovary preparations using a 2,081 bp cDNA for trout P450 17, showed barely detectable expression of a 2.4 kb P450 17 mRNA species during early- and mid-vitellogenic stages of female maturation while abundant levels of this mRNA were expressed in the post-vitellogenic stage and after ovulation.

Following incorporation of a trout P450 17 cDNA expression vector into COS-1 cells, the cells gained characteristic P450 17-related enzyme activities, including the ability to convert progesterone to 17α-hydroxyprogesterone and androstenedione. The cloned trout P450 17 was catalytically similar to the bovine and human P450 17 forms with more 17,20-lyase activity than 17α-hydroxylase activity.

<u>Cytochrome P450 19 (P450$_{arom}$)</u>. Japanese investigators also have cloned, sequenced and characterized a third steroidogenic P450 from trout (Tanaka et al., 1992), P450$_{arom}$ (P450 19), the P450 that converts androgens to estrogens. A cDNA library was prepared from poly(A)$^+$ RNA extracted from trout ovary that was in the early vitellogenic stage and then screened with a radiolabeled human P450 19 cDNA probe. A clone containing a 3 kb insert initially was obtained which, after digestion and subcloning, yielded a second clone with a 1566 bp open-reading frame encoding for a 522 amino acid protein. The deduced protein sequence was 52% homologous with rat, mouse and human P450 19 and 53% homologous with the chicken isoform. Upon expression of the trout P450 19 cDNA in mammalian COS-1 cells, the ability of these cells to convert testosterone to 17β-estradiol was increased approximately 15-fold.

Northern hybridization analysis of the trout P450 19 mRNA demonstrated the presence of a 2.6 kb mRNA in ovaries at the early and late vitellogenic stages. No evidence for expression of the 2.6 kb transcript, however, was seen in the later stages of trout ovulation.

Conclusions

While research on teleost cytochrome P450s has lagged behind that of their mammalian counterparts, characterization of these fish enzymes during the past decade has experienced rapid progress. This has been especially true for the rainbow trout where its P450s have now become the most extensively studied among the fishes. Research conducted in several laboratories have identified many of the P450 isozymes present in trout; determined their properties and structures; and investigated their expression at both the mRNA and protein levels as a function of developmental stage, sex, tissue, chemical exposure and diet. Limited data also has been obtained on the regulation of certain of the trout P450 forms. Published reports on a least 12 different rainbow trout P450s have appeared (Andersson, 1992; Celander et al., 1989; Miranda et al., 1989; Sakai et al., 1992; Takahashi et al., 1993; Tanaka et al., 1992; Williams and Buhler, 1982, 1983a, 1984) and additional studies in progress should further describe the attributes of these and additional P450 forms. Based on the amino acid sequence information provided for trout P450s by these investigations and comparisons with known sequences of mammalian P450 forms, the presence of several orthologs of mammalian P450s have been shown to be present in rainbow trout.

Most of the previous research on trout P450s have involved the classical technique of initial isolation and purification of the enzymes followed by subsequent biochemical and catalytic characterization. This experimental approach, while generally quite successful, is tedious and fraught with difficulties. Moreover, during the purification process, inactivation of P450s could occur with concomitant loss of enzyme activities. Under such circumstances, it was virtually impossible to establish the endogenous substrates for the P450s under study or to determine the role of the P450s in xenobiotic metabolism.

The use of molecular biology methodology, however, offers the promise of markedly reducing the effort involved in isolating and sequencing trout P450s. Hence, screening an appropriate trout cDNA library for orthologs to mammalian forms has been remarkably successful (Sakai et al., 1992; Takahashi et al., 1993; Tanaka et al., 1992) in the case of ovarian P450 forms. However, there are several difficulties associated with the cDNA cloning approach. This method depends on the availability of a cDNA or oligonucleotide probe coding for a P450 ortholog from some other species that has a high sequence homology with the teleost P450 forms so that hybridization can occur between the probe and corresponding P450 clones in an appropriate trout cDNA library. An alternative approach, not yet applied to the cloning of trout P450s, involves the use of the polymerase chain reaction (PCR) with synthetic oligonucleotide primers patterned after known P450 gene sequences from mammals or other species. Following cloning and sequencing of the trout P450 cDNAs, isolated trout P450 cDNAs have been expressed in cultured mammalian COS-1 cells to verify by determining catalytic activities in a reconstituted system that the correct P450 forms had been cloned (Sakai et al., 1992; Takahashi et al., 1993; Tanaka et al., 1992). Unfortunately insufficient quantities of P450 protein are produced by the COS-1 expression system to permit P450 purification and antibody preparation. Consequently, unless antibodies from the orthologous mammalian P450s cross-react with the trout P450s, such investigations have not permitted assessment of both P450 mRNA and protein expression in trout tissues.

Recently, however, Gillam et al. (1993) have inserted human P4503A4 cDNA into a pCW vector which was then used to transform *Escherichia coli* DH5α cells. Sufficient quantities of the human P4503A4 protein were expressed by this system to permit its purification to a high degree (23 nmol P450/mg protein) for use in reconstitution experiments to measure catalytic activity and for raising rabbit polyclonal antibodies. Through the use of the cDNA cloning technique coupled with an expression system permitting P450 purification and antibody preparation, future studies will likely allow examination of P450 expression in rainbow trout at both the mRNA and protein level.

Rapid advances in molecular technology that are occurring in mammalian research (*in situ* hybridization, *in situ* PCR, site-directed mutagenesis, etc.) are slowly being applied to the study of fish P450s. Such investigations will eventually lead a better understanding of trout cytochrome P450 structures, functions and regulation.

Acknowledgements

Supported by grants from the National Institutes of Health, No. ES000210, ES03850, ES04766 AND ESO5533. This manuscript was issued as Technical Paper No. 10369 from the Oregon Agricultural Experiment Station, Oregon State University, Corvallis, OR, U.S.A.

References

Andersson T (1992) Purification, characterization and regulation of a male-specific cytochrome P450 in the rainbow trout kidney. Mar Environ Res 34:109-112

Andersson T, Förlin L (1992) Regulation of the cytochrome P450 enzyme system in fish. Aquatic Toxicol 24:1-20

Aoyama T, Hardwick JP, Imaoka S, Funae Y, Gelboin HV, Gonzalez FJ (1990) Clofibrate-inducible rat hepatic P450s IVA1 and IVA3 catalyze the omega-and (omega-1)-hydroxylation of fatty acids and the omega-hydroxylation of prostaglandins E1 and F2 alpha. J Lipid Res 31:1477-1482

Arınç E, Adalı O (1983) Solubilization and partial purification of two forms of cytochrome P-450 from trout liver microsomes. Comp Biochem Physiol 76B:653-662

Bailey GS, Hendricks JD, Nixon JE, Pawlowski NE (1984) The sensitivity of rainbow trout and other fish to carcinogens. Drug Metab Rev 15:725-750

Brodie BB, Maickel RP (1962) Comparative biochemistry of drug metabolism. In: Brodie BB, Erdos FG (eds) Proc 1st Int Pharmacology Meeting, Macmillan, New York, pp 299-324

Buhl KJ, Hamilton SJ (1991) Relative sensitivity of early life stages of Arctic grayling, coho salmon and rainbow trout to nine inorganics. Ecotoxicol Environ Safety 22:184-197

Buhler DR (1966)Hepatic drug metabolism in fishes. Federation Proc 25:343

Buhler DR, Williams, DE (1988) The role of biotransformation in the toxicity of chemicals. Aquatic Toxicol 11:19-28

Buhler DR, Williams DE (1989) Enzymes involved in metabolism of PAHs by fishes and other aquatic animals. In: Varanasi, U (ed) Metabolism of polynuclear aromatic hydrocarbons in the aquatic environment, CRC Press, Inc, New York, pp 151-184

Buhler DR, Rasmusson ME (1968) The oxidation of drugs by fishes. Comp Biochem Physiol 25:223-239

Buhler DR, Yang Y-H, Dreher TW, Miranda CL, Wang J-L (1993) Cloning and sequencing of the major rainbow trout constitutive cytochrome P450(CYP2K1). Identification of a new cytochrome P450 gene subfamily and its expression in mature rainbow trout liver and trunk kidney. Arch Biochem Biophys, in press

Celander M, Förlin L (1991) Catalytic activity and immunochemical quantification of hepatic cytochrome P-450 in β-naphthoflavone and isosafrol treated rainbow trout (*Oncorhynchus mykiss*). Fish Physiol Biochem 9:189-197

Celander M, Ronis M, Förlin L (1989)Initial characterization of a constitutive cytochrome P-450 isozyme in rainbow trout liver. Marine Environ Res 28:9-14

Dewaide JH, Henderson PT (1968) Hepatic N-demethylation of aminopyrine in rat and trout. Biochem Pharmacol 17:1901-1907

Fong AT, Dashwood RH, Cheng R, Mathews C, Ford B, Hendricks JD, Bailey GS (1993) Carcinogenicity, metabolism, and Ki-ras proto-oncogene activation by 7,12-dimethylbenz(a)-anthracene in rainbow trout embryos. Carcinogenesis 14:629-635

Gillam EMJ, Baba T, Kim B-R, Ohmori S, Guengerich FP (1993) Expression of modified human cytochrome P450 3A4 in *Escherichia coli* and purification and reconstitution of the enzyme. Arch Biochem Biophys 305:123-131

Goksøyr A, Andersson T, Buhler DR, Stegeman JJ, Williams DE, Förlin L (1991) Immunologic chemical cross-reactivity of β-naphthoflavone-inducible cytochrome P-450 (P-450IA) in liver microsomes from different fish species. Fish Physiol Biochem 9:1-13

Goksøyr A, Förlin L (1992) The cytochrome P-450 system in fish, aquatic toxicology and environmental monitoring. Aquatic Toxicol 22:287-312

Guengerich FP (1990) Enzymatic oxidation of xenobiotic chemicals. Critical Rev Biochem Mol Biol 25:97-153

Guengerich FP (1992) Metabolic activation of carcinogens. Pharmacol Ther 54:17-61

Heilmann LJ, Sheen Y-Y, Bigelow SW, Nebert DW (1988) Trout P450IA1: cDNA and deduced protein sequence, expression in liver, and evolutionary significance. DNA 7:379-387

Henderson MC, Zhao X, Miranda CL, Buhler DR (1992) Differential effects of β-naphthoflavone (BNF) and iso-safrole (ISF) on the in vitro metabolism of 7,12 dimethylbenzanthracene (DMBA) by trout liver microsomes The Toxicologist 12:393

Kleinow KM, Haasch ML, Williams DE, Lech JJ (1990) A comparison of hepatic P450 induction in rat and trout (*Oncorhynchus mykiss*): Delineation of the site of resistance of fish to phenobarbital-type inducers. Comp Biochem Physiol 96C:259-270

Klotz AV, Stegeman JJ, Walsh C (1983) An aryl hydrocarbon hydroxylating hepatic cytochrome P-450 from the marine fish *Stenotomus chrysops*. Arch Biochem Biophys 226:578-592

Lech JJ, Vodicnik MJ, Elcombe CR (1982) Induction of monooxygenase activity in fish. In: Weber L (ed) Aquatic Toxicology. Raven Press, New York, vol 1, pp 107-148

Lorenzana RM, Hedstrom OR, Buhler DR (1988) Localization of cytochrome P450 in the head and trunk kidney of rainbow trout (*Salmo gairdneri*). Toxicol Appl Pharm 96:159-167

Lorenzana RM, Hedstrom OR, Gallagher JA, Buhler DR (1989) Cytochrome P450 isozyme distribution in normal and tumor-bearing hepatic tissue from rainbow trout (*Salmo gairdneri*).Exptl Molec Pathol 50:348-361

McKim JM, Bradbury SP, Niemi GJ (1987) Fish acute toxicity syndromes and their use in the QSAR approach to hazard assessment. Environ Health Perspect 71:171-186

Miranda CL, Wang JL, Henderson MC, Buhler DR (1989) Purification and characterization of hepatic steroid hydroxylases from untreated rainbow trout. Arch Biochem Biophys 268:227-238

Miranda CL, Wang J-L, Henderson MC, Buhler DR (1990a) Immunological characterization of constitutive isozymes of cytochrome P-450 from rainbow trout. Evidence for homology with phenobarbital-induced rat P-450s Biochim Biophys Acta 1037:155-160

Miranda CL, Wang J-L, Henderson MC, Williams DE, Buhler DR (1990b) Regiospecificity in the hydroxylation of lauric acid by rainbow trout hepatic cytochrome P-450 isozymes. Biochem Biophys Res Commun 171:537-542

Miranda CL, Wang J-L, Henderson MC, Zhao X, Guengerich FP, Buhler DR (1991)Comparison of rainbow trout and mammalian cytochrome P450 enzymes.Evidence for structural similarity between trout P450 LMC5 and human P450IIIA4. Biochem Biophys Res Commun 176:558-563

Nelson DR, Kamataki T, Waxman DJ, Guengerich FP, Estabrook RW, Feyereisen R, Gonzalez FJ, Coon MJ, Gunsalus IC, Gotoh O, Okuda K, Nebert DW (1993) The P450 superfamily: Update on new sequences, gene mapping, accession numbers, early trivial names of enzymes, and nomenclature.DNA and Cell Biol 12:1-51

Okey AB (1990) Enzyme induction in the cytochrome P-450 system. Pharmacol Therap 45:241-298

Pesonen M, Goksøyr A, Andersson T (1992) Expression of P450 1A1 in a primary culture of rainbow trout hepatocytes exposed to β-naphthoflavone or 2,3,7,8-tetrachlorodibenzo-p-dioxin. Arch Biochem Biophys 292:228-233

Sakai N, Tanaka M, Adachi S, Miller WL, Nagahama Y (1992) Rainbow trout cytochrome P-450c17 (17α-hydroxylase/17,20-lyase) cDNA cloning, enzymatic properties and temporal pattern of ovarian PO-450c17 mRNA expression during oogenesis. FEBS Lett 301:60-64

Soucek P, Gut I (1992) Cytochrome P-450 in rats:structures, functions, properties and relevant human forms. Xenobiotica 22:83-103

Stegeman JJ (1989) Cytochrome P450 forms in fish: catalytic, immunological and sequence similarities. Xenobiotica 19:1093-1110

Stegeman JJ (1994) The diversity and regulation of cytochrome P450 forms in aquatic species. In: Arınç E, Hodgson E, Schenkman JB (eds) Molecular aspects of oxidative drug metabolizing enzymes: Their significance in environmental toxicology, chemical carcinogenesis and health Plenum Press, New York, in press

Stegeman JJ, Lech JJ (1991) Cytochrome P-450 monooxygenase systems in aquatic species: carcinogen metabolism and biomarkers for carcinogen and pollutant exposure. Environ Health Perspect 90:101-109

Takahashi M, Tanaka M, Sakai N, Adachi S, Miller WL, Nagahama Y (1993) Rainbow trout ovarian cholesterol side-chain cleavage cytochrome P450 (P450scc). FEBS Lett 319:45-48

Tanaka M, Telecky TM, Fukada S, Adachi S, Chen S, Nagahama Y (1992) Cloning and sequence analysis of the cDNA encoding P-450 aromatase (P450arom) from a rainbow trout (*Oncorhynchus mykiss*) ovary: relationship between the amount of P450arom mRNA and the production of oestradiol-17β in the ovary. J Mol Endocrinol 8:53-61

Varanasi U, Collier TK, Williams DE, Buhler DR (1986). Hepatic cytochrome P-450 isozymes and aryl hydrocarbon hydroxylase in English sole (*Parophrys vetulus*). Biochem Pharmacol , 35:2967-2971

Watkins PB (1990) Role of cytochrome P450 in drug metabolism and hepatotoxicity. Semin Liver Dis 10:235-250

Williams DE, Buhler DR (1982) Purification of cytochromes P-448 from β-naphthoflavone-treated rainbow trout. Biochim Biophys Acta 717:398-404

Williams DE, Buhler DR (1983a) Comparative properties of purified cytochrome P-448 from β-naphthoflavone-fed rainbow trout. Comp Biochem Physiol 75C:25-32

Williams DE, Buhler DR (1983b) Purified form of cytochrome P-450 from rainbow trout with high activity towards conversion of aflatoxin B_1 to aflatoxin B_1-2,3-epoxide. Cancer Res 43:4752-4756

Williams DE and Buhler DR (1984) Benzo[a]pyrene hydroxylase catalyzed by purified isozymes of cytochrome P-450 from β-naphthoflavone-fed rainbow trout. Biochem Pharmacol 33:3742-3753

Williams DE, Masters BSS, Lech JJ, Buhler DR (1986) Sex differences in cytochrome P-450 isozyme composition and activity in kidney microsomes of   mature rainbow trout. Biochem Pharmacol 35:2017-2023

Williams DE, Okita RT, Buhler DR, Masters BSS (1984) Regiospecific hydroxylation of lauric acid in the (ω-1) position by hepatic and kidney microsomal cytochromes P-450 from rainbow trout. Arch Biochem Biophys 231:503-510

Molecular Aspects of Cytochrome P450 2E1 and Its Roles in Chemical Toxicity

Chung S. Yang and Jun-Yan Hong
Laboratory for Cancer Research
College of Pharmacy
Rutgers University
Piscataway, NJ 08855
U.S.A.

Introduction

Cytochrome P450 2E1, previously known as P450LM$_{3a}$, P450j, P450ac, or P450alc, has received a great deal of attention in recent years because of its vital role in the activation of many toxic chemicals (Yang, *et al.*, 1992, Yang, *et al.*, 1990, Guengerich, *et al.*, 1991, Koop, 1992). In this chapter, the structure, catalytic activities, and regulation of this enzyme are discussed. Examples are used to illustrate this enzyme as a target for the interactions of many environmental chemicals. Although written in the form of a review, it is not intended to be a comprehensive review of the literature. Many of our own studies are used as examples to illustrate pertinent points.

Early Studies on *N*-Nitrosodimethylamine Demethylase

Several seemingly unrelated lines of research converged in the early 1980's in the studies on the structure and catalytic functions of P450 2E1. Our interest in this area derived from investigations on the metabolism of *N*-nitrosodimethylamine (NDMA), a widely occurring environmental carcinogen. Earlier studies had indicated that NDMA was metabolized in an NADPH-dependent reaction in the endoplasmic reticulum, and the reaction had many of the features of a P450-dependent reaction (Lai and Arcos, 1980). However, the role of P450 enzymes in the demethylation (activation) of NDMA was questioned in some early publications, because the activity was not enhanced by the conventional P450 inducers such as phenobarbital, 3-methylcholanthrene and polychlorinated biphenyls; nor was the activity inhibited by common P450 inhibitors such as SKF 525A or metyrapone. The involvement of monoamine oxidase in NDMA

NATO ASI Series, Vol. H 90
Molecular Aspects of Oxidative Drug Metabolizing Enzymes
Edited by E. Arınç, J. B. Schenkman and E. Hodgson
© Springer-Verlag Berlin Heidelberg 1995

metabolism was also suggested because the activity was inhibited by monoamine oxidase inhibitors such as 2-phenethylamine (Lai and Arcos, 1980).

We started to study this problem in 1980 by adopting a hypothesis that the multiple K_m values and a lack of saturation kinetics observed in NDMA demethylation in microsomes are due to the fact that several P450 enzymes can catalyze the oxidation of NDMA and that the low K_m NDMA demethylase activity is due to a unique form of P450 which distincts itself from other P450s in inducibility and susceptibility to inhibitors. In order to test this hypothesis, we started to investigate the inducibility of the low K_m NDMA demethylase. There were some indications in the literature at that time that NDMA demethylase was inducible. What we did was to conduct detailed kinetic analysis to identify the low K_m NDMA demethylase and to examine the induction of this low K_m enzyme systematically. Our first indication was that pyrazole induced a form of NDMA demethylase with a low K_m of 60 μM (Tu, et al., 1981). Subsequent experiments indicated that this low K_m NDMA demethylase also existed in liver microsomes from untreated rats, albeit with a low V_{max}. This form was also inducible by fasting and diabetes as well as by treatment with acetone, isopropanol, or ethanol (Peng, et al., 1983, Peng, et al., 1982, Tu, et al., 1983, Tu and Yang, 1983). In control liver microsomes, in addition to this low K_m form, at least two higher K_m values at approximately 0.25 mM and 28 mM were also observed (Tu, et al., 1983, Tu and Yang, 1983). After treatment with the inducers, the low K_m form became the predoninant enzyme for NDMA demethylation. Gel electrophoresis experiments indicated that the induction of the low K_m NDMA demethylase was associated with the intensification of a protein band with molecular weight around 51,000. The intensity of this protein band was reduced by treating rats with $CoCl_2$, suggesting this band was due to a hemoprotein (Tu, et al., 1983, Tu and Yang, 1983). These results are all consistent with our hypothesis. We therefore initiated the work on the purification of this low K_m NDMA demethylase from acetone- and ethanol-induced microsomes.

In 1982, Koop and Coon purified P450LM$_{3a}$ from ethanol-treated rabbits (Koop, et al., 1982). This enzyme catalyzed the oxidation of alcohols and appeared to have many of the properties of the enzyme we intended to purify. We therefore initiated a collaboration and obtained this enzyme together with purified rabbit P450LM$_2$, LM$_{3b}$, LM$_{3c}$, LM$_4$, and LM$_6$ to conduct metabolic studies to search for the low K_m NDMA demethylase. It was demonstrated that at a low substrate concentration (0.1 mM NDMA), LM$_{3a}$ (2E1) was the only enzyme displaying substantial activity. With 100 μM NADMA, however, other P450s such as LM$_2$, LM$_4$ and LM$_6$, also displayed NDMA demethylase activity (Yang, et al., 1985). This result suggested that P450 2E1 was a low K_m form of NDMA demethylase. However, the K_m observed with LM_{3a} in the reconstituted system, 2.9 mM, was much higher than the 60–70 µM observed in rat liver microsomes (Yang, *et al.*, 1985).

Purification and Characterization of P450 2E1 from Acetone-Induced Rat Liver Microsomes

With a combination of various chromatographic techniques, P450 2E1 (previously referred to as P450ac) was purified to apparent homogeneity from acetone-induced rat liver microsomes at a 0.6% yield (Patten, *et al.*, 1986). It occurred in the ferric form and a portion of this cytochrome was at the high spin state; this fraction increased with the temperature. In the ferrous form, this enzyme showed typical P450 CO-binding spectrum with a maximum at 450.6 nm. Upon reconstitution with NADPH-P450 oxidoreductase, phospholipid, and cytochrome b_5, a turnover number of 28.5 was observed in the demethylation of NDMA, consistent with the theoretical value estimated from the microsomal activity. However, this system displayed a K_m of 0.35 mM, much higher than the microsomal value; without b_5 in the reconstitution, the K_m value was even (8-fold) higher (Patten, *et al.*, 1986). At that time, Ryan *et al.* purified P450j from isoniazid-induced rat liver microsomes (Ryan, *et al.*, 1984). Based on the partial sequence and studies with antibodies, it was established that rat P450j and P450ac are identical and are homologous to the rabbit LM_{3a}; all are referred to as 2E1 in the systematic nomenclature. The human P450 2E1 was subsequently characterized at the gene and protein levels (Song, *et al.*, 1986, Wrighton, *et al.*, 1987), and its was found to be similar to rat P450 2E1 in many aspects.

P450 2E1 efficiently catalyzes the oxidation of aniline, a commonly used substrate (a type II substrate) of P450, and many other small molecules such as alcohols, acetone, hexane, and benzene (Yang, *et al.*, 1990). These activities were inhibited competitively by many commonly used solvents such as ethanol, acetone, dimethyl sulfoxide, hexane, and dimethyl formamide (Yoo, *et al.*, 1987). However, it is not efficient in catalyzing the oxidation of large or charged molecules such as benzo(a)pyrene, testosterone, dimethylamine, or benzphetamine (Patten, *et al.*, 1986, Yang, *et al.*, 1990). Some of the other substrates of P450 2E1 will be discussed in a later section.

Characterization of 2E1 Gene and Transcriptional Regulation

In collaboration with Gonzalez, Gelboin, and coworkers, the cDNA of 2E1 was cloned from the rat and human liver (Song, *et al.,* 1986). The deduced amino acid sequences of the rat and human 2E1 both contained 493 amino acids with calculated molecular weights of 56,634 and 56,916, respectively. Human 2E1 shared 75% nucleotide and 78% amino acid similarities to rat 2E1. Amino acid alignment revealed that 2E1 was 48% similar to 2B1 and 2B2 and 54% similar to 2C6 and 2C7, but had lower similarities to other P450s. Southern blot analyses of rat and human genomic DNAs verified that only a single gene shared extensive homology with 2E1. The human 2E1 gene was mapped to chromosome 10 and spanned 11,413 base pairs with nine exons (Umeno, *et al.,* 1988).

Transcriptional activation of P450 2E1 was demonstrated in rats right after birth (Song, *et al.,* 1986). Nuclear run-on experiments demonstrated that transcription of 2E1 was non-detectable in the livers of new-born rats, but became detectable at day 3 and reached normal level at day 7. Specific cytosine demethylation at the upstream regulatory region was suggested to be related to this activation (Umeno, *et al.,* 1988). Recently, it was reported that hepatocyte-enriched transcription factor-1 α can bind to 2E1 gene between -113 bp to -87 bp and probably controls the transcription of this gene. This transcription factor is fully expressed by day 3 after birth which is coincident to the activation of the 2E1 gene in rats (Gonzalez, 1992).

Posttranscriptional Regulation of 2E1

Transcriptional activation, however, has not been convincingly demonstrated in other conditions which "induce" P450 2E1. In the induction by fasting and diabetes, elevated levels of 2E1 mRNA were observed, suggesting the rate of mRNA degradation may have been retarded under these conditions (Hong, *et al.,* 1987, Song, *et al.,* 1987). In the induction by acetone, isopropanol, ethanol and pyridine, 2E1 mRNA elevation was not demonstrated in most experiments (Hong, *et al.,* 1987, Kim, *et al.,* 1990). The elevated 2E1 protein levels is probably due to enhanced translational activity (Kim, *et al.,* 1990) or increased stability of the 2E1 protein by the binding of these compounds that are also substrates of 2E1 (Song, *et al.,* 1989). It was proposed recently that the binding of substrates to 2E1 switch its degradation from an endoplasmic reticulum proteolytic pathway (with $t_{1/2}$ of 7 h) to a lysosomal proteolytic pathway (with $t_{1/2}$ of 37 h) (Eliasson, *et al.,* 1992). This interesting postulation, however, needs to be further substantiated. Studies on these posttranscriptional mechanisms are hampered by the fact that the extent of induction is generally from 3- to 5-fold, which provides a rather narrow window for observation. Thus definitive conclusions may not be obtained easily.

In rats, hypophysectomy caused a 2-fold increase in hepatic P450 2E1 mRNA and protein levels but no increase in NDMA demethylase activity (Hong, *et al.*, 1990). Under these conditions, the microsomal NADPH-P450 oxidoreductase activity was decreased by 50%, which may explain why the NDMA demethylase activity was not increased. An involvement of pituitary hormones in the regulation of mouse liver microsomal P450 2E1, however, was not observed (Hong, *et al.*, 1990).

The renal P450 2E1 was regulated by testosterone at the posttranscriptional level in mice but not in rats (Hong, *et al.*, 1989). In male mice, P450 2E1 was mainly located in the cortical tubules and low levels were also found in the outer medulla of the kidney. In female mice, the renal P450 2E1 level was very low or nondetectable by immunohistochemistry (Hu, *et al.*, 1990). This sex-related difference in the abundance of renal P450 2E1 is most likely responsible for the much higher renal toxicity or carcinogenicity exhibited in male mice by chloroform, acetaminophen, and NDMA (Pohl, *et al.*, 1984, Hu, *et al.*, 1993).

Roles of P450 2E1 in the Metabolism of Nitrosamines

The important role of P450 2E1 in the metabolism of NDMA is demonstrated by several lines of observations. Among all purified P450 enzymes, P450 2E1 shows the lowest K_m and highest K_{cat}/K_m in catalyzing the oxidation of NDMA (Yang, *et al.*, 1985). The high K_m value observed in the reconstituted system is probably due to the presence of glycerol and other inhibitors in the enzyme preparations. Recent work with heterologously expressed human P450 2E1 in HepG2 cells indicated that the P450 2E1-containing microsomes displayed a low K_m of 20 µM in the NDMA demethylase assay (Patten, *et al.*, 1992), confirming that P450 2E1 is responsible for the low K_m form of NDMA demethylase. In comparison to other P450 forms, P450 2E1 is also the form most active in catalyzing the activation NDMA to a mutagen for the V79 cells (Yoo and Yang, 1985). The predominant role of P450 2E1 in the metabolism of NDMA is suggested by the results that antibodies against P450 2E1 inhibited more than 80% of the NDMA demethylase activity in rat and human microsomes (Yoo, *et al.*, 1990).

P450 2E1 is also very effective in catalyzing the metabolism of *N*-nitrosodiethylamine.

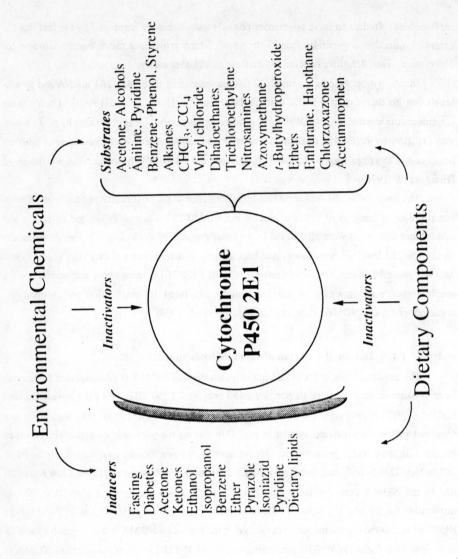

Antibodies against P450 2E1 inhibit the rate of *N*-nitrosodiethylamine deethylation in rat and human microsomes by 50 and 60%, respectively, suggesting that this enzyme also plays an important role in the activation of this carcinogen, but not as dominant as in the activation of NDMA (Yoo, *et al.,* 1990).

P450 2E1 is not responsible for the activation of all nitrosamines. For example, in the bioactivation of the tobacco carcinogen 4-(methylnitrosamino)-1-(3-pyridyl)-1-butanone (NNK), other enzymes are known to play more important roles than P450 2E1 (Smith, *et al.,* 1992, Smith, *et al.,* 1990). It seems that the structures of the alkyl groups are more important than the nitroso group in determining the enzymes specificity in the metabolism of dialkylnitrosamines. Our tentative conclusion is that P450 2E1 is more effective in oxidizing the α-carbon of methyl and ethyl groups whereas other P450 forms, such as P450 2B1, are more effective in catalyzing the dealkylation of larger hydrocarbon chains (Yang, *et al.,* 1993). However, the alkyl group selectivity is dependent on the concentration of the substrate used in the assay. When more sensitive methods become available for assaying the enzyme activity at lower substrate concentrations, this tentative conclusion may change.

P450 2E1 as a Target for Understanding Chemical Interactions

As shown in Figure 1, P450 2E1 is important for the biotransformation of a large number of low molecular weight, noncharged molecules, many of which industrial and environmental chemicals or drugs. Many of these substrates are also inducers of P450 2E1. In addition, this enzyme is induced by high fat diet, fasting, diabetes, and drugs such as isoniazid (Yang, *et al.,* 1992). These properties make P450 2E1 an important site for the following types of chemical interactions: (a) When present together, one substrate may serve as a competitive inhibitor for the metabolic activation of another substrate. This point can be illustrated by the inhibition of NDMA hepatocarcinogenicity by ethanol (Griciute, *et al.,* 1981), the inhibition of styrene oxidation by ethanol (Wilson, *et al.,* 1983) and the inhibition of benzene toxicity by toluene (Sato and Nakajima, 1979). (b) When given prior to the exposure of a toxic chemical, a P450 2E1 inducers may enhance the toxicity. This point is exemplified by the observations that ethanol consumption potentiated the toxicity of CCl_4 and acetaminophen (Krishnan and Brodeur, 1991, Hinson, 1980). Many of the interactions between volatile organic compounds as recently reviewed by Krishnan and Brodeur (1991) are believed to be due to these mechanisms. When rats were pretreated with a

P450 2E1 inducer, acetone or isopropanol, a potentiation of hepatotoxicity was observed only when NDMA was used at doses of 25 mg per kg body weight or higher (Lorr, *et al.*, 1984). The result suggests that at low concentrations of NDMA, the amount of P450 2E1 may not be the rate-limiting factor in NDMA metabolism.

P450 2E1 is inactivated by dietary chemicals such as diallyl sulfide and phenethyl isothiocyanate (Yang, *et al.*, 1992). These actions also point out the importance of this enzyme in mediating the effect of diet on chemical toxicity. Diallyl sulfide has been shown to protect against hepatotoxicity induced by NDMA and CCl_4 (Brady, *et al.*, 1991). The inhibition of acetaminophen toxicity by diallyl sulfide and its metabolite, diallyl sulfone, has been demonstrated in both rats and mice (unpublished results). This approach may be useful for the protection against acetaminophen-induced hepatotoxicity due to overdose or chronic alcohol consumption by human subjects. Although P450 2E1 has been found to activate most of its substrates to toxic or carcinogenic metabolites, it is known to play a detoxification role in the metabolism of 1,2-dihaloethanes and 1,2-dichloropropane. These compounds are known to be activated by a glutathione-dependent pathway (Guengerich, *et al.*, 1991, Igwe, *et al.*, 1986).

Roles of P450 2E1 in the Metabolism of Endogenous and Exogenous Compounds

P450 2E1 is among the most conserved forms in the *CYP2* family and the catalytic activities of P450 2E1 across species are quite similar, suggesting its possible physiological importance. Acetone, a ketone body, is metabolized by P450 2E1 to acetal and then to methylglyoxal (Koop and Casazza, 1985) which can be used for the synthesis of glucose (Landau and Brunengraber, 1987). It has been estimated that under physiological conditions, most of the acetone is metabolized via this oxidative pathway (Landau and Brunengraber, 1987). However, NADPH and ATP molecules are consumed to drive this pathway. One may hypothesize that, during fasting, P450 2E1 is induced for the acceleration of this pathway for making glucose from acetone, for critical physiological functions. However, the physiological significance of this gluconeogenic pathway remains to be determined. Another possible function of P450 2E1 is for the detoxification of acetone. However, the toxicity of the fasting-elevated acetone level is not known. Our recent experiments showed that treating rats with a daily dose of diallyl sulfide (200 mg/kg) elevated the plasma acetone level 4- to 9-fold. Apparent abnormality of the animals was not observed in an experimental period of 4 weeks. The thiobarbiturate reactive products were decreased by this treatment in the kidney and heart, but not in the liver, lung, and brain (unpublished results).

Among all the P450s studied, P450 2E1 is most active in catalyzing the NADPH-dependent formation of H_2O_2 and $\cdot O_2^-$ *in vitro* (Ekstrom and Ingelman-Sundberg, 1989, Gorsky, *et al.*, 1984, Persson, *et al.*, 1990). Antibodies to P450 2E1 almost completely inhibited the NADPH-dependent lipid peroxidation in microsomes (Ekstrom and Ingelman-Sundberg, 1989). It may be postulated that induction of P450 2E1 in the centrilobular region of the liver results in increased oxygen stress by generating higher levels of H_2O_2 and $\cdot O_2^-$. However, this hypothesis remains to be tested *in vivo*. Vaz *et al.* (1990) reported that P450 2E1 catalyzed the efficient reduction of 13-hydroperoxy-9,11-octadecadienoic acid by NADPH to pentane. This raises an interesting possibility that P450 2E1 is involved in both the formation and the reduction of lipid hydroperoxide. Pentane is an oxidative substrate for P450 2E1, showing a low K_m of 8 µM (Terelius and Ingelman-Sundberg, 1986). The possible physiological functions of these pathways remain to be investigated *in vivo*.

Concluding Remarks

The presently discussed molecular properties and catalytic activities of P450 2E1 demonstrate the important role of this enzyme in mediating chemical toxicity and chemical interactions. Because of the high similarity of rodent P450 2E1 to human P450 2E1, many of the interactions observed in animals may be extrapolated to humans. However, this point needs to be further substantiated in human studies. More precise kinetic parameters in the metabolism of different volatile compounds by human 2E1 are needed in order to have reliable parameters in the studies with physiologically based pharmacokinetic models. The physiological functions of P450 2E1 is also an interesting and challenging topic which needs further investigation.

Acknowledgement

This work was supported by NIH grants ES03938 and ES05693. We thank Ms. Dorothy Wong for excellent secretariat assistance.

References

Brady JF, Wang M-H, Hong J-Y, Xiao F, Li Y, Yoo J-SH, Ning SM, Fukuto JM, Gapac JM, Yang CS. (1991) Modulation of rat hepatic microsomal monooxygenase activities and cytotoxicity by diallyl sulfide. Toxicol Appl Pharmacol, 108: 342-354.

Ekstrom G, Ingelman-Sundberg M. (1989) Rat liver microsomal NADPH-supported oxidase activity and lipid peroxidation dependent on ethanol-inducible cytochrome P-450 (P-450IIE1). Biochem Pharmacol, 38: 1313-1319.

Eliasson EE, Mkrtchian S, Ingelman-Sundberg M. (1992) Hormone- and substrate-regulated intracellular degradation of cytochrome P450 (2E1) involving MgATP-activated rapid proteolysis in the endoplasmic reticulum membranes. J Biol Chem, 267: 15765-15769.

Gonzalez FJ. (1992) Control of constitutively-expressed developmentally-activated rat hepatic cytochromes P450 genes. Keio J Med, 41 (2): 68-75.

Gorsky LD, Koop DR, Coon MJ. (1984) On the stoichiometry of the oxidase and monooxygenase reactions catalyzed by liver microsomal cytochrome P-450. J Biol Chem, 259: 6812-6817.

Griciute L, Castegnaro M, Bereziat J-C. (1981) Influence of ethyl alcohol on carcinogenesis with N-nitrosodimethylamine. Cancer Lett, 13: 345-352.

Guengerich FP, Kim D-H, Iwasaki M. (1991) Role of human cytochrome P-450 IIE1 in the oxidation of many low molecular weight cancer suspects. Chem Res Toxicol, 4: 168-179.

Hong J-Y, Ning SM, Ma B-L, Lee M-J, Pan J, Yang CS. (1990) Roles of pituitary hormones in the regulation of hepatic cytochrome P450IIE1 in rats and mice. Arch Biochem Biophys, 281: 132-138.

Hong J-Y, Pan J, Gonzalez FJ, Gelboin HV, Yang CS. (1987) The induction of a specific form of cytochrome P-450 (P-450j) by fasting. Biochem Biophys Res Commun, 142: 1077-1083.

Hong J-Y, Pan J, Ning SM, Yang CS. (1989) Molecular basis for the sex-related difference in renal N-nitrosodimethylamine demethylase in C3H/HeJ mice. Cancer Res, 49: 2973-2979.

Hu JJ, Rhoten WB, Yang CS. (1990) Mouse renal cytochrome P450IIE1: immunocytochemical localization, sex-related difference and regulation by testosterone. Biochem Pharmacol, 40: 2597-2602.

Hu JJ, Lee MJ, Vapawala M, Reuhl K, Thomas PE, Yang, CS. (1993) Sex-related differences in mouse renal and toxicity of acetaminophen. Toxicol Appl Pharmacol, (in press).

Igwe OJ, Que Hee SS, Wagner WD. (1986) Interaction between 1,2-dichloroethane and tetraethylthiuram disulfide (disulfiram). II. Hepatotoxic manifestations with possible mechanism of action. Toxicol Appl Pharmacol, 86: 286-297.

Kim SG, Shehin SE, States JC, Novak, RF. (1990) Evidence for increased translational efficiency in the induction of P450IIE1 by solvents: Analysis of P450IIE1 mRNA polyribosomal distribution. Biochem Biophys Res Commun, 172: 767-774.

Koop DR. (1992) Oxidative and reductive metabolism by cytochrome P450 2E1. FASEB J., 6: 724-730.

Koop DR, Casazza JP. (1985) Identification of ethanol-inducible P-450 isozyme 3a as the acetone and acetol monooxygenase of rabbit microsomes. J Biol Chem, 260: 13607-13612.

Koop DR, Morgan ET, Tarr GE, Coon MJ. (1982) Purification and characterization of a unique isozyme of cytochrome P-450 from liver microsomes of ethanol-treated rabbits. J Biol Chem, 257: 8472-8480.

Krishnan K, Brodeur J. (1991) Toxicological consequences of combined exposure to environmental pollutants. Arch Complex Environ Studies, 3(3): 1-106.

Lai DY, Arcos JC. (1980) Minireview: dialkylnitrosamine bioactivation and carcinogenesis. Life Sci, 27: 2149-2165.

Landau BR, Brunengraber H. (1987) The role of acetone in the conversion of fat to carbohydrate. Trends Biol Sci, 12: 113-114.

Lorr NA, Miller KW, Chung HR, Yang CS. (1984) Potentiation of the hepatotoxicity of N-nitrosodimethylamine by fasting, diabetes, acetone, and isopropanol. Toxicol Appl Pharmacol, 73: 423-431.

Patten C, Aoyama T, Lee MJ, Ning SM, Gonzalez F, Yang CS. (1992) Catalytic activity of human cytochrome P-450 2E1 produced by the vaccinia virus expression system. Arch Biochem Biophys, 299: 163-171.

Patten CJ, Ning SM, Lu AYH, Yang CS. (1986) Acetone-inducible cytochrome P-450: purification, catalytic activity and interaction with cytochrome b_5. Arch Biochem Biophys, 251: 629-638.

Peng R, Tennant P, Lorr NA, Yang CS. (1983) Alterations of microsomal monooxygenase system and carcinogen metabolism by streptozotocin-induced diabetes in rats. Carcinogenesis (Lond.), 4: 703-708.

Peng R, Tu YY, Yang CS. (1982) The induction and competitive inhibition of a high affinity microsomal nitrosodimethylamine demethylase by ethanol. Carcinogenesis (Lond.), 3: 1457-1461.

Persson JO, Terelius Y, Ingelman-Sundberg M. (1990) Cytochrome P-450-dependent formation of reactive oxygen radicals: isozyme-specific inhibition of P-450-mediated reduction of oxygen and carbon tetrachloride. Xenobiotica, 20: 887-900.

Pohl LR, George JW, Satoh H. (1984) Strain and sex differences in chloroform-induced nephrotoxicity. Drug Metab Dispos, 12: 304-310.

Ryan DE, Iida S, Wood AW, Thomas PE, Lieber CS, Levin W. (1984) Characterization of three highly purified cytochromes P-450 from hepatic microsomes of adult male rats. J Biol Chem, 259: 1239-1250.

Sato A, Nakajima A. (1979) Dose dependent metabolic interaction between benzene and toluene *in vivo* and *in vitro*. Toxcol Appl Pharmacol, 48: 249-256.

Smith TJ, Guo Z, Gonzalez FJ, Guengerich FP, Stoner GD, Yang CS. (1992) Metabolism of 4-(methylnitrosamino)-1-(3-pyridyl)-1-butanone (NNK) in human lung and liver microsomes and cytochrome P-450 expressed in hepatoma cells. Cancer Res, 52: 1757-1763.

Smith TJ, Guo Z-Y, Thomas PE, Chung F-L, Morse MA, Eklind K, Yang CS. (1990) Metabolism of 4-(methylnitrosamino)-1-(3-pyridyl)-1-butanone in mouse lung microsomes and its inhibition by isothiocyanates. Cancer Res, 50: 6817-6822.

Song B-J, Gelboin HV, Park S-S, Yang CS, Gonzalez FJ. (1986) Complementary DNA and protein sequences of ethanol-inducible rat and human cytochrome P-450: transcriptional and post-transcriptional regulation of the rat enzyme. J Biol Chem, 261: 16689-16697.

Song BJ, Matsunaga T, Hardwick JP, Park SS, Veech RL, Yang CS, Gelboin HV, Gonzalez FJ. (1987) Stabilization of cytochrome P450j messenger ribonucleic acid in the diabetic rat. Mol Endocrinol, 1: 542-547.

Song BJ, Veech RL, Park SS, Gelboin HV, Gonzalez FJ. (1989) Induction of rat hepatic N-nitrosodiemthylamine demethylase by acetone is due to protein stabilization. J Biol Chem, 264: 3568-3672.

Terelius Y, Ingelman-Sundberg M. (1986) Metabolism of *n*-pentane by ethanol-inducible cytochrome P-450 on liver microsomes and reconstituted membranes. Eur J Biochem, 161: 303-308.

Tu YY, Peng R, Chang Z-F, Yang CS. (1983) Induction of a high affinity nitrosamine demethylase in rat liver microsomes by acetone and isopropanol. Chem Biol Interact, 44: 247-260.

Tu YY, Sonnenberg J, Lewis KF, Yang CS. (1981) Pyrazole-induced cytochrome P-450 in rat liver microsomes: an isozyme with high affinity for dimethylnitrosamine. Biochem Biophys Res Commun, 103: 905-912.

Tu YY, Yang CS. (1983) A High-affinity nitrosamine dealkylase system in rat liver microsomes and its induction by fasting. Cancer Res, 43: 623-629.

Umeno M, McBride OW, Yang CS, Gelboin HV, Gonzalez FJ. (1988) Human ethanol-inducible

P450IIE1: complete gene sequence, promoter characterization, chromosome mapping, and cDNA-directed expression. Biochemistry, 27: 9006-9013.

Vaz ADN, Roberts ES, Coon MJ. (1990) Reductive α-scission of the hydroperoxides of fatty acids and xenobiotics: Role of alcohol-inducible cytochrome P-450. Proc Natl Acad Sci USA, 87: 5499-5503.

Wilson HK, Robertson SM, Waldron HM, Grompertz D. (1983) Effect of alcohol on the kinetics of mandelic acid excretion in volunteers exposed to styrene vapour. Br J Ind Med, 40: 75-80.

Wrighton SA, Thomas PE, Ryan DE, Levin W. (1987) Purification and characterization of ethanol-inducible human hepatic cytochrome P-450HLj. Arch Biochem Biophy, 258: 292-297.

Yang CS, Brady JF, Hong J-Y. (1992) Dietary effects on cytochromes P-450, xenobiotics metabolism, and toxicity. FASEB J, 6: 737-744.

Yang CS, Smith TJ, Hong J-Y, Zhou S. (1993) Kinetics and enzymes involved in the metabolism of nitrosamines. In: The Chemistry and Biochemistry of Nitrosamines, and Ohter N-Nitroso Compounds. (invited paper), ACS Symposium Series, Washington, D. C., (in press).

Yang CS, Tu YY, Koop DR, Coon MJ. (1985) Metabolism of nitrosamines by purified rabbit liver cytochrome P-450 isozymes. Cancer Res, 45: 1140-1145.

Yang CS, Yoo J-SH, Ishizaki H, Hong J-Y. (1990) Cytochrome P450IIE1: Roles in nitrosamine metabolism and mechanisms of regulation. Drug Metab Rev, 22: 147-160.

Yoo J-SH, Cheung RJ, Patten CJ, Wade D, Yang CS. (1987) Nature of N-nitrosodimethylamine demethylase and its inhibitors. Cancer Res, 47: 3378-3383.

Yoo J-SH, Ishizaki H, Yang CS. (1990) Roles of cytochrome P450IIE1 in the dealkylation and denitrosation of N-nitrosodimethylamine and N-nitrosodiethylamine in rat liver microsomes. Carcinogenesis (Lond.), 11: 2239-2243.

Yoo J-SH, Yang CS. (1985) Enzyme specificity in the metabolic activation of N-nitrosodimethylamine to a mutagen for Chinese hamster V79 cells. Cancer Res, 45: 5569-5574.

Problems Associated with Assessment of the Contribution of Individual Forms of Cytochrome P450 to the Metabolism of Xenobiotics

Richard M. Philpot and Lila Overby
Laboratory of Cellular and Molecular Pharmacology
National Institute of Environmental Health Sciences
Research Triangle Park, NC, U. S. A. 27709.

Introduction

One of the broad aims of research in "drug metabolism" is to understand individual drug-metabolizing enzymes well enough to be able to account for the totality of the *in vivo* biotransformation of xenobiotics. If that were possible, differences in metabolism could be related directly to variations in enzyme profiles and, more important, enzyme profiles could be used to predict metabolism. The implications of this knowledge with respect to drug disposition, drug interactions, and formation of reactive products are manifest. A multitude of so-called "phase I" and "phase II" enzymes participate in this process of drug metabolism, the most important arguably being the many isozymes of cytochrome P450.

Over the past ten to fifteen years our ability to define individual forms of cytochrome P450 and to delineate their contributions to the overall activities of crude subcellular preparations -- the start of predicting *in vivo* consequences -- has improved markedly. Certainly, the seminal discovery (realization?) in this evolution of knowledge was of the existence of more than one form of the enzyme. Although this may seem to be a bit trivial today, some of us recall when cytochrome P450 meetings were occupied almost entirely with arguments regarding this point.

Development of a concept of drug metabolism that included multiple forms of cytochrome P450 was somewhat coincident with the genesis of P450 purification, the first of three major technical breakthroughs that account for our present level of expertise. Practical, reproducible procedures for purification of various

NATO ASI Series, Vol. H 90
Molecular Aspects of Oxidative Drug Metabolizing Enzymes
Edited by E. Arınç, J. B. Schenkman and E. Hodgson
© Springer-Verlag Berlin Heidelberg 1995

forms of cytochrome P450 were made possible be a number of methodological advances, most of which hinge on two results that were achieved over twenty years ago. The first, reported in 1968, was the separation of the monooxygenase components on DEAE-cellulose by Anthony Lu while he was in Jud Coon's lab (Lu and Coon, 1968) and the second was the introduction of nonionic detergent to the purification procedure by Imai and Sato (1974). The best of what followed, although important, would not have been possible without these two research milestones.

The advent of reproducible purification procedures greatly facilitated the development of the two areas that have provided the basis for the second and third major advances in the study of drug metabolism -- immunochemistry and molecular biology. Immunochemistry has provided a means for quantitation of individual forms of cytochrome P450 in mixed populations, allowed for the localization of cytochrome P450 within tissues and cells, and made it possible to inhibit reactions catalyzed by specific forms of cytochrome P450 in crude preparations. Molecular biology has given us cytochrome P450 primary structures, insight into regulation of expression, and methods for investigating structure/function relationships at the molecular level.

These three avenues to the study of drug metabolism -- purification, immunochemisty and molecular biology -- theoretically provide the means to assess precisely the contributions of various forms of cytochrome P450 to xenobiotic metabolism. This is not necessarily a straightforward exercise, but with diligence and care it can be accomplished. To achieve this goal a number of items must be examined in detail: the number of cytochrome P450 isozymes involved in a given metabolic pathway; individual substrate specificities, specific activities, and kinetic properties; the extent to which the specific activities are realized *in vivo*; and the concentrations of the individual forms in the relevant tissues and cell types. With respect to these factors, problems encountered in determinations of substrate specificities and specific activities, verification of specific activities and realization of metabolic capacity in crude preparations, quantitation, and localization will be discussed.

Substrate Specificities

The primary source of information about substrate specificities of various forms of cytochrome P450 comes from characterization of purified preparations. Although results obtained in this manner have gained general acceptance, evaluation of interference from small amounts of contaminating isozymes is difficult and, in any case, seldom done in detail. For example, with P450 1A1 being as little as 1% of a preparation, metabolism of benzo(a)pyrene and 7-ethoxyresorufin is guaranteed. Contamination at the level of 1% cannot be detected by protein staining procedures and must be evaluated by immunoblotting. We have experienced this problem with the purification of cytochrome P450 4B1 from rabbit lung (Domin and Philpot, 1986). Prior to the advent of immunochemical methods, we reported that this form of cytochrome P450 was active in the metabolism of benzo(a)pyrene and was highly sensitive to inhibition by a-naphthoflavone (Wolf *et al.*, 1979). Subsequently, we were able to demonstrate that this activity was actually catalyzed by very low levels of cytochrome P450 1A1 present in the preparations, amounts that were not detectable by "silver-staining" but were easily discerned by immunoblotting (Table 1). We have examined over twenty preparations of purified cytochrome P450 from a number of laboratories with only a small armament of antibodies (those to P450s 1A1, 1A2, 2B4, and 4B1) and found the majority of them to contain more than one form of the enzyme with contaminants accounting for as

Table 1.   Catalysis of benzpyrene hydroxylation and 7-ethoxyresorufin O-deethylation by cytochrome P450 1A1 contamination of two purified (> 18nmol P450/mg protein) preparations of cytochrome P450 4B1 from rabbit lung[a].

| | | Rate of Metabolism (nmol product/min) | | | |
| | | Benzpyrene Hydroxylation | | 7-Ethoxyresorufin Deethylation | |
Preparation	Percent P450 1A1	/total P450	/P450 1A1	/total P450	/P450 1A1
Pure P450 1A1	100	3.7	3,7	2.1	2.1
Pure P450 4B1	0.6	0.02	3.1	0.01	2.1
Pure P450 4B1	6.3	0.2	3.5	0.2	2.4

[a]Data taken from Domin and Philpot (1968).

much as 15% of the total cytochrome P450. In addition, every preparation of cytochrome P450 reductase we have examined contains more than one form of P450, contamination that is particularly evident with preparations from animals treated with phenobarbital. Certainly, the contamination involved is usually minor in amount, but it should be remembered that virtually every form of cytochrome P450 catalyzes the metabolism of one or more substrates at rates significantly greater than observed with any other form of the enzyme.

Given that "apparently homogeneous" is, more often than not, less than rigidly verified, the development and application of expression systems for the study of cytochrome P450-mediated metabolism is of substantial importance in establishing substrate specificities that unequivocally reflect the activities of single forms of the enzyme. With these systems, questions regarding intrinsic substrate specificities of cytochromes P450 can now be addressed without fear of the results being confounded by the presence of multiple forms of the enzyme. Although the ability to express and characterize various forms of cytochrome P450 has simplified many problems, it has also added to the list of complications. In the rabbit, for example, microheterogenous forms of cytochrome P450 2B are expressed in in complex patterns in liver, lung, and kidney (Ryan *et al.*, 1993). These forms, which are all greater than 97% identical (Table 2) have not been purified individually and cannot be recognized at the protein level; their distribution can only be assessed at the level of mRNA. Therefore, without mRNA phenotyping, the status of individual animals with respect to cytochrome P450 2B variants is not discernable. This has implications with respect to assessment of the metabolism of some steroids such as androstenedione (Kedzie *et al.*, 1991). One form of cytochrome P450 2B (2B-B2) exhibits a 16a to 16ß hydroxylation ratio that is over two orders of magnitude greater than the ratio observed for the other three forms, and a 16a to 7-ethoxycoumarin O-deethylation ratio that differs by greater than three orders of magnitude (Table 3). Relating these differences to metabolism in microsomal fractions or the intact animal is a difficult proposition.

Characterizing the substrate specificity of a P450 encompasses more than assessing the metabolism of saturating concentrations of various substrates. Highest observed velocities are easily measured but are not necessarily useful in the absence of Km determinations. Unless the information is relevant to expected physiological concentrations of various drugs and other xenobiotics it is not particularly applicable to *in vivo* situations.

Verification of Substrate Specificities

Once the substrate specificity of a given cytochrome P450 has been established (Vmax and Km included), it is appropriate to demonstrate that the intact monooxygenase system behaves in a manner similar to that of the reconstituted

Table 2. Differences among the primary structures of the B0, Bx, B1 and B2 forms of rabbit cytochrome P450 2B.[a]

RESIDUE NUMBER	ENZYME FORM			
	2B-B0	2B-Bx	2B-B1	2B-B2
35	ser	ser	**pro**	ser
39	val	**ile**	val	val
57	arg	**gln**	arg	arg
114	ile	ile	**phe**	ile
120	arg	arg	**his**	arg
174	ile	**val**	**val**	ile
221	pro	pro	pro	**ser**
246	gly	gly	gly	**thr**
248	ser	ser	**thr**	ser
286	gln	gln	**arg**	gln
290	leu	**ile**	leu	leu
294	ser	ser	**thr**	ser
314	met	**leu**	met	met
363	ile	ile	**val**	ile
367	val	val	**ala**	val
370	thr	thr	**met**	thr
403	glu	glu	glu	**lys**
417	asn	asn	**asp**	asn
420	leu	**met**	leu	leu
435	ile	ile	ile	**val**

[a] The amino acid differences (noted in bold type) are listed with cytochrome P450 2B-B0 as the reference sequence. The complete sequence of cytochrome P450 2B-B2 is found in Gasser et al.(1988).

or expressed system. Is turnover limited by the P450 reductase concentration or some other factor? With appropriate antibodies and specific chemical inhibitors, the contribution of a given form of cytochrome P450 to a given microsomal activity can be assessed. In many cases the specific activity of the P450 in a micro-

Table 3. Hydroxylation of androstenedione and deethylation of 7-ethoxy-ethoxycoumarin (ECOD) catalyzed by cytochrome P450 2B forms B0, B1, B2, and Bx expressed in COS Cells.[a]

	Rate (pmol product x 30 min^{-1} x mg $protein^{-1}$)					
	Androstenedione Hydroxylation					
cDNA	16b	16a	15a	16a:16b	ECOD	ECOD:16a
P450 2B-B0	15	69	b	0.22	4260	0.004
P450 2B-B1	11	84	b	0.13	1650	0.007
P450 2B-Bx	25	107	b	0.23	2970	0.008
P450 2B-B2	1950	51	576	38.2	45	43.3

[a]Results represent the average of 2 to 4 assays, data from Ryan et al. (1993).
[b]No activity was detected.

somal preparation approaches that of the purified or expressed enzyme. In other cases, however, the microsomal activity is significantly less. We have found that activities catalyzed by cytochrome P450 1A1 in lungs of rabbits treated with TCDD are only a fraction of that predicted on the basis of the activity of the purified enzyme and the amount of enzyme present as determined by immunoblotting (Figure 1). In addition, the extent to which the predicted activity is approached is substrate dependent. On average, TCDD increases rabbit pulmonary cytochrome P450 1A1 by about 20-fold, increases cytochrome P450 1A1-dependent 7-ethoxyresorufin (7-ERF) O-deethylation activity by about 10-fold, and increases 1A1-dependent benzo(a)pyrene (BP) hydroxylation activity by less than 2-fold (Domin and Philpot, 1968). Addition of purified cytochrome P450 reductase to the incubations serves to increase both 7-ERF and BP activities to the 20-fold level observed for the enzyme. The values obtained indicate that the activities of the purified isozyme accurately reflect the intrinsic catalytic capacity, and that less than predicted activity is, at least in this case, not a reflection of apocytochrome P450 1A1. The reasons for the lack of response of cytochrome P450 1A1 activity

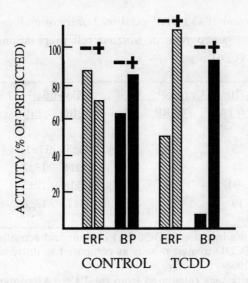

Figure 1. Cytochrome P450 1A1 catalyzed 7-ethoxyresorufin O-deethylation and benzpyrene hydroxylation activities in pulmonary microsomal preparations from control rabbits and rabbits treated with TCDD. Activities are expressed as a percent of predicted based on quantitation of P450 1A1 content by immunoblotting and specific activities determined with the purified enzyme. Activities were determined without (-) and with (+) added cytochrome P450 reductase. These data are taken from Domin and Philpot (1968).

following induction of the enzyme are not known. Obviously, added reductase overcomes the deficiency, but the implications of this are not clear. The primary question to be resolved is whether the lack of activity is due to latent enzyme or to all enzyme turning over at less than maximum efficiency. Our work with isolated pulmonary cell preparation suggested that the reductase/P450 ratio *per se* has little or no bearing on the problem (Domin *et al,*, 1986). The same effects were observed in isolated Clara cells, alveolar type II cells and alveolar macrophages, even though the reductase/total P450 or reductase/P450 1A1 ratios vary significantly among these preparations (Table 4). Also, as seen with preparations from whole lung, added reductase increased activities to expected levels.

Table 4. Cytochrome P450 1A1 catalyzed microsomal metabolism of 7-ethoxyresorufin and benzpyrene in isolated cell preparations from lungs of rabbits treated with TCDD.

Cell Type	Times Control[a]			% Potential[b]		Reductase/[c]	
	P450 1A1	7-ERF	BP	7-ERF	BP	P450	1A1
Clara Cell	25	18	1.3	46±5	17±11	1.1	14
Type II	24	20	3.3	41±14	19±2	0.9	10
Macrophage	>90	66	3.0	41±18	17±5	7.5	10
Whole Lung	19	13	1.7	43±7	13±8	0.5	1.7

[a]These data report the increase in P450 1A1 content and activities in microsomal preparations from TCDD-treated rabbits as compared to untreated rabbits. Data from Domin et al., 1968.
[b]The potential activity was calculated from the P450 1A1 content as determined by immunoblotting and the specific activity of purified P450 1A1.
[c]The reductase to P450 ratios are units of reductase per mg protein divided by nmoles P450 (total or 1A1) per mg protein.

Quantitation of Cytochrome P450

The calculations referred to above (specific activities in microsomal preparations) depend on two things. First, some method of determining the amount of activity associated with a specific P450. Second, some means of assessing the concentration of the specific P450. The latter is normally done by an immunoblotting procedure because it avoids most problems associated with cross-reacting antibodies; the specific antigen is identified both by reactivity and mobility. Cross-reacting species of different mobilities can simply be ignored. The Western blotting procedure can be extremely precise, but cannot be assumed to be accurate without detailed analysis of the behavior of each antibody preparation and each standard.

Usually, investigators are interested in the concentrations of only one or a few isozymes. In such cases it is difficult to rationalize the findings on the basis of total cytochrome P450. and the content calculated will appear reasonable. However, in cases where multiple isozymes have been quantitated it has not been unusual for the sum to exceed significantly that obtained by spectral determination, in some cases by 3 to 5-fold. Such data has been interpreted as evidence of apocytochrome P450 -- some or most of the cytochrome P450 existing without bound heme. This is clearly a possible explanation for immunodetection exceeding spectral detection as the latter determination is dependent on the presence of heme while the former is not.

Our initial attempts to immunoquantitate rabbit lung cytochrome P450 as the sum of 1A1, 2B, and 4B1 gave results that were as much as 4-times the holocytochrome content. This "excess" was later accounted for entirely by the aberrant behavior of the 1A1 and 4B1 standards (Domin and Philpot, 1968). At the concentrations used for the determination standard curves, the extent of detection of 1A1 and 4B1 was found to be inversely related to the time of heating prior to electrophoresis (Table 5).

Table 5. The consequences of heat treatment on the immunodetection of P450 1A1 in microsomal and purified samples.[a]

Treatment (min)	P450 1A1 Detection (% Zero Time)		Apparent Microsomal Concentration
	Pure	Microsomal	
0	100	100	1.00
1	98	100	1.02
3	58	97	1.65
5	27	84	3.10
10	8	51	6.40

[a]These data taken from Domin and Philpot (1968).

As a consequence of this effect the microsomal concentration of cytochrome P450 1A1 in our laboratory is overestimated by 3-fold when standard conditions are used. This situation was even more pronounced for P450 4B1, which was overestimated by about 3-fold with only 3 minutes of heating. In contrast, no effect of heating was observed with samples of cytochrome P450 2B or cytochrome P450 reductase. The large effect on cytochrome P450 1A1 may explain why a lack of agreement between concentrations of cytochrome P450 determined spectrally and immunochemically seems to be most evident following induction with TCDD. It appears that the effect of heating on susceptible isozymes results from the formation of large molecular weight aggregates that do not enter the running gel during electrophoresis.

We have re-examined quantitation of cytochrome P450 isozymes in rabbit lung using revised immunoblotting procedures and the following results have been obtained: the sum of cytochromes P450 1A1, 2B, and 4B1 determined immunochemically is 86% of the spectral total in the microsomal fraction from untreated rabbits and 95% of the spectral total in the fraction from rabbits treated with TCDD (Table 6). These results indicate that the behavior of purified P450 standards in blotting procedures should be investigated, not assumed.

Table 6. Quantitation of cytochrome P450 isozymes in microsomal preparations from untreated and TCDD-treated rabbits by immunoblotting and spectral analysis.

| P450 Isozyme | P450 Isozyme Concentrations (pmol/mg protein) | | | |
| | Control Microsomes | | TCDD Microsomes | |
	Blotting	Spectral	Blotting	Spectral
1A1	5.7±2.7		125±50	
2B[a]	157±45		243±18	
4B1	78±7		119±82	
Total	241	280±100	487	510±100

[a]Several microheterogenous forms of cytochrome P450 2B are expressed in rabbit lung. The value reported is for the sum of all forms present. Immunoblotting was done by the methods of Towbin et al. (1979) as modified by Domin and Philpot (1968). Spectral analysis was done by the method of Omura and Sato (1964) as modified by Estabrook et al. (1972).

The notion that apocytochrome P450 accounts for limited activity following induction and excess immunoquantitation relative to spectral determinations is brought into question by these results. We have shown that in both cases, at least with out research on the cytochrome P450 system of rabbit lung, that all of the cytochrome P450 can be accounted for in terms of active, heme-containing enzyme. It should be noted that hypotheses invoking apocytochrome P450 have never been substantiated by direct measurements of heme-deficient P450. Indeed, no recovery of cytochrome P450 fractions devoid of heme have ever been represented as other than heme loss during purification, and most of those reports are likely due to faulty spectal analysis.. In addition to results suggesting the presence of apocytochrome P450 as an explanation for excessive immunoquantitation and deficient activities, those alluding to the reconstitution of cytochrome P450 by the addition of heme containing compounds to homogenates should be reconsidered. The evidence for successful reconstitution of apocytochrome P450 has been increased activity, It should be remembered that cytochrome P450-mediated activity, particularly following induction, is limited, and that the full potential of the holoenzyme is not expressed. Finally, under normal conditions no direct evidence for the existence of apocytochrome P450, exists, and arguments suggesting that this form of cytochrome P450 explains some apparently anomalous results should be questioned.

In addition to knowing what a cytochrome P450 isozyme does and how much of it there is, it is often necessary to know its precise location. Toxic responses, particularly in extrahepatic tissue, are most often associated with specific cell types, not with tissues. In the lung, for example, a number of xenobiotics selectively damage the nonciliated bronchiolar epithelial (Clara) cell, not the whole lung. This is particularly true in the case of 4-ipomeanol, a pulmonary toxin first described by Boyd et al (1980). With 4-ipomeanol, it is necessary to understand the specific activities, kinetics, and localization of the several cytochromes P450 involved in its activation in order to understand its tissue-selective response (Philpot and Smith, 1984). In such cases studies of whole tissue preparations may produce results of little consequence. Given the problems associated with the isolation of homogeneous cell preparations from various tissues, the method most often used for tissue localization studies is immunohistochemistry at the level of light microscopy. Positive results obtained with well-characterized antibodies can usually be taken as strong evidence for the presence of a given P450 (Serabji-Singh et al., 1980). However, localization at the light level is only marginally adequate with respect to positive identification of most cell types in tissues like lung. More precise information is available from ultrastructural studies (Serabjit-Singh et al., 1988; Overby et al., 1992), which yield data on intracellular localization as well as the opportunity to identify most cell types. A summary of some of our results on the localization of cytochromes P450 1A1, 2B, and 4B1 and cytochrome P450 reductase are shown in table 7.

Table 7. Localization of drug-metabolizing enzymes in lung by immunodetection of protein and *in situ* hybridization of mRNA[a].

	CYTOCHROME P-450						
Cell Type	1 A 1		2 B[a]		4 B 1		Reductase
	p m		p m		p m		p m
(Airway)							
Clara	+ +		+ +		+ n		+ +
Ciliated	+ -		+ +		+ n		+ -
Goblet	- -		+ -		+ n		- -
(Gas Exchange)							
Type 1	- -		- -		- n		- -
Type 2	+ +		+ +		+ n		+ +
(Blood Vessels)							
Endothelium	+ +		+ +		+ n		- -
Smooth muscle	- -		- -		- n		+ -
Collagen	- -		- -		- n		+ -
(Macrophages)							
Alveolar	+ +		+ +		+ n		+ +
Intervasculature	n n		n n		n n		n +
(Other)							
Fibroblasts	- -		- -		- -		- -
Mucous Gland	- -		- -		- -		- -
Immune Cells	- -		- -		- -		- -

[a]Immunodetection of protein is under the heading "p" and *in situ* hybridization of mRNA is under the heading "m". The designation "n" indicates that the experiment has not been carried out.

Although localization results provide valuable information   they do not demonstrate the presence of a an active P450 monooxygenase system.   For example, we have positively identified several cytochromes P450 in endothelial cells of the vasculature, both by ultrastuctural localization and *in situ* hybridization (Overby *et al.*, 1992).   However, we have been unable to demonstrate the presence of cytochrome P450 reductase in these cells.   In addition to tissue sections, we have examined endothelium from dissected pulmonary vasculature systems without positive results.   This is of some interest since marked induction of P450 1A1 by compounds like TCDD appears to take place in the endothelium.   Many investigators have concentrated on the putative relationship between induced cytochrome P450 1A1 and carcinogenesis, but perhaps there is some other "role" for the overexpression of this enzyme. We suggest the possibility that induced cytochrome P450 1A1 acts as a binding protein for some dietary xenobiotics.

Isolation of homogeneous populations of cells is another way to determine the distribution of various forms of cytochrome P450.   This has the advantage of providing preparations that can be used for a number of purposes, but does have several disadvantages.   In particular, procedures that employ proteases
may result in degradation of the enzymes being investigated.   Proteolysis of pulmonary cytochromes P450 explains why activities obtained with isolated Clara cells, which contain the highest concentrations of cytochrome P450 in the lung, have been reported in a number of studies to be less than activities obtained with microsomal preparations from whole lung.

Conclusion

Complete characterization of drug metabolizing enzyme is a formidable task. Many toxic responses, including a multitude of carcinogenic events are thought to be associated with enzyme involved in drug metabolism. While it seems clear that metabolism is a requirement in many of these instances, whether or not it is a determinant remains to be demonstrated. Only after careful determinations of substrate specificities, kinetics, localization, concentrations, and modulation by exogenous and endogenous factors will the answer to this question become clear.

References

Boyd MR (1980) Biochemical mechanisms in pulmonary toxicity of furan derivatives. Hodgson E, Bend JR, and Philpot RM (eds) *Reviews in Biochemical Toxicology* 2:71-102, Elsevier North Holland, New York

Domin BA and Philpot RM (1986) The effect of substrate on the expression of activity catalyzed by cytochrome P-450. Metabolism mediated by rabbit isozyme 6 in pulmonary microsomal and reconstituted monooxygenase systems. *Arch. Biochem. Biophys.* 246:128-142

Domin BA, Devereux TR and Philpot RM (1986) The cytochrome P-450 monooxygenase system of rabbit lung: enzyme components, activities, and induction in the nonciliated bronchiolar epithelial (Clara) cell, alveolar type II cell, and alveolar macrophage. *Mol. Pharmacol.* 30:296-303

Estabrook RW, Peterson J, Baron J, and Hildebrandt A (1972) The spectrophotometric measurement of turbid suspensions of cytochromes associated with drug metabolism. Chignell C (ed) *Methods of Pharmacology* 2:303-350, Appleton-Century-Crofts, New York

Gasser R, Negishi M and Philpot RM (1988) Primary sequences of multiple forms of cytochrome P-450 isozyme 2 derived from rabbit pulmonary and hepatic cDNAs. *Mol. Pharmacol.* 32:22-30

Kedzie KM, Philpot RM and Halpert JR (1991) Functional expression of mammalian cytochromes P450IIB in the yeast *Saccharomyces cerevisiae*. *Arch. Biochem. Biophys.* 291:176-186

Lu, AYH, and Coon, MJ (1968) Role of hemoprotein P-450 in w-hydroxylation in a soluble enzyme system from liver microsomes *J. Biol. Chem.* 243:1331-1332

Omura T and Sato R (1974) The carbon monoxide-binding pigment of liver microsomes *J. Biol. Chem.* 239:2370-2378

Philpot RM and Smith BR (1984) The role of cytochrome P-450 and related enzymes in the pulmonary and metabolism of xenobiotics. *Environ. Health Perspect.* 55:359-367

Ryan R, Grimm SW, Kedzie KM, Halpert JR and Philpot RM (1993) Expression and induction by phenobarbital of cytochrome P450 2B and P450 4B in rabbit, identification of P450 2B-Bx, and functional comparison of four highly related forms of P450 2B. *Arch. Biochem. Biophys.*, in press

Sato R, Satake H, and Imai Y (1973) Partial purification and some spectral properties of hepatic microsomal cytochrome P-450. *Drug Metab. Dispo.* 1:6-13

Serabjit-Singh CJ, Wolf CR, Philpot RM and Plopper C (1980) Cytochrome P-450: Localization in rabbit lung. *Science* 207:1469-1470, 1980.

Towbin H, Staehelin T,Gordon J (1979) Electrophoretic transfer of proteins from polyacrylamide gels to nitrocellulose sheets: procedure and some applications. *Proc. Natl. Acad. Sci. USA* 76:4350-4354

Wolf CR, Smith BR, Ball LM, Serabjit-Singh CJ, Bend JR, and Philpot RM (1979) The rabbit pulmonary monooxygenase system: Catalytic differences between two purified forms of cytochrome P-450 in the metabolism of benzo(a)pyrene. *J. Biol. Chem.* 254:3658-3663

Emergence of the Flavin-Containing Monooxygenase Gene Family: Molecular Biology, Structure, and Function

Richard M. Philpot, Emmanuel Atta-Asafo-Adjei, Kave Nikbakht, Vicki Burnett, and Michael P. Lawton,
Laboratory of Cellular and Molecular Pharmacology
National Institute of Environmental Health Sciences
Research Triangle Park, NC, U. S. A. 27709.

Introduction

The mammalian flavin-containing monooxygenases (FMOs) comprise an important family of "phase I" xenobiotic-metabolizing enzymes. Although not as catalytically or structurally diverse as the cytochrome P450 super-family, the FMOs do metabolize a variety of exogenous compounds, including numerous drugs, pesticides, and other xenobiotics. The FMOs catalyze the oxidation of nitrogen, sulfur, and phosphorous atoms, but cannot carryout the oxidation of carbon (Ziegler, 1980). Like cytochrome P450, the FMOs utilize molecular oxygen as a substrate, NADPH as a source of electrons, produce water as one product, and are localized primarily to the endoplasmic reticulum of cells in a variety of tissues.

FMO activity was described initially in 1972 and for nearly fifteen years was thought to be associated with a single enzyme. However, catalytic and kinetic data, some reported as early as 1977 (Devereux et al., 1977), provided the basis for the eventual purification of two immunochemically distinct forms of the enzyme, referred to as the "liver" and "lung" FMOs (Williams et al., 1984; Tynes et al., 1985). The two FMOs were found to differ with respect to their temperature sensitivities, responses to detergents and certain ions, pH optima, and substrate specificities. Cloning and sequencing of the cDNAs encoding these two FMOs demonstrated conclusively that they are products of distinct genes (Lawton et al., 1990). The "liver" and "lung" FMOs are both proteins of 535 amino acids, and there primary structures are 56% identical.

NATO ASI Series, Vol. H 90
Molecular Aspects of Oxidative Drug Metabolizing Enzymes
Edited by E. Arınç, J. B. Schenkman and E. Hodgson
© Springer-Verlag Berlin Heidelberg 1995

The properties of the "liver" and "lung" FMOs expressed in COS cells, *E. coli* and yeast were found to be the same as those of the native enzyme (Lawton *et al.*, 1991). These findings ruled out the possibility that differences between the membrane environments of liver and lung contribute to differences in the characteristics of the enzymes, and brought into question the conclusion of Cashman and his co-workers (Guan *et al.*, 1991) that the properties of the lung enzyme are somehow dictated by complexation with calreticulin.

Multiple Forms of the FMO

Identification of a Third FMO in Rabbits

Both systematic and accidental findings have contributed to these advances. We hypothesized the existence of additional FMO gene products on the basis of the protein purification work of Ozols (1989) and kinetic considerations. Several investigators has concluded that the non-linear kinetics observed with microsomal and some purified FMO preparations was evidence for substrate activation. However, when we obtained classical, linear kinetics were with the liver and lung FMOs expressed in COS cells, we concluded that the complex kinetics seen previously were a reflection of multiple enzymes, not substrate activation (Lawton *et al.*, 1991). In addition to predicting the existence of additional forms of the FMO, we rationalized that their structural identities would likely be consistent with the two forms already identified. As a consequence, probes and screening conditions suitable for the detection of cDNA clones encoding proteins of approximately 55% identity were developed.

Initially, this approach yielded a single clone that was recognized at low stringency by probes to both the "liver" and "lung" forms of the FMO, but did not hybridize to either at high stringency. This clone was isolated and the insert sequenced. The sequence was characterized by the two GxGxxG/A binding domains present in the sequences of the "liver" and "lung" FMOs and thought to be involved in the binding of FAD and NADPH. This cDNA was used as a probe and a second clone with an identical sequence was isolated from an independent library (Atta-Asafo-Adjei *et al.*, 1993).

The two full-length cDNA clones (2.2 kilobases) encoding the newly discovered FMO encode a polypeptide of 533 amino acids that contains two putative pyrophosphate binding domains characterized by GxGxxG/A sequences and a hydrophobic carboxy-terminus. These properties are also observed with the sequences of the "liver" and "lung" FMOs. The derived sequence of the new FMO is 52% and 57% identical to the sequences of rabbit "liver" and "lung" FMOs, respectively, and 55% identical to the sequence of "liver form 2" published by Ozols (1991). The cDNA encoding the third FMO hybridizes with two species of mRNA (2.6 kb and 5.4 kb) from liver or kidney, but does not detect any mRNA in samples from lung. Guinea pig, hamster, rat and mouse all express this form of FMO in liver, kidney and lung.

Nomenclature for Multiple Forms of the FMO

Comparison of the nucleotide and amino acid sequences of the three forms of the FMO showed that all belonged to the same gene family and that each represented a distinct gene subfamily. A systematic nomenclature based on these comparisons has been proposed as a replacement for the current variety of laboratory-specific classifications, none of which are consistent with current information. This nomenclature is adopted from the one in use for cytochrome P450, and is applied in the same manner (Nebert et al., 1989). A single gene family was recognized and given the designation "1", the gene subfamilies were designated "A, B, and C", and the one gene in each subfamily was designated "1." Thus, the original "liver" FMO becomes FMO 1A1, the "lung" FMO becomes FMO 1B1, and the newly described FMO becomes FMO 1C1.

The proposed nomenclature is not meant to convey any information regarding substrate specificity, enzyme function or mechanism of catalysis. Rather, it provides for the unambiguous identification of orthologous forms of FMOs regardless of the species or tissue. Based on analysis of genomic DNA, each species examined (rabbits, humans, rats, mice, guinea pigs, and hamsters) contains the same set of FMO genes, a factor that supports and greatly simplifies a nomenclature based on structure. Ziegler (1993) has proposed consideration of a nomenclature based on activity. To complete an analog nomenclature, however, would require that each form of the enzyme in each species be characterized functionally before it could be named. Also, it is important to understand that "FMO" defined as a gene family does not include any structurally unrelated enzymes operating by the same mechanism. In contrast, it would include enzymes related by structure, regardless of mechanism.

Characterization of FMO 1C1 expressed in *E. coli.*

Expression of FMO 1C1 in *E. coli* (strain JM109) was monitored by SDS-PAGE analysis of the 100,000xg particulate fraction. The Coomassie blue staining pattern of the fraction from transformed cells showed a significant band corresponding to the molecular weight of FMO 1C1 (~62,000 daltons) that was not seen with the control cells. The 100,000xg particulate fractions from JM109 cells expressing the 62 kd protein and from control cells were analyzed for bound FAD and the amount in the particulate fraction from the transformed cells was significantly higher than in control cells (675 vs 161 pmol/mg protein). Assuming that the difference in FAD content (514 pmol/mg protein) is accounted for by expression of the enzyme (1 mol FAD per mol of enzyme), the FMO 1C1 content averaged 3.2% of the membrane protein for three independent transformations.

Expressed FMO 1C1 showed no activity with methimazole, chlorpromazine, prochlorperazine, imipramine, N'N-dimethylaniline, cysteamine, trimethylamine, triethylamine, n-decylamine as substrates. In contrast, activity (oxidation of NADPH) was observed with n-nonylamine and n-octylamine.Increases in NADPH oxidation with n-octylamine and n-nonylamine (3mM each) were 6.5 and 3.8 nmol x min^{-1} x mg $protein^{-1}$, respectively. The turnover number, approximately 20 nmol product x min^{-1} x nmol FMO $1B1^{-1}$, was similar to that observed for purified FMO 1B1 with thiourea as substrate (21.6) as calculated from the data of Tynes *et al.* (1985).

FMO 1C1 mediated oxidation of NADPH in the presence of n-octyl- amine was examined for a possible uncoupling reaction as well as for substrate metabolism by looking at the formation of H_2O_2 and hydroxylamine equivalents. Expressed FMO 1B1, which is known to metabolize n-octylamine (Tynes *et al.*, 1986), was examined in the same manner. With FMO 1B1, uncoupling and hydroxylamine formation each accounted for approximately 40% of the NADPH oxidized (12.8 nmoles NADPH oxidized x min^{-1} x nmole FMO $1B1^{-1}$) with the remaining 20%

likely due to further metabolism or breakdown of the N-hydroxy metabolite. In contrast, increased rates of formation of H_2O_2 and hydroxylamine equivalents were not observed in association with n-octylamine-induced oxidation of NADPH catalyzed by FMO 1C1. The lack of H_2O_2 formation ruled out an uncoupling reaction (added H_2O_2 could be recovered quantitatively from standard incubations containing expressed FMO 1C1) and indicated that metabolism of n-octylamine must have proceeded to the oxime.

Metabolism of n-octylamine catalyzed by expressed FMO 1C1 was further characterized with respect to a number of parameters. The pH optimum for the reaction was ~9.0 (Fig. 9A), activity was inhibited in a time-dependent manner by 100 mM $MgCl_2$ and 1% Na cholate, and the enzyme was heat labile; complete loss of activity was observed with treatment at 45°C for 5 min.. The Km and Vmax for the reaction, as determined from results obtained with concentrations of n-octylamine between 0.2 and 5mM, were ~1.5 mM and ~11 nmol NADPH oxidized x min^{-1} x protein^{-1}.

Identification of FMO 1D1 and FMO 1E1

Detection of clones encoding two additional forms of the FMO, 1D1 and 1E1, was done with a refinement of the low stringency screening technique used for isolation of cDNA encoding FMO 1C1. This modification took advantage of the fact that the cDNAs encoding FMO 1A1, 1B1 and 1C1 contain an area of relatively high identity over the 5' half of the coding region. Rabbit genomic DNA was probed under low stringency conditions, with a mixture of 5' cDNA fragments of the three cDNAs. Bands associated specifically with FMO 1A1, 1B1 or 1C1 were resolved by analysis at high stringency with individual probes. Several bands were detected that could not be assigned to the known forms of the enzyme. The behavior of the 5' probes at low vs high stringency was used to facilitate the isolation of cDNAs corresponding to the unknown DNA bands. A cDNA library was constructed from rabbit liver mRNA, and screened under low stringency hybridization conditions (37°C, 50% formamide, 1 x SSC, 0.1% SDS) with the probe mixture. Of the 157 clones detected, 95 were identified as FMO 1A1 and 22 as 1C1 by hybridizatiion under high stringency conditions (65°C, 50% formamide, 0.1 x SSC, 0.1% SDS). Of the 40 remaining clones, 36 were characterized by sequence analysis as encoding FMO 1D1, previously identified at the protein level by Ozols (1991) as a second rabbit liver FMO. Four clones were shown to encode an FMO (FMO 1E1) belonging to a fifth subfamily (Burnett and Philpot, personal communication). No clones encoding FMO 1B1 were isolated from the liver library. This is entirely consistent with the lack of FMO 1B1 antibody reactivity with the hepatic microsomal preparation and the negative results obtained upon hybridization of cDNA encoding FMO 1B1 with hepatic mRNA samples.

FMO 1D1 clones 10 and 28 contain open reading frames of 1593 bases; clone 10 has 42 bases of 5′ flanking region and 947 bases of 3′ flanking region, and clone 28 has 30 bases of 5′ flanking region and 1304 bases of 3′ flanking region. The sequences of the two clones are identical prior to their divergence at a NotI-EcoRI site present in clone 28 at position 2080, but not present in clone 10. This discrepancy was resolved by application of the polymerase chain reaction. Rabbit liver mRNA was amplified with primers specific for the 3′ flanking regions of clone 28 or clone 10 located 3′ of base 2080 and a common primer located 5′ of base 2080. A specific band was amplified only when the primer specific for the 3′ end of clone 10 was used, indicating the the 3′ end of clone 28 is an extraneous segment of DNA tandemly ligated to the FMO cDNA.

FMO 1E1 clone 1 has 153 bases of 5′ flanking and 336 bases of 3′ flanking; clone 25 has 183 bases of 5′ flanking region and 341 bases of 3′ flanking. Each clone contains an open reading frame of 1665 bases identical except for position 1131. Clone 1 contains a T at this position and clone 25 contains an A, a difference that results in a change from phenylalanine to isoleucine. Subsequently, FMO 1E1 clones 30 and 42 were partially sequenced and both found to contain a T at position 1131. Clone 25 also differs from clones 1, 30, and 42 in the 5′ flanking region, with the insertion of 3 bases (AGC) beginning 13 bases to the 5′ of the initiation codon. In addition, a single base change, from a C to a T, is found in the 3′ flanking region 13 bases to the 3′ of the termination codon.

Characterization of FMO 1D1 and 1E1 expressed E. coli.

The coding region of the cDNA for FMO 1D1 was amplified by the polymerase chain reaction and ligated into the expression vector pJL-2 (called pJL1D1). The particulate fraction prepared from E. coli strain JM109 transformed with pJL1D1 was analyzed by Coomassie Blue staining following SDS-PAGE and a protein band of approximately 57 kD, not present in cells transformed with vector alone, was clearly visible. Functional properties of FMO 1D1 were examined with methimazole as the substrate. The specific activity of the pJL1D1 particulate fraction was approximately 25 nmol product x min^{-1} x mg^{-1} with 1 mM methimazole at pH 8.4. The responses FMO 1D1 to elevated temperature, sodium cholate, and $MgCl_2$, treatments used to distinguish between FMOs 1A1 and 1B1 were then determined. In each case, the response of FMO 1D1 to these treatments was similar to that of FMO 1A1. For example, the activity of FMO 1D1 decreased when it was heated for 5 min at 45°C, subjected to 1% sodium cholate or incubated with 100 mM $MgCl_2$. However, 3 mM n-octylamine, which increases the activity of FMO 1B1 and inhibits FMO 1A1, had no effect on FMO 1D1.

Following Eadie-Hofstee transformation, the kinetics of methimazole metabolism catalyzed by expressed FMO 1D1 were linear, with a K_m for methimazole of near 30mM (27 µM and 31 µM with two different transformations). Expressed FMO 1D1 was further characterized by monitoring methimazole metabolism in the presence of chlorpromazine, prochlorperazine, and imipramine. These substrates (100 mM) inhibited FMO 1D1-catalyzed metabolism of methimazole (30 mM) much less than metabolism catalyzed by FMO 1A1 or rabbit liver microsomes.

Full length FMO 1E1 cDNA was cloned into the yeast expression vector YEp53 at BamHI-SalI restriction sites and transformed into yeast strain 334. Coomassie blue staining of the proteins did not identify a band unique for FMO 1E1, and methimazole activity was not detected. FMO 1E1 cDNA was then cloned (bases 154 to 2124) into pKKHC *E. coli* expression vector at NcoI-PstI restriction sites and transformed into XL-1 blue strain bacteria. Again, neither FMO 1E1 protein nor methimazole activity could be detected. The FMO 1E1 cDNA was then altered by the polymerase chain reaction to encode a PstI restriction site immediately following the termination codon, and cloned (bases 154 to 1830) into pKKHC at NcoI-PstI restriction sites. The resultant pKK-1E1 was transformed into XL-1 blue or JM109 strain bacteria. No protein or methimazole activity was detected with microsomal preparations from either strain.

Structural Comparisons of FMOs

In the two years since the report that the "liver" and "lung" flavin-containing monooxygenases (FMO 1A1 and FMO 1B1) are products of distinct genes, three additional FMOs, each the product of a distinct gene, have been identified, and, to date, thirteen full-length sequences from cDNAs and three from purified proteins have been reported or made available through GenBank. A list of these sequences is given in Table 1 and a comparison of the structural identities of the derived amino sequences is shown in Table 2.

Table 1. Reported sequences of flavin-containing monooxygenases.

Name	Class	Species	Source	Ref[a].	GenBank
1A1	1A1	Rabbit	cDNA	1	M32030
Form 1	1A1	Rabbit	Protein	2	
Liver	1A1	Pig	cDNA	3	M32031
FMO1	1A1	Human	cDNA	4	M64082
Liver	1A1	Rat	cDNA	GenBank	M84719
Lung	1B1	Rabbit	cDNA	1	M32029
Lung	1B1	Guinea Pig	cDNA	5	L10037
1C1	1C1	Rabbit	cDNA	6	L08449
1D1	1D1	Rabbit	cDNA	7	L10391
Form 2	1D1	Rabbit	Protein	8	
FMOII	1D1	Human	cDNA	9	M83772
1E1	1E1	Rabbit	cDNA	7	L10392
FMO2	1E1	Human	cDNA	10	Z11737

[a]References: 1, Lawton *et al.* (1990); 2, Ozols (1990); 3, Gasser *et al.* (1990); 4, Dolphin *et al.* (1991); 5, Nikbakht *et al.* (1992); 6, Atta-Asafo-Adjei *et al.* (1993); 7, Burnett and Philpot, personal communication; 8, Ozols (1991); 9, Lomri *et al.* (1992); 10, Dolphin *et al.* (1992).

In all, these sequences represent five distinct structural classes (gene subfamilies) in rabbit, five orthologs from species (human, rat, pig, and guinea pig) other than rabbit, and two allelic variants. Results of detailed analysis of genomic DNA indicate that each gene subfamily contains a single gene and that one additional FMO gene product remains to be characterized. If it becomes clear

Table 2. Comparisons of primary structures of flavin-containing monooxygenases.

Species	Enzyme Form	Rabbit				
		FMO 1A1	FMO 1B1	FMO 1C	FMO 1D1	FMO 1E1
		(% Identities)				
Rabbit	FMO 1A1	-	55	52	54	53
Rabbit	1B1	-	-	57	56	56
Rabbit	1C1	-	-	-	55	52
Rabbit	1D1	-	-	-	-	55
Human[a]	1A1,1B1,1C1	86			83	84
Rat[a]	1A1	83				
Pig[a]	1A1	87				
Guinea Pig[a]	1B1		86			

[a]Comparisons of the ortholog sequences with sequences from any other subfamily yields values of between 52% and 57%.

that the FMO gene family is not a member of a gene superfamily (no proteins of ~30% identity are known at present), and that each subfamily does contain only a single gene, the nomenclature could be simplified to include only the subfamily designations, *i.e.* FMOs A, B, C, D, E, etc.

It is interesting that no other flavoproteins exhibit identities with the FMOs that are sufficiently high to suggest the existence of a flavoprotein gene superfamily. This is the case in spite of the fact that putative nucleotide cofactor pyrophosphate binding domains (GxGxxG/A) are present in a number of enzymes in precisely the same positions as in the FMOs. Examples are glutathione reductase, bovine adrenodoxin oxidoreductase, and p-hydroxybenzoate hydroxylase; flavoproteins that are only 19-20% identical to FMO 1B1. Even cyclohexanone monooxygenase, a bacterial enzyme with a broad substrate specificity and a mechanism similar to that of the mammalian FMOs, shares only 23% amino acid identity with FMO 1B1. These identities are not unlike those between FMO 1B1 and totally unrelated proteins.

The amino acid sequences of the rabbit FMOs are aligned in Figure 1 and a schematic representation of structural features common to all five enzymes is shown in Figure 2. In all cases consensus FAD- and NADPH-pyrophosphate-binding sequences are found in identical positions, and several hydrophobic areas are located similarly. All of the FMOs contain an extremely hydrophobic C-terminal sequence. Surprisingly, few highly conserved peptides are found. Although the overall amino acid sequence identity is only 29% (a value that does not take into account the carboxy-terminal extension present only in FMO 1E1), structural conservation of the proteins can be seen by the highly similar locations of hydrophobic and hydrophilic peptides and the locations of the putative FAD and NADPH binding domains. Conservation of structure is also shown by the minimal number of gaps needed for overall alignment: one after residue 2 in 1A1, one after residue 2 in 1B1, one after residue 342 in 1C1, one after residue 1 and two after residue 423 in 1D1 and one each after residues 2, 282, and 429 in 1E1. Also, the 1A1 sequence contains three consecutive residues at positions 317-319 (G, N and A) not present in the other four sequenceAlignment of the FMOs shows only 16 identical peptides of greater than two residues, 15 of which are tripetides. The area of highest identity is residues 1 to 42 (57%) followed by residues 101 to 149 (49%); the regions of lowest identity are residues 400 to 459 (5%) and residues 232 to 285 (11%). Alignment of the derived amino acid sequences of FMO 1A1 from the rat, rabbit, pig, and human demonstrates clearly why these can be classified as products of orthologous genes. For example, the first 95 residues of rabbit and human FMO 1A1 are identical, as are the first 69 residues of the rabbit and pig sequences. The four sequences disagree completely at only a single position, residue 117. Surprisingly, the GNA tripeptide found only in rabbit FMO 1A1 is not present in the FMO 1A1 orthologs from rat, mouse, and human.

Tissue and Species Comparisons

Although the tissue and species distribution of FMO 1A1 and 1B1 proteins has been examined by immunoblotting (Tynes and Philpot, 1987), antibodies to the other FMOs are not available. However, the distribution of mRNAs encoding all five FMOs has been examined in hepatic, renal, and pulmonary samples from rabbits, rats, guinea pig, hamster, and mouse. The distribution of mRNAs encoding FMOs 1A1 and 1B1 is consistent with the reported distribution of the

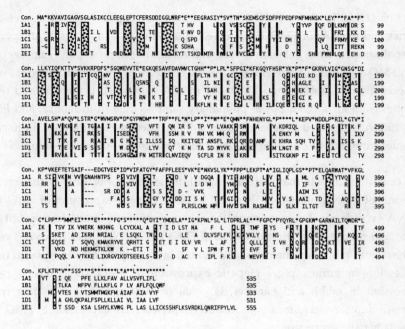

Figure 1. Alignment of the primary sequences of five forms of flavin-containing monooxygenase from rabbit. The cross-hatched boxes show areas of complete agreement.

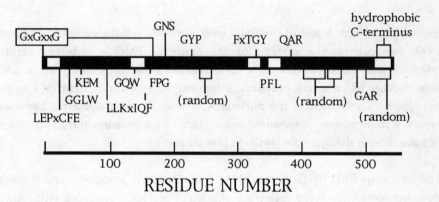

Figure 2. Structural features shared by the five forms of flavin-containing monooxygenase from rabbit. The clear areas show putative areas of membrane association, the GxGxxG sequence indicates the FAD and NADPH binding domains, and the underlined sequences are present in all forms of the enzyme.

proteins (29); mRNA for FMO 1A1 is detected in samples from all tissues and species and mRNA for FMO 1B1 is detected in samples from lungs of all species, in renal samples from hamster, mouse and rabbit, but in no hepatic samples. As with FMO 1A1, mRNA for FMO 1C1 is detected in all tissues and species, with the exception of rabbit lung. FMO 1D1 mRNA is detected in hepatic and renal samples from all species, but in pulmonary samples only from mouse, and FMO 1E1 mRNA is detected in renal samples from all species except mouse, in hepatic samples from guinea pig, hamster, and rabbit, but in no pulmonary samples.

It is important to note that there are significant species differences in the relative amounts of mRNA present in liver and kidney. Hepatic expression of FMO 1A1 mRNA exceeds renal expression in the rabbit, guinea pig, and hamster, but not in rat and mouse. Hepatic expression of FMO 1C1 mRNA greatly exceeds renal expression in rabbit, rat, hamster, and mouse, but is approximately equal to renal expression in guinea pig. Hepatic expression of FMO 1D1 exceeds renal expression in rabbit and hamster, is equal in mouse, and is less than renal expression in rat and guinea pig. In all five species, renal expression of FMO 1E1 exceeds hepatic expression.

Structure/Function Studies

Expression of high levels of the FMO isozymes in *E. coli* (Lawton and Philpot, 1993) has provided a system for the study of FMO structure/function relationships. Aided by site-directed mutagenesis, deletion mutagenesis, and construction of chimeric proteins, a number of structural features can be examined. These include the nucleotide cofactor binding domains, peptides involved in membrane association, and regions of the protein controlling access of substrates to the hydroperoxyflavin intermediate.

The consensus FAD (GxGxxG) and NADP (GxGxxG/A) pyrophosphate-binding domains were first noted following alignment of FMO sequences with other flavin-binding proteins. The importance of these domains to the function of the FMOs is beginning to be investigated. When the third glycine of the FAD-pyrophosphate binding site in FMO 1B1 was changed to valine, the enzyme could no longer oxygenate methimazole, dimethylaniline, thiourea or cysteamine. This deficiency was clearly due to a lack of FAD binding by the modified protein.

A second area investigated in the *E. coli* system has been membrane binding. It has been suggested that the C-terminal hydrophobic peptide serves to anchor the FMO to the endoplasmic reticulum, and we tested this hypothesis. The 26 amino acid hydrophobic tail was deleted from FMO 1B1, and the truncated protein expressed in *E. coli* and tested for membrane association. The truncated protein was detected in the membrane fraction rather than the cytosolic fraction. Sodium carbonate extraction of the membrane-bound truncated protein demonstrated that it behaved like an integral membrane protein. Functional characterization of the truncated protein showed that it was nearly identical to the unmodified enzyme. Additional deletion mutants (removal of 49, 74, 99, and 199 amino acids from the carboxyl terminus) were also expressed and found associated predominantly with the membrane fraction as integral and not peripheral membrane proteins. None of these deletion mutants were active. Preliminary results with two other deletion mutants, one with 30 residues removed and another with 40 residues removed, indicate that the first is active and the second is not, suggesting that an important determinant of enzymatic capability is located between amino acids 495 and 505.

Experiments with chimeric proteins also indicate that the carboxyl terminal region of the FMO is important in substrate association. Hybrids (chimeras) between FMOs 1A1 from pig and rabbit were constructed, expressed and characterized. These two forms of FMO 1A1 were selected because they have similar K_ms for methimazole, but significantly different K_ms for imipramine. The magnitude of the difference is such that imipramine (1 mM) completely inhibits metabolism of methimazole (200 mM) catalyzed by pig FMO 1A1, but has no effect on the same reaction catalyzed by rabbit FMO 1A1. All hybrids that contained 100 or more of the carboxy-terminal amino acids of pig FMO 1A1 exhibited the properties of pig FMO 1A1 -- methimazole activity completely inhibited by imipramine. Again, these results are consistent with the deletion mutagenesis experiments, in which removal of 40 or more amino acids from the carboxy terminus of rabbit 1B1 abolished its activity.

Future Directions

For years the flavin-containing monooxygenase was generally assumed to be a single enzyme. Now it is has been conclusively demonstrated that the FMO is a gene family with at least five members, each with unique tissue and species distribution. The recent and rapid transformation of this field, which has been made possible by by the techniques of molecular biology, suggests that the role of the FMO needs to be reevaluated in some detail. Expression systems will be particularly important in several areas of future work, including production of proteins for development of monospecific antibodies, determination of substrate specificities, and continued examination of structure/function relationships. Another important area of research will be elucidation of the mechanisms of tissue-specific and species-selective expression. This work will be particularly important in the determining which form of the enzyme is involved in trimethyaminuria, a genetic deficiency disease of humans thought to be involved with FMO function.

References

Atta-Asafo-Adjei E, Lawton MP, and Philpot RM (1993) Cloning, sequencing, distribution, and expression of a mammalian flavin-containing monooxygenase from a third gene subfamily. *J. Biol. Chem.* 268:9681-9689

Devereux TR, Philpot RM, and Fouts JR (1977) The effect of Hg2+ on rabbit hepatic and pulmonary solubilized, partially purified N,N-dimethylaniline N-oxidases. *Chem.-Biol. Interactions* 18:277-287

Dolphin CT, Shephard EA, Povey S, Palmer CNA, Ziegler DM, Ayesh R, Smith RL, and Phillips IR (1991). Cloning, primary sequence, and chromosomal mapping of a human flavin-containing monooxygenase (FMO1). *J. Biol. Chem.* 266:12379-12385

Dolphin CT, Shephard EA, Povey S., Smith RL, and Phillips IR (1992) Cloning, primary sequence and chromosomal localization of human FMO2, a new member of the flavin-containing mono-oxygenase family. *Biochem. J.* 287:261-267

Gasser, R., R. E. Tynes, M. P. Lawton, K. K. Korsmeyer, D. M. Ziegler, and R. M. Philpot. The flavin-containing monooxygenase expressed in pig liver: Primary sequence, distribution, and evidence for a single gene. *Biochemistry* 29:119-124 (1990).

Guan S, Falick AM, Williams DE, and Cashman JR (1991) Evidence for complex formation between rabbit lung flavin-containing monooxygenase and calreticulin. *Biochemistry* 30:9892-9900

Lawton MP, Gasser R, Tynes RE, Hodgson E, and Philpot RM (1990) The flavin-containing monooxygenase enzymes expressed in rabbit liver and lung are products of related, but distinctly different genes. *J. Biol. Chem.* 265:5855-5861

Lawton MP, Kronbach T, Johnson EF, and Philpot RM (1991) Properties of expressed and native flavin-containing monooxygenases: Evidence of multiple forms in rabbit liver and lung. *Mol. Pharmacol.* 40:692-698

Lomri N, Gu Q, and Cashman JR (1992) Molecular cloning of the flavin-containing monooxygenase (form II) cDNA from adult human liver. *Proc. Natl. Acad. Sci. USA* 89:1685-1689.

Nebert DW, Nelson DR, Adesnik M, Coon MJ, Estabrook RW, Gonzalez FJ, Guengerich FP, Gunsalus IC, Johnson EF, Kemper B, Levin W, Phillips I, Sato R and Waterman MR (1980) The P450 superfamily: Updated listing of all genes and recommended nomenclature for the chromosomal loci. *DNA* 8:1-13

Nikbakht, K., Lawton MP, and Philpot RM (1992) Guinea pig or rabbit lung flavin-containing monooxygenases with distinct mobilities in SDS-PAGE are allelic variants that differ at only two positions. *Pharmacogenetics* 2: 207-216

Ozols J (1989) Liver microsomes contain two distinct NADPH-monooxygenases with NH_2-terminal segments homologous to the flavin containing NADPH monooxygenase of *Pseudomonas fluorescens. Arch. Biochem. Biophys.* 163:49-55

Ozols J (1991) Multiple forms of liver microsomal flavin-containing monooxygenases: Complete covalent structure of Form 2. *Arch. Biochem. Biophys.* 290:103-115

Tynes RE, and Hodgson E (1985) Catalytic activity and substrate specificity of the flavin-containing monooxygenase in microsomal systems: Characterization of the hepatic, pulmonary and renal enzymes of the mouse, rabbit and rat. *Arch. Biochem. Biophys.* 240:77-93

Tynes, RE, and Philpot RM (1987) Tissue- and species-dependent expression of multiple forms of mammalian microsomal flavin-containing monooxygenase.*Mol. Pharmacol.* 31:569-574

Tynes RE, Sabourin PJ, and Hodgson E (1985) Identification of distinct hepatic and pulmonary forms of microsomal flavin-containing monooxygenase in mouse and rabbit. *Biochem. Biophys. Res. Commun.* 126:1069-1075

Tynes RE, Sabourin PJ, Hodgson E and Philpot RM (1986) Formation of hydrogen peroxide and N-hydroxylated amines catalyzed by pulmonary flavin-containing monooxygenases in the presence of primary alkylamines. *Arch. Biochem. Biophys.* 251:654-664

Williams DE, Ziegler DM, Nordin DJ, Hale SE, and Masters BSS (1984) Rabbit lung flavin-containing monooxygenase is immunochemically and catalytically distinct for the liver enzyme. *Biochem. Biophys. Res. Commun.* 125:116-122

Ziegler DM (1980) Microsomal Flavin-containing monooxygenase: Oxygenation of nucleophilic nitrogen and sulfur compounds. in *Enzymatic Basis of Detoxification* (W. B. Jakoby, ed.), Academic Press, New York 201-225.

Ziegler DM (1993) Recent studies on the structure and function of multisubstrate flavin-containing monooxygenases. *Annual Rev. Pharmacol. Toxicol.* 33:179-199

Flavin-Containing Monooxygenases: Substrate Specificity and Complex Metabolic Pathways

Ernest Hodgson, Bonnie L. Blake,
Patricia E. Levi, Richard B. Mailman[1],
Michael P. Lawton[2], Richard M. Philpot[2]
 and Mary Beth Genter
Department of Toxicology
Box 7633, North Carolina State University
Raleigh NC 27695 USA

Introduction

The flavin-containing monooxygenase (EC 1.14.13.8)(FMO) is located in the endoplasmic reticulum of mammalian cells and is involved in the monooxygenation of a wide variety of xenobiotics. The FMO has a similar distribution and function to many of the isozymes of cytochrome P450 (P450). Originally described as an amine oxidase (Ziegler and Mitchell, 1972), it is now known to catalyze the oxidation of many organic, and some inorganic, chemicals (Ziegler, 1990, 1991). The FAD prosthetic group first reacts with NADPH and then molecular oxygen to give rise to the enzyme-bound hydroperoxyflavin responsible for the oxidation of suitable substrates. These initial reactions occur in the absence of substrate and the enzyme exists primarily in the hydroperoxyflavin form (Poulsen and Ziegler, 1979, Beaty and Ballou, 1981a,b). A consequence of this is that substrates, with few exceptions, have the same Vmax, although Km may vary.

At least five different isoforms of FMO have been demonstrated, some of which appear to be associated with particular tissues or cell types (Lawton and Philpot, 1993).

[1]Brain & Development Research Center
University of North Carolina School of Medicine
Chapel Hill NC 27599-7250 USA

[2]National Institute of
     Environmental Health Sciences
Research Triangle Park NC 27709 USA

NATO ASI Series, Vol. H 90
Molecular Aspects of Oxidative Drug Metabolizing Enzymes
Edited by E. Arınç, J. B. Schenkman and E. Hodgson
© Springer-Verlag Berlin Heidelberg 1995

Classes of Substrate

Substrates for the FMO have very diverse chemical structures, from inorganic ions to organic compounds. (Table 1) All, however, are soft nucleophiles, a category that includes many organic chemicals but particularly those with a sulfur, nitrogen, phosphorus or selenium heteroatom. Although xenobiotic-metabolizing isoforms of cytochrome P450 (P450) appear to prefer hard nucleophiles as substrates, many substrates for FMO are also substrates for P450. However, even when the same substrate is oxidized by both FMO and P450, in addition to rate differences, there may be different products and different stereochemistry. The amount of various isoforms of both FMO and P450 also varies from tissue to tissue.

Alternate Pathways: General Approach

Since P450 and FMO have many substrates in common but, at the same time, these substrates may have different products with different toxic potencies, it is important to know the relative contribution of the two pathways to the metabolism of a particular substrate. Furthermore, in contrast to FMO isozymes, xenobiotic-metabolizing isozymes of P450 are often relatively easily induced, thus making the relative contributions variable with the conditions of exposure. Although it is said that the FMO prefers soft nucleophiles as substrates and P450 preference is for hard nucleophiles, with the exception of compounds oxidized at carbon atoms, this applies only to the relative ability of compounds to serve as substrates for one or the other, since it is difficult to find more than a very small number of FMO substrates that are not also substrates for one or more P450 isozymes.

Such substrates may have complex oxidation patterns and show regioselectivity in the sites attacked, they may yield different products, or different isomers of the same product. A number of methods are available for determining the relative contributions of FMO and P450, including extrapolation from the properties of purified enzymes (or from isozymes cloned and expressed in heterologous expression systems), the use of product specific substrates, the use of enzyme specific substrates or the manipulation of microsomes in which both enzymes are found. This latter technique, (using selective heat treatment to inactivate FMO or an antibody to the NADPH-P450 reductase to inactivate P450) has proven most useful in our hands, particularly in the case of hepatic enzymes.

Table 1

Substrates for the Flavin-Containing Monooxygenases

Chemical Class	Examples
Nitrogen-Containing Organics	
Primary amines	n-Octylamine
Secondary amines	
Acyclic	N-Methylaniline
Cyclic	Desmethyltrifluperazine
Tertiary amines	
Acyclic	Chlorpromazine
Cyclic	Nicotine
Hydroxylamines	N-Hydroxyaminoazobenzene
Hydrazines	
Monosubstituted	Methylhydrazine
Disubstituted (1,1)	1,1-Dimethylhydrazine
Disubstituted (1,2)	1,2-Dimethylhydrazine
Sulfur-Containing Organics	
Thiols	Dithiothreitol
Sulfides	Phorate
Disulfides	Butyl disulfide
Thiocarbamides, etc	Thiourea
Thioamides	Thioacetamide
Dithioacids and Dithiocarbanates	Dithiobenzoate
Mercaptopurines and	
Mercaptopyrimidines	
Phosphorus-Containing Organics	
Phosphines	Diethylphenylphosphine
Phosphonothioates	Fonofos
Selenium-Containing Organics	2-Selenylbenzanilide
Boronic Acids	
Inorganics	HS-, S, I2, I-, IO-, CNS-

Adapted from Ziegler, 1988, 1990, 1991, Ziegler et al., 1992, Hodgson and Levi, 1989, 1992, Tynes and Hodgson, 1985, Hajjar and Hodgson, 1982, Poulsen, 1981.

Alternate Pathways: Specific Compounds

Phorate. The insecticide, phorate, undergoes a complex series of oxidations (Fig 1). The products are generally more toxic than the parent compound and the reaction sequence is, therefore, an activation sequence. This substrate has continued to prove useful in examining the relative importance of FMO and P450.

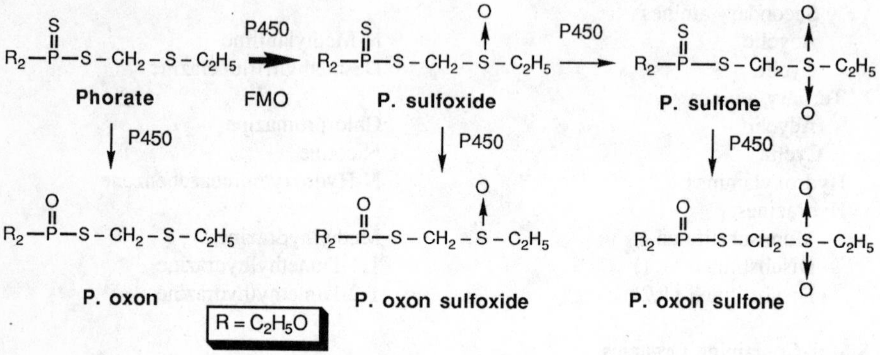

Fig. 1. Oxidation of phorate by FMO and P450.

FMO forms only phorate sulfoxide while P450 yields additional products. The sulfoxidation reaction is stereospecific with FMO producing the (-)-sulfoxide and several P450 isozymes the (+)- sulfoxide. While both sulfoxide isomers are substrates for all P450 isozymes tested, the (+)-sulfoxide is always preferred to the (-)-sulfoxide (Levi and Hodgson, 1988). The relative contribution of FMO to sulfoxide formation is higher in female than in male mice in agreement with the higher activities of FMO toward several other substrates seen in females relative to males of this species. Although overall sulfoxide formation is higher in the liver than in any extra-hepatic tissue, the relative contribution of the FMO is higher in lung, kidney and skin, being as high as 90 percent of the total in renal microsomes from female mice. Furthermore, the contribution of FMO relative to P450 is increased following treatment, in vivo, with P450 inhibitors such as piperonyl butoxide, or decreased following treatment, in vivo, with P450 inducers such as phenobarbital (Kinsler et al., 1988, 1990).

Thioridazine. Many drugs, including antipsychotics, monoamine oxidase inhibitors and antihistamines are substrates for the FMO. The antipsychotic drug, thioridazine (TDZ), is an excellent substrate for examining the relative importance of different oxidative pathways since it is oxidized at multiple sites by both FMO and P450 (Fig 2).

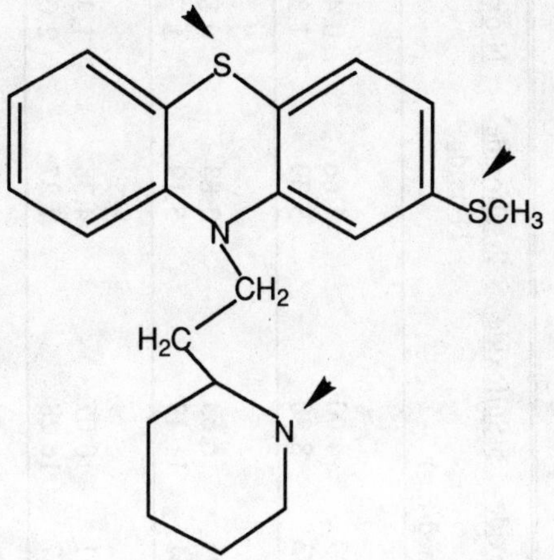

Fig 2. Sites of oxidative attack on thioridazine

Based primarily on examination of urinary and serum metabolite profiles, S-oxidation appears to be the predominant route of metabolism in humans, producing the 2-sulfoxide, the 2-sulfone and the 5-sulfoxide (Hale and Poklis, 1985, Kilts et al., 1981, Widerlöv et al, 1982), the 2 sulfur being in the side chain and the 5 sulfur in the ring. The 2-sulfoxide and the 2-sulfone are known to have greater antipsychotic activity than the parent compound, TDZ, (Kilts et al., 1984, 1982) while the ring sulfoxides may be responsible for the cardiotoxic side effects sometimes seen with TDZ (Hale and Poklis, 1986).

Metabolism by hepatic microsomes from the mouse yielded primarily the 2-sulfoxide of TDZ, along with significant amounts of the 5-sulfoxide, the N-oxide, the N-demethyl derivative and the 2-sulfoxide-N-oxide (Blake et al., 1993) (Table 2). Heat treatment of microsomes to destroy selectively the FMO or treatment with an antibody to the NADPH-cytochrome P450 reductase to inhibit P450 isozymes revealed that the N-oxide was the principal metabolite derived from the FMO, while the 2-sulfoxide and the other products were derived primarily from one or more isozymes of cytochrome P450.

Table 2

Metabolism of Thioridazine by Microsomal Preparations from Livers
of Uninduced and Phenobarbital Induced Mice

	TDZ (μg/ml remaining)	Demethyl	2-Sulfone	2-Sulfoxide	5-Sulfoxide	2-sulfoxide* N-oxide	N-oxide*	Unknown*⊗
				(nmolsformed/mg)				
5 minutes								
Control	10.7	3.72	-	18.29	4.00	1.66	0.43	0.65
PB induced	10.8	5.72	-	34.73	8.28	3.39	1.34	0.69
10 minutes								
Control	10.8	5.74	-	29.95	6.63	2.83	0.85	1.02
PB induced	8.7	9.49	-	47.61	11.48	5.19	1.79	0.78
20 minutes								
Control	9.1	8.43	-	44.51	10.07	4.38	1.34	1.83
PB induced	5.9	13.02	.80	63.47	16.28	7.27	2.05	0.89
30 minutes								
Control	6.5	8.13	-	38.08	12.00	3.70	1.05	1.72
PB induced	4.1	11.48	.80	47.22	12.00	5.44	1.53	0.74

*Calculations based on standard curve for 2-sulfoxide

⊗μg/mg

Studies using FMO purified from mouse liver have shown the N-oxide of TDZ to be the principal and, perhaps the only, product of TDZ metabolism by this enzyme. Similar experiments using cytochrome P450 2D6 reveal that the 2-sulfoxide of TDZ (mesoridazine) is the principal, but not the only metabolite of TDZ and this P450 isozyme. Similar experiments have been carried out with P450 isozymes 2E1 and 2B1.

Since thioridazine and several of its metabolites show similar biological activity, the possible occurrence and role of FMO in the nervous system appears to be important. Previous studies (Duffel & Gillespie, 1984) indicating the possible occurrence of the FMO in rat corpus striatum and whole brain microsomes could not be replicated. Our results, using microsomes prepared from mouse brain, of substrate level oxidations and western blotting with an antibody a form of FMO purified from mouse liver, while not negative, were equivocal. In order to determine the presence of FMO mRNA in rabbit brain we have recently, in conjunction with Drs. Burnett and Lawton, of Richard Philpot's laboratory, utilized PCR techniques to demonstrate FMO in the nervous system of the rabbit. Recently, five forms of FMO (A-E) have been identified in rabbit hepatic and extrahepatic tissues, with most tissues expressing more than one form. PCR amplification of cDNA was performed using primers specific for each of the five forms of FMO found in rabbit tissues. The data suggest that one form, apparently 1E1, is expressed in rabbit brain. This FMO has recently been cloned and sequenced from a human liver cDNA library (Dolphin et al., 1991).

The substrate specificity of FMO 1E1 has not been determined. If it is a metabolically active protein, the difficulty in detecting its presence in brain may be due, in part, to its localization only in certain brain regions or cell types. Studies are in progress to confirm the presence of 1E1 message in brain and to utilize immunocytochemical and in situ hybridization techniques to localize the isozyme in the brain.

Tamoxifen

Studies carried out in David Kupfer's laboratory (Mani et al., 1993) have revealed that the principal metabolites of tamoxifen generated by mammalian liver microsomes are the N-oxide, N-desmethyl, and 4-hydroxy derivatives of tamoxifen. Inhibition of N-oxide formation by mild heat treatment of microsomal preparations or the addition of methimazole to these preparations, implicated the FMO in the formation of this metabolite, an hypothesis further strengthened by the observation that an antibody to the NADPH-cytochrome P450 reductase inhibited the formation of the desmethyl and 4-hydroxy derivatives but not the N-oxide. Only the N-oxide was produced in experiments using purified FMO from mouse liver microsomes, thus providing direct evidence for the role of this enzyme in tamoxifen metabolism. Since tamoxifen N-oxide is readily reduced to the parent compound by liver microsomes in the presence of NADPH, it has been suggested that the N-oxide may serve in vivo as a storage form of tamoxifen.

Alternate Pathways: Portals of Entry and Target Tissues

Alternate pathways and the relative importance of different enzyme systems in nonhepatic tissues are of considerable importance. Portals of entry may not only may be the first line of oxidative attack but in both pulmonary and dermal entry, but first pass through the liver is avoided. In the case of target tissues, it is probable that events taking place close to the site of action, while quantitatively lower than similar events in the liver may, because of their proximity, be more important.

It has been known for some time that the FMO was relatively more important than P450 for the oxidation of phorate and thiobenzamide in lung and kidney. Recently (Venkatesh et al., 1992) we have demonstrated the presence of FMO in the skin of the mouse and the pig using a variety of biochemical techniques and by Western blots with an antibody to a liver form of the enzyme. We have also observed its localization by immunocytochemical methods. Distribution, in both the mouse and the pig, proved to be primarily in the epidermis. The relative contribution of FMO and P450 in the skin was investigated using a variety of techniques to inhibit selectively either FMO or P450, and to measure each one in the absence of the other. The results were compared to those from similar experiments utilizing microsomes from the liver.

As in the lung and kidney, FMO proved to be more important than P450 for the sulfoxidation of phorate. While in liver microsomes P450 was responsible for 68-85% of phorate sulfoxidation, in the skin the FMO was responsible for 66-69% of this activity.

Studies in the laboratory of Dr. Mary Beth Genter (personal communication) utilizing immunocytochemical methods have revealed that at least two isozymes of the FMO are resent in the rodent olfactory mucosa. Isozyme 1B1 is broadly distributed while isozyme 1A1 has a more restricted distribution.

Conclusions

The distribution, molecular biology and function of the FMO are under active investigation but are still not well understood. It is apparent, however, that its role in the oxidation of xenobiotics, particularly relative to that of P450, will have significant toxicological and therapeutic implications.

Acknowledgments

Studies carried out at NCSU were supported, in part, by PHS grants number ES-00044 and ES-07046.

References

Beaty NB, Ballou DP (1981a) The reductive half-reaction of liver microsomal FAD-containing monooxygenase. J Biol Chem 256:4611-4618

Beaty NB, Ballou DP (1981b) The oxidative half-reaction of liver microsomal FAD-containing monooxygenase. J Biol Chem 256:4619-4625

Blake B, Burnette V, Levi P, Hodgson E (1993) Flavin-containing monooxygenase in rabbit brain: amplification and activity. Toxicologist 13:60

Dolphin C, Shephard EA, Povey S, Palmer CN, Ziegler DM, et al. (1991) Cloning, primary sequence, and chromosomal mapping of a human flavin-containing monooxygenase (FMO). J Biol Chem 266:12379-85

Duffel MW, Gillespie SG (1984) Microsomal flavin-containing monooxygenase activity in rat corpus striatu. J Neurochem 42:1350-1353

Hajjar NP, Hodgson E (1982) Sulfoxidation of thioether-containing pesticides by the flavin adenine dinuceotide-dependent monooxygenase of pig liver microsomes. Biochem Pharmacol 31:745-752

Hale PW, Poklis A (1985) Thioridazine-5-sulfoxide diastereoisomers in serum and urine from rat and man following chronic thioridazine administration. J Anal Toxicol 9:179-201

Hale PW, Poklis A (1986) Cardiotoxicity of thioridazine and two sterioisomeric forms of thioridazine-5-sulfoxide in the isolated perfused rat heart. Toxicol Appl Pharmacol 86:44-55

Hodgson E, Levi PE (1991) The flavin-containing monooxygenase (EC 1.14.13.8). pp 11-21 in Molecular Aspects of Monooxygenases and Bioactivation of Toxic Compounds (Arinc, Schenkman and Hodgson, eds) Plenum Press NY

Hodgson E, Levi PE (1992) The role of the flavin-containing monooxygenase (EC 1.14.13.8) in the metabolism and mode of action of agricultural chemicals. Xenobiotica 22:1175-1183

Kilts CD, Mailman RB, Hodgson E, Breese GR (1981) Simultaneous determination of thioridazine and its silfoxidized intermediates by HPLC: use in clinical and preclinical metabolic studies. Fed Proc 40:283

Kilts CD, Patrick KS, Breese GR, Mailman RB (1982) Simultaneous determination of thioridazine and its S-oxidized and N-demethylated metabolites using high performance liquid chromatography on radially compressed silica. J Chromatog 231:377-391

Kilts CD, Knight DL, Mailman RB, Widerlöv E, Breese GR (1984). Effects of thioridazine and its metabaolites on dopaminergic function: drug metabolism as a determinant of the antidopaminergic actions of thioridazine. J. Pharmacol. Exp. Ther. 231:334-342

Kinsler S, Levi PE, Hodgson E (1988) Hepatic and extrahepatic microsomal oxidation of phorate by the cytochrome P-450 and FAD-containing monooxygenase systems in the mouse. Pestic Biochem Physiol 31:54-60

Kinsler S, Levi PE and Hodgson E (1990) Relative contributions of the cytochrome P450 and flavin-containing monooxygenases to the microsomal oxidation of phorate following treatment of mice with phenobarbital, hydrocortisone, acetone and piperonyl butoxide. Pestic Biochem Physiol 23:174-181

Lawton MP, Philpot RM (1993) Molecular genetics of the flavin-containing monooxygenases. Pharmacogenetics 3:40-44

Levi PE, Hodgson E (1988) Stereospecificity in the oxidation of phorate and phorate sulphoxide by purified FAD-containing mono-oxygenase and cytochrome P-450 isozymes. Xenobiotica 18:29-39

Mani C, Hodgson E, Kupfer D (1993) Metabolism of the anti-mammary cancer antiestrogenic agent tamoxifen. II Flavin-containing monooxygenase (FMO) mediated N-oxidation. Drug Metabol. Disp. in press

Poulsen LL (1981) Organic sulfur substrates for the microsomal flavin-containing monooxygenase. Rev Biochem Toxicol 3:33-49

Poulsen LL, Ziegler DM (1979 The liver microsomal FAD-containing mono-oxygenase. Spectral characterization and kinetic studies. J Biol Chem 254:6449-6455

Tynes RE, Hodgson E (1985) Magnitude of involvement of the mammalian flavin-containing monooxygenase in the microsomal oxidation of pesticides. J. Ag. Food Chem. 33:471-479.

Venkatesh K, Levi PE, Inman AO, Monteiro-Riviere NA, Misra R, Hodgson E (1992) Enzymatic and immunohistochemical studies on the role of cytochrome P450 and the flavin-containing monooxygenase of mouse skin in the metabolism of pesticides and other xenobiotics. Pestic. Biochem. Physiol. 43:53-66.

Widerlöv E, Häggström JE, Kilts CD, Anderson U, Breese GR and Mailman RB (1982). Serum concentrations of thioridazine, its major metabolites and serum neuroleptic-like activities in schizophrenics with and without tardive dyskinesia. Acta Psychiatr Scand 66:294-305

Ziegler DM (1988) Flavin-containing monooxygenases: catalytic mechanism and substrate specificities. Drug Metabol Revs 19:1-32

Ziegler DM (1990) Flavin-containing monooxygenases: enzymes adapted for multisubstrate specificity. TIPS Revs

Ziegler DM (1991) Unique properties of the enzymes of detoxication. Drug Metabol Disp 19:847-852

Ziegler DM, Mitchell CH (1972) Microsomal oxidase IV: Properties of a mixed-function amine oxidase isolated from pig liver microsomes. Arch Biochem Biophys 150: 116-125

Ziegler DM, Graf P, Poulsen LL, Stahl W, Sies H (1992) NADPH-dependent oxidation of reduced ebselen, 2-selenylbenzanilide, and of 2-(methylseleno)benzanilide catalyzed by pig liver flavin-containing monooxygenase. Chem Res Toxicol 5:163-166

Developmental Regulation of Biotransformation of Drugs and other Xenobiotics

Wolfgang Klinger
Institute of Pharmacology and Toxicology
Friedrich-Schiller-University of Jena
Loebderstr. 1
D-07743 Jena
Germany

Introduction

Developmental pharmacology (and toxicology) are concerned with the influence of development from the formation of the zygote until the death of an organism on the effect of drugs and xenobiotics as well as on their fate in the organism. Developmental pharmacokinetics must consider the influence of development on absorption, distribution, protein binding, biotransformation, and excretion.

Since the early publications on the insufficiency of newborn animals and human beings to metabolize xenobiotics after second world war more than thousand papers including several reviews have been published concerning the time of appearance and the rates of development of various enzymes associated with the biotransformation of lipid-soluble xenobiotics, the genetic control and regulation and endogenous (e.g.hormonal, immunological) and exogenous (dietary, induction by foreign compounds, regeneration etc.) influences. The developmental aspect of biotransformation has attracted increasing attention especially after the accidents with chloramphenicol in newborns and by the Contergan tragedy. Thus the developmental aspect of biotransformation became the main field of "Developmental Pharmacology", the aims of which can be characterized as follows:

- to protect the newborn from deleterious effects of xenobiotics,
- to optimize drug therapy in the newborn and in prematures,

NATO ASI Series, Vol. H 90
Molecular Aspects of Oxidative Drug Metabolizing Enzymes
Edited by E. Arınç, J. B. Schenkman and E. Hodgson
© Springer-Verlag Berlin Heidelberg 1995

- to bring about the prerequisits for a scientific drug therapy in aged humans, to improve the effectiveness and to reduce the rate of unwanted side effects.

Toxicological aspects thus play a central role. Moreover the importance increased by the finding that these enzymes are also competent for the biotransformation and elimination of endogenous compounds such as bilirubin and steroids. As most biotransformation reactions can be localized in the liver parenchymal cell (predominantly in both main forms of endoplasmic reticulum) especially the morphological, physiological and biochemical maturation of this cell and of the whole organ must be considered. Also the sublobular heterogeneity develops postnatally. Sex differences can be observed mainly in rats and mice and they also develop postnatally. The so-called biogenetic rule that ontogeny is a rapid repetition of phylogeny, as postulated by Ernst Haeckel (1866), proved to be true also in this field, at least with relation to some reactions. Thus the comparison of xenobiotic metabolism in poikilothermic newborn mammals and in different fish species is highly interesting, but shall not be considered in detail in this paper (cp. contributions of R. F. Addison, D. R. Buhler, S. V. Kotelevtsev and J. J. Stegeman in this volume).

In this paper predominantly rat will be considered.

Morphological Basis

Most biotransformation reactions with xenobiotics can be localized in the liver parenchymal cell, predominantly in both main forms of endoplasmic reticulum (ER), smooth (SER) and rough endoplasmic reticulum (RER). But for some enzymes cytosol and according to ER genesis the nuclear membrane are active compartments.
Only some investigators attribute typical phase I reactions or the acetylation mainly to liver mesenchymal cells.

In the rat, the liver is an important hemopoetic organ in fetal and early postnatal life until the 7th day. RER becomes visible at the 14th embryonic day as big vesicles, but free ribosomes are prevalent. The typical picture of the mature liver is reached at the 5th - 7th or 10th postnatal day on the histological and submicroscopical level, but the sublobular heterogeneity of the parenchymal cells has reached maturity not before the 20th day of postnatal life. The growth pattern of rat hepatocytes during postnatal development with regard to ploidy, mitosis rate (hyperplasia) and/or hypertrophy was investigated in detail. The mitotic cycle duration is clearly dependent on age.

Volume, surface and number of microvilli of liver cells as well as the volume of nuclei increase postnatally (David, 1985). But hepatocytes isolated from 6 to 30 month-old female rats did not show a change in cell size with age; in older animals a significant increase in surface folds could be observed, a significant increase in protein content and of the percentage of binuclear cells and cells with higher ploidy.

In fetal liver cells nuclei are bigger, but we find much less mitochondria. In weanling rats 2.6 times more but smaller mitochondria are observed in comparison to newborn rats. In senescent rats cell volume increases, lysosomes are bigger, RER and SER are diminished, only mitochondria do not alter or have even an enlarged surface. The surface of the Golgi apparatus decreases. But there are some discrepancies and conflicting data in literature. Some papers deal with the influence of xenobiotics on differentiation, maturation and aging: The action of CCl_4 in early development and old age and especially the effects of various inducers as spironolactone, pregnenolone-16 alpha-carbonitrile (PCN) and other steroids have been described. Microsomes from liver homogenates of newborn rats spin down already at 200 - 800 g, but with EDTA-buffer high yields of fetal or newborn liver microsomes can be obtained. The microsomal yield determined by the ratio of G-6-Pase in microsomal suspension to that in whole homogenate increased in rats from day 1 to day 21 from 9 % to 39 %. Sonication of the homogenates (10 sec) doubled the microsomal yield of livers (and lungs) from 1 week old rabbits.

The yield is highly dependent on the method used, with 0.25 M sucrose the differential centrifugation according to DeDuve was identical for livers from fetal, 1-, 7-day-old and adult rats. But the differentiation of the mitochondrial fraction into mitochondria, lysosomes and peroxisomes varied considerably. Moreover the whole liver mass must be known for the estimation of the biotransformation capacity of the organism; in the rat liver mass has the highest percentage of whole body mass in 30- to 60-day-old animals. Liver growth is stimulated by the beginning of food uptake after the 15th day of life.

Biochemical Basis

Here only the data relevant for biotransformation reactions shall be reviewed. The ER is not only the site of various hydroxylation reactions, but it is also the main organelle for the synthesis of cellular structural components. During developmental growth faster synthesis and slower degradation of proteins can be observed, whereas in senescent rats microsomal protein as well as protein synthesis (amino acid incorporation) decrease. Also the rate of peptide chain elongation decreased considerable. But a decline in phenylamine incorporation begins as early as in fetal life.

The biosynthesis and transport to the final position of only a few ER membrane proteins have been investigated so far. Cytochrome b5 (b5) for example is synthesized on ribosomes and inserted directly into its final position at the cytoplasmic ER surface. Most senescent rat hepatic microsomal proteins as well as protein synthesis (amino acid incorporation in vivo and in vitro) decrease, due to a general decline in gene expression, transcription and translation, and, more specifically, to a weakening of stimulating and a strengthening of inhibiting factors with age. But also protein degradation decreases during postnatal development.

The lipids of endoplasmic membranes undergo important quantitative and qualitative changes during development, especially in their fatty acid pattern. They strongly influence the fluidity of membranes and by these means the diffusion velocity of xenobiotics, the electron transport from $NADPH_2$ to cytochrome P450 (P450), and moreover they are essential for and decisively influence the binding of xenobiotics to P450 binding sites. Also the changes of ER lipids during development have been investigated. In senescent rats an increase in the cholesterol/phospholipid ratio is remarkable, phospholipids decrease. Microsomal lipids themselves bind to P450 and inhibit monooxygenase reactions in dependence on age. Lipid peroxidation is evidently not only influenced by age dependent differences in lipid composition, but also by age dependent inhibitory and enhancing factors. Variations in the fatty acid pattern of the maternal diet influence the fatty acid composition of microsomal phospholipids, but do not affect enzyme activities in newborn animals and are therefore of minor importance. But acute fasting significantly influences monooxygenase activities, differently in immature and adult rats. Also dietary protein deficiency decreases P450, epoxide hydrolase (EH) and UDP-glucuronosyl transferase (UDPGA-T), more pronounced in young than in older rats; restoration of the diet quickly normalizes these values.

The phospholipid-protein-ratio in microsomes increases with increasing age. Neither in rats nor in guinea pigs of any ages $NADP/NADPH_2$ and $NAD/NADH_2$ concentrations are limiting for the capacity of the microsomal electron transport chain. Microsomal glucose-6-phosphate dehydrogenase activity for the $NADH_2$ and $NADPH_2$ regenerating systems is low in rat hepatocytes at an age of 2 weeks, then an increase can be observed after weaning, more dramatically after 8 weeks of age, maximal values could be detected at an age of 12 - 16 weeks.

Transport Mechanisms

Xenobiotics enter hepatic parenchymal cells by diffusion and are bound to and stored and evtl. excreted unchanged by specific cytosolic proteins which have their own developmental pattern. In newborn rats the overall hepatic excretory function is low and the excretion of organic acids and neutral xenobiotics develops postnatally to reach maximum capacity at an age of 30 days. Thereafter the excretion capacity declines, cp. Fig. 1.

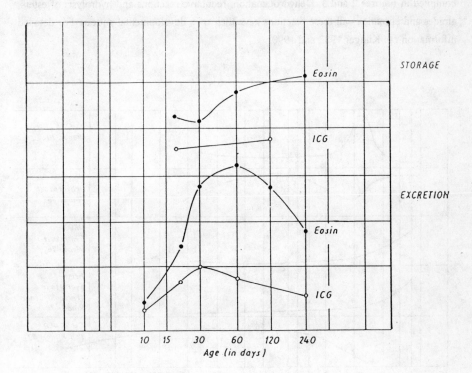

Fig. 1: Influence of age on hepatic storage and biliary excretion of the organic anions and indocyanine green (ICG) in male rats. Data according to Barth et al. (1986).

Biotransformation

Phase I Reactions

The different steps of cytochrome P-450 (P450) formation and breakdown, its steady state concentration, the activities of NADPH-P450 reductase and NADH-b5 reductase activities have very different developmental patterns. Thus also different types of monooxygenation (mixed function oxidation) such as hydroxylation of aliphatic and aromatic compounds, epoxidation, N-dealkylation, O-dealkylation, N-hydroxylation and S-oxidation show different developmental patterns. This holds true also for oxidative deamination and dehalogenation of halogenated aliphates. For comparison some developmental courses are compiled in Figures 2 and 3. Dehydrogenation, reductive reactions and hydrolysis of esters, amides and epoxides - all these reactions have their own developmental pattern. For detailed information cp. Klinger 1982 and 1990.

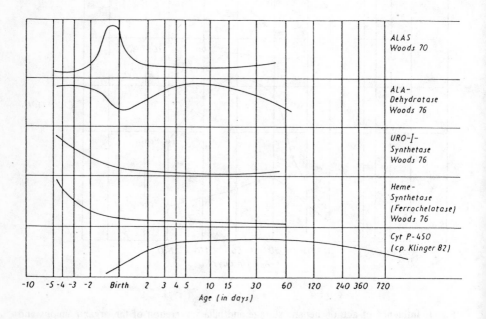

Fig. 2: Influence of pre- and postnatal age on the activities of the enzymes involved in hepatic heme synthesis and of P450 concentration in male rats (Klinger 1990).

Phase II Reactions

Glucuronidation and conjugation with glutathione are each catalyzed by different isozymes which are under different genetic and developmental control. The different isozymes develop in clusters, cp. Figure 4.

Substrates for the fetal cluster of the UDP-glucuronosyltransferases are: 2-aminophenol, 2-aminobenzoate, 4-nitrophenol, 1-naphthol, 4-methylumbelliferone and 5-hydroxytryptamine, as inducers act steroids, mainly glucocorticoids.

Substrates for the postnatal cluster are: bilirubin, testosterone, estradiol, morphine, phenolphthalein and chloramphenicol. The postnatal cluster can be induced by 3-methylcholanthrene (Wishart 1978).

Conjugation with acetate and glycine or other amino acids also needs individual investigation. Much less is known on the ontogenetic development of extrahepatic biotransformation. All reaction types of phase I and phase II need individual investigation in different organs and tissues, respectively, and in various laboratory animals. For more detailed information cp. Klinger 1990.

Regulation of Development

Special interest must be given to factors of regulation, among them hormones and xenobiotics which act as inducers. Finally the interaction of hormones, inducers and other environmental influences (light, imprinting phenomena by hormones and xenobiotics, variation of the immune status, regeneration after impairment of the liver) shall be discussed. In man and also in laboratory animals with long gestation periods (e.g. guinea pigs and rabbits) an early differentiation during pregnancy has been observed, whereas in the rat differentiation and development occur mainly after birth (cp. Klinger et al., 1979). Possible fetal and birth-correlated trigger and control mechanisms have been discussed by Pelkonen (1980). The biogenetic rule that ontogenesis repeats briefly phylogenesis (Ernst Haeckel, 1866), holds in general also true for ontogenetic development of biotransformation reactions. The basic mechanisms which control the changes during ontogeny and especially in old age are in principle unknown. Changes in hormone production, release and efficacy are considered to be part of ageing, but not the reason. Special interrelations between hormone and drug metabolism in the fetal and perinatal period directly influence development of drug metabolism in the fetus and the newborn. There are data available for the hypothesis that

244

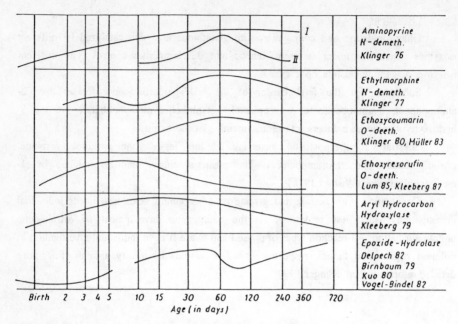

Fig. 3: Influence of age on various monooxygenase reactions (Klinger 1990)

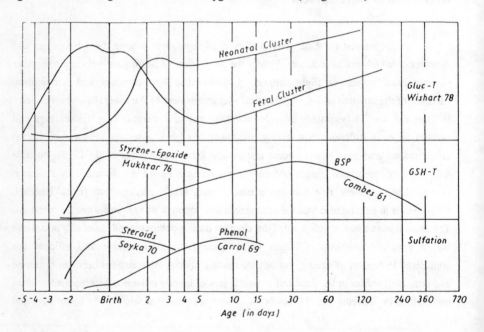

Fig. 4: Influence of age on the activities of various glucuronosyl-, glutathione- and sulfate-transferases (Klinger 1990).

reduction of maternal progesterone inhibits the development of hydroxylation activity in the fetal and perinatal liver. Pregnanolone inhibits MFO activities also directly. Glucocorticoids trigger the development not only of GTs, but also of P450s, and precocious development of P450s in neonatal rat liver is possible by glucocorticoid treatment. But then the question arises what is the mechanism to trigger ACTH and glucocorticoid release.

The postnatal decrease of genetic redundance is at present a hypothesis to explain ageing, but the mechanism remains obscure. We interpreted postnatal development and ageing as a genetically controlled repression-phenomenon, consecutively the postnatal development with increasing activities to be a de-repression of the responsible gene loci and the loss of activities in old age to be an increasing repression. This hypothesis was formulated in connection with investigations on interrelations of development and enzyme induction. It was shown that low activities of various enzymes in infantile animals can be elevated by inducers up to the adult level. But after withdrawel of the inducer these precocious adult values are not stable, they go back to the values which correspond to the age of the animals, and normal development continues (Klinger et al., 1966, 1968, Müller et al., 1971).

Additionally this inducibility in infantile animals shows that the normally low activities are not due to a lack in energy supply, cofactors etc. Until now it has been impossible to demonstrate a fetal or infantile inhibiting factor. And a repressor which should act in vivo could be shown in some experiments with no constant reproducibility. Also an interrelation with the immune system during development could not be confirmed (Klinger et al., 1983). Therefore additional influences are suggested (cp. Klinger, 1982). Nebert also postulates some form of temporal control, associated with the Ah locus (Guenthner and Nebert, 1978; Nebert, 1979; Kahl et al., 1980). Whether the Ah complex includes temporal genes, remains open. The temporal control may influence either the structural gene expression directly or the expression of regulatory genes indirectly. So this temporal control of biotransformation activity and capacity would be a special case, because temporal genes have already been characterized for other enzymes (cp. Kahl et al., 1980). But the biological clock, the time signal, the regulation, the control and mechanism of this temporal control remain an open field of investigation.

Changing environmental conditions such as light and temperature proved to have almost no influence on postnatal development (Häßler 1969). Perinatal and early postnatal impairment of the immunesystem (destruction of the thymus by N-methylnitrosourea or by X-rays, thymectomy or splenectomy) did not significantly influence postnatal development or inducibility by phenobarbital of the monooxygenase reactions (Klinger et al., 1983).

The different P450 forms are not only during postnatal development under different developmental control, but they recover differently in the regenerating liver after 2/3 hepatectomy in 10- and 60-day-old rats: the forms which catalyse ethoxycoumarin O-deethylation recover earlier than the forms which catalyse ethylmorphine N-demethylation. Moreover induction by phenobarbital cannot be demonstrated immediately after hepatectomy whereas the induction effect by beta-naphthoflavone is extraordinarily high and goes back to normal ratios when restoration comes to an end (Klinger and Karge, 1987). Thus it may be concluded that the developmental control of the expression of various P450 genes is different from the influence by factors such as hepatopoietin and hepatotropin.

But the expression of different P450 forms in a given ontogenetic stage is not only developmentally regulated and additionally influenced by hormones, growth factors etc., but the basic level may be additionally fixed by imprinting phenomena in so-called sensitive periods, in general the perinatal and early postnatal period. Cadmium e.g. produces apart from the teratogenic and embryotoxic effects long lasting behavioural alterations in the offspring and also a decreased inducibility of ethoxyresorufin O-deethylation by the inducer beta-naphthoflavone (Jahn and Klinger, 1989). Further examples are given elsewhere (Klinger and Jahn 1985). On the other hand one may observe increased induction responses following the second or repeated administrations of 3-methylcholanthrene, beta-naphthoflavone or phenobarbital, respectively. A time interval of about 3 weeks was inserted between the first inducer stimulus in the early postnatal period and the second exposition of the same animals (Kleeberg et al. 1985).

Thus different concentrations resp. activities of different P450 forms during ontogenetic development are influenced by many factors on the transscriptional and posttransscriptional level. Reviews on concepts and theories of development and ageing have been published by several authors resp. editors, cp. Klinger 1982 and 1990.

References

Barth A, Klinger W, Hoppe H (1986) Hepatic elimination kinetics od organic anions in rats: developmental aspects and influence of phenobarbital. Arch Int Pharmacodyn Ther 283:16-29

Birnbaum LS, Baird MB (1979) Senescent changes in rodent hepatic epoxide metabolism. Chem Biol Interact 26:254-256

Carrol J (1969) Phenolsulfotransferase in the developing rat. Am J Clin Nutr 22:978-985

Combes B, Stakelum GS (1961) Maturation of liver enzyme that conjugates sulfobromophthalein sodium (BSP) and glutathione. J Clin Invest 40:1030-1031

David H (1985) The hepatocyte, Development, Differentiation, and Aging. VEB Gustav Fischer Jena

Delpech I, Kiffel L, Magdalou J, Andre JC, Siest G (1982) 8th Europ. Workshop on Drug Metab, Sart Tilman, Belgium, Sept 5-9, 41

Guenthner TM, Nebert DW (1978) Evidence in rat and mouse liver for temporal control of two polycyclic aromatic inducible forms of cytochrome P-450. Eur J Biochem 91:449-456

Häßler A (1969) The influence of breeding conditions on   individual variation of drug effects in rats of different ages (in German). Thesis Medical Faculty, Friedrich-Schiller-University Jena

Jahn F, Klinger W (1989) Influence of prenatal administration of cadmium on postnatal development and inducibility of hepatic monooxygenases in rats. Pharmacol Toxicol 64:292-292

Kahl GF, Friederici DE, Bigelow SW, Okey AB, Nebert DW (1980) Ontogenetic expression of regulatory and structural gene products associated with the Ah-locus. Devel Pharmac Ther 1:137-162

Kleeberg U, Grohmann G, Volkmann R, Steinert H, Klinger W (1979) In vivo and in vitro inhibition of 3-methylcholanthrene-induced aryl hydrocarbon hydroxylase activity in rat liver by actinomycin D and 7,8-benzoflavone. Pol J Pharmac Pharm 31:675-681

Kleeberg U, Sommer M, Klinger W (1985) Increased response of cytochrome P450 dependent biotransformation reactions in rat liver to repeated administration of inducers. Arch Toxicol Suppl 8:361-365

Kleeberg U (1987) Induktion Cytochrom P-450-abhängiger Biotransformationsreaktionen durch 3-Methylcholanthren. Mechanismen und postnatale Entwicklung. Thesis, University Jena

Klinger W, Ankermann H (1966) Die Hexobarbitalnarkose bei infantilen Ratten. Acta biol med germ 17: 357-359

Klinger W, Kusch T, Neugebauer A, Splinter FK, Ankermann H (1968) Untersuchungen zum Mechanismus der Enzyminduktion. XIV. Der Einfluss des Lebensalters auf die Aktivität und Induzierbarkeit der Phenazon-Hydroxylase, Aminophenazon-N-Demethylase, Kodein-O-Demethylase und Nitro-Reduktase der Rattenleber. Acta biol med germ 21:257-269

Klinger W, Müller D, (1976)   Developmental aspects of xenobiotic transformation. Environm Health Perspect 18:13-23

Klinger W, Müller D (1977) Ethylmorphine N-demethylation by liver homogenate of newborn and adult rats; enzyme kinetics and age course of Vmax and Km. Acta biol med germ 36:1149-1159

Klinger W, Müller D, Kleeberg U (1979) Induction and its dependence on development, In: Estabrook RW, Lindenlaub E (eds) The induction of drug metabolism.   Symp Med Hoechst vol 14. Schattauer Stuttgart New York, pp 517-544

Klinger W (1982) Biotransformation of drugs and other xenobiotics during postnatal development. Pharmac Ther 16: 377-429 (updated version 1992 as manuscript).

Klinger W, Müller D, Danz M, Kob D, Madry M (1983) Influence of impairment of the immune system on hepatic biotransformation reactions, their postnatal development and inducibility. Exp Pathol 24:219-225

Klinger W, Jahn F (1985) Epigenetic imprinting of drug metabolism. In: Magyar T, Scüts T, Vereczkey L (eds) Proc 4th Congr Hung Pharmacol Soc Budapest 1985. vol 2. Sect 4. Pergamon Oxford/ Akademiai Kiado Budapest 1985,   pp 384-397

Klinger W, Karge E (1987) Interaction of induction, ontogenetic development and liver regeneration on the monooxygenase level. Exp Pathol 31:117-124

Klinger W, Müller D, Kleeberg U, Jahn F, Glöckner R (1987) Developmental Pharmacology, In: Pharmacology, Proceedings of the Xth Inteernational Congress of

Pharmacology (IUPHAR), Sydney, 23-28 August 1987, Excerpta Medica Amsterdam New York Oxford, pp 753-763

Klinger W (1987) Developmental Aspects of Enzyme Induction and Inhibition. Pharmac Ther 33:55-61

Klinger W (1990) Biotransformation of Xenobiotics During Ontogenetic Development, In: Ruckpaul K and Rein H (eds): Frontiers in Biotransformation vol 2. Akademie-Verlag. Berlin, pp 113-149

Kuo CH, Hook JB (1980) Postnatal development of renal and hepatic drug-metabolizing enzymes in male and female Fischer 344 rats. Life Sci 27:2433-2438

Lum PY, Walker S, Ioannides C (1985) Foetal and neonatal development of cytochrome P-450 and cytochrome P-448 catalyzed mixed function oxidases in the rat: Induction by 3-methylcholanthrene. Toxicol 35:307-317

Mukhtar H, Bresnick E (1976) Glutathione-S-epoxide transferase activity during development and the effect of partial hepatectomy. Cancer Res 36:937-940

Müller D, Reichenbach F, Klinger W (1971) Die Aktivität der Nitroreduktase und deren Induzierbarkeit durch Barbital in der Leber von Ratten verschiedenen Alters. Acta biol med germ 27:605-609

Müller D, Greiling K, Greiling H, Klinger W (1983) The influence of triiodothyronine (T3) on the postnatal development of drug metabolism in rat liver. Biomed Biochim Acta 42:981-987

Nebert DW (1979) Genetic aspects of enzyme induction by drugs and chemical carcinogens, in: The induction of drug metabolism, Estabrook RW, Lindenlaub E (eds), Symp.Med. Hoechst 14, Schattauer Stuttgart New York, 419-452

Pelkonen O (1980) Biotransformation of xenobiotics in the fetus. Pharmac Ther 10:261-281

Schmucker DL (1985) Aging and drug disposition. Pharmacol Rev 37:133-148

Soyka LF, Gyermek L, Campbell P (1970) A study of the mechanism responsible for the sensitivity of newborn rats to pregnanolone. J Pharmac exp Ther 175: 276-282

Vogel-Bindel U, Bentley P, Oesch F (1982) Ontogenesis, induction, inhibition, tissue distribution, immunological behaviour and purification of microsomal epoxide hydrolase with 16a, 17a-epoxyandrostene-3-one as substrate. Eur Biochem 126:425-431

Wishart GJ (1978) Functional heterogeneity of UDP-glucuronosyltransferase as indicated byits differential development and inducibility by glucocorticoids. Demonstration of two groups within the enzyme's activity towards twelve substrates. Biochem J 114:485-489

Woods JS, Dixon RL (1970) Neonatal differences in the induction of hepatic aminolevulinicacid synthetase. Biochem Pharmacol 19:1951-1954

Woods JS (1976) Developmental aspects of hepatic heme biosynthetic capability and hematotoxicity. Biochem Pharmacol 25:2147-2152

Modulation of Xenobiotic Metabolism and Toxicity by Dietary Chemical

Chung S. Yang and Theresa J. Smith
Laboratory for Cancer Research
College of Pharmacy
Rutgers University
Piscataway, NJ 08855
U.S.A.

Introduction

Dietary and nutritional factors are known to influence the biotransformation of drugs and environmental chemicals. Early studies by Wattenberg (Wattenberg, 1971) demonstrated that rats on commercial rat chow had a 68-fold higher intestinal benzo(a)pyrene hydroxylase activity than those on a purified diet. In human studies, ingestion of cabbage and brussels sprouts was shown to increase the metabolism of antipyrine and phenacetin (Conney, 1982, Pantuck and Pantuck, 1978). Human volunteers on a low-protein diet had lower rates of metabolic clearance of antipyrine and theophylline than those on a high-protein diet (Kappas, et al., 1976). Recent studies demonstrated the transcriptional activation of specific cytochrome P450 genes by dietary chemicals (Pan, et al., 1993, Vang, et al., 1990) and the inhibition of nifedipine metabolism in humans after ingestion of grapefruit juice (Bailey, et al., 1990). Many of these studies have been discussed in several recent reviews (Anderson, et al., 1985, Bidlack, et al., 1986, Hathcock, 1985, Yang, et al., 1992, Yang and Yoo, 1988).

In this communication, some examples of recent studies will be highlighted to illustrate the possible impact of diet on chemical toxicity. Cytochrome P450 2E1, which is modulated by a variety of dietary factors, will receive special attention because of its importance as a site for interactions among environmental chemicals.

Modulation of P450 and Xenobiotic Metabolism by Dietary Lipid and Carbohydrate Ratio

Earlier studies indicated that rats on a low fat or fat-free diet had lower rates of xenobiotic

NATO ASI Series, Vol. H 90
Molecular Aspects of Oxidative Drug Metabolizing Enzymes
Edited by E. Arınç, J. B. Schenkman and E. Hodgson
© Springer-Verlag Berlin Heidelberg 1995

metabolism in liver microsomes (Wade, 1986). It is now known that this is due to the modulation of P450 enzymes by dietary fat. Lipids rich in polyunsaturated fatty acids are more effective than those rich in saturated and monounsaturated fatty acids in increasing the level of P450s. Dietary lipids are also required for the optimal induction of P450s by inducers such as phenobarbital, and in this case the effect is observable at the mRNA level (Yoo, *et al.,* 1990). Although most studies suggested that the factors responsible for the increased drug metabolism were lipids, the lipid to carbohydrate ratio may be a more important factor.

Recent results from our laboratory indicate that, in comparison to a fat-free diet, a 20% corn oil diet produced a 2-fold higher constitutive level of P450 2E1 but did not affect the maximal acetone-inducible level of this enzyme (Yoo, *et al.,* 1990). When diets containing different amounts of corn oil were fed to rats, those on the higher fat diet had higher P450 2E1 levels and higher blood acetone levels (Yoo, *et al.,* 1991), consistent with the hypothesis that ketone bodies and ketosis are key factors for the regulation of P450 2E1 (Miller and Yang, 1984). Menhaden oil and corn oil were more effective than olive oil and lard in maintaining a high level of P450 2E1, suggesting that factors other than ketosis were involved. The total concentration of microsomal P450 was higher in rats on a diet with higher lipid content; specifically, P450s 2E1, 3A and 2A1 were higher, whereas 2B1 and 2C11 were not affected (Yoo, *et al.,* 1992). In addition to P450s, dietary fat is also required for the maximal induction of glutathione *S*-transferase A and B by phenobarbital, and for maintaining the constitutive levels of glutathione *S*-transferase B (Yoo, *et al.,* 1991) and UDPG-glucuronsyl transferase (Dannenberg and Zakim, 1992).

The effect of dietary lipid on P450 2E1-dependent activities was also demonstrsted by studie *in vivo* which showed that the rate of enflurane defluorination was about 2-fold higher in rats on a 25% corn oil diet than rats on a 0.5% corn oil diet (Yoo, *et al.,* 1991). The higher levels of P450 2E1 in rats on a high fat-low carbohydrate diet (in comparison to those on a low fat-high carbohydrate diet) may also, at least partially, be responsible for the observed higher toxicity in these rats induced by halogenated hydrocarbons (Nakajima, *et al.,* 1982), many of which are substrates for P450 2E1 (Yang, *et al.,* 1992).

Induction of P450 and Related Enzymes by Dietary Chemicals

Although the induction of xenobiotic-metabolizing enzymes by dietary chemicals have been observed for many years, the molecular mechanisms of action are just beginning to be understood. Some of the mechanisms will be illustrated by the actions of indole-3-carbinol, diallyl sulfide (DAS), and diallyl sulfone (DASO$_2$) (structures shown in Fig.1). It was demonstrated that after an oral dose of indole-3-carbinol (a compound derived from cruciferous vegetables) to rats, P450 1A1 mRNA was elevated markedly in the liver and colon, and P450 1A2 mRNA was also elevated in the liver (Vang, *et al.*, 1990). It was suggested that under the acidic conditions in the stomach, indole-3-carbinol can be converted to indolo[3,2-*b*]carbazol (Gillner, *et al.*, 1985) or other acid reaction products (Bradfield and Bjeldanes, 1987) which bind to the *Ah* receptor and thereby activate the transcription of the P450 1A1 gene. This suggestion is supported by the observation that the inductive effect was seen when indole-3-carbinol was given to rats orally but not when given intraperitoneally (Bradfield and Bjeldanes, 1987). Acid treated indole-3-carbinol was much more effective than untreated indole-3-carbinol in inducing ethoxyresorufin deethylase activity in primary cultures of rat hepatocytes (Wortelboer, *et al.*, 1990).

A second example of transcriptional regulation is the induction of P450 2B1 by DAS. DAS, a compound derived from garlic, has been shown to induce rat hepatic P450 2B1 based on the increase of immunodetectable protein and pentoxyresorufin dealkylase activity (Brady, *et al.*, 1991). Recent studies in our laboratory indicate that this induction is accompanied by the elevation of 2B1 mRNA levels. Nuclear run-on experiments indicate that after administration of DAS, the transcriptional rate of the P450 2B1 gene was increased markedly, and was observable at 4 h after treatment (Pan, *et al.*, 1993). In primary culture of rat hepatocytes, P450 2B1 was not induced by DAS but by its metabolite $DASO_2$ (Pan, *et al.*, 1993). Thus, the induction is likely to be mediated by $DASO_2$ *in vivo*.

Post-transcriptional mechanisms may also be involved in the induction of P450s by dietary factors. In the induction of P450 2E1 by fasting, elevation of P450 2E1 mRNA was observed (Hong, *et al.*, 1987) but transcriptional activation could not be convincingly demonstrated by nuclear run-on experiments (unpublished results). Stabilization of the mRNA may play a role in this induction. In the induction of P450 2E1 by ethanol and acetone, elevation of the mRNA was not observed but may be observed under certain conditions (Hong, *et al.*, 1987, Song, *et al.*, 1986). Therefore, protein stabilization or increased translation efficiency may be involved in this induction. Protein stabilization due to the binding of substrates or pseudosubstrates may be a very common mechanism by which dietary constituents increase the levels of certain P450 enzymes (Eliasson, *et al.*, 1992).

Dietary induction of other xenobiotic-metabolizing enzymes have also been widely observed, especially for glutathione *S*-transferase and NAD(P)H-quinone oxidoreductase. The induction of these enzymes, mostly known to play roles in the detoxication of xenobiotics, have previously been reviewed (Prestera, *et al.*, 1993, Smith and Yang, 1993, Wattenberg, 1985, Wattenberg, 1992).

Inhibition and Enhancement of Monooxygenase Activities by Dietary Components

Many dietary compounds or their metabolites can bind to the active sites of P450 enzymes, serving as either substrates or competitive inhibitors. For example, DAS and $DASO_2$ are competitive inhibitors of P450 2E1-catalyzed reactions. Phenethyl isothiocyanate, occurring as a glucosinolate in a variety of cruciferous vegetables, competitively inhibits the metabolism of 4-(methylnitrosamino)-1-(3-pyridyl)-1-butanone (NNK), a potent tobacco carcinogen, in lung microsomes (Smith, *et al.,* 1990).

Many dietary flavonoids (Fig. 1) have been shown to inhibit monooxygenase activities. For example, quercetin, kaempferol, morin, and chrysin inhibited benzo(a)pyrene hydroxylase activity in human liver microsomes (Conney, 1982). Quercetin, kaempferol, and naringenin inhibited the P450 3A4-catalyzed oxidation of nifedipine and felodipine in human liver microsomes (Guengerich and Kim, 1991). It appears that flavones having free hydroxy groups on the A ring are inhibitors, whereas flavonoids containing no hydroxy groups (flavone, tangeretin, and nobiletin) are activators (stimulators) of selected monooxygenase activities. In addition, there is enzyme selectivity in the response to these flavonoids. As was demonstrated in the metabolism of acetaminophen and other substrates, the activities of P450 1A2 are more susceptible to the inhibitory effects, whereas the activities of P450 3A4 are more susceptible to the stimulatory effect (unpublished results).

In addition to a reversible inhibition mechanism, many dietary chemicals can inactivate P450 enzymes. In the case of DAS, for example, a metabolism-dependent inactivation of P450 2E1 was demonstrated (Brady, *et al.,* 1991, Hong, *et al.,* 1987). When DAS was given orally to rats, inactivation of microsomal P450 2E1 was observed. The inactivation of P450 2E1 was also demonstrated *in vitro* when $DASO_2$, an oxidative metabolite of DAS, was incubated with microsomes in the presence of an NADPH-generating system (Brady, *et al.,* 1991). The inactivation was time-, concentration-, and NADPH-dependent, showing a typical suicide inhibition pattern. It is believed that $DASO_2$ is converted by P450 2E1 to a reactive intermediate which modifies the heme moiety of this cytochrome and causes inactivation (Brady, *et al.,* 1991). Phenethyl isothiocyanate also inactivated P450 2E1 by a suicide mechanism (Ishizaki, *et al.,* 1990). Psoralens, which are found in edible plants such as figs, celery, parsley and parsnip, were shown to decrease 7-ethoxycoumarin and benzo(a)pyrene hydroxylase activities when added to human liver microsomal incubations in the presence of NADPH (Tinel, *et al.,* 1987). It was suggested that methoxsalen (8-methoxypsoralen), bergapten (5-methoxypsoralen) and psoralen are suicide inhibitors of P450 enzymes; P450s 1A1 or 1A2 may be involved.

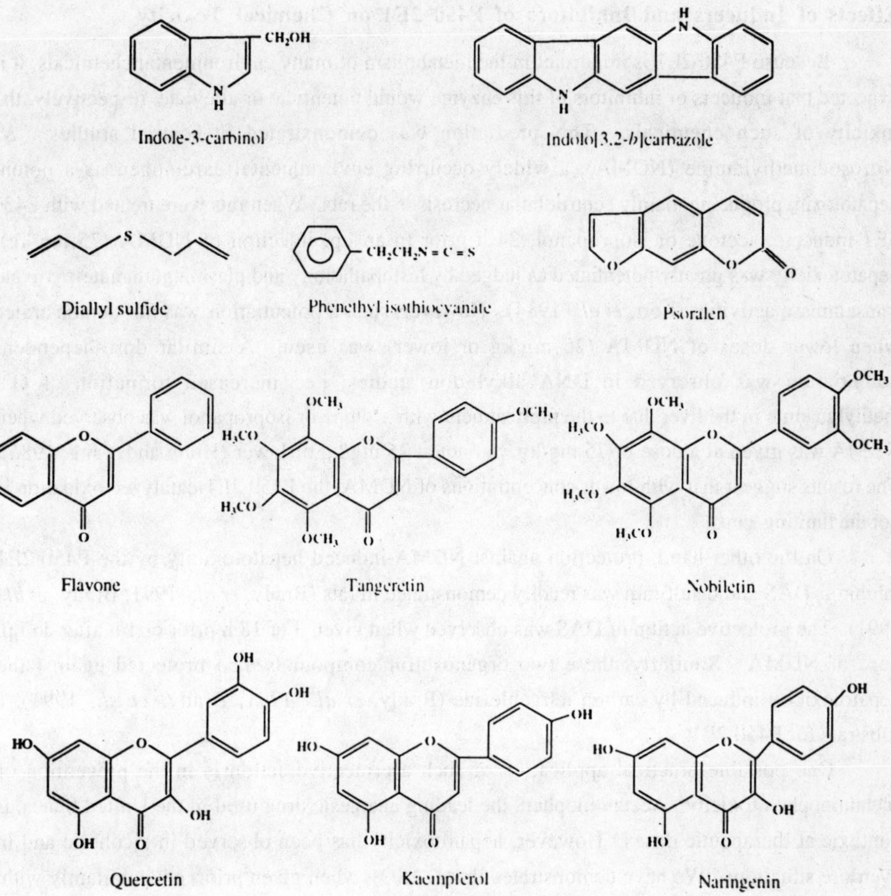
CH₂OH
Indole-3-carbinol
Indolo[3,2-b]carbazole
Diallyl sulfide
CH₂CH₂N = C = S
Phenethyl isothiocyanate
Psoralen
Flavone
OCH₃
H₃CO
Tangeretin
Nobiletin
OH
HO
Quercetin
Kaempferol
Naringenin

Figure 1

Effects of Inducers and Inhibitors of P450 2E1 on Chemical Toxicity

Because P450 2E1 is important in the metabolism of many environmental chemicals, it is expected that inducers or inhibitors of this enzyme would potentiate or alleviate, respectively, the toxicity of such chemicals. This prediction was demonstrated in several studies. *N*-Nitrosodimethylamine (NDMA), a widely occurring environmental carcinogen, is a potent hepatotoxin, producing mainly centrilobular necrosis in the rats. When rats were treated with P450 2E1 inducers, acetone or isopropanol, 24 h prior to an i.p. injection of NDMA (75 mg/kg), hepatotoxicity was greatly potentiated as judged by histopathology and plasma glutamate-pyruvate transaminase activities (Lorr, *et al.,* 1984). However, such a potentiation was not demonstrated when lower doses of NDMA (25 mg/kg or lower) was used. A similar dose-dependent potentiation was observed in DNA alkylation studies; i.e., increased formation of O^6-methylguanine in the liver due to the pretreatment with acetone or isopropanol was observed when NDMA was given at a dose of 75 mg/kg, but not at 25 mg/kg or lower (Hong and Yang, 1985). The results suggest that with lower concentrations of NDMA, the P450 2E1-catalyzed oxidation is not the limiting step.

On the other hand, protection against NDMA-induced hepatotoxicity by the P450 2E1 inhibitors DAS and disulfiram was readily demonstrated in rats (Brady, *et al.,* 1991, Brady, *et al.,* 1991). The protective action of DAS was observed when given 3 or 18 h prior or 1 h after an i.p. dose of NDMA. Similarly, these two organosulfur compounds also protected against the hepatotoxicity induced by carbon tetrachloride (Brady, *et al.,* 1991, Brady, *et al.,* 1991), a substrate for P450 2E1.

One possible practical application of such a protective action is in the prevention of acetaminophen toxicity. Acetaminophen, the leading analgesic drug used in the United States, is nontoxic at therapeutic doses. However, hepatotoxicity has been observed in alcoholic and in overdose situations. We have demonstrated that $DASO_2$, when given prior, concomitantly with, or within 3 h after a high dose of acetaminophen, prevented the development of hepatotoxicity in rats (unpublished results). With mice, a similar protective effect was observed, except that $DASO_2$ had to be given before, or within 20 min of the acetaminophen dose. Diallyl sulfone is water soluble without the pungent garlic odor, effective at low doses (10–25 mg/kg), and rather nontoxic (LD_{50} of 1.8 g/kg). It may have the potential for practical application in the prevention of acetaminophen toxicity, especially in alcoholics whose P450 2E1 levels are usually elevated.

The aforementioned mechanism is only one of the means by which diet affects toxicity. Many dietary constituents are inducers of phase II enzymes, such as glutathione *S*-transferases, which are known to detoxify many compounds. Sulfur containing amino acids, which are precursors of glutathione and 3'-phosphoadenosine-5'-phosphosulfate, are also known to affect the toxicity of chemicals.

Inhibition of Carcinogenesis by Dietary Chemicals

Epidemiological studies showed that frequent consumption of cruciferous vegetables is associated with low incidence of cancer at different sites (Graham, 1983, Graham, *et al.*, 1972, Young and Wolf, 1988). Individuals who frequently consumed garlic and allium vegetables are known to have lower gastric cancer incidence in certain areas (Buiatti, *et al.*, 1989, You, *et al.*, 1989, You, *et al.*, 1988). The aforementioned inhibition of P450 2E1 by phenethyl isothiocyanate and DAS is only one of the mechanisms which contribute to the lowered cancer incidence. Both compounds are known to also inhibit the activation of NNK, which is mainly catalyzed by P450s other than P450 2E1. The inhibition of NNK bioactivation and lung tumorigenesis in mice and rats by phenethyl isothiocyanate and other isothiocyanates has been clearly demonstrated (Guo, *et al.*, 1993, Guo, *et al.*, 1992, Morse, *et al.*, 1989, Morse, *et al.*, 1989, Morse, *et al.*, 1989, Smith, *et al.*, 1993, Smith, *et al.*, 1990) DAS and DASO$_2$ also block the activation of NNK and inhibit lung tumorigenesis in A/J mice (Hong, *et al.*, 1992). In addition to blocking the activation of carcinogens, phenethyl isothiocyanate, DAS, and other organosulfur compounds in cruciferous and allium vegetables can also induce phase II enzymes such as glutathione *S*-transferase which generally plays a role in the detoxification of carcinogens (Smith and Yang, 1993). The effects of indole-3-carbinol and other indoles, which are strong P450 inducers present in cruciferous vegetables, on carcinogenesis have not been clearly demonstrated. It has been suggested that, through its alteration of estrogen metabolism, indole-3-carbinol can decrease mammary tumorigenesis (Bradlow, *et al.*, 1991, Michnovicz and Bradlow, 1990). Dietary chemicals may affect carcinogenesis via mechanisms which are unrelated to xenobiotic metabolism. For example, some garlic related compounds or products such as 1-propenyl sulfide, ajoene, and garlic oil have been shown to inhibit tumor promotion by phorbol-myristate on mouse skin (Belman, 1983). Inhibition of lipoxygenase and ornithine decarboxylase are the possible mechanisms of such an inhibition.

Concluding Remarks

In this chapter, we discussed several mechanisms by which xenobiotic metabolism and toxicity can be affected by diet and dietary components. Although these actions can be readily demonstrated in animal models, it is not clear to what extent these actions occur in humans. There are two major problems in extrapolating results from studies with animals to humans. One is the species differences in the metabolism of carcinogens or toxicants and in how the metabolism is affected by dietary chemicals. Another problem is the extrapolation from the high doses commonly used in animals studies to the low doses of actual human exposure. More comparative studies among different species, more detailed dose-response studies, and additional quantitative epidemiology studies are needed to shed light on the problem of dietary effects on human xenobiotic metabolism and toxicity.

Acknowledgements

This work was supported by NIH grants ES 05673, CA 37037, and CA 46535. We thank Ms. Dorothy Wong for excellent secretarial assistance.

References

Anderson KE, Pantuck EJ, Conney AH, Kappas A. (1985) Nutrient regulation of chemical metabolism in humans. Fed Proc, 44: 130-133.

Bailey DG, Edgar B, Spence JD, Munoz C, Arnold JMO. (1990) Felodipine and nifedipine interactions with grapefruit juice. Clin Pharmacol Ther, 47: 180.

Belman S. (1983) Onion and garlic oils inhibit tumor promotion. Carcinogenesis (Lond.), 4: 1063-1065.

Bidlack WR, Brown RC, Mohan C. (1986) Nutritional parameters that alter hepatic drug metabolism, conjugation, and toxicity. Fed Proc, 45: 142-148.

Bradfield CA, Bjeldanes LF. (1987) Structure-activity relationships of dietary indoles: a proposed mechanism of action as modifiers of xenobiotic metabolism. J Toxicol Environ Health, 21: 311-323.

Bradlow HL, Michnovicz JJ, Telang NT, Osborne MP. (1991) Effects of dietary indole-3-carbinol on estradiol metabolism and spontaneous mammary tumors in mice. Carcinogenesis (Lond.), 12: 1571-1574.

Brady JF, Ishizaki H, Fukuto JM, Lin MC, Fadel A, Gapac JM, Yang CS. (1991) Inhibition of cytochrome P-450IIE1 by diallyl sulfide and its metabolites. Chem Res Toxicol, 4: 642-647.

Brady JF, Wang M-H, Hong J-Y, Xiao F, Li Y, Yoo J-SH, Ning SM, Fukuto JM, Gapac JM, Yang CS. (1991) Modulation of rat hepatic microsomal monooxygenase activities and cytotoxicity by diallyl sulfide. Toxicol Appl Pharmacol, 108: 342-354.

Brady JF, Xiao F, Wang M-H, Li Y, Ning SM, Gapac JM, Yang CS. (1991) Effects of disulfiram on hepatic P450IIE1, other microsomal enzymes, and hepatotoxicity in rats. Toxicol Appl Pharmacol, 108: 366-373.

Buiatti E, Palli D, Decarli A, Amadori D, Avellini C, Bianchi S, Biserni R, Cipriani F, Cocco P, Giacosa A, Marubini E, Puntoni R, Vindigni C, Fraumeni JJ, Blot W. (1989) A case-control study of gastric cancer and diet in Italy. Int J Cancer, 44: 611-616.

Conney AH. (1982) Induction of microsomal enzymes by foreign chemicals and carcinogenesis by polycyclic aromatic hydrocarbons: G. H. A. Clowes Memorial Lecture. Cancer Res, 42: 4875-4917.

Dannenberg AJ, Zakim D. (1992) Dietary lipid regulates the amount and functional state of UDP-glucuronosyltransferase in rat liver. J Nutr, 122: 1607-1613.

Eliasson EE, Mkrtchian S, Ingelman-Sundberg M. (1992) Hormone- and substrate-regulated intracellular degradation of cytochrome P450 (2E1) involving MgATP-activated rapid proteolysis in the endoplasmic reticulum membranes. J Biol Chem, 267: 15765-15769.

Gillner M, Bergman J, Cambillau C, Fernstrom B, Gustafsson J-Ä. (1985) Interactions of indoles with specific binding sites for 2,3,7,8-tetrachlorodibenzo-p-dioxin in rat liver. Mol Pharmacol, 28: 357-363.

Graham S. (1983) Resutls of case-control studies of diet and cancer in Buffalo, New York. Cancer Res, 43 (Suppl.): 2409-2413.

Graham S, Schotz W, Martino P. (1972) Alimentary factors in the epidemiology of gastric cancer. Cancer, 30: 927-938.

Guengerich FP, Kim D-H. (1991) *In vitro* inhibition of dihydropyridine oxidation and aflatoxin B$_1$ activation in human liver microsomes by naringenin and other flavonoids. Carcinogenesis, 11: 2275-2279.

Guo Z, Smith TJ, Wang E, Eklind KI, Chung F-L, Yang CS. (1993) Structure-activity relationships of arylalkyl isothiocyanates for the inhibition of 4-(methylnitrosamino)-1-(3-pyridyl)-1-butanone metabolism and the modulation of xenobiotic-metabolizing enzymes in rats and mice. Carcinogenesis *(Lond.)*, 14: 1167-1173.

Guo Z, Smith TJ, Wang E, Sadrieh N, Ma Q, Thomas PE, Yang CS. (1992) Effects of phenethyl isothiocyanate, a carcinogenesis inhibitor, on xenobiotic-metabolizing enzymes and nitrosamine metabolism in rats. Carcinogenesis (Lond.), 13: 2205-2210.

Hathcock JN. (1985) Metabolic mechanisms of drug-nutrient interactions. Fed Proc, 44: 124-129.

Hong J-Y, Pan J, Dong Z, Ning SM, Yang CS. (1987) Regulation of N-nitrosodimethylamine demethylase in rat liver and kidney. Cancer Res, 47: 5948-5953.

Hong J-Y, Pan J, Gonzalez FJ, Gelboin HV, Yang CS. (1987) The induction of a specific form of cytochrome P-450 (P-450j) by fasting. Biochem Biophys Res Commun, 142: 1077-1083.

Hong J-Y, Wang Z-Y, Smith T, Zhou S, Shi S, Yang CS. (1992) Inhibitory effects of diallyl sulfide on metabolism and tumorigenicity of the tobacco-specific carcinogen 4-(methylnitrosamino)-1-(3-pyridyl)-1-butanone (NNK) in A/J mouse lung. Carcinogenesis (Lond.), 13: 901-904.

Hong J-Y, Yang CS. (1985) The nature of microsomal N-nitrosodimethylamine demethylase and its role in carcinogen activation. Carcinogenesis (Lond.), 6: 1805-1809.

Ishizaki H, Brady JF, Ning SM, Yang CS. (1990) Effect of phenethyl isothiocyanate on microsomal N-nitrosodimethylamine (NDMA) metabolism and other monooxygenase activities. Xenobiotica, 20: 255-264.

Kappas A, Anderson AP, Conney AH, Alvares AP. (1976) Influence of dietary protein and carbohydrate on antipyrine and theophylline metabolism in man. Clin Pharmacol Ther, 20: 643-653.

Lorr NA, Miller KW, Chung HR, Yang CS. (1984) Potentiation of the hepatotoxicity of N-nitrosodimethylamine by fasting, diabetes, acetone, and isopropanol. Toxicol Appl Pharmacol, 73: 423-431.

Michnovicz JJ, Bradlow HL. (1990) Induction of estradiol metabolism by dietary indole-3-carbinol in humans. J Natl Cancer Inst, 82: 947-949.

Miller KW, Yang, CS. (1984) Studies on the mechanisms of induction of N-nitrosodimethylamine demethylase by fasting, acetone, and ethanol. Arch Biochem Biophys, 229: 483-491.

Morse MA, Amin SG, Hecht SS, Chung F-L. (1989) Effects of aromatic isothiocyanates on tumorigenicity, O^6-methylguanine formation, and metabolism of the tobacco-specific nitrosamine 4-(methylnitrosamino)-1-(3-pyridyl)-1-butanone in A/J mouse lung. Cancer Res, 49: 2894-2897.

Morse MA, Eklind KI, Amin SG, Hecht SS, Chung F-L. (1989) Effects of alkyl chain length on the inhibition of NNK-induced lung neoplasia in A/J mice by arylalkyl isothiocyanates. Carcinogenesis, 10: 1757-1759.

Morse MA, Wang C-X, Stoner GD, Mandal S, Conran PB, Amin SG, Hecht SS, Chung F-L. (1989) Inhibition of 4-(methylnitrosamino)-1-(3-pyridyl)-1-butanone-induced DNA adduct formation and tumorigenicity in the lung of F344 rats by dietary phenethyl isothiocyanate. Cancer Res, 49: 549-553.

Nakajima T, Koyama Y, Sato A. (1982) Dietary modification of metabolism and toxicity of chemical substances: with special reference to carbohydrate. Biochem Pharmacol, 31: 1005-1011.

Pan J, Hong J-Y, Li D, Schuetz EG., Guzelian PS, Huang W, Yang CS. (1993) Regulation of cytochrome P450 2B1/2 genes by diallyl sulfone, disulfiram, and other organosulfur compounds in primary cultures of rat hepatocytes. Biochem Pharmacol, 45: 2323-2329.

Pan J, Hong J-Y, Ma B-L, Ning SM, Paranawithana SR, Yang CS. (1993) Transcriptional activation of P-450 2B1/2 genes in rat liver by diallyl sulfide, a compound derived from garlic. Arch Biochem Biophys, 302: 337-342.

Pantuck EJ, Pantuck CB. (1978) Effects of dietary brussels sprouts and cabbage on human drug metabolism. Clin Pharmacol Ther, 25: 88-95.

Prestera T, Holtzclaw WD, Zhang Y, Talalay P. (1993) Chemical and molecular regulation of enzymes that detoxify carcinogens. Proc Natl Acad Sci, 90: 2965-2969.

Smith TJ, Guo Z, Li C, Ning SM, Thomas PE, Yang, CS. (1993) Mechanisms of inhibition of 4-(methylnitrosamino)-1-(3-pyridyl)-1-butanone (NNK) bioactivation in mouse by dietary phenethyl isothiocyanate. Cancer Res, 53: 3276-3282.

Smith TJ, Guo Z-Y, Thomas PE, Chung F-L, Morse MA, Eklind K, Yang, CS. (1990) Metabolism of 4-(methylnitrosamino)-1-(3-pyridyl)-1-butanone in mouse lung microsomes and its inhibition by isothiocyanates. Cancer Res, 50: 6817-6822.

Smith TJ, Yang CS. (1993) Effects of food phytochemicals on xenobiotic metabolism and tumorigenesis. In: Huang MT, Osawa T, Ho CT, Rosen RT (eds) Food Phytochemicals for Cancer Prevention. ACS Symposium Series 546, Washington, D. C., (in press) (invited review).

Song B-J, Gelboin HV, Park S-S, Yang CS, Gonzalez FJ. (1986) Complementary DNA and protein sequences of ethanol-inducible rat and human cytochrome P-450: transcriptional and post-transcriptional regulation of the rat enzyme. J Biol Chem, 261: 16689-16697.

Tinel M, Belghiti J, Descatoire V, Amouyal G, Letteron P, Geneve J, Larrey D, Pessayre D. (1987) Inactivation of human liver cytochrome P-450 by the drug methoxsalen and other psoralen derivatives. Biochem Pharmacol, 36: 951-955.

Vang O, Jensen MB, Autrup H. (1990) Induction of cytochrome P450IA1 in rat colon and liver by indole-3-carbinol and 5,6-benzoflavone. Carcinogenesis, 11: 1259-1263.

Wade AE. (1986) Effects of dietary fat on drug metabolism. J Environ Pathol Toxicol Oncol, 6: 161-189.

Wattenberg LW. (1985) Chemoprevention of cancer. Cancer Res, 45: 1-8.

Wattenberg LW. (1992) Inhibition of carcinogenesis by minor dietary compounds. Cancer Res, 52(suppl.): 2085s-2091s.

Wattenberg LW. (1971) Studies of polycyclic hydrocarbon hydroxylases of the intestine possibly related to cancer: effect of diet on benzpyrene hydroxylase activity. Cancer (Phila.), 28: 99-102.

Wortelboer HM, de Kruif CA, Cassee FR, Falke HE, Noordhoek J, Blaauboer BJ, van Iersel AAJ. (1990) Modification of xenobiotic metabolism by dietary indoles in rats in vivo and in vitro. Stockholm, Karolinska Institutet.

Yang CS, Brady JF, Hong J-Y. (1992) Dietary effects on cytochromes P-450, xenobiotics metabolism, and toxicity. FASEB J, 6: 737-744.

Yang CS, Yoo J-SH. (1988) Dietary effects on drug metabolism by the mixed-function oxidase system. Pharmacol Ther, 38: 53-72.

Yoo J-SH, Hong J-Y, Ning SM, Yang CS. (1990) Roles of dietary corn oil in the regulation of cytochromes P450 and glutathione S-transferases. J Nutr, 120: 1718-1726.

Yoo J-SH, Ning SM, Pantuck CB, Pantuck EJ, Yang CS. (1991) Regulation of hepatic microsomal cytochrome P450IIE1 level by dietary lipids and carbohydrates in rats. J Nutr, 121: 959-965.

Yoo J-SH, Smith TJ, Thomas PE, Ning SM, Lee M-J, Yang CS. (1992) Modulation of cytochromes P450 levels in rat liver and lung by dietary corn oil. Biochem Pharmacol, 43: 2535-2542.

You W-C, Blot WJ, Chang Y-S, Ershow A, Yang ZT, An Q, Henderson BE, Fraumeni JF, Jr., Wang, T-G. (1989) Allium vegetables and reduced risk of stomach cancer. J Natl Cancer Inst, 81: 162-164.

You W-C, Blot WJ, Chang Y-S, Ershow AG, Yang Z-Y, An Q, Henderson B, Xu G-W, Fraumeni, JFJ, Wang, T-G. (1988) Diet and high risk of stomach cancer in Shandong, China. Cancer Res, 48: 3518-3523.

Young TB, Wolf, DA. (1988) Case-control study of proximal and distal colon cancer and diet in Wisconsin. Int J Cancer, 42: 167-175.

Mechanisms of Mutagenicity and Tumour Formation

Diana Anderson
BIBRA Toxicology International
Woodmansterne Road
Carshalton
Surrey SM5 4DS
UK

Introduction

With the continuously enlarging world population, and the necessary associated resources for man's survival, such as energy food, etc., there is a greater chance of exposure of humans to potentially harmful agents. Cancer causation resulting from such exposures and control of these exposures is a primary cause of concern.

Cancer ranks high among the causes of death and current emphasis is on prevention and early detection. Many chemicals are known to cause cancer in experimental animals and according to IARC, 55 agents and processes are proven to do so in man (IARC, 1987). Comparisons of cancer profiles in various ethnic and national groups point to the major role of diet and lifestyle in tumour development (Doll and Peto, 1981). For example, to all cancer deaths, diet contributes around 35% and tobacco 30%, with other factors such as occupation (4%) and industrial products (2%) contributing to a lesser degree.

Despite the low incidence of cancer deaths due to occupation, early epidemiological evidence showed increased cancer risk in workers subjected to high occupational exposures. These studies counted and compared cases of cancer in relation to assessed or measured external exposures and allowed the introduction of preventive measures. Most of the carcinogens identified were potent human carcinogens inducing readily observable increases in cancer yield; animal carcinogenicity bioassays also examined these same chemicals. Nowadays, however, less potent carcinogens are being evaluated and the possibility of using data from studies of putative intermediate effects or cancer-correlated endpoints in the assessment of some exposures,

NATO ASI Series, Vol. H 90
Molecular Aspects of Oxidative Drug Metabolizing Enzymes
Edited by E. Arınç, J. B. Schenkman and E. Hodgson
© Springer-Verlag Berlin Heidelberg 1995

is currently of importance. Data on the mechanisms of action of chemicals are used as adjuncts to and sometimes instead of carcinogenicity bioassays.

Carcinogenesis is a complex biological process and carcinogenic mechanisms extend from the molecular to the population level of investigation. The evidence that a chemical is carcinogenic is drawn from several sources: epidemiological studies, life-time exposure studies in laboratory animals, short-term laboratory tests for genotoxicity, structure-activity relationships, biomarkers of effect, response and susceptibility in humans, etc.

Cancer is thought to be a multistage process. This concept was formulated to account for a variety of epidemiological observations relating cancer incidence to dose and duration of exposure and age at exposure to specific agents. The concepts of multistage carcinogenesis and the genetic mechanisms involved have developed over many years of cancer research (Boveri 1914; Mottram 1935; Rous and Beard 1935; Berenblum 1941, 19785; Case *et al.*, 1954; Foulds 1954; Bishop 1991; Pitot 1993). Other hypotheses based on epidemiological evidence are that immuno-suppression allows the survival of virally induced cancers which would otherwise be eliminated immunologically, that hormones cause certain cancers by inducing cell proliferation and an important mechanism of genetic susceptibility to cancer is the inheritance of one of the changes involved in multistage carcinogenesis which normally occur somatically (IARC 1992). The convergence of molecular, biological and epidemiological perspectives on carcinogenesis is shown in the development of colon cancer (Fearon and Vogelstein 1990) in which a series of early and late stages are identified with changes in specific genes (see later).

Historically, skin pathogenesis has provided much information on the various stages of carcinogenesis (Boutwell, 1964; Berenblum 1974). The process of neoplastic development is often divided into three operationally defined stages, namely initiation, promotion and progression. Furthermore, it is possible to classify agents to which humans are potentially exposed as to whether they are effective at one or the other or

Table 1. Classification of chemical carcinogens in relation to
their action on one or more stages of carcinogenesis

	Examples of chemicals carcinogenic to humans
Initiating Agent (incomplete carcinogen)	A chemical which only initiates cells (at low doses - smokeless tobacco products, cyclophosphamide, bischloromethyl ether)
Promoting agent	A chemical which causes the expansion of initiated cell clones (certain mineral oils, ethanol, caloric intake [high fat diet], oestrogens)
Progressor agent	A chemical which converts an initiated cell or cell in the stage of promotion to a potentially malignant cell (benzene, asbestos, arsenicals)
Complete carcinogen	A chemical which induces cancer from normal cells, usually possessing properties of initiating promoting and progressor agents (vinyl chloride, tobacco smoke, 2-napthylamine and 4-aminobiphenyl)

After Pitot, 1993.

all three stages of carcinogenesis. Such a classification has been outlined by Pitot (1993) in Table 1 with examples of known human carcinogens that exert their effects at the specific stage(s) noted. These three main stages can each consist of multiple stages. In addition to the skin, a wide variety of other tumours exhibit these stages (Hecker *et al.*, 1982), particularly the liver (Pitot and Sirica, 1980). Initiation was defined by treating mouse skin with a subcarcinogenic dose of the chemical carcinogen 7,12-dimethyl benzo(a)anthracene (DMBA). The mouse skin was considered to be initiated because administration of non-carcinogenic promoting treatments, such as croton oil and certain other skin irritants resulted in a high incidence of papillomas, whereas treatment with either the initiator or the promoter alone was ineffective. The order in which the chemicals are applied is important since the tumour response is greatly reduced or absent when the initiator follows the promoter. The effects of initiator treatments are often irreversible, as shown by experiments in which promoter treatments were delayed for over a year with no significant decrease in tumour yield. No exclusive initiating agent has been described (Iversen, 1988) and some potent tumour promoters are also complete carcinogens.

In such experiments, the clearly staged phenomena observed have not been described in human populations but are closely analogous to the concepts of early and late carcinogenesis seen in mathematical models of multistage carcinogenesis (Peto, 1984). These models are based on the assumption that cancer develops from a single cell which passes through more than one stage and that clonal proliferation of cells can occur at each stage. It is difficult to identify single initiated cells but Moore *et al.* (1987) have suggested that single hepatocytes expressing the placental form of glutathione-5-transferase may represent single initiated cells.

Events involved in the multistage process of carcinogenesis

Owing to the discovery of genes and their products involved in the regulation of cell proliferation and differentiation and the identification of genetic events implicated in tumour formation, the understanding of the mechanisms of carcinogenesis is developing rapidly (Varmus, 1989). Genetic damage may range from mutation to gross chromosomal changes which lead to distortion of either the expression or the biochemical function of genes (Bishop, 1987).

Genetic changes induced by carcinogens include gene mutations, gene amplifications, chromosomal rearrangements and aneuploidy. Examples are available of each of these mutational changes in different tumours and provide support for the somatic mutation theory of carcinogenesis. Mutation is unlikely to be the sole mechanism for producing stable changes relevant to carcinogenesis. The expression of genes may be altered permanently by exposure to some agents. Stable changes in gene expression occur during development and differentiation which are often epigenetic or non-genotoxic. An epigenetic change is any modification in a phenotype that does not result from an alteration in a DNA sequence. It may be stable and heritable and includes alterations in DNA methylation, transcription activation, translational control and post-translation modification. In contrast, a genetic or genotoxic change or mutation is any modification in a phenotype resulting from an alteration in a primary DNA sequence. This may be a single base-pair change resulting in a base pair substitution mutation, an addition or deletion of one or more base-pairs resulting in a frameshift mutation. Chromosomal mutation is a morphological alteration in the gross structure of the chromosome resulting from breakage and reunion of chromosomal material during the cell cycle. Genomic mutation is a change in the number of chromosomes from the normal diploid number contained in a complete set of chromosomes from each parent. Only in the last decade have studies in molecular genetics begun to identify particular genes which are altered after carcinogen/mutagen treatment, and genes known as oncogenes and tumour suppressor genes are involved particularly in proliferation and differentiation (P and D genes) (Bos and Kreijl, 1992).

Oncogenes are activated forms of proto-oncogenes. They are activated by chemicals or other agents or spontaneously. Proto-oncogenes are a highly conserved ubiquitous group of genes which encode proteins that regulate normal growth and differentiation. Proto-oncogenes can function as growth factor receptors, transcriptional activators, signal transducers and protein kinases. Normal proto-oncogenes can be converted into oncogenes in various ways by involving modification to DNA structure and or function. They can be activated by point mutation within the oncogene and chromosomal rearrangements can result in the formation of a new gene product, part of which consists of sequences from the proto-oncogene. These two processes can result in qualitative changes to proto-oncogenes by creating new genes which over-ride the functions of the normal genes from which they are derived. Chromosomal rearrangements can also place a proto-oncogene next to an inappropriate promoter of mRNA transcription, or a strong promoter of mRNA transcription can be added from a virus to an oncogene, or there can be an increase in the number of copies of the proto-oncogene in the cell by gene amplification. These last three mechanisms lead to quantitative changes since normal proto-oncogenes make abnormally high levels of their gene products. Activation of proto-oncogenes by all these methods cause the cell to alter genes or gene products which add to the genes or gene products that are expressed normally (IARC, 1992).

In several different tumour types, in humans and laboratory animals, however, the occurrence of many oncogenes is variable within and between tumour categories. In some neoplasms, a given oncogene may occur rarely or never and the occurrence of specific oncogenes is not distinctive for different classes of chemical carcinogens.

Proto-oncogenes may be activated at several different points during the development of tumours such that a succession of different mutagenic events may take place.

Tumour suppressor genes (recessive oncogenes, anti-oncogenes) differ from oncogenes since they are involved in carcinogenesis only when they are deleted or inactivated causing a loss of gene function e.g. the Rb gene which is lost or mutated in the childhood tumour retinoblastoma (Knudson, 1985; Hansen and Cavence, 1987). When inserted into retinoblastoma cells *in vitro*, the gene, which has now been cloned, will restore normal function. In several adult cancers, including carcinomas of the breast, lung and prostate, inactivation of the Rb gene has also been demonstrated. A phosphoprotein known as p-105-rb, the gene product, has been identified and shown to form physical bonds with nuclear oncoproteins coded by nuclear oncogenes found in human DNA viruses. This would suggest ways in which oncogenes and tumour suppressor genes may interact (HMSO, 1991). Genes that do not belong to one of these two categories may also play a role in carcinogenesis: they include those involved in cell recognition (cell adhesion molecules, histocompatibility complex) and metastasis.

A large variety of these genes has been identified, cloned and increased the understanding of the genetic events involved in the evaluation of human and animal tumours (Bishop, 1987; 1991). The development of tumours common in adult humans (e.g. of the lung, colon and breast) involves the loss or inactivation of multiple suppressor genes and/or the activation of proto-oncogenes (Tables 1 & 2). Cancer of the colon is the best characterised human tumour in which three tumour suppressor genes (APC, p53 and DCC) are lost on chromosomes 5, 17 and 18 and one proto-oncogene (K-*ras*) are frequently altered, and specific genetic alterations are associated with the progressive stages of tumour development (Fearon and Vogelstein, 1990; Vogelstein *et al.*, 1988; Vogelstein and Kinzler, 1993)) as shown in Figure 1.

Such findings support the notion that neoplastic development is a multistage progressive process involving multiple genetic changes and demonstrate that genetic changes are required at many stages of the neoplastic process.

Transgeneration transmission of carcinogenic risk of chemicals and radiation has been demonstrated in experimental animals (Tomatis, 1989; Anderson, 1990) and initiating events in multistage carcinogenesis may occur in the germ cells. Studies in humans suggest that germ cell mutation of p53, Rb and other genes is an important determinant of suceptibility to specific tumours (Malkin *et al.*, 1990).

Oncogene	Neoplasm	Activation
abl	CML	Translocation
erb B1	Squamous cell carcinoma; astrocytoma	Amplification
erb B2	Adenoma of breast, ovary and stomach	Point mutations
gip	Carcinoma of ovary and adrenal glands	Point mutations
gsp	Adenoma of pituitary; carcinoma of thyroid	Point mutations
myc	Burkitts lymphoma; carcinoma of lung, breast, cervix	Translocation Amplification
L-myc	Carcinoma of lung	Amplification
N-myc	Neuroblastoma; small cell carcinoma of lung	Amplification
H-ras	Carcinoma of liver, lung, pancreas; melanoma	Point mutations
K-ras	Leukaemia; carcinoma of colon, thyroid, melanoma	Point mutations
N-ras	Carcinoma of genitourinary tract, thyroid	Point mutations
ret	Carcinoma of thyroid	Rearrangement

Table 2. <u>Examples of oncogenes activated in human tumours</u>

(After Davies 1993)

Tumour suppressor gene	Neoplasm	Function	Location
RB	Retinobastoma, osteosarcoma, carcinoma of breast, lung, bladder	Transcription factor	13q14
p53	Astrocytoma, carcinoma of breast, colon, lung, liver, brain, stomach, bone, ovary, lymphoid tumours	Transcription factor	17p13
WT1	Wilms' tumour	Transcription factor	11p13
APC	Colon carcinoma	G-protein activation?	5q21
DCC	Colon carcinoma	Cell adhesion-communication	18q21
MCC	Colon carcinoma	G-protein activation	5q22
NF1	Neurofibromatosis	GAP related	17q11.2
PTP	Renal cell carcinoma	Protein-tyrosine activation	3p

This is not an exhaustive list; other tumour suppressor genes have been located, but not yet cloned.

Table 3.   <u>Tumour suppressor genes found in human neoplasms</u>

(After Davies 1993)

Genotoxic and non-genotoxic carcinogens

A distinction can be drawn between compounds causing cancer which directly damage DNA (genotoxins) and those which damage DNA indirectly (non-genotoxins) e.g. chemicals which generate free radicals or H_2O_2, interfere with topoisomerases or imbalanced DNA precursor pools (Anderson et al., 1981; Anderson, 1985).

However, the division into genotoxic and non-genotoxic carcinogens is not clear-cut since a variety of biological processes is involved in carcinogenesis. Some are related to particular types of carcinogens or to tumours which arise from specific circumstances. The categorisation of carcinogens according to their mechanisms of action is not a simple task since a carcinogen may exert more than one effect and operate by more than one mechanism. Some effects arise in the absence of external exposure suggesting the effect of internal and endogenous stimuli. When tumours in humans arise from complex mixtures, such as tobacco smoke, there may be a contribution from several chemicals. In addition, chemicals interact with other agents including viruses and radiation.

Genetic events play a very important role in the activation of proto-oncogenes and inactivation of tumour suppressor genes which contribute to the development of human and animal cancers. Most known chemical and physical carcinogens are mutagenic and a direct role for carcinogen-induced genetic changes in critical target genes has been demonstrated in tumour induction in various experimental models (Barbacid, 1987; Balmain and Brown 1988). Correlations have been shown in humans between exposure to carcinogens and specific base-pair mutations in tumour oncogenes and tumour suppressor genes (Bos 1989; Hollstein et al., 1991). For such reasons, assays of genotoxicity therefore, have a critical role to play in the identification of potential carcinogenic agents and in the elucidation of mechanisms of action of known carcinogens (IARC, 1987).

It has been suggested (Ashby and Purchase 1992; Ashby et al., 1989; Ashby and Tennant, 1988; 1991) that genotoxic carcinogens produce tumours in multiple species at multiple sites, whereas non-genotoxic carcinogens tend to produce tumours

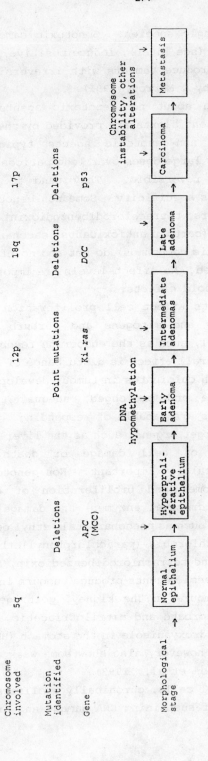

Figure 1. A genetic model for human colorectal cancer.
Based on Vogelstein and Kinzler (1993)

at single sites in single species. Genotoxic carcinogens are structurally- alerting (see later), induce positive responses in short-term tests and produce tumours with irreversible effects with no threshold (Ashby & Morrod, 1991).

Far less is known about non-genotoxic agents and their mechanisms. No consistent alerts are provided by their chemical structure. However, non-genotoxic agents typically exert carcinogenic effects at large doses over long periods of time and effects are initially reversible. They tend to demonstrate target organ and species specificity. Some non-genotoxic agents (hormones, phorbol esters, tetrachlorodibenzodioxin) act through receptors while others (certain antioxidants, saccharin, several hepatoxic and nephrotoxic compounds) do not and no-effect-levels can often be demonstrated. No-effect-levels are important in the determination of threshold effects.

Non-genotoxic agents affect cell proliferation where they may act as cytotoxins and mitogens and perturb the normal processof growth control, such as the endocrine feedback control in the thyroid. The overall effect is disturbance of homeostasis in target tissues, which culminates in tumour development. When stimulation is intense and prolonged, adequate numbers of susceptible target cells are capable of responding. It has been well demonstrated in target organs, such as the liver and kidney, that repeated cycles of cell damage or death and cell regeneration are particularly important. Non-genotoxic agents may stimulate peroxisomes and proliferation of endoplasmic reticulum (affecting microsomal enzymes) and damage lysosomes. Non-genotoxic chemicals such as phenobarbital, ethyl oestradiols, dieldrin, aldrin, trichlorobis-(parachlorphenyl)ethane (DDT), diethylhexylphthalate and tetrachlorodibenzodioxin, halogenated biphenyls and hypolipidaemic agents produce tumours in the liver, chloroform produces tumours in the kidney, goitrogenes in the thyroid, saccharin, ascorbate and nitrilotriacetic acid in the bladder and butylated hydroxyanisole in the stomach (HMSO, 1991). Some of these compounds however, also show some weak activity in short-term tests (Jackson et al., 1993).

Compared with normal cells, chronically proliferating cells are at a greater risk of sustaining DNA damage and are probably less able to repair it by error-free processes. The longer the period of mitotic stimulation the more likely it is that such damage will be incurred. This would suggest that mutations may occur in some instances rather late in the course of tumour development. The nature of the mutagenic stimulus also arises, given that non-genotoxic agents are initially involved. Some non-mutagenic compounds may act partly as indirect genotoxins, such as releasing active oxygen species which can damage DNA by modifying bases and breaking the sugar-phosphate backbone. Genotoxic damage may also arise by endogenous mutagens which form covalently-bound mutagens with DNA. Any imperfectly repaired resulting damage produces a mutation such as is formed with exogenous agents. Endogenous mutations are often referred to as spontaneous events and occur during the ageing process.

Non-genotoxic agents may alter cell proliferation in other ways, such as by causing disturbances in normal cell-to-cell communication. The effects of non-genotoxic agents on cell proliferation are sometimes referred to as promotion. Promotion originally meant the clonal expansion of initiated cells. For non-genotoxic compounds, however, cell proliferation precedes any mutational event which is regarded as equivalent to initiation and proliferation will not be clonal (HMSO 1991). Also some non-genotoxic carcinogens act in other ways than by affecting cell proliferation, such as by immune suppression (Kinlen 1992). This may damage normal cell surveillance and control mechanisms and could even facilitate the expression of oncogenic viruses.

Mechanistic Aspects of Genotoxic and Non-Genotoxic Carcinogens

DNA Binding

The reactivity of carcinogens/mutagens with DNA and its biological consequences have been fundamental to the understanding of chemical carcinogenesis over the last two decades (Lawley and Brookes, 1968; Heidelberger, 1970). Many chemical carcinogens are metabolised to generate electrophilic intermediates capable of binding covalently to the DNA of cells from which tumours arise. DNA adducts have been characterised including those formed by aflatoxin B_1, aromatic amines, polycyclic hydrocarbons and N-nitro-compounds. DNA binding is detected using radiolabelled carcinogens and DNA post-labelling. DNA adducts have shown midcoding properties which suggest how mutation might arise by this means. Belinsky et al. (1987) have reported that carcinogen-DNA adducts differ qualitatively and quantitatively in their mutagenic potential and vary among cell types in a given tumour. These adducts may be eliminated by DNA repair processes or spontaneously and in animals treated with certain N-nitroso-compounds, there has been persistence of O-alkylated guanine in DNA at the site of tumourigenesis.

Proliferation and differentiation genes (P and D genes)   P and D genes (Bos and van Kreijl, 1992) are mutated by chemical carcinogens and evidence came from experiments using animal model systems such as the mouse skin carcinoma model and the rat mammary carcinoma model (Barbacid, 1987; Balmain and Brown, 1988) and from analysis of ras proto-oncogenes and the p53 tumour suppressor gene in human cancers (Bos, 1989; Hollstein et al., 1991).   Initiation has been induced in the mouse skin by 7,12-dimethyl benzo(a)anthracene (DMBA) and N-methyl-N-nitrosoguanidine (MNNG).   After subsequent treatment with the tumour promoter 12-O-tetradecanoylphorbol 13-acetate (TPA) papillomas arise which may progress to carcinomas. Using the NIH3T3 transfection assay system and analysing the carcinomas and papillomas for the presence of activated oncogenes, revealed that many of the tumours initiated by DMBA, contained a mutated H-ras gene with an A→T transversion of the second base in codon 61. This was not the case for tumours initiated by MNNG or benzo(a)pyrene.   This finding of carcinogen-specific mutation would suggest that certain genes are critical for carcinogenesis and mutation is dependent on the reaction of DNA with the carcinogen (Quintanilla et al., 1986; Balmain and Brown, 1988). Similar results were obtained with the rat mammary tumour model. These tumours were induced in female rats by N-methyl-N-nitrosourea (MNU) and DMBA.   Those induced by MNU had a mutated H-ras gene with a mutation in codon 12 in which GAA is substituted for GGA.   In tumours induced by DMBA, the mutation is in codon 61 (Sukumar *et al.*, 1983; Zarbl *et al.*, 1985). The genetic event is again dependent on the initiating carcinogen or its metabolite within the target gene and Kumar *et al.* (1990) have shown the H-*ras* mutation in mammary tissue, two weeks after treatment with the carcinogen. Carcinogen-specific mutations have also been observed in mouse liver carcinomas (Wiseman *et al.*, 1986).

Examples of carcinogen-induced mutations in cancer-related genes also occur in humans. Out of 26 hepatocellular carcinomas from Southern Africa (Bressac *et al.*, 1991) and Quidong, China (Hsu *et al.*, 1991) 11 out of 13 base-substitution mutations in the p53 tumour suppressor gene were at a single site on the third base-position of codon 249 and all but one were G→T transversions. In mainly smoking patients with non-small-cell lung cancer (Bos, 1989; Hollstein *et al.*, 1991), G →T transversions (Slebos *et al.*, 1990) were found in both the p53 and K-*ras* gene (Chiba *et al.*, 1990). Patients with adenocarcinomas of the lung containing the K-*ras* codon 12 mutation were current or ex-smokers (Slebos *et al.*, 1990; 1991). Van't Veer *et al.* (1989) have found mutations in N-ras codon 61 in melanomas induced by sunlight.

Structure Activity Relationships (SAR)

These can be used in the identification of carcinogens and have been reviewed by Phillips and Anderson (1993). Two approaches have been developed: qualitative SAR, based on the identification of likely reactive centres in molecules and quantitative QSAR which attempts to predict the magnitude of the effect. With regard to the latter there are also two fundamental approaches, one that can be applied to congeneric chemicals i.e. with a common skeleton and similar functional groups and one that cannot (Parodi and Waters, 1991). The approach for congeneric groups developed by Hansch in the 1960s can provide a very precise description of an effect for an unknown compound, since a continuous variation in chemical properties within the group determines a continuous variation in biological properties. With non-congeneric chemicals, however, neither feature is present. A different approach therefore, the so-called classification approach, is necessary, where compounds are assigned to one of two classes: active or non-active. There are overlaps between the different approaches and an extension of that of Hansch has been used to classify various groups of carcinogens. The quantitative substructural approach has been used to identify electrophilic reactive centres in chemical groups.

The qualitative approach involves a non-computational chemistry based method for identifying carcinogens (Ashby, 1985; Ashby and Tennant, 1988). It involves recognition of those substituents within an electrophilic carcinogen that are associated with its carcinogenicity (Figure 2). Estimation of the probable electrophile within a DNA reactive carcinogen is based on Miller's electrophilic theory of carcinogenesis (Miller, 1970). Various key (20) structural alerts are recognised (see Figure 2) and possible new ones can be predicted in the absence of data on projected electrophilicity (Tennant and Ashby, 1991). Using the NTP database of 301 compounds tested for carcinogenicity in two rodent species and the Ames test, a high correlation was found between the presence of structural alerts for DNA reactivity in these compounds and their mutagenicity. Nearly 80% of alerting compounds were, for example, found to be mutagenic compared to just 3% of non-alerting compounds. However, the correlation between either property (structurally alerting/mutagenicity) and carcinogenicity was low suggesting that structural alerts are useful but non-definitive indicators of potential carcinogen activity (Ashby and Tennant, 1991). The artificial intelligence technique CASE (Computer Automated Structure Evaluation) has also been used to predict structural alerts both retrospectively and prospectively. In the former (Rosenkranz and Klopman, 1990a) the concordance between CASE-predicted structural alerts and mutagenicity was as good as that achieved by Ashby. In the latter (Rosenkranz and Klopman, 1990b), a concordance of 92% was achieved for the prediction of the presence or absence of alerting features in a group of 39 compounds comprising both genotoxic and non-genotoxic agents. The validity of these predictions with respect to rodent carcinogenicity has yet to be clarified.

Various quantitative approaches have also been used to identify carcinogens -including the classification approach - known as SIMCA (Soft Independent Modelling of Class Analogy) which has characterised mutagenic and non-mutagenic N-nitroso compounds, 4-nitroquinoline-1-oxides and polyaromatic hydrocarbons (Dunn and Wold, 1981). COMPACT (Computer Optimised Molecular Parametric Analysis of Chemical Toxicity) is another method which uses discriminant analysis and is based on the idea that carcinogens and mutagens could be identified by their affinity for specific macromolecular receptor sites. The accumulated experience with over 1,000 compounds has suggested that genotoxins and non-genotoxins can be distinguished by their molecular shape and the difference in energy between frontier molecular orbitals. Such compounds are known to be substrates of a particular cytochrome P450 isoenzyme family. The degree of correlation between COMPACT and rodent carcinogenicity was 92% (Lewis et al., 1992).

Cell proliferation

Cell proliferation is an important feature for both genotoxic and non-genotoxic carcinogens (Rajewsky, 1972; Ames and Gold, 1990a & b; Baserga, 1990; Butterworth, 1990; Cohen and Ellwein, 1990; Preston-Martin et al., 1990; Weinstein 1991a&b; Butterworth et al., 1992). Cell proliferation acts at each stage of the carcinogenic process and is dependent on cell kinetics. Carcinogens may act differently on the various stages of the multistage process of carcinogenesis. Hyperplasia is a factor in tumour promotion in mouse skin carcinogenesis but not all skin hyperplasia-inducing compounds exert tumour promoting activity (Slaga, 1985; Marks, 1990).

Enhanced cell replication may increase the frequency of mutations by inducing errors in replication or by converting DNA adducts to mutations before DNA repair can occur (Loeb and Cheng, 1990). However, according to Craddock (1976) a genotoxic agent will be more effective as a carcinogen if it is given at a dose which does not induce cell proliferation. During cell division non-disjunction and mitotic recombination can also take place and

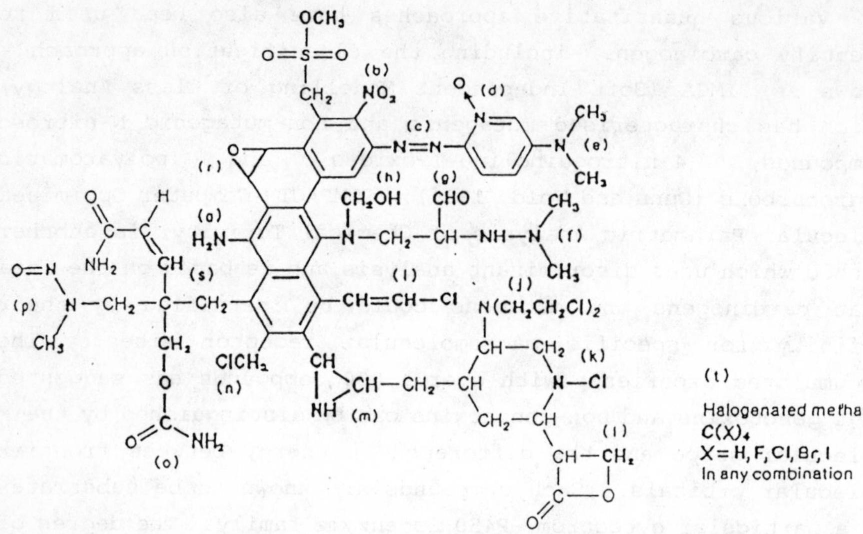

Fig. 2. Modified version of the model compound upon which structural alerts to genotoxicity are based. The structure shown earlier in Ashby and Tennant (1988) has been supplemented with a centre of Michael reactivity(s). Halomonocarbons(t) have been added as a separate structure because their potential genotoxicity will probably not become evident when attached to a larger molecule. Thus a,a-dichlorotoluene (PhCHCl$_2$) should not be considered as an active analogue of dichloromethane (CH$_2$Cl$_2$), just as aniline (PhNH$_2$) is not usefully considered as an analogue of ammonia (NH$_3$). The structural subunits a–r were identified in Fig. 1 of Ashby and Tennant (1988), and are as follows: (a) alkyl esters of either phosphonic or sulphonic acids; (b) aromatic nitro groups; (c) aromatic azo groups, not per se, but by virtue of their possible reduction to an aromatic amine; (d) aromatic ring N-oxides; (e) aromatic mono- and di-alkylamino groups; (f) alkyl hydrazines; (g) alkyl aldehydes; (h) N-methylol derivatives; (i) monohaloalkenes; (j) a large family of N and S mustards (β-haloethyl); (k) N-chloramines (see below); (l) propiolactones and propiosultones; (m) aromatic and aliphatic aziridinyl derivatives; (n) both aromatic and aliphatic substituted primary alkyl halides; (o) derivatives of urethane (carbamates); (p) alkyl N-nitrosamines; (q) aromatic amines, their N-hydroxy derivatives and the derived esters; (r) aliphatic epoxides and aromatic oxides. Qualifications or refinements of these units are discussed in Ashby and Tennant 1988. The N-chloramine substructure (k) has not yet been associated with carcinogenicity, but potent genotoxic activity has been reported for it (discussed in Ashby and Tennant, 1988).

mutation is enhanced because of greater exposure to chemical agents (Mead and Fausto, 1989).

There are two types of cell proliferation: namely, regenerative cell proliferation, which occurs after cell death in a tissue and mitogen-induced cell proliferation. Examples of agents which act in the former case are carbon tetrachloride and chloroform in the rodent liver, α-limonene in the male rat kidney and sodium saccharin in the bladder (Doolittle *et al.*, 1987; Flamm and Lehmann-McKeeman, 1991). Preneoplastic hepatic foci have been shown to proliferate in response to the regenerative stimulus provided by partial hepatectomy (Rabes, 1988). The carcinogenic activity of many mitogenic agents depends on the continued administration of the chemical or other agents to stimulate clonal expansion (Barrett, 1992). Not all cytotoxic agents that cause compensating cell proliferation induce neoplasms in affected tumour tissue (Huff, 1992).

Hormones

These can produce mutagenesis in normal cells or clonal expansion of genetically altered cells. The mechanism by which this occurs is unclear, but receptor-mediated events are often involved (Lucier, 1992). Receptor-mediated stimulation of mitogenesis operates through a different mechanisms than cell proliferation arising from lethality (Roth and Grunfeld, 1985). For some receptor mediated responses, a relationship exists between receptor occupancy and biochemical/carcinogenic response (Roth and Grunfeld (1985). Hormones may interact with receptors located in the plasma membrane or in the cytosol and/or nucleus.

Factors affecting multistage mechanisms

Metabolic capacity

Cancer is dependent on the rate at which the chemical reaches the target tissue and this is then dependent on the metabolic capacity of the animal. Most chemicals which damage DNA act by forming reactive intermediates generated by metabolic reactions (Miller, 1970). Many of these reactions convert chemically stable precarcinogens to electrophilic reactive species carried out by cytochrome P450 enzymes (Guengerich, 1988). Metabolic differences are an important cause of species variation in carcinogenicity e.g. the metabolism of vinyl chloride is fifteen and twelve times faster in mice and rats respectively on the basis of body weight than humans (Buchter et al., 1978) and this rank order is maintained for vinyl chloride induced cancer (Gold et al., 1989).

Physiological differences

Such differences may also account for species differences since formaldehyde causes very different tumourigenic responses in rats and mouse (Dybing and Huitfeldt, 1992).

Inter-individual variability

Inter-individual variability amongst species exists and this has been studied in human populations where polymorphic variation is known to exist. A link has been reported between debrisoquine metabolic polymorphism and risk from lung cancer (Ayesh et al., 1984; and Caporaso et al., 1990). As association between aryl hydrocarbon hydroxylase activity and the risks for lung and pharyngeal cancer has also been suggested (Kellerman et al., 1973; Kouri et al., 1982). The N-acetyl-transferase polymorphism has been related to a risk for bladder cancer among individuals exposed to arylamines (Cartwright et al., 1982; Mommsen et al., 1985). Slow acetylators poorly deactivate arylamines which are bladder carcinogens (Caparoso et al., 1992). Carcinogen DNA adduct studies in humans in vivo have shown inter-individual variation which may be due to DNA repair rates in addition to difference in metabolism (Perera et al., 1991).

Individual variability sources in human susceptibility to carcinogens include not only genetic and/or acquired differences in activation and detoxification but also variation in nutritional and health status and DNA repair.

DNA repair

From an historical viewpoint, the relevance of DNA repair to carcinogenesis was related to defective repair of ultraviolet-induced damage to DNA in cells from xeroderma pigmentosum (XP) patients, with increased susceptibility to sunlight and induced skin cancer. The relevance of DNA repair was also related to defective repair in cells from in ataxia telangiectasia patients. This condition, however, is also associated with pathological conditions other than cancer. In both bacterial and mammalian cells there is excision repair and other independent enzymic processes through which the structural and biological integrity of damaged DNA is restored. Molecular damage to DNA may be subject to a sequential series of enzymic reactions which constitute the excision repair process. This damage may include altered bases, the covalent binding of 'bulky' adducts, intrastrand or interstrand cross-links and generation of strand breaks, as well as damage caused by ultraviolet (UV) radiation. The enzymology of excision repair was initially characterised in bacteria with a subsequent demonstration of a corresponding process in mammalian cells (Friedberg, 1984). The process involves breakage of the damaged DNA strand adjacent to the site of damage, excision of the damaged nucleotide(s) along with a variable number of undamaged nucleotides by an exonuclease, with synthesis of a DNA repair patch to fill the gap created and ligation of the free 3' end of the repair patch to the 5' end of the pre-existing DNA strand to complete the repair (Sancar and Sancar, 1988).

Different types of damage initiate slightly different reaction pathways. Base excision repair involves cleaving the bond between a modified base and its associated sugar by an N-glycosylase and then schission of the phosphodiester backbone of the DNA, catalysed by an apurinic/apyrimidric endonuclease. A number of N-glyosylases have been isolated from mammalian cells including those which recognise 3-methyladenine, 7-methylguanine, uracil and hypoxanthine and also including a formamidopyrimidine-DNA glycosylase which recognises the ring-opened derivative of 7-methylguanine. A glycosylate that recognises GT mismatches has been demonstrated in human cell extracts (Wiebauer and Jiricny, 1989).

Direct schission of the phosphodiester backbone of DNA by an exonuclease acting at the site of UV damage or a bulky adduct is termed nucleotide excision repair. The late stages of DNA repair including the respective roles of different polymerases and parameters which determine the length of repair patches are not fully understood (Perrino and Loeb, 1990).

A number of clinical conditions, apart from XP are associated with hypersensitivity to DNA damaging agents, such as patients with Bloom's syndrome who suffer from photosensitivity, immunodeficiency and dwarfed development and Fanconi's anaemia who have developmental abnormalities and increased incidences of acute leukaemia. These diseases provide evidence that a relationship exists between DNA repair and increased likelihood of malignant transformation.

Altered intercellular communication

The importance of altered intercellular communication in human carcinogenesis is known (Yamasaki, 1990). Gap junctions are specialised membrane components that permit the direct exchange of ions and small molecules between cells. Intercellular communication through gap junctions is involved in homeostasis and differentiation (Loewenstein, 1979). Oyamada et al. (1990) showed aberrant expression of the gap junctional protein (connexin) genes in surgically removed hepatocellular carcinomas. Cell adhesion molecules have been reported to be prerequisite for the function of gap junctions (Mege et al., 1988; Jongen et al., 1991) and a loss of function of known cell adhesion molecules is associated with the malignancy of human tumours (Shimoyama et al., 1989). Studies in vivo of various stages of hepatocarcinogenesis have shown a sequential decrease in gap junctional intercellular communication in preneoplastic foci and carcinomas of the liver in treated rats (Krutovskikh et al., 1991).

Immune factors

There are known to be involved in cancer etiology (Thomas, 1959; Burnet, 1965). It was thought that cancer arose in antigenically distinct and aberrant cells which were formed during cell replication. Such cells would normally be eliminated in a healthy person but in an immunocompromised person, cancer cells would continue to develop. Only a minority of cancers have a

viral origin such as malignant non-Hodgkin's lymphoma and
Kaposi's sarcoma (Kinlen, 1992). The former is increased among
transplant patients receiving immunosuppression. The latter
occurs in people with acquired immunodeficiency syndrome (AIDS).
Patients with AIDS also have increased risks for non-Hodgkin's
lymphomas as do patients with rare genetic diseases involving
deficiencies of the immune system such as ataxia telangiectasia.
AIDS patients are also at risk for Epstein-Barr virus associated
B-cell lymphoproliferative disease and Burkitt's-type lymphoma.

Hormones
Sex hormones can both increase and decrease the risk for cancer
in certain groups of humans (Key and Beral, 1992). Oestrogen
replacement therapy increases the risk of endometrial cancer
(IARC, 1979), since oestrogens, in the absence of progestins,
stimulate the initiation of endometrial cells. Combined oral
contraceptives, however, containing both oestrogen and progestin
reduce the risk of cancer. Such alterations in cancer risk can
be observed within a few years of and many years after the change
in hormone production indicating that both early and late stages
of carcinogenesis are affected. Experimental data have shown
that the administration to rats of testosterone propionate, a
synthetic androgen, increases the development of spontaneous and
carcinogen induced carcinomas of the prostate and seminal
vesicles (Shirai et al., 1991) but the mechanisms are unclear.

Carcinogenicity Bioassays and Short-term tests

Long-term carcinogenicity assays are carried out identifying
chemicals and other agents that may cause cancer in humans.
Long-term studies in laboratory animals are the most accepted
means for determining carcinogenic hazards to public health.
Such studies use one or both sexes of one or more species of
rodents involving control and exposure groups. Duration of
exposure generally ranges from 18-30 months. They are generally
conducted to identify carcinogens rather than determine
mechanisms. Exposure concentrations are generally selected on
the basis of shorter-term exposures, for example, in ninety-day

experiments. The top level is chosen so as not to affect unduly the normal well-being or growth and survival patterns of the animals. The levels should be lower than those that cause life-threatening conditions other than neoplasia (benign and malignant tumours) and should not affect adequate body weight gain in order that suitable numbers of control and exposed animals are available for pathological analysis at the end of the study. Results obtained provide some qualitative and quantitative information and may be useful for suggesting a mode of action.

A mechanistic insight into carcinogenesis might be gained from phenotypic cellular changes observed in preneoplastic lesions in different tissues (Bannasch, 1986). These may be changes in the amount and/or activity of enzymes, particularly of carbohydrate and drug metabolism; accumulation of certain metabolites such as glycogen, ribonucleoproteins and lipids, changes in the organisation of cytoplasmic organelles such as mitochondria, peroxisomes etc; and increase in nuclear alterations and cell proliferation. These changes indicate early changes in energy metabolism during carcinogenesis and in rodent liver such changes suggest a clonal origin of the focal lesions (Bannasch and Zerban, 1990, Bannasch *et al.*, 1992).

A number of short-term tests for detecting mutagenicity and genotoxicity of chemicals are in general use. They have been examined in international trials *in vitro* (Ashby *et al.*, 1985) and *in vivo* (Ashby *et al.*, 1988). They measure point mutations, chromosome damage, DNA repair and aneuploidy. Whilst there are tests to measure the first three categories, no generally available assay for gene amplification or aneuploidy exists, although such information can be obtained.

Genetic Biological Markers in Humans

DNA and protein adducts and oncoprotein levels

Bartsch and Malaveille (1990) analysed the data on genetic and related effects of chemicals evaluated by the IARC (1985) and found that 80-93% of the Group 1A human carcinogens were considered to exert genetic and related effects and 84-95% and 50-90% respectively of Groups 2A and B. For many of these compounds, DNA binding products have been described. A smaller proportion of known carcinogens are inactive in the short-term tests, and for most of them no DNA binding has been demonstrated (Hemminki, 1992).

DNA binding and damage inflicted by carcinogens occur in the early initiating stages of carcinogenesis (Berenblum, 1975), but could also occur at later stages of the multistage process particularly for clinically manifest cancers (Cerutti, 1985; Weinberg, 1989).

Exposure to a chemical may be estimated by environmental monitoring (external dose) biological monitoring (internal dose) and biochemical effect monitoring (tissue dose) (European Chemical Industry, Ecology and Toxicology Centre, 1989). The tissue dose can be assessed from the amount of DNA adducts in surrogate tissue such as white blood cells or directly in target tissue. Neumann (1986) has shown in animal experiments that DNA adduct levels in target and non-target tissues are proportional to the external dose. The extent of binding is illustrated by the covalent binding index, which spans orders of magnitude depending on the chemical (Lutz, 1979).

Binding of a chemical to haemoglobin or albumin is also proportional to the external dose and this is the reason that protein adducts can be used in the estimation of tissue dose (Neumann, 1986; Farmer et al., 1987). Protein adduct measurements offer advantages over DNA adduct measurements as haemoglobin adducts are stable over the lifetime of the erythrocyte and are readily available.

In humans there has been a development of sensitive methods based on ^{32}P-post-labelling (Randerath et al., 1981) and immune assays (Poirier, 1981). Interlaboratory comparisons carried out on the post-labelling technique and/or immune assays have shown reasonable correlations (Hemminki et al., 1988).

As yet no direct causal link has been established between the formation of DNA adducts and cancer; however there is a great deal of circumstantial evidence.

Recently, there has also been shown to be an increase in oncoprotein levels in serum of iron foundry workers exposed to polycyclic aromatic hydrocarbons (Brand-Rauf 1991; Brand-Rauf et al., 1990)). However, smoking (2-30 cigarettes per day) has not been shown to be a confounding variable (Brinkworth et al., 1992).

Gene Mutations

Various methods are available for the study of gene mutations arising in human somatic cells in vivo. They allow determination of the frequency of mutant lymphocytes and erythrocytes in blood and characterisation of mutation at the molecular level in lymphocytes. They are used to determine biological dosimetry of human exposure to mutagens in vivo and for identifying molecular alterations characteristic of chemicals and other agents.

Four systems have been developed for biomonitoring of humans exposed to carcinogenic agents in which gene mutation is the end point. These can be used to detect mutations in T lymphocytes at the X-linked locus for the purine salvage enzyme hypoxanthine phosphoribosyltransferase (hprt) (Albertini et al., 1982; Morley et al., 1983) and at the autosomal locus for human leukocyte antigen A-A (HLA-A) (Janatipour et al., 1988; Turner et al., 1988). The other two make use of red blood cells for detecting haemoglobin variants (Stamatoyannapoulos et al., 1984; Tates et al., 1989) and loss of the cell surface glycoprotein glycophorin A (Langlois et al., 1986; 1987).

The induction of an increased frequency of gene mutation in any of the above mentioned assays in vivo indicates the possibility that mutation may also be induced in proto-oncogenes and tumour-suppressor genes in the relevant cells to produce cancer (Lambert, 1992).

Cytogenetic Damage

Most of the IARC (1987) human carcinogens give positive results in routinely used tests such as the Salmonella and clastogenicity assays (Ennever et al., 1987; Shelby, 1988; Bartsch and Malaveille, 1989, 1990; Shelby and Zeiger, 1990; Shelby et al., 1993). An observed absence of genotoxicity may simply be because of the limitations of the test system (Ramel, 1992) or because the substances may be acting by an indirect or non-genotoxic mechanisms. However, it is known that some agents classified as non-genotoxins do give positive results in chromosome assays *in vitro* or *in vivo* (Barrett, 1992; Jackson *et al.*, 1993).

Malignant cells often feature chromosomal rearrangements and are known to activate proto-oncogenes. That chromosomal breakpoints tend to cluster at these sites supports the association with cancer indirectly (Heim and Mitelman, 1987; de Klein, 1987). In the IARC data base (IARC, 1987), 55 agents are carcinogenic to humans (Group 1), 45 agents are probably carcinogenic to humans (Group 2A) and 191 agents that are possibly carcinogen to humans (Group 2B). Cytogenetic data are available for 27 compounds in Group 1, 10 in Group 2A and 15 in Group 2B (Sorsa *et al.*, 1992). Cytogenicity in humans was found in 19/27, 6/10 and 5/15 cases respectively. This would suggest that carcinogenic chemicals tend to be clastogenic, and that clastogenicity tends to be associated with known human carcinogens.

Thirty-one agents are listed in the IARC data base at exposures which induce chromosomal aberrations in humans *in vivo*. Fifteen belong to Group 1 carcinogens, four to Group 2A and five to Group 2B. Seven agents have no evaluation of carcinogenicity (Group 3). The concordance of the endpoint of chromosomal aberrations *in vitro* and *in vivo* is good in most cases. In five cases, (alcoholic beverages, welding fumes, bis(chloromethyl) ether, cyclosporin and radon) there is a discrepancy between chromosomal aberrations induced *in vivo* in animals and in humans. For alcoholic beverages and welding fumes, the qualitative exposure of humans and animals may not be similar. For others there may be confounding of the human studies by other clastogenic agents.

Most experience with cytogenetic surveillance has involved high-exposure occupational situations.

Conclusion

Thus, the various mechanisms of mutagenicity and tumour formation have been briefly reviewed. According to IARC (1992), when the data on available mechanisms are thought to be relevant to the evaluation of the carcinogen i.e. risk of an agent to humans, they should be used in making the overall evaluation, together with the combined evidence for animal and/or human carcinogenicity.

References

Albertini RJ, Castle KL, Borcherding WR (1982). T-cell cloning to detect the mutant 6-thioguanine resistant lymphocytes present in human peripheral blood. Proc. Natl. Acad. Sci. USA 79:6617-6621.

Ames BN, Gold LS (1990a). Chemical carcinogenesis. Too many rodent carcinogens. Proc. Natl. Acad Sci. USA 87:7772-7776.

Ames BN, Gold LS (1990b). Too many rodent carcinogens. Mitogenesis increases mutagenesis. Science 249:970-971.

Anderson D. (1990). Ed. Special Issue Male-Mediated F_1 abnormalities. Mutation Res. 229:103-249.

Anderson D. (1985). Induction of mutation and chromosome damage by excess bases and nucleosides. In: de Serres FJ (ed). Genetic Consequences of Nucletoide Pool Imbalance, Plenum Press, New York pp.283-295.

Anderson D, Richardson CR, Davies PJ (1981). The genotoxic potential of bases and nucleosides. Mutat. Res. 91:265-272.

Ashby J. (1985). Fundamental structural alerts to potential carcinogenicity or non-carcinogenicity. Environ. Mutagen 7:919-921.

Ashby J. (1992). Use of short-term tests in determining the genotoxicity or non-genotoxicity of chemicals. In: Vainio H, Magee PN, McGregor DB, McMichael AJ (eds). Mechanisms of Carcinogenesis in Risk Identification. International Agency for Research on Cancer, Lyon pp.135-164.

Ashby J, de Serres FJ, Draper M, Ishidate M Jr, Margolin B, Matter BE, Shelby MD (1985) (Eds). Evaluation of short-term tests for carcinogens. Report of the International Programme on Chemical Safety's Collaborative Study in *In Vitro* Assays. UNEP/ILO/WHO, Progress in Mutation Research Volume 5. Elsevier.

Ashby J, de Serres FJ, Shelby MD, Margolin BH, Ishidate M Jr, Becking GC (1988) (eds). Evaluation of Short Term Tests for

Carcinogens. Report of the International Programme on Chemical Safety's Collaborative Study in *In Vitro* Assays. UNEP/ILO/WHO, Progress in Mutation Research. Volumes I & II. Cambridge University Press.

Ashby J, Morrod RS (1991). Detection of human carcinogens. Nature 352:185-196.

Ashby J, Purchase IFH (1992). Non-genotoxic carcinogens : an extension of the perspective provided by Perera. Environ. Hlth. Perspect. 98:223-236.

Ashby J, Tennant RW, (1988) Chemical structure, *Salmonella* mutagenicity and extent of carcinogenicity as indicators of genotoxic carcinogenesis among 222 chemicals tested in rodents by the US NCI/NTP. Mutat Res 204:17-115

Ashby J, Tennant RW, Zeiger E, Stascewicz S (1989). Classification according to chemical structure, mutagenicity to *Salmonella* and level of carcinogenicity of a further 42 chemicals tested for carcinogencity by the US National Toxicology Program. Mutat. Res. 223:73-103.

Ashby J, Tennant RW (1991) Definitive relationships among chemical structure, carcinogenicity and mutagenicity for 301 chemicals tested by the US NCI/NTP. Mutat Res. 257:229-306.

Ayesh R, Idle JR, Ritchie JC, Crothers MJ, Hetzel MR (1984). Metabolic oxidation phenotypes as markers for susceptibility to lung cancer. Nature 312:169-170.

Balmain A, Brown K (1988). Oncogene activation in chemical carcinogenesis. Adv. Cancer Res. 51:147-182.

Bannasch P, Moore MA, Klimek F, Zerban H (1982). Biological markers of preneoplastic foci and neoplastic nodules in rodent liver. Toxicol. Pathol. 10:19-34.

Bannasch P, Zerban H (1990b). Animal models and renal carcinogens. In: Eble JN (ed) Tumors and Tumor-like Conditions of the Kidneys and Ureters, New York, Churchill Livingston, pp.1-34.

Barbacid M (1987). *ras* Genes. Ann. Rev. Biochem. 56:779-827.

Barrett JC (1992). Mechanisms of action of known humam mutagens. In: Vainio H, Magee PN, McGregor DB, McMichael AJ (eds). Mechanisms of Carcinogenesis in Risk Identification. International Agency for Research on Cancer Lyon pp.115-134.

Bartsch H, Malaveille C (1989). Prevalence of genotoxic chemicals among animal and human carcinogens evaluated in the IART Monograph series. Cell Biol. Toxicol. 5:115-127.

Bartsch H, Malaveille C (1990). Screening assays for carcinogenic agents and mixtures : an appraisal based on data in the IARC monograph series. In: Vainio H, Sorsa M, McMichael A.J. (eds). Complex Mixtures and Cancer Risk. IARC Scientific Publications No.104 Lyon pp.65-74.

Baserga R (1990). The cell cycle: myths and realities. Cancer Res. 50:6969-6771.

Belinsky SA, White CM, Devereux TR, Swenberg JA, Anderson MW (1987). Cell selective alkylation of DNA in rar lung following low dose exposure to the tobacco specific carcinogen 4-(N-methyl)-N-nitrosamine)-1-(3-pyridyl)-1-butanone. Cancer Res. 47:1143-1148.

Berenblum I (1941). The mechanism of carcinogenesis : a study of the significance of carcinogenic action related phenomena. Cancer Res. 1:807-814.

Berenblum I (1974). Carcinogenesis as a Biological Problem (Frontiers of Biology Vol.4), Amsterdam, North Holland.

Berenblum I (1975). Sequential Aspects of Chemical Carcinogenesis : Skin. In: Pecker FF (ed) Cancer: A Comprehensive Treatise. Vol.1 New York, Plenum Press, pp.323-344.

Bishop JM (1987). The molecular genetics of cancer. Science 235:305-311.

Bishop JM (1991). Molecular themes in oncogenesis. Cell 64:235-248.

Bos JL (1989). *ras* Oncogenes in human cancer: a review. Cancer Res.49:4682-4689.

Bos JL, van Kreijl CF (1992). Genes and gene products that regulate proliferation and differentiation : critical targets in carcianogenesis. In: Vainio H, Magee PN McGregor DB, McMichael AJ (eds). Mechanisms of Carcinogenesis in Risk Identification International. Agency for Research on Cancer Lyon, pp.57-65.

Boutwell RK (1964). Some biological aspects of skin carcinogenesis. Prog. Exp. Tumor Res. 4:207-250.

Boveri T (1914) Zur Frage der Entstehung Maligner Tumoren Jena, Gustave Fischer.

Brandt-Rauf PW (1991). Advances in cancer biomarkers as applied to chemical exposures. The *ras* oncogene and p21 protein and pulmonary carcinogenesis. J. Occ. Med. 33:951-955.

Brandt-Rauf PW, Smith S, Perera FP, Niman HL, Yohanna W, Hemminki K, Santella R (1990) Serum oncogene proteins in foundry workers. J. Soc. Occup. Med. 40:11-14.

Bressac B, Kew M, Wands J, Ozturk M (1991). Selective G to T mutations of p53 gene in hepatocellular carcinoma from Southern Africa. Nature, 350: 429-431.

Brinkworth MH, Yardley-Jones A, Edwards AJ, Hughes JA, Anderson D (1992) A comparison of smokers and non-smokers with respect to oncogene products and cytogenetic parameters. J Occup. Med. 34:1181-1188.

Buchter A, Bolt HM, Filser JG, Georgens HW, Laib RJ, Bolt W. (1978). Pharmakokinetic und Karzinogenese von Vinylchlord, Arbeitsmedizinische Risikobeurteilung. In: Loskant, H. ed. Möglichkeiten und Grenzen des Biological Monitoring, Arbeitzsmedizinische Probleme des Diensteleistungsgewerbes, Stuttgart, Gentner Verlag, pp.111-124.

Burnet FM (1965). Somatic mutation and chronic disease. Br. Med. J., i, 338-342.

Butterworth BE (1990). Consideration of both genotoxic and non-genotoxic mechanisms in predicting carcinogenic potential. Mutat. Res. 239:117-132.

Butterworth BE, Popp JA, Connolly, RB, Goldsworth TL (1992). Chemically induced cell proliferation in carcinogenesis. In: Vainio H, Magee PN, McGregor DB, McMichael AJ (eds). Mechanisms of Carcinogenesis in Risk Identification. International Agency for Research on Cancer Lyon, pp.279-306.

Caporaso NE, Tucker MA, Hoover RN, Hayes RB, Pickle LW, Issaq HJ, Muschink, GM, Green-Gallo L, Buivys D, Aisner S, Resau JH, Trump BF, Tollerud D, Weston A, Harris CC (1990). Lung cancer and the debrisoquine metabolic phenotype. J. Natl. Cancer Inst. 82: 1264-1272.

Cartwright RA, Glashan RW, Rogers HJ, Ahmad RA, Barham-Hall D, Higgins E, Kahn MA (1982). Role of N-acetyltransferase phenotypes in bladder carcinogenesis: a pharmacogenetic epidemiological approach to bladder cancer. Lancet, **ii**, 842-845.

Case RAM, Hosker ME, McDonald DB, Pearson JT (1954) Tumours of the urinary bladder in workmen engaged in the manufacture and use of certain dyestuff intermediates in the British chemical industry. I. The role of aniline, benzidine, alpha-naphthylamine and beta-naphthylamine. Br. J. Ind. Med. 11:75-104.

Cerutti PA (1985). Prooxidant states and tumour promotion. Science, 227:375-381.

Chiba I, Takahashi T, Nau MM, D'Amico D, Curiel DT, Mitsudomi T, Buchhagen DL, Carbone D, Piantadosi S, Koga H, Reissman PT, Slamon DJ, Holmes EC, Minna JD (1990). Mutations in the p53 gene are frequent in primary, resected non-small cell lung cancer. Oncogene 5:1603-1610.

Cohen SM, Ellwein LB (1990). Cell proliferation in carcinogenesis. Science, 249:1007-1011.

Craddock VM (1976). Cell proliferation and experimental liver cancer. In: Cameron HM, Linsell CA, Warwick GP (eds). Liver Cell Cancer, Amsterdam, Elsevier/North-Holland Biomedical Press, pp.153-201.

Davies MJ (1993). PhD Thesis. Molecular analysis of chemically induced mutations in mammalian cells. University of Surrey.

de Klein A (1987). Oncogene activation by chromosomal rearrangements in chronic myelocyte leukaemia. Mutat. Res. 186:161-172.

Doll R, Peto R, (1981). The Causes of Cancer. Oxford, New York. University Press.

Doolittle DJ, Muller CJ, Scribner HE (19897). Relationship between hepatotoxicity and induction of replicative DNA synthesis following single or multiple doses of carbon tetrachloride. J. Toxicol. Environ. Health. 22:63-78.

Dunn WJ, Wold S (1981). The carcinogenicity of N-nitrosocompounds: a SIMCA pattern recognition study. Bioorg. Chem., 10:29-45.

Dybing E, Huitfeldt (1992). Species differences in carcinogenic metabolism and interspecies extrapolation. In: Vainio H, Magee

PN, McGregor DB, McMichael AJ (eds). Mechanisms of Carcinogenesis in Risk Identification. International Agency for Research on Cancer Lyon, pp.501-524.

Ennever FK, Noonan TJ, Rosenkranz HS (1987). The predictivity of animal bioassays and short-term genotoxicity tests for carcinogenicity and non-carcinogenicity in humans. Mutagenesis, 2:73-78.

European Chemical Industry, Ecology and Toxicology Centre (1989). DNA and Protein Adducts : Evaluation of their use in exposure monitoring and risk assessment. Monograph No.13, Brussels.

Farmer PB, Neumann, HG, Henzchler D (1987). Estimation of exposure of man to substances reacting covalently with macromolecules. Arch. Toxicol. 60:251-260.

Fearon, E.R. and Vogelstein, B. (1990). A genetic model for colorectal tumorigenesis. Cell, 61:759-767.

Flamm, G.W. and Lehman-McKeeman, L.D. (1991). The human relevance of the renal tumor-inducing potential of d-limonene in male rats: implications for risk assessment. Regul. Toxicol. Pharmacol. 13:70-86.

Foulds L (1954) The experimental study of tumor progression : a review. Cancer Res. 14:327-339.

Friedberg, E.C. (1984). DNA Repair. San Francisco. W.H. Freeman.

Gold LS, Slone TH, Bernstein L (1989). Summary of carcinogen potency and positivity for 492 rodent carcinogens in the carcinogenic potency database. Environ. Health Perspect. 79:259-272.

Guengerich FP (1988). Roles of cytochrome P-450 enzymes in chemical carcinogenesis and cancer chemotherapy. Cancer Res. 48:2946-2954.

Hansch C, Fujita T (1964). r-s-p Analysis. A method for the correlation of biological activity and biogical structure. J. Am. Chem. Soc. 86 1616-1626.

Hansen MF, Cavenee WK (1987). Genetics of cancer predisposition. Cancer Res. 47:5518-5527.

Hecker F, Kunz W, Füsenig NE, Marks F, Thielmann HW eds (1982). Carcinogenesis - A Comprehensive Survey, Vol.7. Carcinogenesis and Biological Effects of Tumor Promoters, New York, Raven Press.

Heidelberger C (1970). Chemical carcinogenesis, chemotheraphy : cancer's continuing core challenges - G.H.A. Clowes Memorial Lecture. Cancer Res. 30:1549-1569.

Heidelberger C, Freeman AE, Pienta RJ, Sivak A, Bertram JS, Casto BC, Dunkel VC, Francis MW, Kakunaga T, Little JB, Schechtman LM (1983). Cell transformation by chemical agents - a review and analysis of the literature. A report of the US Environmental Protection Agency Gene-Tox Program. Mutat. Res. 114:283-385.

Heim S, Mitelman F (1987). Nineteen of 26 cellular oncogenes precisely localised in the human genome map to one of the 83 bands involved in primary cancer. Specific rearrangements. Hum. Genet. 75:70-72.

Hemminki K (1992). Significance of DNA and protein adducts. In: Vainio H, PN, Magee DB, McGregor, AJ McMichael (eds). Mechanisms of Carcinogenesis in Risk Identification. International Agency for Research on Cancer, Lyon, pp.525-534.

Hemminki K, Perera FP, Phillips DH, Randerath K, Reddy MV, Santella RM (1988). Aromatic DNA adducts in white blood cells of foundary workers. In: Bartsch, H., Hemminki, K. and O'Neill, I.K. (eds). Methods for Detecting Agents in Humans. Application in Cancer Epidemiology and Prevention. IARC Scientific Publications No.89. Lyon, pp.190-195.

HMSO (1991). Report of Health and Social Subjects 42. Guidelines for the Evaluation of CLhemicals for Carcinogenicity. Committe on Carcinogenicity of Chemicals in Food, Consumer Products and the Environment. Department of Health.

Hollstein M, Sidransky D, Vogelstein B, Harris CC (1991). p53 Mutations in human cancers. Science 253:49-53.

Hsu IC, Metcalf RA, Sun T, Welsh JA, Wang NJ, Harris CC (1991). Mutational hotspot in the p53 gene in human hepatocellular carcinomas. Nature 350:427-428.

Huff JE (1992). Chemical toxicity and chemical carcinogenesis. Is there a causal condition? A comparative morphological examination of 1500 experiments. In: Vainio H, Magee PN, McGregor DB, McMichael AJ (eds). Mechanisms of Carcinogenesis in Risk Identification. International Agency for Research on Cancer, Lyon. pp.437-476.

IARC (1979). Monographs on the Evaluation of the Carcinogenic Risks of Chemicals to Humans, Vol.21 Sex Hormones (11), Lyon.

IARC (1987). IARC Monographs on the Evaluation of Carcinogenic Risks to Humans. Suppl.7. Overall Evaluations of Carcinogenicity: An updating of IARC Monographys Volumes 1 to 42, Lyon.

IARC (1992) Scientific Publications No.116. Mechanisms of Carcinogenesis in Risk Identification eds. Vaino H, Magee P, McGregor D, McMichael AJ (eds). International Agency for Research on cancer, Lyon.

Iversen OH (1988). Initiation, promotion: Critical remarks on the two-stage theory. In: Iversen OH (ed). Theories of Carcinogenesis, Washington DC, Hemisphere Publishing, pp.119-126.

Jackson MA, Stack HF, Waters MD (1993). The genetic toxicology of putative non-genotoxic carcinogens. Mutation Res. 296:241-277.

Janatipour M, Trainor KJ, Kutlaca R, Bennet G, Hay J, Tunrer DR Morely AA (1988). Mutations in human lymphocytes studied by an HLA system. Mutat. Res. 198:221-226.

Jongen WM, Fitzgerald DJ, Asamoto M, Piccoli C, Slaga TJ, Gros D, Takeichi M, Yamasaki H (1991). Regulation of connexin 43-mediated gap junctional intercellular communication by Ca^{2+} in mouse epidermal cells is controlled by E-cadherin. J. Cell Biol. 114:545-555.

Kellerman G, Shaw CR, Luyten-Killerman M (1973). Aryl hydrocarbon hydroxylase inducibility and bronchogenic carcinoma. New Engl. J. Med. 289:934-937.

Key TJA, Beral V (1992). Sex hormones and Cancer. In: H Vainio, PN Magee, DB McGregor, AJ, McMichael (eds). Mechanisms of Carcinogenesis in Risk Identification. International Agency for Research on Cancer, Lyon. pp.255-270.

Kinlen LJ (1992). Immunosuppression and Cancer. In: H Vainio, PN Magee, DB McGregor, AJ, McMichael (eds). Mechanisms of Carcinogenesis in Risk Identification. International Agency for Research on Cancer, Lyon. pp.237-253.

Kouri RE, McKinney CE, Slomiany DJ, Snodgrass DR, Wray NP, McLemore TL (1982). Positive correlation between high aryl hydrocarbon hydroxylase activity and primary lung cancer as analysed in cryopreserved lymphocytes. Cancer Res. 42:5030-5037.

Knudson AG (1985). Heriditary cancer, oncogenes and antioncogenes. Cancer Res.45:1437-1443.

Krutovskikh VA, Oyamada M, Yamasaki H. (1991). Sequential changes of gap-junctional intercellular communications during multistage liver carcinogenesis: direct measurement of communication in vivo. Carcinogenesis, 12:1701-1706.

Kumar R, Sukumar S, Barbacid M (1990). Activation of ras oncogenes preceding the onset of neoplasia. Science, 248:1101-1104.

Lambert B (1992). Biological markers in exposed humans: gene mutations. In: H Vainio, PN Magee, DB McGregor, AJ McMichael (eds). Mechanisms of Carcinogenesis in Risk Identification. International Agency for Research on Cancer, Lyon pp.535-542.

Langlois RG, Bigbee WL, Jensen RH (1986). Measurement of the frequency of human erythrocytes with gene expression loss phenotypes at the glycophorin A locus. Hum. Genet. 74:353-362.

Langlois RG, Bigbee WL, Kyoizumis, Naskamura N, Bean MA, Akiyama M, Jensen RH (1987). Evidence for increased somatic mutations at the glycophorin A locus in atomic bomb survivors. Science, 236:445-448.

Lawley PD, Brookes P (1968). Cytotoxicity of alkylating agents towards sensitive and resistant strains of Escherichia coli in relation to extent and mode of alkylation of cellular macromolecules and repair of alkylation lesions in deoxyribonucleic acids. Biochemi. J. 109:433-447.

Lewis DFV, Ioannides C, Parke DV (1992). Validation of a novel molecular orbital approach (COMPACT) to the safety evaluation of chemicals by comparison with Salmonella mutagenicity and rodent carcinogenicity data evaluatedby the US NCI/NTP. Mutat. Res. 291:61-77.

Loeb LA, and Cheng, K.C. (1990). Errors in DNA synthesis: a source of spontaneous mutations. Mutat. Res. 238:297-304.

Loewenstein WR (1979). Junctional intercellular communication and the control of growth. Biochim. Biophys. Acta. 560:1-65.

Lucier GW (1992). Receptor-mediated carcinogenesis. In: Vainio H, Magee PN, McGregor DB, McMichael AJ (eds). Mechanisms of Carcinogenesis in Risk Identification. International Agency for Research on Cancer, Lyon. pp.87-114.

Lutz WK (1979). *In vivo* covalent binding of organic chemicals to DNA as a quantitative indicator in the process of chemical carcinogenesis. Mutat. Res. 65:289-356.

Malkin D, Li FP, Strong LC, Fraumeni JF, Nelson CE, Kim DH, Massel J, Gryka MA, Bischoff FZ, Tainsky MA, Freind SH (1990). Germ line p53 mutations in a familial syndrome of breast cancer, sarcomas and other neoplasms. Science, 250:1233-1238.

Marks F (1990). Hyperplastic transformation: the response of the skin to initiation and injury. In: Galli CL, Hensby CN, Marinovich M (eds). Skin Pharmacology and Toxicology: Recent Advances (Series: Life Sciences, Vol.181), New York, Plenum Press, pp.121-145.

Mead JE, Fausto N (1989). Transforming growth factor alphas may be a physiological regulator of liver regeneration by means of an autocrine mechanism. Proc. Natl. Acad. Sci. USA, 86:1558-1562.

Mege R-M, Matsuzaki F, Gallin WJ, Goldberg JI, Cunningham BA, Edelman GM (1988). Construction of epithelioid sheets by transfection of mouse sarcoma cells with cDNAs for chicken cell adhesion molecules. Proc. Natl. Acad. Sci. USA, 85:7274-7278.

Miller JA (1970). Carcinogenesis by chemicals: an overview - G.H.A. Clowes Memorial Lecture. Cancer Res. 30:559-576.

Mommsen S, Barford DW, Heidelberger C (1976). Two-stage chemical oncogenesis in cultures of C3H/10T1/2 cells. Cancer Res. 36:2254-2260.

Moore MA, Nakagawa K, Satoh K, Ishikawa T, Sato K (1987) Single GSTP-positive liver cells - putative initiated hepatocytes. Carcinogenesis 8:483-486.

Morley AA, Trainor KJ, Seshadri R, Ryall RG (1983). Measurement of *in vivo* mutations in human lymphocytes. Nature 302:155-156.

Mottram JC (1935) The origin of tar tumours in mice, whether from single or many cells. J. Pathol. 40:407-409.

Neumann HG (1986). The role of DNA damage in chemical carcinogenesis of aromatic amines. J. Cancer Res. Clin. Oncol. 112:100-106.

Oyamada M, Krutovskikh VA, Mesnil M, Partensky C, Berger F, Yamasaki H (1990). Aberrant expression of gap junction gene in primary human carcinomas: increased expression of cardiac-type gap junction gene connexin 43. Mol. Carcinog. 3:273-278.

Parodi S, Waters MD (1991). Introduction and summary. Genotoxicity and carcinogenicity databases: an assessment of the present situation. Environ. Hlth. Perspectives 96:3-4.

Perera F, Mayer J, Santella RM, Brenner D, Jeffrey A, Latriano, L, Smith S, Warburton D, Young TL, Tsai WY, Hemminki K, Brandt-Rauf P. (1991). Biological markers in risk assessment for environmental carcinogens. Environ. Health Perspectives 90:247-254.

Perrino FW, Loeb LA (1990). Animal cell DNA polymerases in DNA repair. Mutat. Res. 236:289-300.

Peto J (1984). Early- and late-stage carcinogenesis in mouse skin and in man. In: Börzsönyi M, Day NE, Lapis K, Yamasaki H. (eds). Models, Mechanisms and Etiology of Tumour Promotion. IARC Scientific Publications No.56 Lyon, IARC, pp.359-371.

Pitot HC (1993) The dynamics of carcinogenesis : Implications for human risk. CITT Activities 13 No.6:1-6.

Pitot HC, Sirica AE (1980). The stages of initiation and promotion in hepatocarcinogenesis. Biochim. Biophys. Acta. 605:191-215.

Phillips JC. Anderson D (1992). Predictive Toxicology. Occupational Health Review, 41:27-30.

Poirier MC (1981). Antibodies to carcinogens - DNA adducts. J. Natl. Cancer Inst. 67:515-519.

Preston-Martin S, Pike MC, Ross RK, Jones PA. Henderson BE (1990). Increased cell division as a cause of human cancer. Cancer Res. 50:7415-7421.

Quintanilla M, Brown K, Ramsden M, Balmain A (1986). Carcinoge-specific mutation and amplification of Ha-ras during mouse skin carcinogenesis. Nature, 322:78-80.

Rabes HM (1988). Cell proliferation and hepatocarcinogenesis. In: Roberfroid MB, Préat V, eds. Experimental Hepatocarcinogenesis, New York, Plenum Press, pp.121-132.

Rajewsky MF (1972). Proliferative parameters of mammalian cell systems and their role in rumor growth and carcinogenesis. Krebsforsch. 78:12-30.

Ramel C (1992). Genotoxic and non-genotoxic carcinogens. In: H Vainio, PN Magee, DB McGregor, AJ McMichael (eds). Mechanisms of Carcinogenesis in Risk Identification. International Agency for Research on Cancer, Lyon. pp.195-210.

Randerath K, Reddy MV, Gupta (1981). ^{32}P-Labelling test for DNA damage. Proc. Natl. Acad. Sci. USA, 78:6126-6129.

Rosenkranz HS, Klopman G (1990a). Structural alerts for genotoxicity: the interaction of human and artificial intelligence. Mutagenesis, 5:333-361.

Rosenkranz HS, Klopman G (1990b). Evaluating the ability of CASE, an artificial intelligence structure-activity relational system, to predict structural alerts for genotoxicity. Mutagenesis, 5:525-527.

Roth J, Grunfeld C (1985). Mechanism of action of peptide hormones and catecholamines. In: Wilson JD, Foster DW (eds). Textbook Endocrinology, Philadelphia, W.B. Saunders, pp.76-122.

Rous P, Beard JW (1935) The progression to carcinoma of virus-induced rabbit papillomas (shape). J. Exp. Med. 62:523-548.

Sancar A, Sancar GB (1988). DNA repair enzymes. Ann Rev. Biochem, 57:29-67.

Shelby MD (1988). The genetic toxicity of human carcinogens and its implications. Nutat. Res. 204:3-15.

Shelby MD, Erexson EL, Hook GJ, Tice PR (1993). Evaluation of three-exposure mouse bone marrow micronucleus protocol: results with 49 chemicals. Envir. Molec. Mutagen 21:160-179.

Shelby MD, Zeiger E (1990). Activity of human carcinogen in the human Salmonella and rodent bone-amrrow cytogenetics test. Mutat. Res. 234:1264-1269.

Shimoyama Y, Hirohashi S, Harano S, Noguchi M, Shimosato Y, Takeichi M, Abe O (1989). Cadherin cell-adhesion molecules in human epithelial tissues and carcinomas. Cancer Res. 49:2128-2133.

Shirai T, Tamano S, Kato T, Iwaski S, Takahaski S, Ito N (1991). Induction of invasive carcinomas in the accessory six organs other than the ventral prostate of rats given 2,3'-dimethyl-4-aminobiphenyl and testosterone propionate. Cancer Res. 51:1264-1269.

Slaga TJ (1985). Mechanisms involved in multistage skin tumorigenesis. In: Huberman E, Barr SH eds. Carcinogenesis - A Comprehensive Survey, Vol.10. The Role of Chemicals and Radiation in the Etiology of Cancer, New York, Raven Press, pp.189-199.

Slebos RJ, Hruban RH, Dalesio O, Mooi WJ, Offerhaus GJ, Rodenhuis S (1991). Relationship between K-*ras* oncogene activation and smoking in adenocarcinoma of the human lung. J. Natl. Cancer Inst, 83:1024-1027.

Slebos RJ, Kibbelaar RE, Dalesio O, Kooistra A, Stam J, Meijer CJ Wagenaar S, Vanderschueren RG, van Zandwijk N, Mooi WJ, Bos JL, Rodenhuis S (1990). K-*ras* Oncogene activation as a prognostic marker in adenocarcinoma of the lung. New Engl. J. Med. 323:561-565.

Sorsa M, Wilbourne J, Vainio H (1992). Human cytogenetic damage as a predictor of cancer risk. In: Vainio H, Magee PN, McGregor DB, McMichael AJ (eds). Mechanisms of Carcinogenesis in Risk Identification. International Agency for Research on Cancer, Lyon. pp.543-556.

Stamatoyannopoulos E, Nute PE, Lindsley D, Farquhar M, Brice M, Nakamato N, Papayannopoulou T (1984). Somatic cell mutation monitoring system based on human haemoglobin mutants. In: Ansari AA, deSerres FJ (eds). Single Cell Monitoring Systems. Topics in Chemical Mutagenesis pp.1-35.

Sukumar S, Notorio V, Martin-Zanca D, Barbacid M (1983). Induction of mammary carcinomas in rats by nitroso-methyl urea involves malignant activation of H-*ras* 1 locus by single point mutations. Nature, 306:658-661.

Swierenga SHH, Yamasaki H (1992). Performance of tests for cell transformation and gap junction-intercellular communication for detecting non-genotoxic activity. In: Vainio H, Magee PN, McGregor DB, McMichael AJ (eds). Mechanisms of Carcinogenesis in Risk Identification. International Agency for Research on Cancer, Lyon. pp.165-194.

Tates AD, Berninih F, Natarajan AT, Ploem JS, Verwoerd NP (1989). Detection of somatic mutants in man HPRT mutations in lymphocytes and hemoglobin mutations in erythrocytes. Mutat. Res. 213:73-82.

Tennant RW, Ashby J (1991). Classification according to chemical structure, mutagenicity to Salmonella and level of carcinogenicity of a further 39 chemicals tested for carcinogenicity by the US National Toxicology Program. Mutat. Res, 257:209-227.

Thomas L (1959). Mechanisms involved in tissue damage by the endotoxins of gram negative bacteria. In: Lawrence HS (ed). Cellular and Humoral Aspects of the Hypersensitive States, London, Cassel, pp.451-468.

Tomatis L (1989). Overview of perinatal and multigeneration carcinogenesis. In: Napalkov NP, Rice JM, Tomatis L, Yamasaki H (eds). Perinatal and Multigeneration Carcinogenesis (IARC Scientific Publications No.96), Lyon, IARC, pp.1-5.

Turner DR, Grist SA, Janatipour M, Morely AA (1988). Mutations in human lymphocytes commonly involved gene duplication and resemble those seen in cancer cells. Proc. Natl. Acad. Sci. USA, 85:3189-3192.

Van't Veer LJ, Burgering BMT, Versteeg R, Boot AJM, Ruiter DJ, Osanto S, Schrier PI, Bos JL (1989). N-ras Mutations in human cutaneous melanoma from sun-exposed body sites. Mol. Cell. Biol, 9:3114-3116.

Varmus HE (1989). Oncogenes and the Molecular Origins of Cancer, Cold Spring Harbor, NY, CSH Press.

Vogelstein B, Fearon ER, Hamilton SR, Kern SE, Preisinger AC, Leppert M, Nakamura Y, White R, Smits AM, Bos JL (1988). Genetic alterations during colorectal-tumor development. New Engl. J. Med, 319:525-532.

Vogelstein B, Kinzler KW (1993). The multistep nature of cancer. Trends in Genetics 9:138:141.

Weinstein BI (1991a). Mitogenesis is only one factor in carcinogenesis. Science, 251:387-388.

Weinstein BI (1991b). Cancer prevention : recent progress and
  future opportunities. Cancer Res, 51:5080s-5085s.

Wiebauer K, Jiricny J (1989). *In vitro* correction of G.T.
  misrepairs to G.C. pairs in nuclear extracts from human cells.
  Nature, 339:234-236.

Weinberg (1989). Oncogenes, antioncogenes and the molecular
  basis of multistep carcinogenesis. Cancer Res, 49:3713-3721.

Wiseman RW, Stowers SJ, Miller EC, Anderson MW, Miller JA (1986).
  Activating mutations of c-Ha-*ras* proto-oncogene in chemically
  induced hepatomas of the male B6C3F$_1$ mouse. Proc. Natl Acad.
  Sci. USA. 83:5825-5829.

Yamasaki H (1990). Gap junctional intercellular communication
  and carcinogenesis. Carcinogenesis 11:1051-1058.

Zarbl H, Sukumar S, Arthur AV, Martin-Zanca D, Barbacid M (1985).
  Direct mutagenesis of Ha-*ras*-1 oncogenes by *N*-nitroso-N-
  methylurea during initiation of mammary carcinogenesis in rats.
  Nature, 315:382-385.

Genotoxicity assays

Diana Anderson
BIBRA Toxicology International
Woodmansterne Road
Carshalton
Surrey SM5 4DS

1. Introduction

A long standing goal of genetic toxicologists has been to predict the carcinogenic potential of chemicals by means of short-term tests.

Genetic toxicology is a branch of general toxicology which addresses the problems of toxicity to the DNA. It is a discipline which has arisen from the early studies of Müller (1927) and Auerbach et al. (1947) in which irradiation and chemicals were shown to induce mutations in Drosophila, the fruit fly. When it was shown that mutations could be induced in mammals, the possibility arose that some of the hereditary diseases in man might be induced by environmental agents. Such materials include not only man-made chemicals, but natural carcinogens and mutagens found in fungi and plants and produced by cooking processes (Anderson and Purchase, 1983; Felton and Knize, 1991; IPCS 1992; Rowland et al., 1984; Sugimura, 1978; Wakabayashi et al., 1993). The impact of these agents on human health of present and future generations could be of consequence if the extent of exposure is sufficient to permit expression of genotoxic properties in somatic or germ cells. Over the last generation many substances shown to be mutagenic were also shown to be carcinogenic, and early correlations as high as 90% between mutagenicity and carcinogenicity were claimed (McCann et al., 1975a and b; Purchase et al., 1978; Sugimura et al., 1976, 1977). Such correlations stimulated a new interest in the somatic theory of cancer of the 1950's and has resulted in intensive study of the genetic effects of chemical substances.

It has been estimated that there is a genetic element in at least 10% of all human pathological conditions; thus genetic

NATO ASI Series, Vol. H 90
Molecular Aspects of Oxidative Drug Metabolizing Enzymes
Edited by E. Arınç, J. B. Schenkman and E. Hodgson
© Springer-Verlag Berlin Heidelberg 1995

changes, if produced in man, could be of serious consequence. Abnormalities could arise as a result of either gene or chromosome mutation (Carter, 1977). McKusick (1975) segregated 'traits' inherited at the level of the gene as dominant or recessive, and as autosomal or sex-linked. A dominant gene is immediately expressed in the next generation (e.g. Huntingdon's chorea and achondroplasia). A recessive mutation may take many generations to be expressed (e.g. phenylketonuria) except for sex-linked recessives (e.g. haemophilia). Many constitutional and degenerative diseases such as epilepsy or schizophrenia could be caused by irregularities of gene expression or arise from multiple genes.

Chromosome abnormalities may arise either by errors in the distribution of chromosomes leading to abnormalities of chromosome numbers, such as non-disjunction [Down's syndrome (mongolism) and Klinefelter's and Turner's syndromes], where the effect is seen in the next generation; or chromosome breakage (ataxia telangiectasia and Fanconi's anaemia) which arise from recessive autosomal gene changes. Some chromosome changes are thought to give rise to early embryonic loss and thus have little genetic impact on society. Deleterious mutations that are compatible with survival may be a heavy burden to society if the affected person requires medical or institutional care (mongolism, for example, which occurs in 1 in 1000 births).

It is not possible to give a quantitative estimate of the contribution of chemical agents to the incidence of genetic disease, but that they could constitute an aetiological factor in such disease which must be accepted. Hence, in toxicological assessments, there is a need to define by the appropriate tests the mutagenic activity of chemical substances.

This is a complex and difficult task and no single method gives conclusive information about genetic risk. A variety of test systems have been put forward, and some governmental agencies have produced guidelines for recommended methods (e.g. Department of Health and Social Security, UK, 1981; Department of Health, 1989; Official Journal of the EEC, 1979; and the OECD, 1983).

Structurally DNA, the target of mutagenic action, is identical in all living organisms but there are differences in the organisation of the genetic material in different species so that some organisms may be considered to constitute better predictive models for man. There are many interactions between chemicals and organisms which may determine whether genetic damage is expressed - unlike ionising irradiation, which is immediately active in terms of mutagenic potential. Irradiation generates extremely short-lived free radicals, some of which are close to or within the genetic target molecules, whereas there are many factors that influence whether a chemical compound may reach or react with such targets. Among such factors are the chemical structure of the compound, the duration of treatment, the route of administration, metabolic transformation (which may all be dependent on species, strain and sex), the membrane barriers, and the presence of pertinent defence mechanisms.

Other factors determine whether the damage is expressed as genetic damage. Such factors include the innate susceptibility of the cell (i.e. genetic repair capacity), the numbers of susceptible cells, the type of genetic target, the selection processes involved and the mode of inheritance (Matter, 1976).

Many compounds that require testing may not bind covalently with biologically important molecules unless they are first metabolised to highly reactive forms. Others disturb other functions and cause a genetic effect indirectly. Thus bases and nucleosides can themselves be mutagenic, an effect thought to arise as a result of unbalanced DNA precursor pools (Anderson *et al.*, 1981b; Clode *et al.*, 1986; Clode and Anderson *et al.*, 1988a and b; Meuth, 1984). Clearly the opportunities that theoretically exist for a given compound to attack genetic material are widespread, as are the obstacles and defences against such an attack being effective. As a consequence, the methods available for assessing such effects are numerous and in many respects very specialised.

2. Genotoxicity tests

The mutagenic activity of chemical substances has been studied in a number of test systems including micro-organisms, plants, insects, mammalian and human cells using *in vitro* and *in vivo* techniques. The mutagenic effect detectable in one system may not be found in another or even in different tissues of the same system.

The main sub-mammalian test systems currently available are those which reflect the present stage of knowledge and may be improved with the progress of experimental work in this field. Not all of the available tests are described here. Three categories are (a) tests for gene mutations; (b) tests for chromosome damage including aneuploidy or non-disjunction; (c) tests for DNA repair. The tests and cells and organisms used in them are listed in Table 1.

A.   Gene Mutation Tests

In molecular terms, gene mutations consist of the substitution, deletion or insertion of one or more nucleotide base-pairs in DNA. Mutations due to substitution are known as base-pair substitutions and those due to deletions or insertions as frameshifts. When a base-pair substitution mutation occurs, a wrong base is inserted, which then pairs with its natural partner during replication (adenine with thymine, cytosine with guanine) so that a new pair of incorrect bases is inserted into the DNA. When a frameshift mutation occurs, if there is base-pair loss, the messenger DNA, which reads the DNA in triplet codons, incorporates the first base-pair from the next triplet codon. Thus the subsequent code of the DNA becomes scrambled until a 'nonsense' or terminating codon is reached. A similar process happens with a base-pair gain.

Gene (or point) mutations cannot be detected by cytological methods but can be distinguished as variants of characteristics controlled by specific gene loci or as recessive lethal conditions (generally linked to sex chromosomes). Mutations of specific sites can arise in either a 'backward' (from mutant to normal 'wild' type) or a 'forward' direction (from a wild type to mutant).

TABLE 1   Tests, cells and organisms used in mutagenicity tests

Gene mutation tests	Chromosome damage tests	DNA repair tests
Microbial tests	Microbial tests	Microbial tests
Bacteria	Nondisjunction in fungi	Spot tests for
Salmonella typhimurium	Chromosome loss	inhibition zones
Histidine independence	*Saccharomyces cerevisiae*	*Salmonella tyhimurium*
(reverse mutation)	*Sordaria breviocollis*	*Bacillus subtilis*
8-Azaguanine resistance		*Escherichia coli*
arabinose resistance	Plants	Saccharomyces cerevisiae
(forward mutation)	*Vicia faba*	mitotic recombination
Escherichia coli	*Allium cepa*	gene conversion
Tryptophan independence	*Tradescantia*	Saccharomyces cerevisiae
(reverse mutation)	*Zea Mays*	Schizosaccharomyces
5-Methyltryptophan	*Tricitum vulgare*	pombe
Streptomycin	*Hordeum vulgare*	Unscheduled DNA
(forward mutation)	*Lycopersium exulentum*	synthesis
		in mammalian cells
		in vitro and in vivo
prophage reduction	*Nicotiania tobacum*	synthesis
fluctuation test	*Pisum sativum*	Breakage of single chain
		DNA
		using gradient or
		elution techniques
		Single strand DNA breaks
Funqi	Insects	and alkali labile damage
		individual cells using
Saccharomyces cerevisiae	*Drosophila melanogaster*	single cell gel
Schizosaccharomyces pombe	*Bombyx mori*	electrophoresis
Neurospora crassa		(COMET assay)
Aspergillus	**Mammalian cells** *in vitro*	
	Chinese hamster ovary cells	
Insect test	human fibroblasts	
Drosophila	human lymphocytes	
Plants	**Mammalian cells** *in vivo*	
Tradescantia hair	the rat bone marrow metaphase	
	assay	
Mammalian cells *in vitro*	the micronucleus test	
HPRT ase locus	dominant lethal assay	
Chinese hamster V79	the heritable translocation	
Chinese hamster ovary	test	
human lymphocytes	direct spermatocyte test	
human fibroblasts	sex chromosome losses	
TK ase locus	the fertilised oocyte test	
L5178Y		
P388F		

Mammalian cells *in vivo*
the spot test
the specific locus test
the dominant skeletal test
the cataract test

Others
Biochemically based tests
involving protein change,
histocompatability loci,
pigment loci
Inversion techniques to
measure recessive lethals
Sperm morphology changes
Transgenic mice

Backward (reverse) mutation is generally studied in mutant cells for which the base-pair substitutions or frameshift mutations are known. Normal functional activity is re-established following a new substitution, or a second deletion and insertion near the first one. Backward mutation is highly specific in the type of DNA interaction required, and a chemical compound that does not increase the frequency of this type of mutation may nevertheless still cause other genetic effects.

Forward mutations may arise from effects which range from the substitution, insertion, or deletion of the nucleotide bases of a gene, to the deletion of the entire gene and neighbouring genes. They are detected when there is a loss of an enzyme function (auxotrophy) or from acquired resistance to various toxic chemicals, substrate analogues and agents such as heat. In some instances where only one specific locus may be affected, the forward mutation can be just as specific as the backward mutation.

The specificity of some of these mutation systems, together with secondary factors such as the effect of the particular nucleotide sequence near the original mutation, makes a quantitative comparison or a between-species extrapolation of the mutagenicity of different chemicals very difficult. For a list of other factors that can influence effects *in vitro*, see Ashby and Styles (1978) and Brusick (1987).

Bacterial Tests

Point mutation can be detected in various bacterial species. Those most commonly used are *Salmonella typhimurium* and *Escherichia coli*.

In addition to its use in the reverse mutation assay of Ames, detecting independence to histidine as a medium nutrient (Ames *et al.*, 1973; 1975; Ashby and Tennant, 1988; Maron and Ames, 1983; McCann *et al.*, 1975a and b; McCann and Ames, 1976; Tennant *et al.*, 1987; Zeiger, 1987) *S. typhimurium* can be used to detect forward mutations to 8-azaguanine resistance (Skopek *et al.*, 1978) and arabinose resistance (Pueyo, 1978). The Ames test has been reviewed in the EPA Gene-Tox Programme (Kier *et al.*, 1986), and by Gatehouse *et al.*, 1990. Strains most commonly used are TA1535, TA1537, TA1538, TA97, TA98, TA100 and TA102.

E. coli has also been reviewed in the EPA Gene-Tox Programme (Brusick *et al.*, 1980) and can detect (a) the induction of mutations from arginine dependence to independence (base-pair substitution reverse mutation) (Mohn *et al.*, 1974) or from tryptophan dependence to independence (frameshift reverse mutation) (Green and Muriel, 1976; Sleigh, 1981; Venitt and Crofton-Venitt *et al.*, 1983); (b) the induction of forward mutations from inability to ability to ferment galactose (Saedler *et al.*, 1968) and from sensitivity to resistance to 5-methyltryptophan (Mohn, 1973) or streptomycin (Wild, 1973). These forward systems allow the detection or base-pair substitutions, frameshift mutations, or small deletions; (c) the induction of prophage (inductest) which induces lysis of bacterial cells harbouring a latent bacteriophage through changes that lead to the destruction of deactivation of the phage repressor. These effects can be observed in a single strain which carry the necessary genetic markers and the bacteriophage in the prophage form (Ho and Ho, 1981; Moreau *et al.*, 1976; Speck *et al.*, 1978); and, (d) the reverse fluctuation test (Green *et al.*, 1976).

Tests using DNA-repair deficient bacterial strains have been reviewed by Leifer *et al.* (1981 (Gene-Tox).

Fungal Tests

The yeasts *Saccharomyces cerevisiae* and *Schizosaccharomyces pombe* and the Ascomycetes *Neurospora crassa* and *Aspergillus nidulans* are among the fungi most used for the detection of point mutations.

In the yeasts *Saccharomyces cerevisiae* and *Schizosaccharomyces pombe*, mutations at each of the two genetic loci that control adenine biosynthesis cause a red pigmentation of the colonies. In *Saccharomyces cerevisiae*, the effect has been used to develop forward mutation tests either in haploid or diploid strains (Mortimer and Manney, 1971). In this system it is possible to distinguish (by established morphological criteria) the effect of the mutation from that of mitotic recombination (Zimmerman, 1975; Zimmerman *et al.*, 1984, Gene-Tox).

Schizosaccharomyces pombe has been used to detect forward mutation in wild or adenine-dependent haploid strains where mutations have been introduced that increase sensitivity to chemical mutagens (Mortimer and Manney, 1971). The use of these two yeasts for mutagenicity assays has been extensively reviewed (Loprieno *et al.*, 1974; Loprieno *et al.*, 1983, Gene-Tox; Zimmerman, 1975). A system of forward mutation from canavine sensitivity to resistance (Brusick, 1972) has also been developed in *Saccharomyces cerevisiae* and the organism can also be used in a fluctuation test, for example, the D-7 strain is used for detecting isoleucine dependence (Parry, 1977a and b; Parry *et al.*, 1984).

Strains of *Neurospora crassa* can also detect the induction of forward mutations in each of two genetic loci which control adenine biosynthesis (Ong and de Serres, 1972). In addition, strains have been developed that allow identification of recessive point mutations at each of the two genetic loci mentioned, dominant lethal mutations in the genetic region in which the two loci are situated and recessive mutations in the whole genome (Brockman *et al.*, 1984, Gene-Tox; de Serres and Malling, 1971; Ong and de Serres, 1972). Strains of aspergillus can be used for the induction of forward mutation to 8-azaguanine and *p*-fluoropheny-lalanine and back mutation to methionine independence (Roper, 1971). Aspergillus haploid and diploid strains, respectively, have been reviewed in the gene-tox programme (Kafer *et al.*, 1976; Kafer *et al.*, 1982; Scott *et al.*, 1982).

Insect Tests

The most widely used point mutation test in *Drosophila melanogaster* is that which detects the induction of sex-linked recessive lethals. These become evident in the second generation of treated individuals and can be observed at much lower doses than those needed to induce chromosome loss or dominant lethal mutations in this system. The presence of the X chromosome (which is about 20% of the genome) makes it possible to classify most of the sex-linked recessive lethals as gene mutations or multi-gene deletions. Autosomal recessive lethal mutations can be observed in flies of second generation. The induction of visible mutations is measured by crossing wild males (treated or control) with females that are homozygous for the visible mutations. The function of *Drosophila* in genetic toxicology testing has been described by Abrahamson and Lewis (1971), Vogel (1976; 1977 and 1987) and others (Auerbach, 1977; Baker *et al.*, 1976; Bootman and Kilbey, 1983; Lee *et al.*, 1983, Gene-Tox; Mollet and Wurgler, 1974; Valencia *et al.*, 1984, Gene-Tox; Wurgler *et al.*, 1977).

Mammalian Cell Tests In Vitro

There are various genetic systems for the detection of point mutations in cultures of mammalian cells (Cole *et al.*, 1990). They are based on the use of phenotypic markers generally resulting from resistance to antimetabolites or temperature sensitivity. In addition to gene mutation, variants of somatic cells in culture can be generated by other effects such as gene suppression. Hamster, mouse and human diploid fibroblasts are among the cell lines most often used to detect genetic changes. For some markers many cell lines may be used. For example, the locus controlling the hypoxanthine guanine phosphoribosyl transferase (HPRT) enzyme occurs in Chinese hamster embryo lung cells [V-79 cells] (Arlett, 1977; Bradley *et al.*, 1981, Gene-Tox; Chu *et al.*, 1976; Cole *et al.*, 1983; Fox *et al.*, 1976; Huberman and Sachs, 1974; Krahn and Heidelberg, 1977) Chinese hamster ovary cells [CHO cells] (Hsie *et al.*, 1981, Gene-Tox; Neill *et al.*, 1977) and primary culture of lung cells (Dean and Senner, 1977), human fibroblast (Jacob and Mars, 1977) and lymphoblast cells (Thilly *et al.*, 1976). The locus controlling adenine phosphoribosyl transferase (APRT) is present in human fibroblasts (de Mars, 1974). The thymidine kinase (TK) locus occurs in L5178Y (Clive, 1973; Clive *et al.*, 1972; Clive *et al.*, 1983, Gene-Tox; Clive and Spector, 1975; Cole and Arlett, 1973; Cole *et al.*, 1983; Knaap and Simons, 1975; Nakamura *et al.*, 1977) and P388F mouse lymphoma cells (Anderson, 1975a; Anderson and Cross, 1982; Anderson and Cross, 1985; Anderson and Fox, 1974; Fox and Anderson, 1976).

A fluctuation test (Cole *et al.*, 1976) and a host-mediated assay (Fischer *et al.*, 1974) have been developed using L5178Y cells, and other cell lines from the mouse have been employed for mutagenesis assays (Anderson, 1975b).

Mammalian Cell Tests In Vivo

The 'spot' test (Fahrig, 1978; Russel et al., 1981b, Russel et al., 1984, Gene-Tox) This test detects somatic mutations. Mice that are heterozygous for certain coat colour genes are treated in utero by administration of the test compound to the mother. If a mutation occurs during the development of the hair follicles, this may be observed in the mouse when the coat is fully developed. Thus, the female is treated during pregnancy and the offspring grown until the coat colour has fully developed. Each mouse is then examined for the coat markings indicative of a clone arising from a single mutant cell. This system has the advantage of being one of the few assays for gene mutations in mammals. The main disadvantages are that the somatic mutations for coat colour may occur as a consequence, not of gene mutations, but of chromosomal mutations, and the coloured spots can be difficult to distinguish from the natural pigmentation zones of the coat.

The specific locus test (Russel, 1951; Russel et al., 1981a, Gene-Tox; Searle, 1975). This test in mice is at present the only one available that can detect heritable mutations in mammalian germ cells.   The test uses mouse strains, homozygous for certain dominant wild-type alleles and strains that are homozygous for recessive alleles at these loci. The dominant, wild-type strains are treated and then mated with the recessive strains. Changes induced at any of the loci cause visible changes in the offspring, in respect of eye colour, ear size and coat colour. Although it is recognised that these events are gene mutations, their exact nature cannot be identified in most instances and they are assumed to be small multi-locus deletions. The main disadvantage of the specific locus test is that the spontaneous rate for these loci is low. Hence, for a negative result to be recorded, large numbers of offspring (up to 30,000) need to be scored, making the test extremely costly and laborious.

Biochemically-based tests There are a number of biochemical tests for detecting gene mutations in somatic and germ cells in mammals. Enzyme activity can be used as a genetic marker to detect induced microlesions in mammals. They depend on variations in specific proteins caused by mutations in the genes controlling the synthesis of the enzymes of proteins (Mays *et al.*, 1978; Neel, 1979; Valcovic and Malling, 1973).

Other approaches to detection of mutants in mice include changes in mandible shape (Festing and Wolff, 1979), skeletal mutations (Ehling, 1970; Selby and Selby, 1977), histocompatibility loci (Kohn, 1973), recessive lethals measured by the inversion technique (Evans and Phillips, 1975; Roderick, 1979) and pigment loci (Searle, 1979).

Transgenic animal tests
Transgenic animals provide efficient systems for detecting gene mutations *in vivo*. Such systems rely on the use of a bacteriophage lambda shuttle vector which has been integrated into the genome of an inbred mouse via microinjection so that as division occurs every somatic and germ cell in the animal contains this gene which is quickly recovered using standard laboratory techniques. Two systems are commercially available - Mutamouse and the Big Blue mouse. Mutamouse was created by incorporating DNA from the bacteriophage lambda gt10 lacZ into the genome of the CD2 hybrid mouse. Each Mutamouse cell contains approximately 40 copies of the transgene arranged head to tail at a single insertion site. Mutamouse has become homozygous for the transgene and contains approximately 80 lac Z sites within every normal diploid cell and 40 per haploid gamete.

In the case of the Big Blue mouse the target organ is the Lac1^q gene in C57Bl6/6 mouse strain.

These Lac genes are susceptible to mutation like any other endogenous genes in the cell. After the transgenic mouse is exposed to a test compound, genomic DNA is isolated from the desired tissues such as bone marrow, germ tissue and liver. The integrated lambda vectors are then easily and efficiently recovered by subjecting the genomic DNA to an *in vitro* packaging extract (proprietorial) which excises the vector DNA and packages it into a lambda phage head. Each packaged phage is then used to infect *E. coli* cells. The packaging extract is used to give high target gene rescue efficiencies.

When *E. coli* infected with the lambda vectors are plated and incubated (18 hr) on indicator agar dishes containing the chromagenic substrace X-gal, mutations with the Lac1^q target gene cause the formation of blue plaques, while unmutated or intact Lac1^q targets result in clear plaques. In the case of the gt10 LacZ gene (Mutamouse) mutations are the opposite from those in the Big Blue mouse, i.e. they form clear plaques while the unmutated ones are blue.

Details of the transgenic systems have been described in several publications (Kohler *et al.*, 1990; Kohler *et al.*, 1991a and b).

It has been found, for example, that for male germ cells the spontaneous mutation frequency is lower in these cells than in somatic tissues (Kohler *et al.*, 1991a).

Transgenic rats containing the same lambda/Lac1 shuttle vector have been developed for inter-species comparison of mutagenesis testing results, which may offer a better understanding of the specific mechanisms involved in mutagenesis at the molecular level *in vivo* (Provost *et al.*, 1993).

3. The Use of Metabolic Activation Systems

Microbial and mammalian cell systems *in vitro* generally do not possess the metabolic capability of cells of the intact animal. Several supplementary have been developed for use with microbial and mammalian test systems.

The most widely used systems for metabolic activation is the rat liver S-9 fraction supplemented with the co-factors NADP and glucose-6-phosphate (Ames *et al.*, 1973a and b; Ames *et al.*, 1975; Gatehouse *et al.*, 1990; Kier *et al.*, 1986, Gene-Tox; Maron and Ames, 1983). The S-9 fraction is the supernatant fraction of the liver homogenate obtained by centrifugation at 9000 g for 10 min. It contains a reponderance of microsomes and the attendant oxidases. Before the fraction is prepared, the animals are commonly treated with inducing agents such as polycholorinated biphenyls (PCB), phenobarbital (PB), or 3-methylcholanthrene. PCB can induce microsomal enzymes capable of metabolising many different types of environmental mutagens and carcinogens, but a combination of β-naphthoflavone and PB is to be preferred (Matsushima *et al.*, 1976).

Rat liver is the most widely used source but the livers from other species, including man, have also been used. The use of human liver might be important in evaluating the risk of environmental mutagens and carcinogens to man (Bartsch *et al.*, 1975; Tang and Friedman, 1977).

Several experimental procedures exist for microbial systems: (a) *the liquid method*. The bacteria are incubated with the S-9 mix and the test substance, washed, and the revertants and surviving bacteria counted; (b) In *the plate method* (Ames *et al.*, 1975; Maron and Ames, 1983), the mixture of bacteria, S-9 mix and test substance is added directly to molten-soft agar and poured onto hard agar. This method is relatively simple and quick; (c) In a variant of this procedure (Nagao *et al.*, 1978; Sugimura *et al.*, 1976) the mixture of bacteria, test substance and S-9 mix is incubated for 20 min at 37°C before adding molten-soft agar. With this *pre-incubation method*, dimethylnitrosamine, a definite mutagen and carcinogen, gives positive results (Sugimura *et al.*, 1976) which are not readily detected by other methods.

An alternative approach, where mammalian cells in culture constitute the test system, utilises 'feeder' layers of mammalian cells (Huberman, 1975). These are usually of embryonic origin, which their enzyme system fully intact but with cell replication inactivated by radiation or a high dose of an antimitotic agent such as mitomycin C.

A method involving *in vivo* metabolic activation of a drug has been reported (Legator *et al.*, 1977; Legator *et al.*, 1982, Gene-Tox; Legator and Malling, 1971). The test substance is administered to animals by gastric tube or subcutaneous injection and the test bacteria (indicator organisms) are injected intraperitoneally, withdrawn after several house, and examined for mutation using an *in vitro* assay. This 'host-mediated assay' has been used to investigate the mutagenicity of undefined or unsterilised material thought to contain mutagens and carcinogens. The test organism may also be given intravenously or into the testes. This assay is relatively insensitive by comparison with the plate incorporation assay. High doses of test compound have to be used to produce a response but it is conceivable that this method gives a realistic measure of the mutagenic impact *in vitro*. Clearly the antibacterial activity of the host may be a confounding factor. *Salmonella typhimurium* (Legator and Malling, 1971), *Escherichia coli* (Mohn and Ellenberger, 1973), *Saccharomyes cerevisiae* (Fahring, 1974; 1977), *Schizosaccharomyces pombe* (Legator and Malling, 1971), *Neurospora crassa* (Legator and Malling, 1971), and mouse lymphoma cells (Fischer *et al.*, 1974) have all been used as indicator organisms.

Chemical substances which have undergone metabolism are excreted in the urine as degradation or conjugated products and as such could be identified by suitable microbial indicators, after deconjugation (Commoner *et al.*, 1974; Durston and Ames, 1974). Although of little value in the detection of new mutagens, analysis of the mutagenic activity of the urine or body fluids (Legator *et al.*, 1976; Legator *et al.*, 1982, Gene-Tox; Siebert, 1973) or faeces (Bruce *et al.*, 1977; Combes *et al.*, 1984) might be of value to screen workers with known exposures.

Metabolic activation systems have been described with tests for gene mutations, but can also be applied with tests for chromosome damage and DNA repair.

B.   Chromosome Damage Tests
Chromosome damage is defined as a modification of the number or structure of chromosomes. It can be detected both by cytological and genetic methods.

Variations in the number of chromosomes (aneuploid and polyploidy) may result from endoreduplication (continued chromosome division), metaphase arrest, anaphase retardation and non-disjunction in mitosis and meiosis.

Structural changes (Figure 1) are mainly the results of breaks in the chromatid arms. Depending on the number of breaks and the way in which they may be rejoined a series of unstable structural modifications may arise that are not transmitted to successive cellular generations. These include 'gaps' (achromatic interruptions, but see Anderson and Richardson, 1981), breaks of one or both the chromatids, chromatid interchanges, acentric fragments, ring and dicentric chromosomes. Stable structural modifications that are transmissible may also arise, such as inversions, translocations and deletions (Buckton and Evans, 1973; Evans, 1976; Preston *et al.*, 1981, Gene-Tox; Scott *et al.*, 1983; Scott *et al.*, 1990).

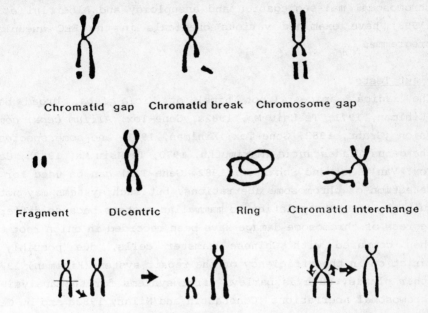

Figure 1 *Diagram of different categories of chromosome damage*

Microbial Tests

It is possible to measure non-disjunction in micro-organisms. Tests for detecting mitotic non-disjunction have been developed using the fungus *Aspergillus nidulans* (Bignami *et al.*, 1974; Kafer *et al.*, 1976; Kafer *et al.*, 1982, Gene-Tox; Scott *et al.*, 1982, Gene-Tox) and the yeast *Saccharomyces cerevisiae* (Parry, 1977b; Parry *et al.*, 1984; Parry and Zimmermann, 1976; Sora and Magni, 1988; Zimmermann, 1975) meiotic non-disjunction can be measured in the fungus *Sordaria brevicollis* (Bond, 1976).

Mitotic chromosome loss has also been examined in *Saccharomyces cerevisiae* (D61M) (Zimmerman *et al.*, 1985). The test relies upon the recovery and expression of multiple recessive markers reflecting the presumptive loss of the chromosome VII homologue carrying the corresponding wild type alleles.

Deviations from the conventional aneuploid or diploid state often causes serious perturbations in normal growth and development. Whittaker *et al.* (1989; 1990) have carried out an inter-laboratory assessment of this assay. Albertini (1989) has examined chromosomal mal-segregation and aneuploidy and Albertini *et al.* (1993) have examined various chemicals in the EEC aneuploidy programme.

Plant Tests

The radical apices (root tips) of *Vicia faba*, broad bean, (Kihlman, 1971; Te-Hsiu Ma, 1982a, Gene-Tox) *Allium cepa*, common onion (Grant, 1982, Gene-Tox; Kihlman, 1971) and some species of the genus *Tradescantia* (Marimuthu, 1970; Te-Hsiu Ma, 1982b, Gene-Tox; Van't Hof and Schairer, 1982, Gene-Tox) can be used for the detection of chromosome aberrations, but such systems may not be suitable for extrapolation to mammalian systems because different degrees of chromosome damage have been recorded in onion root tips when compared with Chinese hamster cells, due possibly to variation in the efficiency of the repair system (Kihlman, 1971). Other plants, such as barley, offer systems for the analysis of chromosomal aberrations (Constantin and Nilan, 1982a and b, Gene-Tox; Ehrenberg, 1971; Nilan and Vig, 1976; Plewa, 1982, Gene-Tox; Vig, 1982). The role of plants for genetic and cytogenetic screening has been reviewed by Constantin and Owens (1982, Gene-Tox).

Insect Tests

Effects at the chromosome level such as non-disjunction, loss of the X chromosome, deletions, translocations, dominant lethal mutations and mitotic and meiotic recombination have been observed in *Drosophila melanogaster* (see earlier references, also Valencia *et al.*, 1984, Gene-Tox; Wurgler *et al.*, 1977). Changes are generally shown by observing the phenotypes of the progeny from the appropriate crosses. For example, reciprocal translocations may be detected in the second generation of treated individuals carrying defined recessive markers in the autosome. The absence of any of the expected phenotypes in the progeny indicates that the translocations occurred in the parent reproductive cells. The loss or acquisition of a chromosome, or the loss of part of a chromosome, can be detected by observing phenotypes in the progeny of crosses in which only one of the parents have been treated.

Mammalian Cell Tests In Vitro

Chromosome damage can easily be observed by cytological methods in mammalian cells in culture (Preston *et al.*, 1982, Gene-Tox; Scott *et al.*, 1983; Scott *et al.*, 1990). Human fibroblasts, lymphocytes and rodent cells (hamster fibroblasts and mouse lymphocytes) are used. Translocations, inversions and other stable rearrangements indicate genetic damage that can be inherited. Gaps probably indicate early toxic effects unless accompanied by other types of damage (Anderson and Richardson, 1981) but have in fact persisted for three months when other categories of damage have been eliminated after vinyl chloride treatment.

Sister chromatid exchange Sister chromatid exchange in somatic mammalian cells is a measure of the reciprocal exchange in segments of homologous loci between sister chromatids, i.e. the number of crossovers during replication that occur between paired chromatids following treatment with a test agent (Latt *et al.*, 1981, Gene-Tox; Perry *et al.*, 1984; Perry and Evans, 1975; Stetka and Wolff, 1976). Cells are exposed to 5-bromodeoxyuridine for one round of replication. A second round without it follows, so that one arm of the chromosome is a substituted chromatid (these are recognised by differential staining techniques) and the frequency of such chromatid exchanges is increased by mutagens. It is considered to be a sensitive method for detecting chromosome damage but the significance of the test is not yet fully understood. It is considered as a test that indirectly measures DNA damage which involves repair.

Mammalian Cell Tests In Vivo

Methods exist for the assay of chromosomal aberrations in somatic cells after the administration of a test chemical to the intact animal (Albanese *et al.*, 1984; Preston *et al.*, 1981, Gene-Tox; Richold *et al.*, 1990; Topham *et al.*, 1983). Peripheral lymphocytes (Lilly *et al.*, 1975) and bone marrow cells (Cohen and

Hirschorn, 1971; Legator *et al.*, 1973; Nichols *et al.*, 1973; Richold *et al.*, 1990; Schmid *et al.*, 1971; Tijo and Whang, 1962) are commonly used. Bone marrow cells are a naturally proliferating cell population but peripheral lymphocytes are a synchronised population of G_o cells that rarely proliferate and have to be stimulated by mitogens for analysis. Such systems have large numbers of cells available and show good potential for analysis of human cells, provided solid baseline data are available (Anderson *et al.*, 1988).

The micronucleus test In anaphase chromosomes, abnormal chromosome or chromatid fragments lag behind the other migrating chromosomes, and deformed anaphase bridges occur. At telophase, these lagging elements form micronuclei (Heddle, 1973; Heddle *et al.*, 1983, Gene-Tox; Maier and Schmid, 1976; Salamone *et al.*, 1980; Schmid, 1973, 1976) because they do not become incorporated into the nuclei of the daughter cells. Micronuclei are most often observed in polychromatic erythrocytes but they can be observed in embryonic and other cells in culture by suitable staining techniques. The test is also of value in the detection of mitotic spindle poisons. There has been debate about the number of cells to be analysed, up to 2000 are recommended for suitable sensitivity (Ashby and Mirkova, 1987).

Dominant lethal test Basically, this assay depends on an increase in the foetal abortion rate due to defective sperm or ova, measured as a decrease in the numbers of uterine implants or an increase in early foetal deaths, i.e. pre-and post-implantation loss, respectively (Anderson *et al.*, 1983; Bateman, 1977, Bateman and Epstein, 1971, Ehling *et al.*, 1978; Epstein and Rohrborn, 1970; Green *et al.*, 1985, Gene-Tox). An effect on fertility may be measured in the same study. Pre-implantation losses can occur due to other than genetic reasons (including lack of fertility) but the early implantation death is thought to be due to chromosomal abnormalities produced in germ cells. In order to sample all stages of sensitivity of the mating germ cell, an 8-week mating period is required in the mouse and a 10-week period in the rat (Epstein and Rohrborn, 1970), these being the times taken to produce mature sperm for the germinal cells.

The assay may be extended by examining surviving offspring for congenital malformations (Anderson *et al.*, 1993; Frances *et al.*, 1990; Jenkinson *et al.*, 1987; Jenkinson *et al.*, 1990; Knudsen *et al.*, 1977).

The dominant lethal assay can also be carried out using *Drosophila* (Wurgler *et al.*, 1977) (see earlier references).

Sperm-head morphology   Some mutagens give rise to abnormally shaped spermatocytes (Topham, 1980b, Wyrobek and Bruce, 1975) and this can result in inherited sperm abnormalities after parental treatment in animals (Topham, 1980; Wyrobek *et al.*, 1983a, Gene-Tox).   An evaluation has been made of human sperm as indicators of chemically-induced alterations of spermatogenic function (Wyrobek *et al.*, 1983b, Gene-Tox; Wyrobek *et al.*, 1984).

Heritable translocation test   The presence of a heritable translocation (i.e. a balanced chromosome translocation) is thought to confer sterility or partial sterility on the F_1 offspring.   Its presence can be confirmed either cytogenetically or by mating on several occasions suspect males who will continue to produce litters of reduced numbers (Cachiero *et al.*, 1974; Generoso *et al.*, 1977, Gene-Tox; Generoso *et al.*, 1980).

Direct spermatocyte test   Spermatogonia are examined at or near metaphase 1 for the abnormal mitotic figures that result from reciprocal translocation (Léonard, 1975; 1977).

Sex chromosome losses   The test also measures the heritability of chromosomal damage by scoring the progeny of exposed animals for infertility (Russel, 1979).

The fertilised oocyte test   The first cleavage embryos from treated male and female germ cells are scored for structural chromosome damage.   As the male and female pronuclei condense at different times, both genomes can be examined.   Chromosome aberrations in the male pronuclei have been shown to correlate with dominant lethality (Albanese, 1987; Brewen *et al.*, 1975).

The various germ cell methods have been reviewed by Albenese (1987).

C.   DNA Damage and Repair Tests

When DNA is damaged, repair normally follows (Cleaver, 1975; 1977; Larsen *et al.*, 1982, Gene-Tox). The initial lesions in DNA may be lethal, may remain without being repaired, may be repaired correctly to restore a normal genome, or incorrectly to produce errors and an abnormal genome. There are several types of repair possibilities.

Excision repair   Cut and patch or pre-replication repair occurs when a lesion, recognised by endonuclease, is excised by an exonuclease and the missing part re-synthesised by a polymerase to reconstitute the original strand. The new part is joined to existing DNA by a ligase. Excision repair is normally error-free and is known to occur in human cells. Hydroxyurea has been used to inhibit normal DNA replication, which then allows the detection of excision repair, but it is suspected that this compound itself may have some effect on excision repair so that the results from experiments using hydroxyurea should be considered with caution (Pearson and Styles, 1984).

Post-replication (by-pass) repair   The lesion is by-passed in newly synthesised daughter DNA, thus leaving a gap that is sealed by insertion of a DNA segment by recombination into the new daughter DNA. This process is error-prone. The post-replication gap may be filled by DNA synthesis *do novo* and thus correct errors copied by replication.

Recombination repair is a post-replication repair process in bacteria which involves the recombination of daughter strands of DNA to reconstruct the correct genome. This process is error-free, but error can occur if the repair requires *de novo* synthesis. This has been established in bacteria but has not been demonstrated conclusively in mammalian cells.

Damage to the DNA molecule may be considered as a primary lesion that could be involved in the process of mutation and the extent of the repair is an indicator of the amount of damage that has occurred to DNA.

Several methods exist for detecting DNA repair phenomena. These include differential zones of inhibition or killing in bacterial strains with and without repair processes (Ishinotsubo *et al.*, 1977; Leifer *et al.*, 1981, Gene-Tox; Tanooka, 1977), gene conversion in strains of yeast (Zimmerman *et al.*, 1984, Gene-Tox), sister chromatid exchange in mammalian cells (Latt *et al.*, 1981, Gene-Tox; Perry *et al.*, 1984; Perry and Evans, 1975; Steka and Wolff, 1976) and the direct measurement of DNA damage and repair (Cleaver, 1975; 1977; Larsen *et al.*, 1982, Gene-Tox).

'Spot' tests measuring differential zones of inhibition (Tweats *et al.*, 1984). A Petri dish is seeded or streaked with the test organism (*Salmonella typhimurium*) (Ames *et al.*, 1975; Leifer *et al.*, 1981, Gene-Tox), *Bacillus subtilis* (Tanooka, 1977), *Escherichia coli* (Leifer *et al.*, 1981, Gene-Tox; Sugimura *et al.*, 1977), *Saccharomyces cerevisiae* (Zimmerman, 1975; Zimmerman *et al.*, 1984, Gene-Tox) or *Aspergillus nidulans* (Kafer *et al.*, 1976; Kafer *et al.*, 1982, Gene-Tox; Roper, 1971). The test compound is placed in the dish and the inhibition zone or lethal effect produced by the compound is evaluated in two different strains of the test organism, one being the wild-type strain and one being deficient in a DNA repair system (e.g. pol A⁻ in *B. subtilis* and rec A⁻ and uvr⁻ in *E. coli*). When a greater zone of inhibition is produced in the repair-deficient strain than in the wild-type strain, the compound is considered to be capable of affecting DNA. The assay can be carried out with and without metabolic activation (S-9 mix) incorporated in the agar. If minimal medium is used, both mutation and inhibition zones can be detected.

Fungal Tests Measuring Mitotic Recombination or Gene Conversion
In eukaryotes, it is possible to measure an increase in the frequency of mitotic recombination or gene conversion when a recessive phenotype is expressed in the transition from a

heterozygote to a homozygote situation. These tests are thought to be related to the exchange following breakage of the chromatids of two homologous chromosomes. Such changes may allow for the expression of recessive mutation, as it is known that meiotic recombination allows the expression of recessive gene mutations in man.

Such tests are carried out in the yeast *Saccharomyces cerevisiae* and *Schizosaccharomyces pombe* (Schiestl, 1989). A fluctuation test can also be used to detect mitotic-gene conversation (Zimmerman *et al.*, 1984). Other fungi such as *Aspergillus nidulas* (Kafer *et al.*, 1982; Gene-Tox) have also been used.

Mammalian Cell Tests In Vitro and In Vivo
Tests to measure sister chromatid exchange in mammalian cells have already been described.

Test to measure DNA repair by unscheduled synthesis. DNA damage can be detected by determining unscheduled DNA synthesis which occurs as a result of DNA excision repair. One of the techniques reveals repair synthesis by determining radioactivity (tritiated thymidine) incorporated into DNA during the repair process (Mitchell *et al.*, 1983, Gene-Tox; Waters *et al.*, 1984). Tests to measure unscheduled synthesis in liver cells *in vivo* are now suggested if a negative result has been obtained in the bone marrow assay (see DH 1989 Guidelines - Fig.2).

The radioactivity can be measured either by autoradiography or by direct counting of the incorporated thymidine by liquid scintillation. Metabolic activation can also be included with such systems.

Another technique measures *the breakage of the single DNA chain in alkali*, using gradient or elution techniques. The gradient technique is the most sensitive but alkaline elution is simple and faster, measuring the rate of elution of DNA through a filter, a function of the relative molecular mass of the DNA (Cleaver, 1975; 1977).

The COMET or single cell gel electrophrosesis assay is more recently becoming established as a useful technique for detecting

DNA damage (e.g. Green and Lowe, 1992). It detects single strand breaks and alkali labile damage in individual cells. Cells are electrophoresed under alkaline conditions. Stained nuclei with increased DNA damage display increased migration of single stranded DNA towards the anode. The length of migration of the tail can be measured with a graticule Density profiles can also be measured using image analysis. This method has been reviewed by McKelvey-Martin *et al.* (1993).

Conclusion

Not all of the systems described have been equally well-studied and some are used more for specialised purposes than for screening chemicals. The most widely used screening system today is the Ames test. Its use is widespread because of its extensive validation and its potential to detect genotoxic carcinogens.

4. Usefulness of some of the systems for screening purposes

<u>Microbial assays</u>. The major advantage of the microbial methods is that they are rapid, inexpensive and relatively simple to carry out for the experienced scientist, though not for the novice.

A. Bacterial Assays

The plate incorporation test of *Salmonella typhimurium* (the Ames test) is the test used routinely for screening purposes. It was shown in early blind trials to have correlations with known carcinogens and non-carcinogens as high as 90% and also to detect carcinogen and non-carcinogen pairs equally well (Purchase *et al.*, 1978a). It is a test which can be carried out by many laboratories (Ashby *et al.*, 1985; de Serres and Ashby, 1981). It has a stable phenotype which is demonstrated by its lack of genetic drift (Anderson *et al.*, 1984).

Correlations with known carcinogens have been much lower in recent years (Tennant *et al.*, 1987; Zeiger, 1987) but this is primarily because non-genotoxic carcinogens have been included in the testing programme. The assay still has high predictivities for genotoxic carcinogens (Ashby and Tennant, 1988). This will be discussed in more detail later - see Prediction of Human Carcinogens.

The liquid incubation or the pre-incubation method has been used for detecting those mutagens difficult to determine in the plate incorporation assay. The fluctuation assay has been suggested as being more sensitive than the plate incorporation assay, in that it can detect compounds at lower dose levels of the test compound. A forward mutation assay is also available for *Salmonella* but these deviations from the standard Ames test method have not been validated. The repair-deficient/proficient microbial tests and spot tests, where a 'spot' of the test compound is placed in the centre of the bacterial plate, are really only suitable as pre-screens. The Ames test can be completed in about three days.

B.   Yeast Assays

Both *Saccharomyces cerevisiae* and *Schizosaccharomyces pombe* are suitable for use in routine screening assays. Strains can be cultivated in both the diploid and haploid phases, which allows for the detection of a wide range of mutation events. Yeasts can be used to detect both point mutations and DNA repair events in terms of mitotic recombination as evaluated by crossing-over and gene conversion. A disadvantage is that the chromosomes are too small for direct cytological observation but chromosome damage can be measured by tests for non-disjunction. Yeast systems take a few days longer than bacterial systems for colony growth but the overall time scale involved is not greatly different. They tend to be used as supplementary assays.

C.   Plant Assays

Plant systems can detect most types of damage. They have short generation times and the cost, handling and space requirements are relatively small; genetics of seeds can be investigated under a wide range of environmental conditions such as pH, water content, and temperature, and chromosomal organisation is similar to the human system (Nilan and Vig, 1976). Difficulty is experienced, however, in extrapolating the results to mammalian systems, including man. Several agents, such as cytosine arabinoside, daunomycin, and adriamycin, are known to be ineffective on the plant genomes and yet cause severe genetic damage to mammalian calls. This may be because the cell wall inhibits absorption and because of greater ability to repair DNA lesions. Such systems are probably not satisfactory for predicting potential human mutagens. They are used as supplementary assays but may be useful for testing chemicals which are sprayed on plants such as pesticides.

D. Insect Assays

Drosophila has a short generation time of 10-12 days and is cheap and easy to breed in large numbers with relatively simple facilities. Extensive studies on the metabolism of insecticides performed over 15 years (Würgler, 1977) revealed that insect microsomes are capable of similar enzymic activity to those of the mammalian liver, but insects do not have any specific organ in which the enzymes are predominantly located.

In *Drosophila*, mutagenic activity can be tested at different germ cell stages which is important where mutagens have specificity of action. *Drosophila*, permits the scoring for the whole spectrum of genetic effects. The observation that the lowest effective concentration (LEC) values (and therefore the highest mutagenic effectiveness) have been recorded for recessive lethals indicates the superior discriminating power of this test. (The X chromosome represents a fifth of the whole genome.) By comparison, the test for dominant lethality is of limited value. High doses are required, and dominant lethals sometimes fail to arise when agents cause the induction of recessive lethals. Changes in hatchability sometimes produce false-positives. However, *Drosophila* is a good 'catch-all' system owing to the variety of genetic and end-points that can be detected.

Vogel (1987) points out that, based on Gene-Tox Report data and two international collaborative trials, the sex-linked recessive lethal test does not have a high ability to detect carcinogens when genotoxins other than direct acting and simple pro-mutagens are included. However, it has a high predictability for non-carcinogens. The tests detecting somatic mutation/mitotic recombination (SMART) have higher predictibilities than the sex-linked recessive lethal assays. *Drosophila* assays, although they were included in some regulatory guidelines, e.g. DHSS, 1981, are not currently included in those of DH, 1989 (Figure 2).

E.   Mammalian Cell Assays In Vitro

Mammalian cell systems are generally considered more valid than non-mammalian systems in terms of extrapolation to man, because mammalian DNA is more similar to that of man. Few chemicals are known which are detected exclusively in mammalian cell mutation assays, and are not detected by the similar (in terms of cost) mammalian cell *in vitro* cytogenetic assays. For this reason the use of mammalian cell mutation assays is questioned. In addition, the V79 and the CHO cell mutation systems, whilst suitable for detecting a positive response are basically inadequate in terms of cell numbers for detecting a negative response. By increasing cell numbers, whilst the mutant fraction remains constant, there would be sufficient cell numbers available for statistical purposes. Suitable cell numbers are only available at present in the cell suspension assay of mouse lymphoma - the L5178Y system. The L5178Y cell system of Clive (Clive *et al.*, 1983, Gene-Tox) detects both large and small colonies. The large are thought to arise from gene mutation and the small from chromosome damage. Since the genetic end-points are measurable, the system is very sensitive but it has been suggested that this system may be too sensitive and lack discriminating power for the detection of carcinogens and non-carcinogens. The thymidine kinase (TK) and hypoxanthine guanine phosphoriboxyl transferase (HPRT) loci are commonly used in mammalian cell assays but the latter, which is used in V79 and CHO cells, is less sensitive than the TK locus (McGregor, personal communication). If mutation cell systems are to be used, however, cells grown in suspension are easier to handle.   They do not require trypsinisation, are easily sub-cultured and are not subject to metabolic co-operation, so do not suffer from the reduced sensitivity that occurs through metabolic co-operation when mutated cells are in close contact with non-mutated cells.

Figure 2 Flow diagram for testing strategy for investigating mutagenic properties of a substance*

STAGE 1

Initial Screening

Two tests required (a + b) except where human exposure would be expected to be extensive and/or sustained, and, difficult to avoid, when all three tests are necessary

In Vitro Tests

(a) Bacterial assay for gene mutation

(b) Test for clastogenicity in mammalian cells for example metaphase analysis

(c) Test for gene mutation in mammalian cells (for example the L5178Y TK +/− assay)

STAGE 2

Tests for:
Compounds positive in one or more tests in Stage 1, and
All compounds where high, or moderate, prolonged levels of human exposure are anticipated

In Vivo Tests

(a) Bone marrow assay for chromosome damage (metaphase analysis or micronucleus test)

Plus, if above negative, and any in vitro test positive

(b) Test(s) to examine whether mutagenicity or evidence of DNA damage can be demonstrated in other organs (eg liver, gut etc)

STAGE 3

If risk assessment for germ cell effects is justified (on basis of properties including pharmacokinetics, use and anticipated exposure).

Quantitative studies need strong justification in view of their complexity, long duration, costs and use of large numbers of animals.

In Vivo Tests for Germ Cell Effects

(a) Tests to show interaction with DNA

Dominant lethal assay (most useful)

Cytogenetics in spermatogonia

One cell embryo test

(b) Tests to show potential for inherited effects

Dominant lethal assay gives indication of likelihood of inherited effects

Cytogenetics in spermatocytes for reciprocal translocations

Non-disjunction in the mouse (10-day embryo)

(c) Test for quantitative assessment of heritable effects

Mouse heritable translocation test

Mouse specific locus test.

* General guidance only is given in this flow diagram. Decisions regarding, for example, whether a specific compound is expected to produce high or moderate, but prolonged, exposure would normally be taken by Regulatory Authorities having regard to other relevant data, and on a case-by-case basis.

There may be instances where alternative tests to those specified might be more appropriate, and it is important that a flexible approach is adopted. Each compound should be considered on a case-by-case basis with regard to the selection of tests as well as the interpretation of results.

(After Department of Health Mutagenicity Guidelines, HMSO, 1989)

Figure 2. Flow diagram for testing strategy for investigating mutagenic properties of a substance

STAGE 1

Initial Screening

Two tests required (a + b) except where human exposure would be expected to be extensive and/or sustained, and, difficult to avoid, when all three tests are necessary

STAGE 2

Tests for:
Compounds positive in one or more tests in Stage 1,
and
All compounds where high, or moderate, prolonged levels of human exposure are anticipated

STAGE 3

If risk assessment for germ cell effects is justified (on basis of properties including pharmacokinetics, use and anticipated exposure).

Quantitative studies need strong justification in view of their complexity, long duration, costs and use of large numbers of animals.

In Vitro Tests

(a) Bacterial assay for gene mutation

(b) Test for clastogenicity in mammalian cells (for example metaphase analysis

(c) Test for gene mutation in mammalian cells (for example the L5178Y TK +/− assay)

In Vivo Tests

(a) Bone marrow assay for chromosome damage (metaphase analysis or micronucleus test)

Plus, if above negative, and any in vitro test positive

(b) Test(s) to examine whether mutagenicity or evidence of DNA damage can be demonstrated in other organs (eg liver, gut etc)

In Vivo Tests for Germ Cell Effects

(a) Tests to show interaction with DNA

Dominant lethal assay (most useful)

Cytogenetics in spermatogonia

One cell embryo test

(b) Tests to show potential for inherited effects

Dominant lethal assay gives indication of likelihood of inherited effects

Cytogenetics in spermatocytes for reciprocal translocations

Non-disjunction in the mouse (10-day embryo)

(c) Test for quantitative assessment of heritable effects

Mouse heritable translocation test

Mouse specific locus test.

* General guidance only is given in this flow diagram. Decisions regarding, for example, whether a specific compound is expected to produce high or moderate, but prolonged, exposure would normally be taken by Regulatory Authorities having regard to other, relevant data, and on a case-by-case basis.

There may be instances where alternative tests to those specified might be more appropriate, and it is important that a flexible approach is adopted. Each compound should be considered on a case-by-case basis with regard to the selection of tests as well as the interpretation of results.

(After Department of Health Mutagenicity Guidelines, HMSO, 1989)

(Current DH (1989) guidelines suggest that chemicals giving negative responses in 3 *in vitro* systems (bacterial mutation, chromosomal and cellular point mutation assay) with and without metabolic activation do not require further testing in animal systems).

F.   Mammalian Assays In Vivo

The advantage of *in vivo* studies is that the test compound is metabolised in the animal.   Both the bone marrow metaphase and micronucleus assays thus have this advantage.   Positive results in such assays indicate that the test chemical is a mutagen to somatic mammalian cells *in vivo*.   Results correlate well with carcinogenicity studies, particularly human carcinogens (Shelby, 1988).

The dominant lethal assay in rodents has been claimed to be insensitive to detecting chemicals, but the lack of sensitivity may reflect the real situation because of the so-called 'testes barrier' formed by the Sertoli cells surrounding the germ cells. In addition to the pharmacokinetic hurdles and organ specifity etc. and the short half-life of the chemical, the blood-testes barrier may be important.   Even if the compound reaches the testes, it may not be metabolised. With dimethylnitrosamine, for example, there is less alkylation of the DNA in the testes than in any other organ (Swann and Magee, 1968) and the compound gives a negative result in the dominant lethal assay (Propping *et al.*, 1972).

These considerations, of course, also affect the other assays concerned with the germ cells, such as the specific locus and heritable translocation assays.   The drawbacks to the latter two assays are that they require vast numbers of animals in order to detect a response and thus are very costly in terms of resources and time.   However, the mammalian *in vivo* assays are required for determining genetic hazard and risk estimates.   The dominant lethal assay although not useful for predicting carcinogens (Green *et al.*, 1985, Gene-Tox) could be useful for predicting heritable hazard and the specific locus and heritable translocation assays are useful for quantitative risk assessment.

Kirkland (1987) discusses the implications of germ cell cytogenetic tests in the regulatory process. One of the responses to a questionnaire sent to regulators was that germ cell tests are rarely requested as a matter of course. Where germ cell tests are indicated, a dominant lethal test (most often) of a heritable translocation test (sometimes) would be seen as helpful in elucidating germ cell effects. The mouse specific locus test is rarely requested due to the large number of animals involved and the small number of laboratories with the relevant experience.

The current DH (1989) guidelines (Figure 2) recommend germ cell assays only for risk assessment as the last tier. In previous guidelines (DHSS, 1981) germ cell assays were included in a battery of assays.

G.   DNA Repair Assays

Measurements of excision repair are determined as mean values for a population of cells, whereas mutation is a rare event in individual cells. The amount of excision repair after exposure to an agent will depend on several factors, such as the extent of reaction with the DNA (the total number of damaged sites), the number of sites that can be excision repaired, the size of the repaired regions, the kinetics of excision repair as functions of time and dose, and the extent to which chemical interactions modify other sites, and possibly inhibit excision repair (Cleaver, 1975; 1977). The amount of excision repair will therefore be greatest for mutagens that induce the greatest proportion of extensively damaged sites requiring repair by the large substitution. The number of mutations depends on the severity of pre- and post-replication damage.

Studies of the relationship between DNA damage, excision repair, post-replication repair, and mutagenesis must take account of the numbers and varieties of lesions involved in mutagenesis and the modes of repair. Exclusive reliance on any one measurement is not recommended. It is best to consider DNA repair, for example, only in conjunction with some other parameters before assessing the possible mutagenicity of an agent. However, the COMET assay is proving useful for the rapid measurement of genetic damage (McKelvey-Martin, 1993).

Cell Transformation Assays

These assays are thought to 'bridge the gap' between mutagenicity and carcinogenicity. Cell transformation has been defined as the induction in cultured cells of certain phenotypic alterations that are related to neoplasia (Barrett *et al.*, 1986). There are several types of endpoints for cell transformation, including loss of anchorage dependence and alterations in morphology, viral dependence and altered growth in agar (McGregor and Ashby, 1985). Morphological transformation and altered growth have been used most extensively. These systems are the Syrian hamster (SHE) assay (Barrett and Lamb, 1985; Berwald and Sachs, 1963; 1963, Di Poalo *et al.*, 1969a and b; 1971; 1972; Huberman and Sachs, 1966) and the mouse C3H/IOT1/2 and mouse BALB/c3T3 assays (Heidelberger *et al.*, 1983; Kakunaga, 1973; Reznikoff *et al.*, 1973). Of these, the SHE cell transformation assay is unique in that it uses normal diploid cells (Berwald and Sachs, 1963, 1965). The other two systems are based on established cell lines which have undergone some adaptive changes in culture, resulting in an aneuploid karyotype and a potentially preneoplastic phenotype. A recommended protocol for all three assays based on a survey of current practice has been suggested (Dunkel *et al.*, 1991). SHE cells have a limited lifespan in culture and rarely become tumorigenic unless treated with carcinogens (Barrett *et al.*, 1977). Therefore the cellular events underlying the morphological transformation of SHE cells might be indicative of earlier neoplastic changes compared with those underlying cell transformation observed in the other cell lines. The acquisition of a fully neoplastic phenotype in these cells in a multistep process analogous to that *in vivo* and this system is therefore particularly useful for studying the cellular and molecular events involved in neoplastic development (Barrett *et al.*, 1986; Koi and Barrett, 1986). Fitzgerald and Yamasaki (1990) have addressed the issue of tumour promotion describing models and assay systems (1990).

In the SHE transformation assay morphologically transformed colonies are identified by a disorientated pattern of piled up cells (Berwald and Scales, 1965). The cells have a considerable range of metabolic activities but chemicals requiring further metabolic activation can be studied with incorporation of appropriate sub-cellular fractions or a second cell type.

5. Test significance and interpretation

The available sub-mammalian test systems, used without mammalian *in vivo* studies, would not be acceptable by any governmental authority for estimating risk to man (except perhaps for chemicals to which humans are exposed at insignificant levels). They do provide a useful tool for a preliminary screening of possible human mutagens. A positive result in a well constructed and validated system is generally regarded as a warning sign. However, when considering the many and diverse chemicals which are of unknown mutagenic potential which require metabolic activation and which may react differently with different cells and organs before producing genotoxic effects, it is not surprising that many give equivocal data that cannot be resolved from experience gained from classical studies.

By comparison, the handling of positive data is much more clear-cut, but it is desirable that dose-response curves should be established in routine testing and results should be reproducible. Difficulties may arise when mutagens have a strong killing effect so that a genetic effect is obviously not as readily detected. This can be exemplified by mammalian cell mutation assay systems where the activity of a chemical can produce an absolute increase in the number of mutants per treated cell or an absolute decrease if there is a strong killing effect. In the former case the rate of increase of mutants (over the spontaneous) is then greater than the rate of inactivation, per unite dose; in the latter, this is not so. Weak mutagens are more difficult to evaluate when there is an increase only in mutants per surviving cell (that is, after correction for survival) and not per treated cell. The apparently weak positive effect could be due to the induction of new mutants but may be due to a greater resistance of spontaneous mutants to inactivation by the agent used. Reconstruction experiments (where known numbers or mutant and wild-type cells are mixed together) can solve the former problem, but to the latter there is no good solution.

Dose-response curves may have linear or diphasic shape, or may be diphasic at high dosage with a linear function at low doses. Thresholds may exist at some chemical concentrations when the chemical is without effect below a certain concentration. At higher doses, dose-response curves tend to flatten or plateau or decline when the killing effect overrides the mutagenic effect. When a dose-response relationship is established it is easier to reach a conclusion regarding the mutagenicity of a chemical.

In *in vitro* mammalian cell systems, particular emphasis should be placed on using doses of chemicals which are not too high and so do not alter the pH or affect the osmolality of the test system. The problems arising in such instances, where false positives can be generated, have been highlighted in a special issue of Mutation Research (Brusick, 1987).

6. Strategies for the protection of man

Auerbach (1975) stated that the procedures for the estimation of possible hazards from genotoxins are full of uncertainties. This is still the case. It is, nevertheless, important to continue to develop approaches that will allow such hazard identification. The *first* approach is to use a group of short term tests to detect possible mutagen/carcinogenic activity while recognising that such an approach is subject to limitations in terms of test variability, species sensitivity and interspecies extrapolation (but this is true for toxicological tests in general).

The *second* is to measure induced genetic damage directly in man.

The *third* is to use results from an evaluation of effects in the gonads which can be combined with the pattern of expected human exposure from which a judgement as to the amount of risk can be estimated. This can be attempted by combining human exposure data with (a) the known dose-response observed in animal studies, or (b) measured target/germ cell concentrations combined with mutagenic responses defined by the best understood *in vitro* systems. Concepts such as radiation-equivalents, doubling doses and parallelogram considerations may be of value in this approach.

A. Approach 1 - Use of Short Term Tests for Mutagenicity Prediction of Human Mutagens

Developments in short term test procedures for genotoxicity have generally focused on their ability or inability to identify potential carcinogens. However, in regulatory practice in the UK and some other European countries, such tests are conducted to determine mutagenicity *per se* with a view to identifying potential human mutagens.

There are no examples of induced mutations in man with proven causality. Cigarette smoke, vinyl chloride, lead and anaesthetic gases are the agents which are best documented but no effects in man are yet unequivocally determined. Thus a validated mutational assay for detecting human mutagens is not yet attainable, although attempts are being made to address this issue.

Before this can be done there is a need to examine the data on the performance of the test systems and to establish how consistent these data are when derived from different sources.

Within test variation. An assessment of the variation that is obtained within a test is most easily achieved by comparing the performance of tests which independently assay the same chemical. Over a decade ago, three such studies were completed (de Serres and Ashby, 1981; Dunkel, 1979; Poirier and de Serres, 1979). All showed errors of about 10% in detecting positive effects and about the same in detecting negative effects. It is worth considering the consequence of a 10% discrepancy when 6 test systems are used. Thus if a single test is used, 90% of the mutagens will be correctly identified. If it is assumed that each one of the six tests provides an independent assessment of the mutagenicity of the chemical, and that the error rate of each test for both positive and negative results is 10%, six tests will identify 99.999% of the mutagens. At the same time, however, only 56% of the non-mutagens will be negative in all six test systems.

Between test variation. When considering test systems which, although they may assess the same genetic end-point, use different organisms, the range of variability is wide. One study (Poirier and de Serres, 1979) using three assay systems found agreement for fifty-five chemicals (twenty-five all negative and thirty all positive) and disagreement for forty-four. Similar findings occurred in an international study (de Serres and Ashby, 1981).

The use of such schemes for regulatory purposes was much discussed (Brusick, 1982; Purchase, 1980) but as yet has still not been satisfactorily resolved. The reproducibility of the *Salmonella typhimurium* and *Escherichia coli* mutagenicity assays was examined by Dunkel *et al.* (1985). It is generally recognised that a single positive is not sufficient to define mutagenicity, but is was proposed that a single response from some test systems might have a greater weighting than from others (Brusick, 1982). However, as more work has been done with these tests it is realised that some of these weightings may not apply. Assessment panels have addressed the evaluation of mutagenicity assays for genetic risk assessment (Brusick *et al.*, 1992; Russel *et al.*, 1984), whilst Ray *et al.* (1987) have examined the various assays for identifying classes of chemicals.

Evaluation of the performance of short term tests in identifying germ cell mutagens

Waters *et al.* (in press), evaluated the performance of various STTs in identifying germ cell mutagens. Using a combined data set derived from the US EPA/IARC Genetic Activity Profile (GAP) database and the US EPA Gene-Tox database, a total of 56 germ cell mutagens were identified. These chemicals had given positive results in one or more of the following assays: the mouse specific locus test; *in vivo* tests for chromosomal aberrations in the germ cells; the dominant lethal test in mice or rats; and the mouse heritable translocation test. The same two databases were used to provide information (where available) on the activity of these chemicals in bacterial mutagenicity assays, in two *in vitro* mammalian cell assays (one for chromosome aberrations, another for gene mutation), and in two *in vivo* tests on the bone marrow (chromosome aberrations or micronucleus formation). The performance of the various STTs is summarised in Figure 3. Although the sample size was only small, the data indicated that the two *in vitro* assays with mammalian cells were able to identify 86% (gene mutation) and 93% (chromosome aberration) of the germ cell mutagens, whilst bacterial mutagenicity assays were slightly less sensitive (75%). [Unfortunately, in the absence of any data on chemicals that are not germ cell mutagens, it is not possible to assess the specificity of the assays (i.e. their ability to correctly identify such chemicals as non-mutagenic to the germ cells)] The sensitivity of the individual assays was increased when the results from two or three of the assays were combined. Of the 36 germ cell mutagens that had been tested in a bone marrow assay (for chromosomal aberrations or micronucleus formation), 33 gave positive results, and the evidence for germ cell mutagenicity of two of the three not identified was called into question. The problem of strain variability among rodents was raised as a possible source of discrepancy. This study indicates that STTs can provide valuable information on the potential of a chemical to induce germ cell mutations.

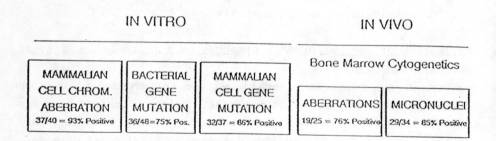

Figure 3 Test performances are given for the germ cell mutagens from the combined EPA/IARC GAP and GENE-TOX databases. Performance is indicated by the fraction of agents with positive test results divided by the number of agents tested and is expressed also as the percentage positive.

(After Waters et al (in press)

Ashby (1986) proposed a testing strategy to detect genotoxic agents *in vivo*, where after a positive response in *Salmonella* or *in vitro* cytogenetics, a chemical should be tested in a mouse micronucleus or a rat liver unscheduled DNA synthesis assay *in vivo*. This strategy was debated by Garner and Kirkland (1986) and Gatehouse and Tweats (1986). However, the guidelines for mutagenicity testing (DH, 1989) use a similar scheme, but *in vivo* tests for germ cell effects may be justified on the basis of properties including pharmacokinetics, use and anticipated exposure (Figure 2). Guidelines in most countries and internationally are constantly under scrutiny (e.g. OECD Guidelines Harmonisation Meeting at the 6th International Environmental Mutagen Meeting in Melbourne 1993). As a result of such moves, it is hoped that better models for evaluating human mutagenic risk may emerge.

Prediction of human carcinogens
The use of genetic toxicology assays in predicting the carcinogenicity of chemicals to rodents or humans has come under a great deal of scrutiny. In a 1975 publication in which 300 chemicals were tested in the Ames test, McCann *et al.* found that 90% of the 174 carcinogens were mutagenic. Similar figures were provided by other workers (see introduction), and it seemed only a matter of time before complementary short-term tests (STTs) would be developed to detect the remaining 10% of carcinogens that were 'missed' by the Ames test.

Through the late 1970s and early 1980s there was a period of activity as the matrix of STTs and chemicals tested increased, and various groups of workers endeavoured to show that their favoured assays could identify known mammalian carcinogens. A series of internationally coordinated (through WHO/IPCS, UNEP and ILO) validation studies (funded partly by the US National Institute of Environmental Health Sciences and the UK Medical Research Council) were conducted (Ashby *et al.*, 1985, 1988). The various STTs were assessed on their <u>sensitivities</u> (the percentage of known carcinogens correctly detected), their <u>specificities</u> (the percentage of non-carcinogens correctly identified) and their <u>concordances</u> (the overall accuracy in their identifications).

These studies revealed that the predictive values of the tests were no longer as high as they had been in earlier investigations. As more and more rodent carcinogens were identified, mainly under the National Toxicology Program (NTP), it was found that the predictive power of the Ames test, in particular, declined. While the activity of most of the long-standing and well-established rodent carcinogens could be rationalised in terms of their electrophilic properties (because this was the primary stimulus for their initial selection for carcinogenicity testing), an increasing proportion of the newly-identified carcinogens (not pre-selected in this way) were both without a supporting chemical rationale for their activity and were non-mutagenic to Salmonella. The fact that these chemicals were appearing positive in the animal carcinogenicity studies but were consistently negative in STTs was of great concern, and had regulatory implications.

This can now be explained in terms of the two types of carcinogen that are thought to exist, those acting by a genotoxic mechanism (that would have produced positive results in the STTs), and those acting by a non-genotoxic mechanism (generally negative in the STTs).

Analysis of concordance between STT results and carcinogenicity data

The inconsistency between STT results and carcinogenicity findings prompted an investigation by the Cellular and Genetic Toxicology Branch of the NTP, the aim of which was to assess the ability of prokaryotic and eukaryotic STTs to detect rodent carcinogens. The resultant publication (Tennant *et al*. 1987) revealed that four of the most used STTs (the Ames test, the mouse lymphoma L5178Y mutagenicity assay, and tests for chromosomal aberrations and sister chromatid exchange in Chinese hamster ovary cells) were poor predictors of carcinogenic activity. The assessment of 73 chemicals that had recently been tested for carcinogenicity by the NTP revealed that the concordances between the carcinogenicity results and the genotoxicity findings were only about 60% for each STT. The individual tests exhibited different data profiles, the mouse lymphoma and SCE assays giving more positive results than the Ames or chromosomal aberration assays. Although some chemicals always gave consistent results, there were eleven chemicals which showed only a single positive STT result. No combination of the tests improved the accuracy of the cancer prediction: for instance, defining a chemical as positive if it gave a positive result in any of the four tests increased the overall sensitivity but decreased the specificity. The most difficult problem from the genotoxicity viewpoint was the failure of six carcinogens to show any positive results at all, despite the fact that three of these (dioxin, reserpine and a polybrominated biphenyl mixture) were the most potent carcinogens in the whole group, at least on the basis of the dose producing statistically significant increases in tumour incidences. In an update of this study, Zeiger et al. (1990) tested a further 41 NTP chemicals and found a wider variation in the level of concordance for the four STTs (ranging from 54% for the SCE assay to 73% for the Ames test), but this was not considered to represent any substantial improvement in the predictive power of the Ames test. Again, no combinations of STT improved upon the concordance and predictivity of the Ames test alone.

The 1987 publication by Tennant and colleagues provoked immediate responses and counter-responses in the genetic toxicology journals (e.g. Ashby, 1988a; Ashby and Purchase, 1988; Auletta and Ashby, 1988; Brockman and DeMarini, 1988; Haseman et al. 1988; Kier, 1988; Trosko, 1988; Young, 1988). The 73 compounds tested were said to be a distorted sample, as they represented compounds that had only recently been tested by the NTP and had been selected for testing on the basis of production volumes, degrees of human exposure and suspicion of carcinogenic potency. They were certainly more representative of the more subtle 1980s type of carcinogen than the classic potent carcinogens of the 1950s and 1960s. Brockman and DeMarini (1988) criticised the approach of blind testing, because no account was taken of what might already be known about a chemical's properties or those of structurally related compounds. Individually 'customized' protocols, it was argued, might have given results that more closely matched the carcinogenicity findings. They also considered that the results of animal carcinogenicity bioassays (particularly negative ones) did not deserve the exalted position that the scientific community had generally assigned to them because they had low statistical power, were rarely replicated, were seldom done under different sets of experimental conditions, and had many limitations that were unlikely to be overcome in the near future. Other investigators have also questioned the scientific validity of lifetime animal feeding studies in which the chemical is administered at the maximum tolerated dose (see later for further discussion) (Ames and Gold, 1990a and b; Ashby and Morrod, 1991). The problem with the numerical approach of Tennant et al. (1987) is that the mathematical sophistication does not make up for the inherent limitations of the carcinogenicity and genotoxicity assays. The expression of carcinogenicity bioassay findings as a simple positive or negative result is, in many cases, an over-simplification or misrepresentation of a chemical's true carcinogenic activity. Similarly, the results of a number of STTs for a particular chemical often include conflicting or equivocal findings which do not allow a simple positive or negative categorisation. Mammalian cell assays are said to present a particular problem in producing so-called 'false positive' results (Adler et al. 1989; Scott et al. 1991). Such limitations can seriously affect the interpretation of the data obtained, although various attempts may subsequently be made to rationalise the discordant results in the different STTs, as occurred for the 1987 study by Tennant et al. (e.g. Ashby, 1988b; Myhr and Caspary, 1991; Prival and Dunkel, 1988).

Despite the limitations, the mathematical approach has been used by a number of other investigators (e.g. Auletta and Ashby, 1988; Benigni and Giuliani, 1988; Ennever and Rosenkranz, 1989; Klopman and Rosenkranz, 1991; Kuroki and Matsushima, 1987; Loprieno et al. 1991; Parodi et al. 1991). Kuroki and Matsushima (1987) evaluated the performance of a range of STTs in detecting 71 established, probable or possible human carcinogens (classified in IARC Groups 1, 2A and 2B respectively) and concluded that the chromosome aberration test in mammalian cells in vitro provided results that were complementary to those obtained in the Ames test. No figures could be derived to assess the specificity of the chromosome aberration test because of the lack of clearly identified human non-carcinogens. Sorsa et al. (1992) evaluated the available cytogenetic data on 27 proven, ten probable and 15 possible human carcinogens and found that 19/27, 6/10 and 5/15 induced chromosomal aberrations, sister chromatid exchanges and/or micronuclei in humans. A large prospective cohort study suggested that chromosomal aberrations but not sister chromatid exchanges are significant for prospective cancer risk (Sorsa et al. 1992).

Klopman and Rosenkranz (1991) selected 253 compounds that had been tested by the NTP for carcinogenicity to animals, and used probability calculations to assess the predictivity of the Ames test and of assays for sister chromatid exchange (SCE) and chromosomal aberrations (CA). The unscheduled DNA synthesis (UDS) assay was similarly evaluated for 130 chemicals, the corresponding cancer bioassay data being taken from Williams et al. (1989). The analysis revealed concordances of 62% (Ames), 61.3% (SCE), 56.5% (CA) and 56.2% (UDS). The probabilities of a carcinogen being positive in individual tests were 56% (Ames), 88% (SCE), 75% (CA) and 53% (UDS), while the probabilities of a non-carcinogen being positive in individual tests were 27% (Ames), 89% (SCE), 78% (CA) and 33% (UDS). Although the former might suggest that the SCE and CA assays were better than the Ames and UDS assays at detecting carcinogens, the latter indicate that they may in fact be too sensitive to be of use in predicting carcinogenicity, as a high proportion of non-carcinogens also gave positive results in these assays. The concordance figures reflect the fact that none of the four assays is particularly good at predicting carcinogenicity.

As noted previously, Tennant et al. (1987) and Zeiger et al. (1990) found that various combinations of in vitro STTs (Ames, CA, SCE and the mouse lymphoma assay) were no better at predicting carcinogenicity than was the Ames test alone. However, Jenssen and Ramel (1980) and Shelby (1988) proposed that a combination of two genotoxicity assays, the Ames test and the (in vivo) bone marrow micronucleus test, could be used for the detection of genotoxic chemicals that might be predicted to be carcinogens. In an analysis of 23 chemicals designated by the International Agency for Research on Cancer (IARC) as Group 1 compounds (carcinogenic to humans), Shelby (1988) found that 20 of the 23 carcinogens (87%) were active in one or both STTs. Seventeen of the 22 that were tested in the Ames assay gave positive results, and the untested chemical (treosulphan) was considered likely to be active on structural grounds (subsequently confirmed by Zeiger and Pagano, 1989); 12 of these 17 were also active in the *in vivo* bone marrow test, and four of the remaining five were considered as likely positives due to structural similarities to known bone marrow clastogens. Three other chemicals (benzene, diethylstilboestrol and arsenic) were inactive in the Ames test but gave positive results in the bone marrow test, while of the remaining three chemicals not tested in the bone marrow test, treosulphan was again considered a likely positive (as was subsequently demonstrated by Gulati *et al.*, 1990 and Shelby *et al.*, 1989), and asbestos and conjugated oestrogens (both negative in the Ames test) were not expected to affect the bone marrow. Thus, the latter two carcinogens would not be detected by this combination of assays, nor would their carcinogenicity be anticipated on structural grounds (Shelby, 1988).

Since that time, numerous studies have been conducted to assess the various possible protocols for the *in vivo* bone marrow test, including single or multiple exposures and different sampling times (e.g. Adler and Kliesch, 1990; Čihák and Vontorková, 1990; George *et al.*, 1990; Gulati *et al.*, 1990; Mavournin *et al.*, 1990; Mirkova, 1990; Tice *et al.*, 1990). The various investigators reached different conclusions on which protocol was most effective in detecting carcinogens, but as there was some evidence that a three-exposure protocol might be more effective than a single-exposure protocol (e.g. Gulati *et al.* 1990; Tice *et al.* 1990), Shelby *et al.* (1993) went on to test 49 NTP chemicals (25 rodent carcinogens and 24 non-carcinogens) using a three-exposure intraperitoneal protocol with a single sampling time. Only five of the 25 rodent carcinogens gave positive results, although a further two were found to be positive in a single-exposure protocol. Two of the seven (benzene and monuron) would not have been suspected from Ames test data or from their chemical structure. Four of the 24 chemicals that had shown no evidence of carcinogenicity in rodent bioassays (ascorbic acid, phenol, titanium dioxide and 2,6-toluenediamine) were found to be active in the micronucleus test; all but 2,6-toluenediamine were non-mutagenic in the Ames test. This study indicates that the

three-exposure (ip) protocol with single sampling time is not
satisfactory for distinguishing between carcinogens and non-
carcinogens; further work is needed before the micronucleus test
can provide additional meaningful information on a chemical's
genotoxic activity.

In an evaluation of a mammalian gene mutation assay, the
Chinese hamster ovary/ hypoxanthine guanine phosphoribosyl
transferase (CHO/HGPRT) assay, Li *et al.* (1988) found that 40 out
of 43 reported animal carcinogens (93%) gave positive results,
while the only definitive non-carcinogen, caprolactam, was
negative. Since then, a further nine chemicals that gave negative
results in NTP bioassays have been tested in this assay; seven
gave negative results, one (2-chloroethanol) was positive and one
(benzoin) was equivocal (Li *et al.* 1991). The investigators
concluded that the CHO/HGPRT assay seemed to have high specificity
as well as high sensitivity, and that it could be used to
complement the Ames test and cytogenetic assays. Oshiro *et al.*
(1991) tested ten compounds deemed non-carcinogenic in the
literature, and concluded that the CHO/HGPRT assay produced more
relevant results than other genotoxicity tests in mammalian cells:
only two of the chemicals (dichlorvos and 2-(chloromethyl)
pyridine), were positive in this assay, and the non-carcinogenic
status of both was considered questionable.

In an analysis by Loprieno *et al.* (1991), a test battery of
two *in vitro* assays (Ames and CA tests) and one *in vivo* assay (the
rodent bone marrow micronucleus test) was reported to have greater
predictivity than the Ames test alone. In an analysis of 716
chemicals that had been tested in cancer bioassays, the accuracy
of the three STTs was 68.6% for the Ames test (number of test
chemicals, n=544), 64.3% for the CA test (n=445) and 70.6% for the
in vivo micronucleus test (n=163). This was increased to 71.6%
for Ames + CA (n=310), 85.0% for Ames + micronucleus (n=113),
87.9% for CA + micronucleus (n=107) and 92.5% for a combination of
all three STTs (n=93). Thus the three STTs together correctly
identified 43 out of 45 carcinogens and 43 out of 48 non-
carcinogens. The 716 chemicals that had been tested for
carcinogenicity were part of a much larger data set of 3389
chemicals, 2898 of which had been tested in the Ames test (85.5%),

1399 (41.3%) in the CA test, and 319 (9.4%) in the *in vivo* micronucleus test. For the 270 chemicals that had been tested in all three STTs, 107 were positive in both *in vitro* assays, of which just over half (56) were also positive *in vivo*; a smaller proportion (25% or less) of the 16 or 71 chemicals that were positive only in the Ames test or only in the CA test (respectively) also gave positive results *in vivo*; and eleven of the 76 chemicals (14.5%) that gave no indication of genotoxicity *in vitro* were found to be active in the micronucleus test. Five of the eleven (chlorobenzene, ortho- and para-dichlorobenzene, toluene and trichloroethylene) have also demonstrated carcinogenic activity in rodents, while the other six (isoxaben, 1,3,5-, 1,2,4- and 1,2,3-trichlorobenzene, trimethoprim and vincristine) have not been adequately tested. The eleven chemicals were thought by the investigators to represent a class of compound for which the mechanism of genotoxicity could not be fully applied, and more information was being collected on them (Loprieno *et al.* 1991).

Various supplementary assays have been developed, including tests for recombination, gene conversion and aneuploidy (chromosome loss) in different strains of the yeast, *Saccharomyces cerevisiae*. These have not been well validated, and inconsistencies have been reported, for example, in an aneuploidy test system using strain D61.M (Albertini, 1991). Nevertheless, there is some indication that tests for recombination (Schiestl, 1989; Schiestl *et al.* 1989) or for gene mutation (Morita *et al.* 1989) in yeast cells might be able to detect carcinogens that are 'missed' by the Ames test.

Ashby and Leigibil (1992) authored a commentary on the use of transgenic mouse mutation assays in the context of genotoxic and non-genotoxic mutagens. Recommendations for dose levels and treatment protocols for these mice were presented and the authors asked for further discussion of dosing regimens. Mirsalis (1993) and Ashby and Leigibil (1993) debated the proposed dosing approach for the detection of genotoxic and non-genotoxic carcinogens. Gunz *et al.* (1993) further examined whether non-genotoxic carcinogens can be detected with the lac1q transgenic mouse assays.

Use of the Ames test and/or structural alerts in predicting carcinogenicity

A non-computational chemistry based method for identifying carcinogens has been proposed (Ashby, 1985). (This has been progressed in parallel with developments in quantitative structure activity relationships (QSAR) - see review by Phillips and Anderson (1993). This chemistry based approach utilises the electrophilic theory of carcinogenesis propounded by the Millers in the 1970's and relies on personal skill to identify any electrophilic centre(s) within the molecule under consideration. Such a feature constitutes a "structural alert" for reaction with a nucleophilic site in DNA.

Following the 1987 study by Tennant *et al.*, Ashby and Tennant (1988) and Ashby *et al.* (1989) examined the Ames data, carcinogenicity verdicts and chemical structures of a set of 264 compounds tested in the NTP's rodent cancer bioassays. Chemicals were classified by the breadth of their cancer activity. For instance, 66 were assigned to Group A because they caused cancer in both rats and mice at one or more sites. Group B chemicals (19 in all) produced tumours at multiple sites in once species. Group C (27 compounds) and Group D (26 compounds) resulted in cancer at a single site either in both sexes or a single sex (respectively) in a single species. The two remaining classes consisted of 94 chemicals where there was no indication of carcinogenicity in either species (Group F) and 32 where there was an equivocal or uncertain result (Group E). Aspects of the chemical structure that were suggestive of electrophilic activity either of the compound or its metabolites were then identified. The investigators assessed how well the results of the Ames test and the presence or absence of structural alerts predicted the NTP carcinogenicity results. The concordance between the four carcinogenic and non-carcinogenic classes (A,B,C,D and F) and the Ames test was 64%, as low as Tennant's earlier results. The concordance with the structural alerts was also 64%. The number of mismatches between the structural alerts and the Ames test results was 31 (12%). Thus, although there was a high correlation between the presence or absence of structural alerts in the 264 chemicals and their mutagenic activity in the Ames test, neither of these was particularly effective in predicting carcinogenicity. The observed sensitivity of the Ames test (in correctly identifying carcinogens) was only 58%, but this increased to 72% when the 66 two-species carcinogens (Group A) were considered on their own (Ashby *et al.*, 1989). This compared with only about 50% (Groups B and D) or 32% (Group C) for the 72 single-species carcinogens. The evaluation of a further 39 chemicals had little effect on these figures for sensitivity (Tennant and Ashby, 1991).

These differences between the four classes of carcinogens provided the impetus for a further analysis of all the chemicals that had been included in the three earlier surveys (i.e. Ashby *et al.*, 1989; Ashby and Tennant, 1988; Tennant and Ashby, 1991). The combined data set consisted of 301 chemicals, which were split into broad classes based, this time, on their known or projected chemical reactivity (Ashby and Tennant, 1991). Roughly half of the chemicals (154) were structurally-alerting chemicals, and these were further subdivided into aromatic amino/nitro-type compounds (84 chemicals), natural electrophiles including reactive halogens (46) and miscellaneous structurally-alerting groups (24). The 147 non-alerting chemicals were broadly categorised as compounds devoid of actual or potentially electrophilic centres (61), compounds containing a non-reactive halogen (50), or compounds previously classed as non-alerting in structure but with minor concerns about a possible structural alert (36). When the six classes were each analysed separately, Ashby and Tennant reported a clear distinction between the three groups of structurally-alerting chemicals and the three groups of non-alerting chemicals. Thus 65-70% of the structurally-alerting chemicals were carcinogenic compared with 25-52% of the non-alerting chemicals, and 65-86% were mutagenic in the Ames test compared with only 0.02-9% of the non-alerting chemicals. The sensitivity of the Ames test in predicting carcinogenicity was 93% for the nitro/amino compounds, 83% for the electrophiles and 56% for the miscellaneous structurally-alerting compounds; however, for all three groups there was a high false-positive rate, as 71%, 55% (6/11) and 75% (3/4) of the non-carcinogens in the respective groups gave positive results in the Ames test. For the non-alerting chemicals, the sensitivity was low (13%) or non-existent, while the high specificity (with only one non-carcinogen in each group showing mutagenic activity) was of little value.

Genotoxic and non-genotoxic carcinogens

Ashby and Tennant (1991) distinguished between structurally-alerting carcinogens, i.e. chemicals with one or more electrophilic groups that could potentially interact with and damage the cell's DNA in some fairly immediate way ("genotoxic carcinogens"), and non-alerting carcinogens that had no obvious electrophilic centre and were thought to induce tumours by some mechanism that did not involve early attack of the chemical or a direct metabolite on the cell's DNA (putative "non-genotoxic carcinogens"). Some examples of non-genotoxic (or epigenetic) routes to tumour formation include enhanced cell proliferation resulting from cytotoxicity or mitogenesis (induced cell division), hormonal changes, and peroxisome proliferation. At a 1990 meeting on early indicators of non-genotoxic carcinogenesis, several possible mechanisms were described (Anon., 1991).

Since the various STTs are specifically designed to detect genotoxic activity, any chemical that induces tumours by a non-genotoxic mechanism will not (or at least in theory ought not to) be active in these STTs.

Jackson *et al.* (1993) indicated that the situation is not that simple. Many chemicals that are described as non-genotoxic carcinogens in the literature have in fact shown evidence of genotoxic activity in appropriate STTs. In an analysis of 39 putative non-genotoxic carcinogens which had been tested in five or more STTs for gene mutation, chromosomal aberrations and/or aneuploidy, 14 showed evidence of activity (seven *in vivo* and seven *in vitro*) and a further ten showed limited evidence (*in vivo* and/or *in vitro*). Only two were considered to have been sufficiently tested to warrant the description non-mutagenic (i.e. non-genotoxic). The remaining 13 chemicals had not been adequately tested *in vitro* and *in vivo* for each of the three broad categories of DNA alterations (i.e. gene mutation, chromosome aberration and aneuploidy). Thus, although all but one of the 39 chemicals gave predominantly negative results in the Ames test, and most were also devoid of structural alerts, over half showed genotoxic potential in other STTs, and another third might also do so if they were to be adequately tested. Whether their genotoxic activity plays an important role in their carcinogenicity is questionable, but the investigators noted that four out of six compounds that induce peroxisome proliferation, two out of five cytotoxic carcinogens and one out of three mitogens all demonstrated some mutagenic activity.

Distinguishing between genotoxic and non-genotoxic carcinogens on the basis of their tumour profile

Ashby *et al.* (1989) and Ashby and Tennant, (1988, 1991) distinguished between two types of carcinogen: multi-species (and usually multi-site) carcinogens which could generally be predicted by either a positive result in the Ames test or (equally well) by a structural alert (the so-called genotoxic carcinogens); and the single-species (often single-site, and/or single-sex) carcinogens which were less likely to be mutagenic or structurally-alerting (the so-called non-genotoxic carcinogens). Analysis of the patterns of carcinogenic response provided by these two classes of carcinogen had revealed that for many tissues (notably the mouse liver), equal sensitivity to genotoxic and non-genotoxic carcinogens was observed, but that certain tissues (notably the Zymbal's gland in the rat) appeared only to be sensitive to genotoxic carcinogens (Ashby *et al.*, 1989; Ashby and Tennant, 1988). When Ashby and Tennant (1991) evaluated the target sites for 59 carcinogenic nitro/amino compounds (93% of which were Ames-positive) and 30 carcinogenic electrophilic agents (83% Ames-positive), they were all said to have been previously associated with genotoxic carcinogens. In contrast, the target sites for the 57 carcinogenic non-alerting chemicals (only two of which were mutagenic in the Ames test) were said to be almost exclusively confined to tissues connected with non-genotoxic carcinogens. This in fact may be an over-simplification. Whilst the 1988 study had identified 16 tissues that were affected only by genotoxic carcinogens, the 1991 study revealed a marked reduction in this number, as some non-genotoxic carcinogens also induced tumours in these tissues. Only the Zymbal's gland (rat) and the lungs (both species) appeared to be exclusively targeted by genotoxic carcinogens. The authors' post-analysis rationalisation of certain putative non-genotoxic carcinogens which were active in tissues previously only associated with genotoxic carcinogens did not explain away all of the discrepancies.

In a further analysis of the pattern of the carcinogenic response among tissues and test groups, Ashby and Paton (1993) aimed to test the tentative conclusions of the 1991 paper on a larger data set, restricting their analysis of genotoxicity to a simple consideration of chemical structure. The data for the analysis were taken from the carcinogen database compiled by Gold *et al.* (1991), which detailed the site of carcinogenesis for 522 chemical carcinogens. Of the 511 rodent carcinogens, 251 had been tested in both rats and mice, 168 in rats only and 92 in mice only. One chemical, benzene, was eliminated from the analysis because of its 'exceptional' carcinogenic effects (it induces an extensive range of tumours in both sexes of rats and mice often at unusual sites, and yet no useful structural alert can be derived without condemning all benzenoid chemicals). Around 70% of the remaining 510 rodent carcinogens were structurally-alerting, i.e. 300 of the 418 chemicals tested in rats and 236 of the 342 tested in mice. When these were analysed in terms of the number of tissues affected, 27.7% (rat) and 18.6% (mouse) of the structurally-alerting chemicals were active in more than two tissues, compared with only 5.9% (rat) and 0% (mice) of the non-alerting chemicals. In contrast, a much higher percentage of the non-alerting chemicals affected only a single site (rat, 58.8%; mouse 65.9%), compared with the structurally-alerting chemicals (rat, 34.2%; mouse, 34.6%). Of the 250 chemicals tested in both species, roughly half of the 43 rat-specific carcinogens and of the 64 mouse-specific carcinogens were structurally-alerting chemicals, compared with 77% of the 143 two-species carcinogens. The investigators considered this a confirmation of their earlier hypothesis, that structurally-alerting (or genotoxic) carcinogens tend to be active in more than one species and at multiple sites, whereas the non-alerting (putatively non-genotoxic) carcinogens are more likely to affect only a single species and a single tissue.

In an earlier paper, Ashby and Purchase (1992) had selected five "genotoxic carcinogens" (positive in the Ames test and with structural alerts) and five "non-genotoxic carcinogens" (negative Ames and no structural alerts), and had shown that all five genotoxins were carcinogenic to both species and both sexes, and that four of the five were multi-site carcinogens; in contrast, the five non-genotoxins were active in only one species, three of the five affected only a single sex, and four were single-site carcinogens.   These ten compounds well supported their earlier hypothesis.   However, when Ashby and Paton (1993) tried to demonstrate the same effect amongst over 500 carcinogens, the findings were not so clear cut.   Thus, around 34% of the structurally-alerting carcinogens were active in only a single tissue and 35-40% of the non-alerting chemicals were active in more than one tissue.   Furthermore, although 110 of the 143 two-species carcinogens were structurally-alerting, this still left 33 which had no structural alerts.   For the 107 carcinogens that were tested in both species and found to be active in only one, 53 were structurally-alerting and 54 were not.   In a separate study by Gold *et al.* (1993), involving an analysis of 351 mutagenic (in the Ames test) or non-mutagenic rodent carcinogens from the Carcinogenic Potency Database, it was found that 42% of the single-site, single-species carcinogens were mutagenic while 31% of the two-species carcinogens were not.   This does not greatly support the distinction between genotoxic carcinogens (multi-species, multi-site) and non-genotoxic carcinogens (single-species, single-site).

Gold *et al.* (1993) also compared the target organs for the mutagenic and non-mutagenic carcinogens, and concluded that both groups induced tumours in a wide variety of sites, that most organs were target sites for both, and that the same sites tended to be the most common targets for both.   When the more unusual tumour sites had been excluded (because the number of mutagenic and/or non-mutagenic chemicals that affected them was too small to be meaningful), Gold *et al.* found that only the Zymbal's gland was targeted exclusively by mutagenic carcinogens.   The one discordant chemical was benzene, which Ashby and Paton (1993) had eliminated from their own analysis because it was clastogenic but non-alerting and non-mutagenic. The latter investigators also reported the Zymbal's gland to be the sole tissue that was affected only by structurally-alerting chemicals (after benzene had been excluded). Ashby and Tennant (1988) attributed this 'property' to 15 specific sites in 1988, but ten of the sites were subsequently shown also to be susceptible to non-alerting chemicals (Ashby and Paton, 1993); it remains to be seen whether the other sties (including the Zymbal's gland, subcutaneous tissues, clitoral gland, spleen and ovary) will also be shown, eventually, to be susceptible to non-genotoxic carcinogens. Gold *et al*. (1993) noted that out of 351 rodent carcinogens, 20 induced Zymbal's gland tumours in rats and two induced them in mice (these two being included in the 20 that were active in the rat). The 20 chemicals were all multi-site carcinogens. Of the 230 carcinogens in the overall Carcinogenic Potency Database that were tested in both rats and mice, only 42 (18%) affected multiple sites in both species; but of the 14 that were tested in both species and that induced Zymbal's gland tumours, ten (71%) affected multiple sites in both species. This indicates that chemicals that induce Zymbal's gland tumours are generally multi-site, multi-species carcinogens.

It can be seen that successive attempts to distinguish between genotoxic and non-genotoxic carcinogens on the basis of their tumour profile have become less convincing as the number of carcinogens analysed has increased.

Screening for non-genotoxic carcinogens

Whilst there are a number of STTs that can be used to detect genotoxic chemicals, the problem remains of how to screen for non-genotoxic carcinogens when the actual mechanisms involved are not known in any detail. Carcinogenesis is a complex, multi-stage process, and there are many ways in which its onset and progress may be affected (Green, 1991). The various stages of cancer development have been defined operationally as initiation, promotion, progression and metastasis. Initiation appears to represent damage to key genes involved in the regulation of cell growth, and genotoxic chemicals are believed to contribute to tumour initiation as a result of their damaging effect on DNA.

Other factors such as viruses, UV light, ionising radiation and error-prone DNA replication may also lead to initiation. The clonal expansion of these initiated cells may result from the action of, for example, growth factors, hormones or many non-genotoxic carcinogens.

1. *Cell proliferation*

The induction of cell proliferation appears to be a key factor in non-genotoxic carcinogenicity, and may involve either a direct mitogenic effect on the target tissue without apparent cytolethality or a cytotoxic effect which produces cell death in the target tissue followed by regenerative cell proliferation (Butterworth, 1990). Cell proliferation is a key factor in genotoxic as well as non-genotoxic carcinogenicity, as mutagenic activity may occur as a secondary event in the carcinogenic process (Butterworth *et al.* 1992). As noted previously, use of the maximum tolerated dose (MTD) in animal carcinogenicity bioassays has been the subject of much debate; such a high dose may cause cell death and subsequent cell proliferation, the development of tumours being secondary to this excessive organ-specific toxicity. Thus, tumours that are induced only at the MTD are of questionable relevance to humans who are generally exposed at much lower levels (Ames and Gold, 1990a and b; Ashby and Morrod, 1991). Information on a chemical's capacity to induce cell proliferation may therefore be very useful in setting bioassay doses or in evaluating bioassay results. Two measures that have been used to assess the extent of cell proliferation are the mitotic index and the labelling index. These indicate, respectively, the fraction of cells that are in the process of mitosis and the percentage that have taken up a radiolabelled DNA precursor (Butterworth *et al.* 1992). More complex studies may be conducted to identify the specific receptors that mediate the mitogenic action of some non-genotoxic carcinogens, or to understand how the various proto-oncogenes and tumour suppressor genes that regulate cell growth are influenced by non-genotoxic carcinogens (Green, 1991). These types of study are based on mechanistic considerations and have not been applied to any great extent in testing and validation programmes.

2. Cell transformation assays

The most well-established assays for detecting non-genotoxic (and also genotoxic) carcinogens are the various cell transformation assays, which are based on carcinogen-induced loss of contact inhibition of cultured cells resulting in a piling up of transformed cells in a criss-cross fashion. These assays have been critically evaluated by various investigators and expert groups, who considered them highly relevant for the process of carcinogenesis *in vivo* because they involve the same endpoint (i.e. the transformed cells are tumorigenic in appropriate hosts) and because they may well share many of the same cellular and molecular mechanisms (Dunkel *et al.* 1981; Heidelberger *et al.* 1983; IARC/NCI/EPA Working Group, 1985). Studies evaluating the performance of cell transformation tests in predicting animal or human carcinogens have generally given promising results (Barrett and Lamb, 1985; DiPaolo *et al.* 1972; Fitzgerald *et al.* 1989; Jones *et al.* 1988; Pienta *et al.* 1977; Swierenga and Yamasaki, 1992), although the lack of a dose response, the low transformation frequency and the difficulties in obtaining consistently reproducible results in repeat assays have impeded the development of this system for routine use (Jones *et al.* 1988; Tu *et al.* 1986).

In an assessment of the Syrian hamster embryo (SHE) cell transformation assay, Jones *et al.* (1988) tested 18 coded chemicals in three different laboratories using the same basic protocol. Rodent carcinogenicity data were available for 16 of the 18 chemicals. When the four chemicals that gave discordant transformation responses in two laboratories were counted as positive, the investigators found the results of the two test systems (for carcinogenicity and cell transformation) to be in agreement for 14 of the 16 chemicals. Four rodent carcinogens that were inactive in the Ames test gave positive results in the SHE assay. However, two of the eight purported rodent non-carcinogens, caprolactam and geranyl acetate, were found to be active in the SHE assay in all three laboratories. A non-carcinogen might turn out to be carcinogenic, however, if tested in the appropriate species or strain and under appropriate conditions, but it is interesting to note that caprolactam is the only chemical classified by the International Agency for Research on Cancer (IARC, 1987) as a Group 4 compound (probably not carcinogenic to humans).

Swierenga and Yamasaki (1992) evaluated the performance of transformation tests in identifying IARC Group 1 compounds (carcinogenic to humans) and Group 2A compounds (probably carcinogenic to humans). The data from cell transformation tests were considered collectively, so that a chemical was positive if it gave a positive response in any such assay. Out of 28 Group 1 carcinogens for which data were available, 25 (89%) gave a positive response in the cell transformation assay, while for the three chemicals that were negative, only a single test result was identified. For comparison, the Ames test and the chromosome aberration test identified 25 out of 41 (61%) and 25 out of 30 (83%) Group 1 carcinogens, respectively. For Group 2A compounds, the cell transformation assay again showed a high level of sensitivity, with 21 of the 25 (84%) giving positive results; the corresponding figures for the Ames test and the chromosome aberration test were 34 out of 41 (83%) and 30 out of 31 (96%), respectively. The specificity of the cell transformation assays (in identifying human non-carcinogens) is more difficult to assess because of the lack of clearly identified non-carcinogens, but Ennever et al. (1987) reported that only four out of twelve probable human non-carcinogens (33%) gave negative results in at least one cell transformation assay. According to Swierenga and Yamasaki (1992), some of these probable human non-carcinogens have since shown evidence of carcinogenic activity in animals, but the exclusion of these from the analysis apparently does not increase the specificity of the cell transformation test. These investigators nevertheless concluded that the importance of cell transformation tests may have been underestimated in previous IARC evaluations while that of the Ames test may have been overestimated. They considered the various assays for cell transformation to have generally shown good agreement, and suggested that further refinements were needed to include the use of human cells, human xenobiotic metabolism and appropriate tissue-specific target cells (Swierenga and Yamasaki, 1992).

3. Gap-junction intercellular tests

It has been suggested that, since normal cells surrounding those that contain transforming oncogenes are able to suppress the transformed phenotype, the disruption of intercellular communication via gap junctions may play a role in carcinogenesis (Green, 1991; Yamasaki, 1990). In an evaluation of the performance of gap-junction intercellular tests, Swierenga and Yamasaki (1992) reported that three out of four Group 1 compounds (established human carcinogens) and four out of five Group 2A compounds (probable human carcinogens) gave positive results in this test. For the two compounds that were not detected, one, crystalline silica, was thought to exert its carcinogenic effects without disturbing cell-to-cell communication via the gap junction, and the other, diethylstilboestrol was thought to be cell-type specific, as two steroidal oestrogens gave positive results in this assay. Four out of five organochlorine pesticides also gave positive results, and three of the four were classified by IARC as Group 2B compounds (possible human carcinogens). Whilst these findings are encouraging, the data set is obviously far too small for any definitive conclusions on this type of test. Even so, Swierenga and Yamasaki (1992) considered that the results of such tests should be accorded greater weight than they currently receive.

4. Use of toxicity data

An alternative approach might be to use existing toxicological data on a chemical as an indication of its carcinogenic potential. A relationship between carcinogenicity and systemic toxicity might be expected, given that the highest dose used in carcinogenicity tests is selected on the basis of toxicity. The possible tautologous nature of this relationship has been discussed by Bernstein et al. (1985) and Crouch et al. (1987). In fact, only limited correlations have been shown between LD_{50} values and carcinogenic potencies of known animal carcinogens (McGregor, 1992; Metzger et al. 1989; Zeise et al. 1984). McGregor (1992) suggested that better correlations might be obtained if the maximum tolerated dose (MTD) were used as a measure of toxicity rather than the LD_{50} value, but that even this assumed that the mechanisms of death and of carcinogenicity were the same in animals exposed to a particular chemical. Haseman and Seilkop (1992) looked for such an association between the MTD and carcinogenicity in an analysis of 326 NTP studies conducted in rodents, and they found the overall concordance between toxicity and carcinogenicity to be only 56%. When 130 NTP carcinogenicity studies, involving around 1500 sex-species-exposure groups, were analysed to see if there was a direct causal relationship between organ toxicity and carcinogenicity, it was concluded that the available data did not support a correlation between chemically-induced toxicity and carcinogenicity (Hoel *et al.* 1988; Huff, 1992; Tennant *et al.* 1991). Some chemicals caused organ toxicity without cancer, others induced site-specific cancer with no associated toxicity, and a third group caused toxicity and cancer in the same organ; only seven of the 53 positive carcinogenicity studies were said to have exhibited the types of target organ toxicity that could have been the cause of all observed carcinogenic effects (Hoel *et al.* 1988). Huff (1992) concluded that it would be scientifically premature to make any inference about the influence of toxicity or of cell replication on chemical carcinogenesis.

A wider approach suggested by Travis *et al.* (1990a and b, 1991) involves making use of all available information on a compound's biological activity; they found that a combination of acute and reproductive toxicity data, mutagenicity data and subchronic and chronic tumorigenicity data provided a far better prediction of carcinogenic potency than did mutagenicity data alone. Unfortunately the method cannot be used to distinguish between carcinogens and non-carcinogens, as it can only predict the carcinogenic potency of known mouse carcinogens.

It can be seen that, for non-genotoxic carcinogens, mechanistic considerations must play a major role in any assessment of risk. Until the various mechanisms of action are more fully understood, it is difficult to see how a suitable battery of short-term tests can be developed to screen against the carcinogenic effects of such chemicals. This area merits intensive study if it is to be systematised to the extent necessary to be of use in carcinogen regulation (Clayson and Arnold, 1991).

Despite all the problems identified, the short term tests do have a role in predicting carcinogens. IARC is currently using data from short term tests, alongside any epidemiological evidence to reclassify carcinogens. Until 1987, there were 23 human chemical carcinogens and 7 from industrial processes. Now as a result of reclassification, there are 55 recognised human carcinogens.

In addition to their use in predicting carcinogens, short term tests when used with due consideration, have a role to play in mechanistic studies, for examining complex mixtures and air pollutants and for the early identification of the genetic toxicity of new chemical products.

B.   Approach 2 - The Direct Measurement of Genetic Damage in Man
An approach to measuring genetic damage in man has been to investigate the incidence of chromosome damage in peripheral lymphocytes of subjects exposed to a given chemical (Anderson, 1988; Kilian et al., 1975; Purchase et al., 1978b). It has been reported by several authors, for example, that workers occupationally exposed to vinyl chloride have an increase in the incidence of chromosome damage (Ducatman et al., 1975; Fleig and Theiss, 1977; Hansteen et al., 1978; Kilian et al., 1975; Natarajan et al., 1978; Szentesi et al., 1976). In the study of Purchase et al. (1978b), 81 workers were investigated (57 VC exposed and 24 controls) and effects found in the exposed group. Eighteen months later, the incidence of chromosome damage was still present, though it had decreased in those workers who had changed jobs. At 42 months the incidence of damage had returned to control values (Anderson et al., 1980). Many other chemicals have been investigated in a similar way, e.g. ethylene oxide, epichlorohychin, styrene, butadiene, acrylonitrile, asbestos, benzene, chloroprene, cyclophosphamide, chromium, lead etc.

Such studies have to be well-controlled with age- and sex-matched individuals and confounding factors, such as drinking and smoking, taken into account (Anderson et al., 1985; Anderson, et al., 1990; Brinkworth et al., 1992; Dewdney et al., 1986). As yet, the studies yield results which can only be interpreted on a group basis and not used for individual counselling. In these circumstances, chromosome analysis is useful for determining whether exposure levels of a chemical do or do not induce chromosome damage. At a meeting in Luxembourg in 1987 (convened by the EEC, IPCS, WHO, IARC, and Institute of Occupational Health, Helsinki) on Human Monitoring, a consensus opinion was that increased levels of chromosome damage were indicative of an increased risk of cancer even though no causal relationship had yet been established between chromosome damage and cancer. However, most of the agents causing chromosome damage have been shown to be carcinogenic to animals. Sorsa *et al.* (1990) in a preliminary study, however, did suggest a causal relationship between chromosome damage and cancer, but not between SCE and cancer.

Other techniques for direct application to man are available, such as the use of concentrated urine or blood plasma from exposed workers in combination with a microbial assay (Legator *et al.*, 1982, Gene-Tox), electrophoretic monitoring of enzymatic markers in man (Neel, 1979) detection of variants in haemoglobin molecules (Nute *et al.*, 1976; Popp *et al.*, 1979), determining increases in the formation of haemoglobin adducts (Farmer, 1982; Farmer *et al.*, 1986; Tornquist *et al.*, 1988) and DNA adducts (Pfeifer *et al.*, 1993; Phillips *et al.*, 1988; Weston, 1993), investigations of sperm morphology (Wyrobek *et al.*, 1983b, Gene-Tox), increases in YY bodies in sperm (Kapp *et al.*, 1979), thioguanine mutant frequency in human lymphocytes (Albertini *et al.*, 1988; Cole *et al.*, 1988), oncoproteins in plasma (Brinkworth *et al.*, 1992).

A special issue of Mutation Research reviews Human Monitoring methods (Anderson, 1988) and there is another issue in preparation (Anderson, 1994).

C. **Approach 3 - Measurement of Gonadal Effects in Mammalian Systems and Extrapolation to Man Based on Principles Determined from Radiation Genetics**

The virtual universality of DNA as the genetic material furnishes a rationale for using various sub-mammalian and non-human test systems for such predicting mutagenic potential. Nevertheless, some organisms may be considered more similar to man than others and thus more suitable for the purpose of evaluating human genetic risk. In man, however, there are many phamacokinetic factors which affect the mutagenic efficiency of a compound.

The assessment of heritable mutagenic risk to humans involves determining gonadal effects and acknowledging the influence of many variables (Brinkworth *et al.*, 1992). Results can be assessed on the basis of expected human exposure and known mutagenic potential from such studies as dominant lethal, heritable translocation, or specific locus assays. Similarly, the concentration of the mutagen in germ cells can be assessed biochemically, extrapolated to human exposure levels, and combined with mutagenicity data from the most-used and well-understood *in vitro* test systems.

It has been suggested that the population effects of chemical mutagens should taken radiation as an equivalent and equate the population dose of those mutagens to the radiation dose admissible for that population (Bridges, 1974; Crow, 1973); or should not exceed a dose which doubles the spontaneous mutation rate of that population. Both concepts, however, have been criticised (Auerbach, 1975; Schantel and Sankaranarayanan, 1976, Sobels, 1977). An approach currently regaining favour is the parallelogram approach of Sobels (1977) where with data from rodent germ cells and human and rodent somatic cells, human germ cell data can be predicted/estimated.

7. General conclusions

The present methods available for the testing for mutagenicity are not equally reliable or reproducible. Even the systems most frequently used (e.g. Ames test) give different results in different circumstances.

Since no one test system satisfactorily detects all genetic end-points, a combination of tests is required. Such a 'battery' should preferentially consist of a microbial test and a test detecting chromosome damage. This combination would be the minimum required. Currently in some countries an *in vitro* assay for gene mutation is also carried out (Arlett and Cole, 1988). If a larger 'battery' is to be used to identify those chemicals which are potentially hazardous to man, a mammalian *in vivo* system should be considered. By making the test battery too large a greater number of false results may be generated (Purchase *et al.*, 1978a; Purchase, 1980; Tennant *et al.*, 1987). It is possible that the 'gap' that exists between mutagenicity and carcinogenicity may be partially bridged at an early stage of testing with a cell transformation assay (Dunkel, 1991; COC Guidelines, HMSO, 1991; Fitzgerald and Yamasaki, 1990; Heidelberger *et al.*, 1983; Meyer *et al.*, 1984).

The need for safety evaluation in general toxicological testing is well recognised and this is certainly true in the field of genetic toxicology. However, it is not certain that positive or negative results in laboratory model test systems are relevant to man because of man's unique metabolism and because of the absence of any convincing 'no-effect' level data for animals or man. Epidemiological evidence for germ cell mutation after chemical exposure (or, in fact, any agent) is lacking. In the industrial situation, it is often difficult to identify the exact chemical or agent that may be causing a problem. Unbiased abortion rates are difficult to determine by comparison with control or unexposed populations. Not all abortions are recorded.

The limitations of the simpler short term tests, which are more concerned with the concept of somatic mutation than of heritable genetic damage, are becoming better understood (Ashby and Paton, 1993; Ashby and Tennant, 1988; Purchase, 1980; Tennant *et al.*, 1987; Waters *et al.*, 1993) but to extrapolate to man in terms of heritable damage, *in vivo* animal studies are required. The logistics of tests for this purpose (the specific locus test and heritable translocation test), however, are difficult to satisfy.

There is a school of opinion which holds that just as mutagenicity tests might detect genotoxic carcinogens, so might carcinogenicity tests detect mutagens. This opinion is based on the concept that cancer may be an easily observable phenotype of DNA mutation. If this is so, then the carcinogenicity and mutagenicity data for a chemical should be considered together.

This is probably true for genotoxic carcinogens and IARC is using this approach.

It is hoped that as research progresses, our understanding and techniques will improve so that the results generated in our model systems will become unequivocal in terms of hazard to man. To achieve this aim, attention will have to be given to studies aimed at assessing the significance to man of positive mutagenic responses produced by a test system for a given chemical, in addition to the search for better assay procedures and the understanding and interpretation of effects for chemicals producing mutations by indirect means.

References

Abrahamson, S. and Lewis, S.B. (1971). The detection of mutations in *Drosophila*: In: Chemical Mutagens, Principles and Methods for their Detection. Plenum Press, New York, London, Vol.2, pp.461-489.

Adler, I.D., Ashby, J. and Würgler, F.E. (1989). Screening for possible human carcinogens and mutagens: a symposium report. Mutat. Res. 213:27-39.

Adler, I.-D. and Kliesch, U. (1990). Comparison of single and multiple treatment regimens in the mouse bone marrow micronucleus assay for hydroquinone (HQ) and cyclophosphamide (CP). Mutat. Res. 234:115-123.

Albanese, R. (1987). Mammalian male germ cell cytogenetics. Mutagenesis 2:79-87.

Albanese, R., Topham, J.C., Evans, E., Clare, M.G. and Tease, C. (1984). Mammalian germ-cell cytogenesis. The report of the UKEMS Sub-Committee on Guidelines for Mutagenicity Testing. Part 2, pp.145-172.

Albertini, S. (1989). Influence of different factors on the induction of chromosome malsegregation in *Saccharomyces cerevisiae* D61.M by baviston and assessment of its genotoxic property in the Ames test and in *Saccharomyces cerevisiae* D7. Mutat. Res. 216:327-340.

Albertini, S. (1991). Reevaluation of the 9 compounds reported conclusive positive in yeast *Saccharomyces cerevisiae* aneuploidy test systems by the Gene-Tox Program using strain D61.M of *Saccharomyces cerevisiae*. Mutat. Res. 260:165-180.

Albertini, S., Brunner, M. and Würgler, F.E. (1993). Analysis of six additional chemicals for *in vitro* assays of the European Economic Communities (EEC) aneuploidy programme using *Saccharomyces cerevisiae* D61.M and the *in vitro* Porcine Brain Tubulin Assembly assay. Environ. Molec. Mutagen. 21:180-192.

Albertini, R.J., Sullivan, L.M., Berman, J.K., Greene, C.J., Stewart, J.A., Silveira, J.M. and O'Neill, J.P. (1988). Mutagenicity monitoring in humans by autoradiographic assay for mutant T lymphocytes. Mutat. Res. 204:481-492.

Ames, B.N., Durston, W.E., Yamasaki, E. and Lee, F.D. (1973a). Carcinogens are mutagens: a simple test system combining liver homogenates for activation and bacteria for detection. Proc. natn. Acad. Sci. (USA) 70:2281-2285.

Ames, B.N. and Gold, L.S. (1990a). Chemical carcinogenesis: too many rodent carcinogens. Proc. natn. Acad. Sci. (USA) 87:7772-7776.

Ames, B.N. and Gold, L.S. (1990b). Too many rodent carcinogens: mitogenesis increases mutagenesis. Science NY 249:970-971.

Ames, B.N., Lee, F.D. and Durston, W.E. (1973b). An improved bacterial test system for the detection and classification of mutagens and carcingoesn. Proc. Natl. Acad. Sci. (USA) 70:782-786.

Ames, B.N., McCann, J. and Yamasaki, E. (1975). Methods for detecting carcinogens and mutagens with the Salmonella/mammalian microsome mutagenicity test. Mutat. Res. 31:347-364.

Anderson, D. (1975a). The selection and induction of 5-iodo-2-deoxyuridine and thymidine variants of P388 mouse lymphoma cells which agents with are used for selection. Mutat. Res., 33:399-406.

Anderson, D. (1975b). Attempts to produce systems for isolating spontaneous and induced variants in various mouse lymphoma cells using a variety of selective agents. Mutat. Res. 33:407-416.

Anderson, D. (1988). Human Monitoring - Special Issue. Mutat. Res. 204: No.3.

Anderson, D. and Cross, M.F. (1982). Studies with 4-chloromethylbiphenyl in P388 cells resistant to 5-iodo-2-deoxyuridine. Mutat. Res. 100:257-261.

Anderson, D. and Cross, M.F. (1985). Suitability of the P388 mouse lymphoma system for detecting potential carcinogens and mutagens. Food Chem. Toxicol. 23:115-118.

Anderson, D., Dewdney, R.S., Jenkinson, P.C., Lovell, D.P., Butterworth, K.R. and Conning, D.M. (1985). Sister chromatid exchange in 106 control individuals. In Monitoring of Occupational Genotoxicity. Proceedings of the Fourth ICEM Satellite Meeting, Helskini, June 30-July 2. Eds. M. Sorsa and H. Norppa, pp.38-58. Alan R. Liss, and abstract in Human Toxicol. 4:49-116.

Anderson, D. and Fox, M. (1974). The induction of thymidine and IUdR resistant variants in P388 mouse lymphoma cells by X-rays, UV and mono- and bifunctional alkylating agents. Mutat. Res. 25:107-122.

Anderson, D., Francis, A.J., Godbert, P., Jenkinson, P.C. and Butterworth, K.R. (1990). Chromosome aberrations (CA) sister chromatid exchanges (SCE) and mitogen induced blastogenesis in cultured peripheral lymphocytes from 48 control individuals sampled 8 times over 2 years. Mutat. Res. 250:467-476.

Anderson, D., Green, M.H.L., Mattern, I.E. and Godley, M.J. (1984). An international collaborative study of 'Genetic drift' in Salmonella typhimurium strains used in the Ames test. Mutat. Res. 130:1-10.

Anderson, D., Jenkinson, P.C., Dewdney, R.S., Francis, A.J., Godbert, P. and Butterworth, K.R. (1988). Chromosome aberrations, mitogen-induced blastogenesis and proliferative rate index in peripheral lymphocytes from 106 control individuals of the UK population. Mutat. Res. 204:407-420.

Anderson, D., Jenkinson, P.C., Edwards, A.J., Evans, J.G. and Lovell, D.P. (1993). Male-mediated F$_1$ abnormalities. In: Reproductive Toxicology, Eds. M. Richardson, VCH Verlagsgesellschaft Weinheim, pp.101-116.

Anderson, D., McGregor, B.D. and Bateman, A.J. (1983). Dominant lethal mutation assays. Report of the UKEMS sub-committee on guidelines for mutagenicity testing, pp.43-164.

Anderson, D. and Purchase, I.F.H. (1983). The mutagenicity of food. In: Toxic Hazards in Food, Eds. D.M. Conning and A.B.G. Lansdown. Croom Helm Ltd, London, pp.145-182.

Anderson, D. and Richardson, C.R. (1981a). Issues relevant to the assessment of chromsome damage in vivo. Mutat. Res. 90:261-272.

Anderson, D., Richardson, C.R. and Davies, P.J. (1981b). The genotoxic potential of bases and nucleosides. Mutat. Res. 91:265-272.

Anderson, D., Richardson, C.R., Purchase, I.F.H., Weight, T.M. and Adams, W.G.F. (1980). Chromosomal analyses in vinyl chloride exposed workers. Mutat. Res. 70:151-162.

Anon. (1991). Early indicators of non-genotoxic carcinogenesis. Mutat. Res. 248:211-376 (Special Issue).

Arlett, C.F. (1977). Mutagenicity testing with V79 Chinese hamster cells. Handbook of Mutagenicity Test Procedures, Eds. B.J. Kilbey, M. Legator, W. Nichols and C. Ramel. Elsevier Biomedical Press, North-Holland, Amsterdam, pp.175-191.

Arlett, C.F. and Cole, J. (1988). The role of mammalian cell mutation assays in mutagenicity and carcinogenicity testing. Mutagenesis 3:455-458.

Ashby, J. (1985). Fundamental structural alerts to potential carcinogenicity or non-carcinogenicity. Environ. Mutagen. 7:919-921.

Ashby, J. (1986a). The prospects of simplified and internationally harmonized approach to the detection of possible human carcinogens and mutagens. Mutagenesis 1:3-16.

Ashby, J. (1986b). Letter to the Editor. Mutagenesis 1:309-317.

Ashby, J. (1988a). The separate identities of genotoxic and non-genotoxic carcinogens. Mutagenesis 3:365-366.

Ashby, J. (1988b). An opinion on the significance of the 19 non-clastogenic gene-mutagens reported by Tennant et al. (1987). Mutagenesis 3:463-465.

Ashby, J., de Serres, F.J., Draper, M., Ishidate, M., Jr., Margolin, B.H., Matter, B.E. and Shelby, M.D. (1985). Evaluation of short term tests for carcinogens. Report of the International Programme on chemical safety's collaborative study on in vitro assays. Vol.5. Elsevier Science Publishers, Amsterdam, Oxford, New York.

Ashby, J. de Serres, F.J., Shelby, M.D., Margolin, B.H., Ishidate, M. and Becking, G. (1988). Evaluation of short term tests for carcinogens. Report of the International Programme on Chemical Safety's Collaborative Study on in vitro assay. Vol.1. and in vivo assays. Vol.11. Cambridge University Press on behalf of the World Health Organisation.

Ashby, J. and Liegibel, U. (1992). Transgenic mouse mutation assays. Potential for confusion of genotoxic and non-genotoxic carcinogenesis. A proposed solution. Environ. Mol. Mutagen. 20:145-147.

Ashby, J. and Liegibel, U. (1993). Dosing regimens for transgenic animals mutagenesis assays. Environ. Mol. Mutagen. 21:120-121.

Ashby, J. and Mirkova, E. (1987). The activity of MNNG in the mouse bone marrow micronucleus assay. Mutagenesis 2:199-205.

Ashby, J. and Morrod, R.S. (1991). Detection of human carcinogens. Nature, London 352:185-186.

Ashby, J. and Paton, D. (1993). The influence of chemical structure on the extent and sites of carcinogenesis for 522 rodent carcinogens and 55 different human carcinogen exposures. Mutat. Res. 286:3-74.

Ashby, J. and Purchase, I.F.H. (1988). Reflections on the declining ability of the Salmonella assay to detect rodent carcinogens as positive. Mutat. Res. 205:51-58.

Ashby, J. and Purchase, I.F.H. (1992). Non-genotoxic carcinogens: an extension of the perspective provided by Perera. Environ. Hlth Perspect. 98:223-226.

Ashby, J. and Styles, J.A. (1978). Factors influencing mutagenic potency *in vitro*. Nature, London 274:20-22.

Ashby, J. and Tennant, R.W. (1988). Chemical structure, Salmonella mutagenicity and extent of carcinogenicity as indicators of genotoxic carcinogenesis among 222 chemicals tested in rodents by the U.S. NCI/NTP. Mutat. Res. 204:17-115.

Ashby, J. and Tennant, R.W. (1991). Definitive relationships among chemical structure, carcinogenicity and mutagenicity for 301 chemicals tested by the U.S. NTP. Mutat. Res. 257:229-306.

Ashby, J., Tennant, R.W., Zeiger, E. and Stasiewicz, S. (1989). Classification according to chemical structure, mutagenicity to Salmonella and level of carcinogenicity of a further 42 chemicals tested for carcinogenicity by the U.S. National Toxicology Program. Mutat. Res. 223:73-103.

Auerbach, C. (1975). The effects of six years of mutagen testing on our attitude to the problems posed by it. Mutat. Res. 33:3-10.

Auerbach, C. (1977). The role of *Drosophila* in mutagen testin. In: Topics in Toxicology, Mutagenesis in Sub-mammalian Systems. Status and Significance, Ed. G. Paget. MTP Press Ltd. Lancaster, pp.13-20.

Auerbach, C., Robson, J.M. and Carr, J.G. (1947). The chemical production of mutations. Science 105:243.

Auletta, A. and Ashby, J. (1988). Workshop on the relationship between short-term test information and carcinogenicity; Williamsburg, Virginia, January 20-23, 1987. Environ. Molec. Mutagen. 11:135-145.

Baker, B.S., Boyd, J.B., Carpenter, A.T.C., Green, M.M., Nguyen, T.D., Tipoll, P. and Smith, P.D. (1976). Genetic controls of meiotic recombination and somatic DNA metabolism in *Drosophila melanogaster*. Proc. Natl. Acad. Sci. (USA) 73:4140-4144.

Barrett, J.C. (1985). Cell culture models of multistep carcinogens. In: Age-related factors in carcinogenesis. Ed. A. Likhacker, V. Arisimov and R. Montesano. IARC, Lyon, pp.181-202.

Barrett, J.C., Crawford, B.D., Grady, D.L., Hester, L.D., Jones, P.A., Benedict, W.F. and Ts'o, P.O.P. (1977). Temporal acquisition of enhanced fibrinolytic activity by Syrian hamster embryo cells following treatment with benzo(a)pyrene. Cancer Res. 37:3815-3823.

Barrett, J.C., Kakunaga, T., Kuroki, T., Neubert, D., Troske, J.E., Vasilieu, J.M., Williams, G.M. and Yamaski, H. (1986). Mammalian Cell Transformation in Culture. In: Long and short term test assays for carcinogens. A critical appraisal. Eds. R. Montesano, H. Bartsch, H. Vainio, J. Wilbourn and H. Yamasaki. IARC Sci. Publ. No.83, Lyon, pp.267-286.

Barrett, J.C. and Lamb, P.W. (1985). Tests with the Syrian hanster embryo cell transformation assay. In: Progress in Mutation Research. Volume 5. Evaluation of Short-Term Tests for Carcinogens. Report of the International Programme on Chemical Safety's Collaborative Study on In Vitro Assays. J. Ashby, F.J. de Serres, M. Draper, M. Ishidate Jr., B.H.. Margolin, B.E. Matter and M.D. Shelby (eds). pp.623-628, Elsevier, Amsterdam.

Bateman, A.J. (1977). The dominant lethal assay in the mouse. In: Handbook of Mutagenicity Test Procedures, Eds. B.J. Kilbey, M. Legator, W. Nichols and C. Ramel. Elsevier Biochemical Press, North-Holland, Amsterdam, pp.225-255.

Bateman, A.J. and Epstein, S.S. (1971). Dominant lethal mutations in mammals. In: Chemical Mutagens, Principles and Methods for their Detection. Plenum Press, New York, Vol.2, pp.541-568.

Bartsch, H., Malaveille, C. and Montesano, R. (1975). Human rat and mouse liver-mediated mutagenicity of vinyl chloride in S. typhimurium strains. Int. J. Cancer 15:429-437.

Benigni, R. and Giuliani, A. (1988). Statistical exploration of four major genotoxicity data bases: an overview. Environ. Molec. Mutagen. 12:75-83.

Bernstein, L., Gold, L.S., Ames, B.N., Pike, M.C. and Hoel, D.G. (1985). Some tautologous aspects of the comparison of carcinogenic potency in rats and mice. Fund. appl. Toxicol. 5:79-86.

Berwald, Y. and Sachs, L. (1963). In vitro transformation with chemical carcinogens. Nature, London 200:1182-1184.

Berwald, Y. and Sachs, L. (1965). In vitro transformation of normal cells to tumour cells by carcinogenic hydrocarbons. J. Natn. Cancer Inst. 35:641-661.

Bignami, M., Morpugo, C., Pagliani, R., Carere, A., Conte, G. and Di Guiseppe, G. (1974). Non-disjunction and crossing-over induced by pharmaceutical drugs in *Aspergillus nidulans*. Mutat. Res. 26:159-170.

Bond, D.J. (1976). A system for the study of meiotic non-disjunction using *Sodaria brevicollis*. Mutat. Res. 39:213-220.

Bootman, J. and Kilbey, B.J. (1983). Recessive lethal mutations in *Drosophila*. The report of the UKEMS sub-committee on guidelines for mutagenicity testing. Part 1.

Bradley, M.O., Bhuyan, B., Francis, M.C., Langenbach, R., Peterson, A. and Huberman, E. (1981). Mutagenesis by chemical agents in V79 Chinese hamster cells: A review and analysis of literature. A report of the US Environmental Protection Agency Gene-Tox program. Mutat. Res. 87:81-142.

Brewen, J.C., Payne, H.S., Jones, K.P. and Preston, R.J. (1975). Studies on chemically-induced dominant lethality. 1. The cytogenetic basis of MMS-induced dominant lethality in post-meiotic male germ cells. Mutat. Res. 33:239-250.

Bridges, B.A. (1974). The three-tier approach to mutagenicity screening and the concept of the radiation equivalent dose. Mutat. Res. 26:335-340.

Brinkworth, M.H., Yardley-Jones, A., Edwards, A.J., Hughes, J.A. and Anderson, D. (1992). A comparison of smokers and non-smokers with respect to oncogene products and cytogenetic parameters. J. Occup. Med. 34:1181-1188.

Brockman, H.E., de Serres, F.J., Ong, T.-M., DeMarini, D.M., Katz, A.J., Griffiths, A.J.F. and Stafford R.S. (1984). Mutation tests in *Neurospora crassa*. A report of the US Environmental Protection Agency Gene-Tox Program. Mutat. Res. 133:87-134.

Brockman, H.E. and DeMarini, D.M. (1988). Utility of short-term tests for genetic toxicity in the aftermath of the NTP's analysis of 73 chemicals. Environ. Molec. Mutagen 11:421-435.

Bruce, W.R., Varghese, A.J., Furrer, R. and Land, P.C. (1977). A mutagen in the faeces of normal human. In: Origins of Human Cancer, Eds. H.H. Hiatt, J.D. Watson and J.A. Winsten. Cold Spring Harbor Laboratory, New York, Vol.C, pp.1641-1646.

Brusick, D.J. (1972). Induction of cyclohexamide resistant mutants in *Saccharomyces cerevisiae* with *N*-methyl-*N*-nitronitrosoguanidine and ICR-70. J. Bacteriol. 109:1134.

Brusick, D.J. (1982). Genetic toxicology. In: Principles and Methods of Toxicology, Ed. A.W. Hayes. Raven Press, New York, pp.223-272.

Brusick, D.J. (1987). Genotoxicity produced in mammalian cell assays by treatment conditions. Special issue. Mutat. Res. 189.

Brusick, D.J. and Auletta, A. (1985). Developmental status of bioassays in genetic toxicology. A report of Phase II of the US Environmental Protection Agency Gene-Tox Program. Mutat. Res. 153:1-10.

Brusick, D.J., Gopalan, H.N.B., Heseltine, E., Huismans, J.W. and Lohman, P.H.M. (1992). Assessing the risk of genetic damage. *UNEP/ICPEMC*, Hodder and Stoughton, pp.1-52.

Brusick, D.J., Simmon, V.F., Rosenkranz, H.S., Ray, V.A. and Stafford, R.S. (1980). An evaluation of the *Escherichia coli* WP2 and WP2 *uvrA* reverse mutation assay. Mutat. Res. 76:169-190.

Buckton, K.E. and Evans, H.J. (1973). Methods for the analysis of human chromosome aberrations. WHO Publication, Geneva.

Butterworth, B.E. (1990). Consideration of both genotoxic and nongenotoxic mechanisms in predicting carcinogenic potential. Mutat. Res. 239:117-132.

Butterworth, B.E., Popp, J.A., Conolly, R.B. and Goldsworthy, T.L. (1992). Chemically induced cell proliferation in carcinogenesis. In: Mechanisms of Carcinogenesis in Risk Identification. H. Vainio, P.N. Magee, D.B. McGregor and A.J. McMichael (eds). pp.279-305, IARC, Lyon.

Cachiero, N.L.A., Russel, L.B. and Swarthout, M.S. (1974). Translocation, the predominant cause of total sterility in sons of mice treated with mutagens. Genetics 76:73-91.

Carter, C.O. (1977). The relative contribution of mutant genes and chromosome aberrations in genetic ill health. In: Progress in Genetic Toxicology, Eds. D. Scott, B.A. Bridges and F.H. Sobels. Elsevier Biomedical Press, North-Holland, Amsterdam, pp.1-14.

Chu, E.G.Y., Brimer, P., Jacobson, K.B. and Merriam, E.V. (1976). Mammalian cell genetis. 1. Selection and characterisation of mutations auxotrophic for 1-glutamine or resistance to 8-azaguanine in Chinese hamster cells *in vitro*. Genetics 62:359-377.

Čihák, R. and Vontorkova, M. (1990). Activity of acrylamide in single-, double-, and triple-dose mouse bone marrow micronucleus assays. Mutat. Res. 234:125-127.

Clayson, D.B. and Arnold, D.L. (1991). ICPEMC Publication No.19. The classification of carcinogens identified in the rodent bioassay as potential risk to humans: What type of substance should be tested next? Mutat. Res. 257:91-106.

Cleaver, J.E. (1975). Methods for studying repair of DNA damaged by physical and chemical carcinogens. In: Methods of Cancer Research, Ed. H. Busch. Academic Press, New York, Vol.9, pp.123-165.

Cleaver, J.E. (1977). Methods for studying excision repair of DNA damaged by physical and chemical mutagens. In: Handbook of Mutagenicity Test Procedures, Eds. B.J. Kilbey, M.S. Legator, W. Nichols and C. Ramel. Elsevier Biomedical Press, North-Holland, Amsterdam, pp.19-47.

Clive, D.W. (1973). Recent developments with the L5178TK heterozygote mutagen assay system. Environ. Hlth. Perspect. 6:119-125.

Clive, D.W., Flamm, W.G., Machesko, M.R. and Bernheim, N.H. (1972). Mutational assay system using the thymidine kinase locus in mouse lymphoma cells. Mutat. Res. 16:77-87.

Clive, D.W., McCuen, R., Spector, J.F.S., Piper, C. and Mavournin, K.H. (1983). Specific gene mutations in L5178Y cells in culture. A report of the US Environmental Protection Agency Gene-Tox Program. Mutat. Res. 155:225-251.

Clive, D.W. and Spector, J.F.S. (1975). Laboratory procedure for assessing specific locus mutations at the TK locus in cultured L5178Y mouse lymphoma cells. Mutat. Res. 31:17-29.

Cohen, M.M. and Hirschhorn, K. (1971). Cytogenetic studies in animals. In: Chemical Mutagens, Principles and Methods for their Detection, Ed. A. Hollaender. Plenum Press, New York, Vol.2, pp.515-534.

Clode, S.A. and Anderson, D. (1988a). High performance liquid chromatographic analysis of bases and nucleosides in mouse testes following in vivo thymidine administration. Mutat. Res. 200:63-66.

Clode, S.A. and Anderson, D. (1988b). Germ and somatic cell abnormalities following in vivo administration of thymidine and adenine. Mutat. Res. 200:249-254.

Clode, S.A., Anderson, D. and Gangolli, S.D. (1986). Studies with unbalanced precursor pools in vivo. In: C. Ramel, B. Lambert and J. Magnusson (Eds). Genetic toxicology of Environmental Chemicals, Part A, Basic principles and mechanisms of action. Liss, New York, pp.551-561.

Cole, J. and Arlett, C.F. (1976). Ethyl methanesulphonate mutagenesis with L5178Y mouse lymphoma cells. A comparison of ouabain, thioguanine, and excess thymidine resistance. Mutat. Res. 24:507-531.

Cole, J., Arlett, C.F. and Green, M.H.L. (1976). The fluctuation test as a more sensitive system for determining induced mutation in L5178Y mouse lymphoma cells. Mutat. Res. 41:377-386.

Cole, J., Fox, M., Garner, R.C., McGregor, D.B. and Thacker, J. (1983). Gene mutation assay in cultured mammalian cells. The report of the UKEMS sub-committee on guidelines for mutagenicity testing. Part 1, pp.65-102.

Cole, J., Green, M.H.L., James, S.E., Henderson, L. and Cole, H. (1988). A further assessment of factors influencing measurements of thioguanine-resistant mutant frequency in circulating T-lymphocytes. Mutat. Res. 204:493-507.

Cole, J., McGregor, D.B., Fox, M., Thacker, J., Garner, R.C. (1990). Gene mutation assays in cultured mammalian cells. In: Basic Mutagenicity Tests, UKEMS Recommended Procedures. Ed. D.J. Kirkland, Cambridge University Press, Cambridge, pp.87-114.

Combes, R.D., Anderson, D., Brooks, T.M., Neale, S. and Venitt, S. (1984). Mutagens in urine, faeces and body fluids. The report of the UKEMS sub-committee on guidelines for mutagenicity testing. Part 2, pp.203-243.

Commoner, B., Vithayathal, A.J. and Henry, J.I. (1974). Detection of metabolic carcinogen intermediates in urine of carcinogen fed rats by means of bacterial mutagenesis. Nature, London 249:850.

Constantin, M.J. and Nilan, R.A. (1982a). Chromosome aberration assays in barley (*Hordeum vulgare*). A report of the US Environmental Protection Agency Gene-Tox Program, Mutat. Res. 99:13-36.

Constantin, M.J. and Nilan, R.A. (1982b). The chlorophyll-deficient mutant assay in barley (*Hordeum vulgare*). A report of the US Environmental Protection Agency Gene-Tox Program. Mutat. Res. 99:37-49.

Constantin, M.J. and Owens, E.T. (1982). Introduction and perspectives of plant genetic and cytogenetic assays. A report of the US Environmental Protection Agency Gene-Tox Program. Mutat. Res. 99:1-12.

Crouch, E., Wilson, R. and Zeise, L. (1987). Tautology or not tautology? J. Toxicol. Envir. Hlth. 20:1-10.

Crow, J. (1973). Impact of various types of genetic damage and risk assessment. Environ. Hlth Perspect. 6:1-5.

Dean, B.J. and Senner, K.R. (1977). Detection of chemically induced mutation in Chinese hamsters. Mutat. Res. 46:403-405.

DeMars, R. (1974). Resistance of cultured human fibroblasts and other cells to purine and pyrimidine analogues in relation to mutagenesis detection. Mutat. Res. 24:335-364.

de Serres, F.J. and Ashby, J. (Eds), (1981). Evaluation of short term tests for carcinogenesis. Chemical Mutagens, Principles and Methods for their Detection, Elsevier Biomedical Press, North-Holland, Amsterdam.

de Serres, F.J. and Malling, H.V. (1971). Measurement of recessive lethal damage over the entire genome and at two specific loci in the Ad-3 region of a two component heterokaryon of *Neurosopora crassa*. In: Chemical Mutagens, Principles and Methods for their Detection, Ed. A. Hollaender, Plenum Press, New York, London, Vol.2, pp.311-341.

Dewdney, R.S., Lovell, D.P., Jenkinson, P.C. and Anderson, D. (1986). Variation in sister chromatid exchange using 106 members of the general UK population. Mutat. Res. 171:43-51.

DHSS (1981), No.24. Guidelines for the testing of chemicals for mutagenicity, Report on Health and Social Subjects. Committee on mutagenicity of chemicals in food, consumer products and the environment. Department of Health and Social Security (July 8, 1981), London, HMSO.

DHSS (1989), No.35. Guidelines for the testing of chemicals for mutagenicity. Committee on mutagenicity of chemicals in food, consumer products and the environment. Department of Health and Social Security (April 1989), London, HMSO.

DHSS (1991), No.42. Guidelines for the evaluation of chemicals for carcinogenicity. Committee on carcinogenicity of chemicals in food, consumer products and the environment. Department of Health and Social Security (July 1991), London, HMSO.

DiPaolo, J.A., Donovan, P.J. and Nelson, R.L. (1969a). Qualitative studies of *in vitro* transformation by chemical carcinogens. J. Natn. Cancer Inst. 42:867-876.

DiPaolo, J.A., Nelson, R.L. and Donovan, P.J. (1969b). Sarcoma producing cell clones derived from clones transformed *in vitro* by benzo(a)pyrene. Science, NY 167:917-918.

DiPaolo, J.A., Nelson, R.L. and Donovan, P.J. (1971). Morphological, oncogenic and karyological characteristics of Syrian hamster embryo cells transformed *in vitro* by carcinogenic polycyclic hydrocarbons. Cancer Res. 31:1118-1127.

DiPaolo, J.A., Nelson, R.L. and Donovan, P.J. (1972). *In vitro* transformation of Syrian hamster embryo cells by diverse chemical carcinogens. Nature, London 235: 278-280.

Ducatman, A., Hirshhorn, K. and Selikoff, I.J. (1975). Vinyl chloride exposure and human chromosome aberrations. Mutat. Res. 31:163-168.

Dunkel, V. (1979). Collaborative studies on the *Salmonella* microsome mutagenicity assay. J. Assoc. Anal. Chem. 62:874-882.

Dunkel, V.C., Pienta, R.J., Sivak, A. and Traul, K.A. (1981). Comparative neoplastic transformation responses of Balb/3T3 cells, Syrian hamster embryo cells, and Rauscher murine leukemia virus-infected Fischer 344 rat embryo cells to chemical carcinogens. J. Natn. Cancer Inst. 67:1303-1315.

Dunkel, V.C., Rogers, C., Swierenga, S.H.H., Brillinger, R.L., Gilman, J.P.W. and Nestmann, E.R. (1991). Recommended protocols based on a survey of current practice in genotoxicity testing laboratories. III. Cell transformation in C3H/10T½ mouse embryo cell, BALB/c3T3 mouse fibroblast and Syrian hamster embryo cell cultures. Mutat. Res. 246:285-300.

Dunkel, V., Zeiger, E., Brunswick, D., McCoy, E., McGregor, D., Mortelmans, K., Rosenkranz, H.S. and Simmon, V.F. (1985). Reproducibility of microbial mutagenicity assays. II. Testing of carcinogens and non-carcinogens in *Salmonella typhimurium* and *Escherichia coli*. Environ. Mutagenesis 77:(Suppl.5), 1-248.

Durston, W. and Ames, B.N. (1974). A simple method for detection of mutagens in urine, studies with the carcinogen 2-acetylaminofluorene. Proc. Natl. Acad. Sci. (USA) 71:737-741.

Ehling, U.H. (1970). Evaluation of presumed dominant skeletal mutations. In: Chemical Mutagenesis in Mammals and Man, Eds. F. Vogel and G. Rohrborn. Springer-Verlag, Berlin, Heidelberg, New York, pp.162-166.

Ehling, U.H., Machemer, L., Buselmaier, W., Dycka, J., Frohbert, H., Kratochvilova, J., Lang, R., Lorke, D., Muller, K.D., Pehn, J., Rohrborn, G., Rollm, R., Schulze-Schencking, M. and Wiseman, H. (1978). Standard protocol for the dominant lethal test on male mice set up by the work group 'Dominant Lethal Mutations of the *ad hoc* committee chemogenetics'. Arch. Toxicol. 39:173-185.

Ehrenberg, L. (1971). Higher plants. In: Chemical Mutagens, Principles and Methods for their Detection, Ed. A. Hollaender, Plenum Press, New York, London, Vol.2, pp.365-386.

Ennever, F.K., Noonan, T.J. and Rosenkranz, H.S. (1987). The predictivity of animal bioassays and short term genotoxicity tests for carcinogenicity and non-carcinogenicity to humans. Mutagenesis 2:73-78.

Ennever, F.K. and Rosenkranz, H.S. (1989). Application of the carcinogenicity prediction and battery selection method to recent National Toxicology Program short-term test data. Environ. Molec. Mutagen. 13:332-338.

Ennever, F.K. and Rosenkranz, H.S. (1987). Prediction of carcinogenic potency by short term genotoxicity tests. Mutagenesis 2:39-44.

Epstein, S.S. and Rohrborn, G. (1970). Recommended procedures for testing genetic hazard from chemicals based on the induction of dominant lethal mutations in mammals. Nature, London 230:264.

Evans, E.P. and Phillips, R.J.S. (1975). Inversion heterozygosity and the origin of XO daughters of Bpa/+ female mice. Nature, London 256:40-41.

Evans, H.J. (1976). Cytological methods for detecting chemical mutagens. In: Chemical Mutagens, Principles and Methods for their Detection, Ed. A. Hollaender, Plenum Press, New York, London, Vol.4, p.1.

Fahrig, R. (1974). Development of host-mediated mutagenicity tests. 1. Differential response of yease cells injected into testes of rats and peritoneum of mice and rats to mutagens. Mutat. Res. 26:29-36.

Fahrig, R. (1977). Host-mediated mutagenicity tests - yeast systems. Recovery of yeast cells out of testes, liver, lung and peritoneum of rats. In: Handbook of Mutagenicity Test Procedures, Eds. B.J. Kilbey, M. Legator, W. Nichols and C. Ramel. Elsevier Biomedical Press, North-Holland, Amsterdam, pp.135-147.

Fahrig, R. (1978). The mammalian spot test: A sensitive in vivo method for the detection of genetic alterations in somatic cells of mice. In: Chemical Mutagens, Principles and Methods for their Detection, Eds, A. Hollaender and F. de Serres, Plenum Press, New York, Vol.5, 152-176.

Farmer, P.B. (1982). Monitoring of human exposure to carcinogens. Chem. Brit. 18:790-794.

Farmer, P.B., Bailey, S.M., Gorf, M., Tornqvist, S., Osterman-Golkar, A., Kautianen and Lewis-Enright, D.P. (1986). Monitoring human exposure to ethylene oxide by the determination of haemoglobin adducts using gas chromatography-mass spectrometry. Carcinogenesis 7:637-640.

Festing, M.F.W. and Wolff, G.L. (1976). Quantitative characteristics of potential value in studying mutagenesis. Genetics 92: No.1, Part 1, S.173.

Felton, S.J. and Knize, M.G. (1991). Occurrence, identification, and bacterial mutagenicity of heterocyclic amines in cooked food. Mutat. Res. 259:205-217.

Fielding, R. (1987). Letter to the Editor. Mutagenicity Testing. Mutagenesis 2:315-316.

Fischer, G.A., Lee, S.Y. and Calabresi, P. (1974). Detection of chemical mutagens using a host-mediated assay L5178Y mutagenesis system. Mutat. Res. 16:501-511.

Fitzgerald, D.J., Piccoli, C. and Yamasaki, H. (1989). Detection of non-genotoxic carcinogens in the BALB/c3T3 cell transformation/mutation assay system. Mutagenesis 4(4):286-291.

Fitzgerald, D.J. and Yamasaki, H. (1990). Tumour promotion : models and assay systems. Teratogen Carcinogen Mutagen 10:89.

Fleig, I. and Theiss, A.M. (1977). External chromosome studies undertaken on persons and animals with VC Illness (Abstract 2nd Int. Conf. Environ, p.219). Mutat. Res. 52:187.

Fox, M. and Anderson, D. (1974). Induced thymidine and 5-iodo-2-deoxyuridine resistant clones of mouse lymphoma cells. Mutat. Res. 25:89-105.

Fox, M., Boyle, J.M. and Fox, B.W. (1976). Biological and biochemical characterisation of purine analogue resistant clones of V79 Chinese hamster cells. Mutat. Res. 35:289-310.

Francis, A.J., Anderson, D., Evans, J.G., Jenkinson, P.C. and Godbert, P. (1990). Tumours and malformations in the adult offspring of cyclophosphamide-treated and control male rats. Preliminary communcation 229:239-246.

Funes-Cravioto, F.B., Lambert, B., Linsten, L., Ehrenberg, L., Natarajan, A.T. and Osterman-Golkar, S. (1975). Chromosome aberrations in workers exposed to vinyl chloride. Lancet, i, 459.

Garner, R.C. and Kirkland, D.J. (1986). Reply - Letter to the Editor. Mutagenesis 1:233-235.

Gatehouse, D.G., Rowland, I.R., Wilcox, P., Callander, R.D. and Forster, R. (1990). Bacterial mutation assays in basic mutagenicity tests, UKEMS Recommended Procedures. Ed. D.J. Kirkland, Cambridge University Press, Cambridge, pp.13-61.

Gatehouse, D.G. and Tweats, D.J. (1986). Letter to the Editor. Mutagenesis 1:307-309.

Generoso, W.M., Bishop, J.B., Gosslee, D.G., Newell, G.W., Sheu, C.-J. and von Hallé, E. (1980). Heritable translocation test in mice. A report of the US Environmental Protection Agency Gene-Tox Program. Mutat. Res. 76:191-215.

Generoso, W.M., Cain, K.T., Huff, S.W. and Gosslee, D.G. (1977). Heritable translocation in mice. In: Chemical Mutagens, Principles and Methods for their Detection, Ed. A. Hollaender, Plenum Press, New York, Vol.5, p.55.

George, E., Wootton, A.K. and Gatehouse, D.G. (1990). Micronucleus induction by azobenzene and 1,2-dibromo-3-chloropropane in the rat: evaluation of a triple-dose protocol. Mutat. Res. 234:129-134.

Gold, L.S., Slone, T.H., Manley, N.B. and Bernstein, L. (1993). Comparison of target organs of carcinogenicity for mutagenic and non-mutagenic chemicals. Mutat. Res. 286:75-100.

Gold, L.S., Slone, T.H., Stern, B.R. and Bernstein, L. (1991). Target organs in chronic bioassays of 533 chemical carcinogens. Environ. Hlth Perspect 93:233-246.

Grant, W.F. (1982). Chromosome aberration assays in *Allium*. A report of the US Environmental Protection Agency Gene-Tox Program. Mutat. Res. 99:273-291.

Green, M.H.L. and Lowe, J.E. (1992). Dietary vitamin C modulates the response of human lymphocytes to ionising radiation : Studies with the Comet Assay. *6th International Conference of Environmental Mutagens, Melbourne*. Abstract. 27-1.

Green, M.H.L. and Muriel, W.J. (1976). Mutagen testing using tryp + reversion in *Escherichia coli*. Mutat. Res. 38:3-32.

Green, M.H.L., Muriel, W.J. and Bridges, B.A. (1976). Use of a simplified fluctuation test to detect low levels of mutagens. Mutat. Res. 38:33-42.

Green, S. (1991). The search for molecular mechanisms of non-genotoxic carcinogens. Mutat. Res. 248:371-374.

Green, S., Auletta, A., Fabricant, J., Kapp, R., Manandhar, M., Sheu, C.-J., Springer, J. and Whitfield, B. (1985). Current status of bioassays in genetic toxicology - the dominant lethal assay. A report of the US Environmental Protection Agency Gene-Tox Program, Mutat. Res. 154:49-67.

Gulati, D.K., Wojciechowski, J.P. and Kaur, P. (1990). Comparison of single-, double-, or triple-exposure protocols for the rdoent bone marrow/peripheral blood micronucleus assay using 4-aminobiphenyl and treosulphan. Mutat Res. 234:135-139.

Gunz, D., Shephard, S.E. and Lutz, W.K. (1993). Can non-genotoxic carcinogens be detected with the lacl transgenic mouse mutation assay? Environ. Molec. Mutagen 21.

Hansteen, I., Hillestad, L., Thiis-Evensen, E. and Heldaas, S.S. (1978). Effect of vinyl chloride in man. A cytogenic follow-up study. Mutat. Res. 51:172-178.

Haseman, J.K., Margolin, B.H., Shelby, M.D., Zeiger, E. and Tennant, R.W. (1988). Do short-term tests predict rodent carcinogenicity? Response. Science 241:1233.

Haseman, J.K. and Seilkop, S.K. (1992). An examination of the association between maximum-tolerated dose and carcinogenicity in 326 long-term studies in rats and mice. Fund. Appl. Toxicol. 19:207-213.

Heddle, J.A. (1973). A rapid *in vivo* test for chromosomal damage. Mutat. Res. 18:187-190.

Heddle, J.A., Hite, M., Kirkhart, B., Mavournin, K., MacGregor, J.T., Newell, G.W. and Salamone, M.F. (1983). The induction of micronuclei as a measure of genotoxicity. A report of the US Environmental Protection Agency Gene-Tox Program. Mutat. Res. 123:61-118.

Heidelberger, C., Freeman, A.E., Pienta, R.J., Sivak, A., Bertram, J.S., Casto, B.C., Dunkel, V.C., Francis, M.W., Kakunaga, T., Little, J.B. and Schechtman, L.M. (1983). Cell transformation by chemical agents - a review and analysis of the literature. A report of the US Environmental Protection Agency Gene-Tox Program. Mutat. Res. 114:283-385.

Ho, Y.L. and Ho, S.K. (1981). Screening of carcinogens with the prophage clts 857 induction test. Cancer Res. 41:532-536.

Hoel, D.G., Haseman, J.K., Hogan, M.D., Huff, J. and McConnell, E.E. (1988). The impact of toxicity on carcinogenicity studies: implications for risk assessment. Carcinogenesis 9:2045-2052.

Hsie, A.W., Casciano, D.A., Couch, D.B., Krahn, D.F., O'Neill, J.P. and Whitfield, B.L. (1991). The use of Chinese hamster ovary cells to quantify specific locus mutation and to determine mutagenicity of chemicals. A report of the US Environmental Protection Agency Gene-Tox Program. Mutat. Res. 86:193-214.

Huberman, E. (1975). Mammalian cell transformation and cell-mediated mutagenesis by carcinogenic polycyclic hydrocarbons. Mutat. Res. 29:285-291.

Huberman, E. and Sachs, L. (1966). Cell susceptibility to transformation and cytotoxicity by the carcinogenic hydrocarbon benzo(a)pyrene. Proc. Natl. Acad. Sci. USA 56:1123-1129.

Huberman, E. and Sachs, L. (1974). Cell-mediated mutagenesis of mammalian cells with chemical carcinogens. Int. J. Cancer 13:326-333.

Huff, J.E. (1992). Chemical toxicity and chemical carcinogenesis. Is there a causal connection? A comparative morphological evaluation of 1500 experiments. In: Mechanisms of Carcinogenesis in Risk Identification. H. Vainio, P.N. Magee, D.B. McGregor and A.J. McMichael (Eds). pp.437-475, IARC, Lyon.

IARC (1987). IARC Monographs on the Evaluation of Carcinogenic Risks to Humans. Overall Evaluations of Carcinogenicity: An updating of IARC Monographs Volumes 1 to 42. Supplement 7. p.55. International Agency for Research on Cancer, Lyon.

IARC/NCI/EPA Working Group (1985). Cellular and molecular mechanisms of cell transformation and standardization of transformation assays of established cell lines for the prediction of carcinogenic chemicals: overview and recommended protocols. Cancer Res. 45:2395-2399.

IPCS/WHO (1992). Final Report of the Steering Groups on Naturally Occurring toxins of plant anigen.

Ishinotsubo, D., Mower, H.F., Setliff, J. and Mandel, M. (1977). The use of REC bacteria for testing carcinogenic substances. Mutat. Res. 46:53-62.

Jackson, M.A., Stack, H.F. and Waters, M.D. (1993). The genetic toxicology of putative non-genotoxic carcinogens. Mutat. Res. 296:241-277.

Jacob, L. and de Mars, R. (1977). Chemical mutagenesis with diploid human fibroblasts. In: Handbook of Mutagenicity Test Procedures. Eds. B.J. Kilbey, M. Legator, W. Nichols and C. Ramel. Elsevier Biomedical Press, North-Holland, Amsterdam, pp.193-220.

Jenkinson, P. and Anderson, D. (1990). Malformed foetuses and karyotypic abnormalities in the offspring of cyclophosphamide and allyl alcohol-treated male rats. Mutat. Res. 229:173-184.

Jenkinson, P., Anderson, D. and Gangolli, S.D. (1987). Increased incidence of abnormal foetuses in the offspring of cyclophosphamide treated mice. Mutat. Res. 188:57-62.

Jenssen, D. and Ramel, C. (1980). The micronucleus test as part of a short-term mutagenicity test program for the prediction of carcinogenicity evaluated by 143 agents tested. Mutat. Res. 75:191-202.

Jones, C.A., Huberman, E., Callahan, M.F., Tu, A., Halloween, W., Pallota, S., Sivak, A., Lubet, R.A., Avery, M.D., Kouri, R.E., Spalding, J. and Tennant, R.W. (1988). An interlaboratory evaluation of the Syrian hamster embryo cell transformation assay using eighteen coded chemicals. Toxic. in Vitro, 2:103-116.

Kafer, E., Marshall, P. and Cohen, G. (1976). Well marked strains of *Aspergillus* for tests of environmental mutagens. Identification of induced recombination and mutation. Mutat. Res. 38:141-148.

Kafer, E., Scott, B.R., Dorn, G.L. and Stafford, R. (1982). *Aspergillus nidulans*: Systems and results of tests for chemical induction of mitotic segregation and mutation. 1. Diploid and duplication assay systems. A report of the US Environmental Protection Agency Gene-Tox Program. Mutat. Res. 98:1-48.

Kakunaga, T. (1973). A quantitative system for assay of a malignant transformation by chemical carcinogens using a clone derived from Balb/3T3. Inst. J. Cancer 12:463-473.

Kapp, R.W., Picciano, D.J. and Jacobson, C.B. (1979). Y-chromosomal non-disjunction in dibromochloropropane exposed workmen. Mutat. Res. 64:47-51.

Kier, L.D. (1988). Comments and perspective on the EPA Workshop on "The relationship between short-term test information and carcinogenicity". Environ. Molec. Mutagen 11:147-157.

Kier, L.D., Brusick, D.J., Auletta, A.E., von Hallé, E.S., Simmon, V.F., Brown, M.M., Dunkel, V.C., McCann, J., Mortelmans, K., Prival, M.J., Rao, T.K. and Ray, V.A. (1986). The *Salmonella typhimurium*/mammalian microsome mutagenicity assay. A report of the US Environmental Protection Agency Gene-Tox Program. Mutat. Res. 168:69-240.

Kihlman, B.A. (1971). Root tips for studying the effects of chemicals on chromsomes. In: Chemical Mutagens, Principles and Methods for their Detection, Ed. A. Hollaender, Plenum Press, New York, Vol.2, pp.365-514.

Kilian, D.J., Picciano, D.J. and Jacobson, C.B. (1975). Industrial monitoring: a cytogenetic approach. Ann. N.Y. Acad. Sci. 249:411.

Kirkland, D. (1987). Regulatory aspects of genotoxicity testing. Implications of germ cell cytogenetic tests in the regulatory process. Mutagenesis 2:61-67.

Klopman, G. and Rosenkranz, H.S. (1991). Quantification of the predictivity of some short-term assays for carcinogenicity in rodents. Mutat. Res. 253:237-240.

Knaap, A.G.A.C. and Simmons, J.W.I.M. (1975). A mutational assay system for L5178Y mouse lymphoma cells using hypoxanthine guanine phosphoribosyl transferase deficiency as a marker. The occurrence of a long expression time for mutations induced by X-rays and EMS. Mutat. Res. 30:97-109.

Knudsen, I., Hansen, E.V., Meyer, O.A. and Poulsen, E. (1977). A proposed method for the simultaneous detection of germ cell mutations leading to foetal death/dominant lethality and malformations (Male Teratogenicity) in mammals. Mutat. Res. 48:267-270.

Kohler, S.W., Provost, G.S., Fieck, A., Kretz, P.L., Bullock, W.O., Sorge, J.A., Putnam, D.L. and Short, J.M. (1991a). Spectra of spontaneous and mutagen induced mutations in the lacI gene in transgenic mice. Proc. Natl. Acad. Sci. USA 17:4707-4716.

Kohler, S.W., Provost, G.S., Fieck, A., Kretz, P.L., Bullock, W.O., Putnam, D.L., Sorge, J.A. and Short, J.M. (1991b). Analysis of spontaneous and induced mutations in transgenic mice using a Lambda ZAP/lacI shuttle vector. Environ. Mol. Mutagen 18:316-321.

Kohler, S.W., Provost, G.S., Kretz, P.L., Dycaico, M.J., Sorge, J.A. and Short, J.M. (1990). Development of a short term *in vivo* mutagenesis assay. The effects of methylation on recovery of a Lambda phage shuttle vector from transgenic mice. Nucleic Acids Res. 18:3007-3013.

Kohn, H.I. (1973). H-gene (Histocompatibility) mutations induced by triethylene-melamine in the mouse. Mutat. Res. 20:235-247.

Koi, M. and Barrett, J.C. (1986). Loss of tumour-suppressive function during chemically induced neoplastic progression of Syrian hamster embryo cells. Proc. Natn. Acad. Sci. USA 83:5992-5996.

Krahn, D.F. and Heidelberger, C. (1977). Liver homogenate-mediated mutagenesis in Chinese hamster V79 cells by polycyclic 'Aromatic' hydrocarbons and aflatoxins. Mutat. Res. 46:27-44.

Kuroki, T. and Matsushima, T. (1987). Performance of short term tests for detection of human carcinogens. Mutagenesis 2:33-37.

Larsen, K.H., Mavournin, K.H., Brash, D., Cleaver, J.E., Hart, R.W., Maher, V.M., Painter, R.B. and Sega, G.A. (1982). DNA repair assays as tests for environmental mutagens. A report of the US Environmental Protection Agency Gene-Tox Program. Mutat. Res. 98:287-318.

Latt, S.A., Allen, J., Bloom, S.E., Carrano, A., Falke, E., Kram, D., Schneider, E., Schreck, R., Tice, R., Whitfield, B. and Solff, S. (1981). Sister-chromatid exchanges: A report of the US Environmental Protection Agency Gene-Tox Program. Mutat. Res. 87:17-62.

Lee, W.R., Abrahamson, S., Valencia, R., von Hallé, E.S., Wurgler, F.E. and Zimmering, S. (1983). The sex-linked recessive lethal test for mutagenesis in Drosophila melanogaster. A report of the US Environmental Protection Agency Gene-Tox Program. Mutat. Res. 123:183-279.

Legator, M.S., Bueding, E., Batzinger, R., Connor, T.H., Eisenstadt, E., Farrow, M.G., Fiscor, G., Hsie, A., Seed, J. and Stafford, R.S. (1982). An evaluation of the host-mediated assay and body fluid analysis. A report of the US Environmental Protection Agency Gene-Tox Program. Mutat. Res. 98:319-374.

Legator, M.S. and Malling, H.V. (1971). The host-mediated assay, a practical procedure for evaluating potential mutagenic agents in mammals. In: Chemical Mutagens, Principles and Methods for their Detection, Ed, A. Hollaender, Plenum Press, New York,, Vo.2, pp.569-589.

Legator, M.S., Palmer, K.A. and Adler, I.A. (1973). A collaborative study of in vivo cytogenetic analysis. 1. Interpretation of slide preparations. Toxicol. Appl. Pharmacol. 24:337.

Legator, M.S., Pullin, T.G. and Connor, T.H. (1977). The isolation and detection of mutagenic substances in body fluid and tissues of animals and body fluid of human subjects. In: Handbook of Mutagenicity Test Procedures, Eds. B.J. Kilbey, M. Legator, W. Nichols and C. Ramel, Elsevier Biomedical Press, North-Holland, Amsterdam, pp.149-159.

Legator, M.S., Zimmering, S. and Connor, T.H. (1976). The use of indicator systems to detect mutagenic activity in human subjects and experimental animals. In: Chemical Mutagens, Principles and Methods for their Detection, Ed, A. Hollaender, Plenum Press, New York, Vol.4, pp.171-192.

Leifer, Z., Kada, T., Mandel, M., Zeiger, E., Stafford, R. and Rosenkranz, H.S. (1981). An evaluation of tests using DNA repair-deficient bacteria for predicting genotoxicity and carcinogenicity. A report of the US Environmental Protection Agency Gene-Tox Program, Mutat. Res. 87:211-297.

Léonard, A. (1975). Tests for heritable translocation in male mammals. Mutat. Res. 31:291-298.

Léonard, A. (1977). Observations on meiotic chromosomes of the male mouse as a test of the potential mutagenicity of chemicals. In: Chemical Mutagens, Principles and Methods for their Detection. Ed, A. Hollaender, Plenum Press, New York, Vol.3, pp.21-56.

Li, A.P., Aaron, C.S., Auletta, A.E., Dearfield, K.L., Riddle, J.C., Slesinski, R.S. and Stankowski, Jr., L.F. (1991). An evaluation of the roles of mammalian cell mutation assays in the testing of genotoxicity. Regul. Toxic. Pharmac. 14:24-40.

Li, A.P., Gupta, R.S., Heflich, R.H. and Wassom, J.S. (1988). A review and analysis of the Chinese hamster ovary/hypoxanthine guanine phosphoribosyl transferase assay to determine the mutagenicity of chemical agents. A report of Phase III of the US Environmental Protection Agency Gene-Tox Program. Mutat. Res. 196:17-36.

Lilly, L.J., Bahner, B. and Magee, P.N. (1975). Chromosome aberrations induced in rat lymphocytes by N-nitroso compounds as a possible basis for carcinogen screen. Nature, London 258:611-612.

Loprieno, N., Barale, R., Bauer, C., Baroncelli, S., Bronzetti, G., Cammellini, A., Cini, A., Corsi, G., Leporini, C., Nieri, R., Mozzolini, M. and Serra, C. (1974). The use of different test systems with yeasts for the evaluation of chemically-induced gene conversion and gene mutations. Mutat. Res. 25:197-217.

Loprieno, N., Barale, R., von Hallé, E.S. and von Borstel, R.C. (1983). Testing of chemicals for mutagenic activity with Schizosaccharomyces pombe. A report of the US Environmental Protection Agency Gene-Tox Program. Mutat. Res. 115:215-223.

Loprieno, N., Boncristiani, G., Loprieno, G. and Tesoro, M. (1991). Data selection and treatment of chemicals tested for genotoxicity and carcinogenicity. Environ. Hlth Perspect 96:121-126.

Maier, P. and Schmid, W. (1976). Ten model mutagens evaluated. Mutat. Res. 50:325-338.

Marimuthu, K.M., Sparrow, A.H. and Schairer, L.A. (1970). The cytological effects of space flight factors, vibration, clinostat and radiation on root tip cells of Tradescantia Radiat. Res. 42:105.

Maron, D. and Ames, B.N. (1983). Revised methods for the *Salmonella* mutagenicity test. Mutat. Res. 113:173-215.

Matter, B.E. (1976). Problems of testing drugs for potential mutagenicity. Mutat. Res. 38:243-258.

Matsushima, T., Sawamura, M., Hara, K. and Sugimura, T. (1976). A safe substitute for polychlorinated biphenyls as an inducer of metabolic activation systems. In: *In vitro* Metabolic Activation of Mutagenesis Testing, Eds. F.J. de Serres, J.R. Fout, J.R. Bend and R.M. Philpot. Elsevier Biochemical Press, North-Holland, Amsterdam, p.85-88.

Mavournin, K.H., Blakey, D.H., Cimino, M.C., Salamone, M.F. and Heddle, J.A. (1990). The *in vivo* micronucleus assay in mammalian bone marrow and peripheral blood. A report of the US Environmental Protection Agency Gene-Tox Program. Mutat. Res. 239:29-80.

Mays, J., McAninch, J., Feurs, R.J., Burkhart, J., Mohrenweiser, H. and Casciano, D.A. (1978). Enzyme activity as a genetic marker to detect induced microlesions in mammals. Mutat. Res. 58: No.1. Abstract 58.

McCann, J. and Ames, B.N. (1976). Detection of carcinogens as mutagens in the *Salmonella*/microsome test: Assay of 300 chemicals: Discussion. Proc. Natl. Acad. Sci. (USA) 73:950-954.

McCann, J., Choi, E., Yamasaki, E. and Ames, B.N. (1975a). Detection of carcinogens as mutagens in the Salmonella/microsome test: assay of 300 chemicals. Proc. Natn. Acad. Sci. USA 72:5135-5139.

McCann, J., Spingarn, N.E., Kobori, J. and Ames, B.N. (1975b). Detection of carcinogens as mutagens: bacterial tester strains with R factor plasmids. Proc. Natl. Acad. Sci. (USA) 72:979-983.

McGregor, D.B. (1992). Chemicals classified by IARC: their potency in tests for carcinogenicity in rodents and their genotoxicity and acute toxicity. In: Mechanisms of Carcinogenesis in Risk Identification. H. Vainio, P.N. Magee, D.B. McGregor and A.J. McMichael (eds). pp.323-352, IARC, Lyon.

McGregor, D. and Ashby, J. (1985). Summary report on the performance of the cell transformation assays. In: J. Ashby, F. de Serres, M. Draper, M. Ishidate, B.H. Margolin, B.E. Matter and M.D. Shelby. Evaluation of short-term tests for carcinogens. Elsevier, New York. pp.103-115.

McKelvey-Martin, V.J., Green, M.H.L., Schmezer, P., Pool-Zobel, B.L., De Méo, M.P. and Collins, A. (1993). The single cell gel electrophoresis assay (Comet assay). Mutat. Res. 288:47-63.

McKusick, V. (1975). Mendelian inheritance in Man, 4th Edn. John Hopkins Press, Baltimore.

Metzger, B., Crouch, E. and Wilson, R. (1989). On the relationship between carcinogenicity and acute toxicity. Risk Analysis 9, 169-177 (cited in Travis et al. 1990a).

Meuth, M. (1984). The genetic consequences of nucleotide precursor pool imbalance in mammalian cells. Mutat. Res. 126:107-112.

Meyer, A.L., McGregor, D.B. and Styles, J.A. (1984). In vitro cell transformation assays. The report of the UKEMS Sub-Committee on Guidelines for Mutagenicity Testing. Part 2, 123-145.

Mirkova, E. (1990). Activity of the human carcinogens benzidine and 2-naphthylamine in triple- and single-dose mouse bone marrow micronucleus assays: results for a combined test protocol. Mutat. Res. 234:161-163.

Mirsalis, J. (1993). Dosing regimes for transgenic animal mutagenesis assays. Environ. Mol. Mutagen 21:118-119.

Mitchell, A.D., Casciano, D.A., Meltz, M.L., Robinson, D.E., San, R.H.C., Williams, G.M. and von Hallé, E.S. (1983). Unscheduled DNA synthesis tests. A report of the US Environmental Protection Agency Gene-Tox Program. Mutat. Res. 123:363-410.

Mohn, G. (1973). 5-Methyl-tryptophan resistance mutations in Escherichia coli K-12. Mutagenic activity of monofunctional alkylating agents, including organophosphorus insectides. Mutat. Res. 20:7-15.

Mohn, G. and Ellenberger, J. (1973). Mammalian blood-mediated mutagenicity tests using a multipurpose strain of Escherichia coli K-12. Mutat. Res. 19:187-196.

Mohn, G. and Kerklaan, R. and Ellenberger, J. (1984). Methodologies for the direct and animal mediated determination of various genetic effects in derivatives of strain 343/113 of E. coli K-12. In: Handbook of Mutagenicity Test Procedures, 2nd Edition, Ed, B.J. Kilbey, M. Legator, W. Michels and C. Ramel, Elsevier, Amsterdam, pp.189-214.

Mollet, P. and Wurgler, F.E. (1974). Detection of somatic recombination and mutations in *Drosophila*. A method for testing genetic activity of chemical compounds. Mutat. Res. 25:421-424.

Moreau, P., Bailone, A. and Devoret, R. (1976). Prophage induction in *Escherichia coli* K-12 enVA uvrB: A highly sensitive test for potential carcinogens. Proc. Natl. Acad. Sci. (USA) 73:3700-3704.

Morita, T., Iwamoto, Y., Shimizu, T., Masuzawa, T. and Yanagihara, Y. (1989). Mutagenicity tests with a permeable mutant of yeast on carcinogens showing false-negative in Salmonella assay. Chem. Pharm. Bull. 37:407-409.

Mortimer, R.K. and Manney, T.R. (1971). Mutation induction in yeast. In Chemical Mutagens, Principles and Methods for their Detection, Ed, A. Hollaender, Plenum Press, New York, Vol.1, pp.209-237.

Müller, H.J. (1927). Artificial transmutation of the gene. Science 66:84-87.

Myhr, B.C. and Caspary, W.J. (1991). Chemical mutagenesis at the thymidine kinase locus in L5178Y mouse lymphoma cells: results for 31 coded compounds in the National Toxicology Program. Environ. Molec. Mutagen 18:51-83.

Nagao, M., Yahagi, T. and Sugimura, T. (1978). Differences in effects of Norharman with various classes of chemical mutagens and amounts of S-9. Biochem. Biophys. Res. Commun. 83:373-378.

Nakamura, J., Suzuki, N. and Okada, S. (1977). Mutagenicity of furylfuramide, a food preservative tested by using alanine-requiring mouse L5178Y cells *in vitro* and *in vivo*. Mutat. Res. 46:355-364.

Natarajan, A.T., van Buul, P.W. and Raposa, T. (1978). An evaluation of the use of peripheral blood lymphocyte systems for assessing cytological effects induced *in vivo* by chemical mutagens. In: Mutagen Induced Chromosome Damage in Man. Eds, H.C. Evans and D.C. Lloyd. Edinburgh University Press, pp.268-276.

Neel, J.V. (1978). Some trends in the study of spontaneous and induced mutation in man. Genetics, 92, No.1, part 1, S25.

Neel, J.V., Kohrenweiser, H., Satoh, C. and Hamilton, B.H. (1979). A consideration of two biochemical approaches to monitoring human populations for a change in germ cell mutation rats. In: Genetic Damage in Man Caused by Environmental Agents, Ed. K. Berg. Academic Press, New York, London, pp.29-47.

Neill, J.P., Brimer, P.A., Machanoff, R., Hirsch, G.P. and Hsie, A.W. (1977). A quantitative assay of mutation induction at HGPRT locus in Chinese hamster ovary cells: development and definition of the system. Mutat. Res. 45:91-101.

Nesnow, S., Argus, M., Bergman, H., Ohu, K., Firth, C., Helmes, T.,McGaughey, R., Ray, V., Staga, T.J., Tennant, R., and Weisburger, E. (1983). Chemical carcinogens: A review and analysis of the literature of selected chemicals and the establishment of the Gene-Tox carcinogen data-base. A report of the US Environmental Protection Agency Gene-Tox Program. Mutat. Res. 185:1-195.

Nicholls, W.W., Moorhead, P., and Brewen, G., (1973). Chromosome methodologies in mutation testing. Toxicol. Appl. Pharmacol. 24:337.

Nikaido, O., and Fox, M., (1976). The relative effectiveness of 6-thioguanine and 8-azaguanine in selecting resistant mutants from two V79 Chinese hamster clones in vitro. Mutat. Res. 35:279-288.

Nilan, R.A., and Vig. B.K. (1976). Plant test systems for detection of chemical mutagens. In: Chemical Mutagens, Principles and Methods for their Detection. Ed. A. Hollaender. Plenum Press, New York, Vol. 4, pp.143-170.

Nute, P.E., Wood, N.E., Stammatoyannopoulos, G., Olivery, C., and Failkaw, P.J. (1976). The Kenya form of hereditary persistence of foetal haemoglobin: Structural studies and evidence for homogeneous distribution of haemoglobin F using fluorescent F antibodies. Br. J. Haemotol. 32:55.

OECD (1983). Organisation for Economic and Cultural Development. Guidelines for testing of chemicals. Genetic Toxicology Guidelines.

Official Journal of the European Communities. Council Directive 18 September 1979 Amending for the Sixth Time Directive 67/548/EEC on the approximation of the Laws, Regulations, and Administrative Provisions relating to the classification, packaging, and labelling of dangerous substances.

Ong, T.M., and de Serres, F.J. (1972). Mutagenicity of chemical carcinogens in Neurospora Crassa. Cancer Res. 32:1890-1893.

Oshiro, Y., Piper, C.E., Balwierz, P.S. and Soelter, S.G. (1991). Chinese hamster ovary cell assays for mutation and chromosome damage: data from non-carcinogens. J. Appl. Toxicol. 11:167-177.

Parodi, S., Malacarne, D. and Taningher, M. (1991). Examples of uses of databases for quantitative and qualitative correlation studies between genotoxicity and carcinogenicity. Environ. Hlth Perspect. 96:61-66.

Parry, J.M. (1977a). The use of yeast cultures for the detection of environmental mutagens using a fluctuation test. Mutat Res. 46:165-176.

Parry, J.M. (1977b). The detection of chromosome non-disjunction in the yeast *Saccharomyces cerevisiae*. In: 'Progress in Genetic Toxicology', *Eds* D. Scott, B.A. Bridges, and F.H. Sobels. Elsevier Biomedical Press, North-Holland, Amsterdam, pp.223-229.

Parry, J.M., Brooks, T.M., Mitchell,I.de G., and Wilson, P. (1984). Genotoxicity studies using yeast cultures. Report of the UKEMS Sub-Committee on Guidelines of Mutagenicity Testing. Part 2, pp.27-62.

Parry, J.M. and Zimmerman, F.K. (1976). The detection of monosomic colonies produced by mitotic chromosome non-disjunction in the yeast *Saccharomyces cerevisiae*. Mutat. Res. 36:49-66.

Pearson, C.M. and Styles, J.A (1984). Effects of hydroxyurea consistent with the inhibition of repair synthesis in UV irradiated HeLa cells. Cancer Lett. 21:247-252.

Perry, P. and Evans, H.J. (1975). Cytological detection of mutagen-carcinogen exposure by sister chromatid exchange. Nature, London 258:121-125.

Perry, P.E., Henderson, L. and Kirkland, D.J. (1984). Sister chromatid exchange in cultured cells. In: Report of the UKEMS Sub-Committee on Guidelines of Mutagenicity Testing. Part 2, pp.89-123.

Pfeifer, G.P., Drouin, R. and Holmqvist, G.P. (1993). Detection of DNA adducts at the DNA sequence level by ligation-mediated PCR. Mutat. Res. 288:39-46.

Phillips, D.H., Hemminki, K., Alhonen, A., Hewer, A. and Grover, P.L. (1988). Monitoring occupational exposure to carcinogens. Detection by ^{32}P-postlabelling of aromatic DNA adducts in white blood cells from iron foundry workers. pp.531-541.

Phillips, J.C. and Anderson, D. (1993). Predictive toxicology Structure Activity relationships and carcinogens. Occup. Hlth. Rev. 41:27-30.

Pienta, R.J., Poiley, J.A. and Lebherz, W.B., III (1977). Morphological transformation of early passage golden Syrian hamster embryo cells derived from cryopreserved primary cultures as a reliable *in vitro* bioassay for identifying diverse carcinogens. Int. J. Cancer 19:642-655.

Plewa, M.J. (1982). Specific-locus mutation assays in *Zea mays*. A report of the US Environmental Protection Agency Gene-Tox Program. Mutat. Res. 99:317-337.

Poirier, L.A. and de Serres, F.J. (1979). Initial National Cancer Institute studies on mutagenesis as a pre-screen for chemical carcinogens: an appraisal. J. Natl. Cancer Inst. 62:919-926.

Popp, R.A., Hrisch, G.P. and Bradshaw, B.S. (1979). Amino acid substitution: Its use in detection and analysis of genetic variants. Genetics, 92, No.1, Part 1:S.39.

Preston, R.J., Au, W., Bender, M.A., Brewen, J.G., Carrano, A.V., Heddle, J.A., McFee, A.F., Wolff, S. and Wassom, J.S. (1981). Mammalian *in vivo* and *in vitro* cytogenetic assays. A report of the US Environmental Protection Agency Gene-Tox Program. Mutat. Res. 87:143-188.

Prival, M.J. and Dunkel, V.C. (1989). Reevaluation of the mutagenicity and carcinogenicity of chemicals previously identified as "false positives" in the *Salmonella typhimurium* mutagenicity assay. Environ. Molec. Mutagen 13:1-24.

Propping, P., Rohrborn, G. and Buselmaier, W. (1972). Comparative investigations on the chemical induction of point mutations and dominant lethal mutations in mice. Mol. Gen. Genet. 117:197.

Provost, G., Kretz, P.L., Hammer, R.T., Matthews, C.D., Rogers, B.J., Lundberg, K.S., Dycaico, M.J. and Short, J.M. (1993). Transgenic systems for *in vivo* mutation analysis. Mutat. Res. 288:123-131.

Pueyo, C.A. (1978). A forward mutation assay using 1-Arabinose sensitive strain. Abstracts (Ab^{-2}). American Environmental Mutagen Society.

Purchase, I.F.H. (1980). Appraisal of the merits and shortcomings of tests of mutagenic potental. In: Mechanisms of Toxicity and Hazard Evaluation, Eds. B. Holmstedt, R. Lauwerys, M. Mercier and M. Roberfroid. Elsevier Biochemical Press, North-Holland, Amsterdam, pp.105-119.

Purchase, I.F.H., Longstaff, E., Ashby, J., Styles, J.A., Anderson, D., Lefevre, P.A. and Westwood, F.R. (1978a). Evaluation of six short-term tests for detecting organic chemical carcinogens and recommendations for their use. Nature, London 263:624-627; Br. J. Cancer 37:873-903.

Purchase, I.F.H., Richardson, C.F., Anderson, D., Paddle, G.M. and Adams, W.G.F. (1978b). Chromosomal analysis in vinyl-chloride exposed workers. Mutat. Res. 57:325-334.

Ray, V.A., Kier, L.D., Kannan, K.L., Haas, R.T., Auletta, A.E., Wassom, J.S., Nesnow, S. and Waters, M.D. (1987). An approach to identifying specialised batteries of bioassays for specific causes of chemicals and analysis using mutagenicity and carcinogenicity relationships and phylogenetic concordance and discordance patterns. 1. Comparison and analysis of the overall data base. A report of the US Environmental Protection Agency Gene-Tox Program. Mutat. Res. 185:197-241.

Reznikoff, C.A., Bertram, J.S., Brankow, D.W. and Heidelberger, C. (1973). Quantitative and qualitative studies of chemical trasnformation of cloned C34 mouse embryo cells sensitive to post-confluence inhibition of cell division. Cancer Res. 33:3239-3249.

Richold, M., Chandleg, A., Ashby, J., Gatehouse, D.G., Bootman, J. and Henderson, L. (....). *In vivo* cytogenetics assays, UKEMS Recommended Procedures, Ed, D.J. Kirkland, Cambridge University Press, Cambridge, pp.115-141.

Roderick, J.H. (1979). Chromosomal-inversions - Studies of mammalian mutagenesis. Genetics, 92, No.1, Part 1, S.121.

Roper, J.A. (1971). *Aspergillus*. In: Chemical Mutagens, Principles and Methods for their Detection. Ed. A. Hollaender, Plenum Press, New York, Vol.2, pp.343-365.

Rowland, I.R., Robery, E.D. and Walker, R. (1984). Bacterial assays for mutagens in food. In Report of the UKEMS Sub-Committee on Guidelines of Mutagenicity Testing, Part 2 173-202.

Russell, L.B. (1979). *In vivo* somatic mutations in the mouse. Genetics, 92, No.1, Part 1, S.153.

Russell, L.B., Aaron, C.S., de Serres, F.J., Generoso, W.M., Kannan, K.L., Shelby, M.P., Spring, J. and Voytek, P. (1984). Evaluation of mutagenicity assays for purposes of genetic risk assessment. A report of the US Environmental Protection Agency Gene-Tox Program. Mutat. Res. 134:143-157.

Russell, L.B., Selby, P.B., von Hallé, E., Sheridan W. and Valcovic, L. (1981a). The mouse specific-locus test with agents other than radiation: interpretation of data and recommendations for future work. Mutat. Res. 86:329-354.

Russell, L.B., Selby, P.B., von Hallé, E., Sheridan, W. and Valcovic, L. (1981b). Use of the mouse spot test in chemical mutagenesis: interpretation of past data and recommendations for future work. Mutat. Res. 86:355-359.

Russell, W.L. (1951). Specific locus mutations in mice. Cold Spring Harbor Laboratory. Quant. Biol. 16:237.

Saedler, H., Gullon, A., Fiethen, L. and Starlinger, P. (1968). Negative control of galactose operon in *Escherichia coli*. Molec. Gen. Genet. 102:79.

Salamone, M., Heddle, J., Stuart, E. and Katz, M. (1980). Towards an improved micronucleus test. Studies on 3 agents, Mitomycin C, cyclophosphamide and dimethylbenzanthracene. Mutat. Res. 35:347-370.

Schantel, A.P. and Sankaranarayanan, K. (1976). Evaluation and re-evaluation of genetic radiation hazards in man. Mutat. Res. 35:341-370.

Schiestl, R.H. (1989). Nonmutagenic carcinogens induce intrachromosomal recombination in yeast. Nature 337:285-288.

Schiestl, R.H., Gietz, R.D., Mehta, R.D. and Hastings, P.J. (1989). Carcinogens induce intrachromosomal recombination in yeast. Carcinogenesis 10:1445-1455.

Schmid, W. (1973). Chemical mutagen testing on *in vivo* somatic cells. Agents and Action 3:77-85.

Schmid, W. (1976). The micronucleus test for cytogenetic analysis. In: Chemical Mutagens, Principles and Methods for their Detection. Ed. A. Hollaender. Plenum Press, New York, Vol.4, pp.31-53.

Schmid, W., Arakaki, D.T., Breslau, N.A. and Culbertson, J.C. (1971). Chemical mutagenesis. The Chinese hamster bone marrow as an *in vivo* test system. Humangenetik 11:103-118.

Scott, D., Danford, N.D., Dean, B.J., Kirkland, D.J. and Richardson, C.R. (1983). *In vitro* chromosomal aberration assays. In: Report of the UKEMS sub-committee on Guidelines of Mutagenicity Testing, Part 1, pp.41-64.

Scott, D., Dean, B.J., Danford, N.D. and Kirkland, D.J. (1990). Metaphase chromosome aberration assays *in vitro*. In: Basic Mutagenicity Tests UKEMS Recommend Procedures, Cambridge University Press, Cambridge, pp.62-86.

Scott, B.R., Dorn, G.L., Kafer, E. and Stafford, R. (1982). *Aspergillus nidulans*: Systems and results of tests for induction of mitotic segregation and mutation. II. Haploid assay systems and overall response of all systems. A report of the US Environmental Protection Agency Gene-Tox Program. Mutat. Res. 98:49-94.

Scott, D., Galloway, S.M., Marshall, R.R., Ishidate, M., Brusick, D., Ashby, J. and Myhr, B.C. (1991). Genotoxicity under extreme culture conditions. Mutat. Res. 257:147-206.

Searle, W.G. (1975). The specific locus test in the mouse. Mutat. Res. 31:277-290.

Searle, W.G. (1979). The use of pigment loci for detecting reverse mutations in somatic cells of mice. Arch. Toxicol. 38:105-108.

Selby, P.B. and Selby, P.R. (1977). Gamma-ray induced dominant mutations that cause skeletal abnormalities in mice. Mutat. Res. 43:357-376.

Shelby, M.D. (1988). The genetic toxicity of human carcinogens and its implications. Mutat. Res. 204:3-15.

Shelby, M.D., Erexson, G.L., Hook, G.J. and Tice, R.R. (1993). Evaluation of a three-exposure mouse bone marrow micronucleus protocol: results with 49 chemicals. Environ. Molec. Mutagen 21:160-179.

Shelby, M.D., Gulati, D.K., Tice, R.R. and Wojciechowski, J.P. (1989). Results of tests for micronuclei and chromosomal aberrations in mouse bone marrow cells with the human carcinogens 4-aminobiphenyl, treosulphan and melphalan. Environ. Molec. Mutagen. 13:339-342.

Siebert, D. (1973). Induction of mitotic conversions of *Saccharomyces cerevisiae* by lymph fluid and urine of cyclophosphamide-treated rats and human patients. Mutat. Res. 21:202.

Skopek, T.R., Liber, J.L., Krowleski, J.J. and Thilly, W.G. (1978). Quantitative forward mutation assays in *Salmonella typhimurium* using 8-azaguanine resistance as a genetic marker. Proc. Natl. Acad. Sci. (USA) 75:410-411.

Sobels, F.H. (1977). Some thoughts on the evaluation of environmental mutagens. From the Address presented to the 7th Annual EMS Meeting in Atlanta, Georgia.

Sora, S. and Magni, G.E. (1988). Induction of meiotic chromosomal malsegregation in yeast. Mutat. Res. 201:375-384.

Sorsa, M., Ojajarui, A. and Salomaa, S. (1990). Cytogenetic surveillence of workers exposed to genotoxic chemicals : preliminary experiences from a prospective cancer study in a cytogenetic cohort. Teratogen Cacinogen Mutagen 10:215-221.

Sorsa, M., Wilbourn, J. and Vainio, H. (1992). Human cytogenetic damage as a predictor of cancer risk. In: Mechanisms of Carcinogenesis in Risk Identification. H. Vainio, P.N. Magee, D.B. McGregor and A.J. McMichael (eds). pp.543-554. IARC, Lyon.

Speck, W.T., Santella, R.M. and Rozenkranz, H.S. (1978). An evaluation of the prophage induction (inductest) for the detection of potential carcinogens. Mutat. Res. 54:101-104.

Stetka, E.G. and Wolff, G. (1976). Sister chromatid exchange as an assay for genetic damage by mutagen-carcinogens: *In vivo* test for compounds requiring metabolic activation. Mutat. Res. 41:333-342.

Stich, H.F., San, R.H.C., Lam, P., Koropatnick, J. and Lo, L. (1977). Unscheduled DNA synthesis of human cells as a short term assay for chemical carcinogens. In: Origins of Human Cancer. Eds. H.H. Hiatt, J.D. Watson and J.A. Winsten. Cold Spring Harbor Laboratory, New York, Vol.C. pp.1499-1512.

Sugimura, T. (1978). Let's be scientific about the problems of mutagens in cooked foods. Mutat. Res. 55:149-152.

Sugimura, T., Kawachi, T., Matsushima, T., Nagao, M., Sato, S. and Yahai, T. (1977). A critical review of sub-mammalian systems for mutagen detection. In: Progress in Genetic Toxicology, Eds. D. Scott, B.A. Bridges and F.H. Sobels. Elsevier Biochemical Press, North-Holland, Amsterdam, pp.25-140.

Sugimura, T., Yahagi, T. Nagao, M., Takuichi, M., Kawacki, T., Hara, K., Yamasaki, E., Matsushima, T., Hashimoto, Y. and Okada, M. (1976). Validity of mutagenicity testing using microbes as a rapid screening method for environmental carcinogens. In: Screening Tests in Chemical Carcinogens, Eds. R. Montesano, H. Bartsch and T. Tomatis. IARC, Lyon, 12:81-101.

Swann, P.F. and Magee, P.N. (1968). Nitrosamine-induced carcinogenesis. The alkylation of nucleic acids of the rat by N-methyl-Mnitrosourea, dimethyl-nitrosamine dimethyl sulphate, and methyl methanesulphonate. Biochem. J. 110:39-47.

Swierenga, S.H.H. and Yamasaki, H. (1992). Performance of tests for cell transformation and gap-junction intercellular communication for detecting nongenotoxic carcinogenic activity. In: Mechanisms of Carcinogenesis in Risk Identification. H. Vainio, P.N. Magee, D.B. McGregor and A.J. McMichael (eds). pp.165-193. IARC, Lyon.

Szentesi, I., Hornyak, E., Unvary, G., Czeizel, A., Bognar, Z. and Timor, M. (1976). High rate of chromosomal aberrations in PVC workers. Mutat. Res. 37:313-316.

Tang, T. and Friedman, M.A. (1977). Carcinogen activation by human liver enzymes in the Ames mutagenicity test. Mutat. Res. 46:387-394.

Tanooka, J. (1977). Development and applications of Bacillus subtilis test system for mutagens involving DNA repair deficiency and suppressible auxotrophic mutations. Mutat. Res. 48:367-370.

Te-Hsui, Ma. (1982a). Vicia cytogenetic test for environmental mutagens. A report of the US Environmental Protection Agency Gene-Tox Program. Mutat. Res. 99:293-302.

Te-Hsui, Ma. (1982b). Tradescantia cytogenetic tests (root-top mitosis, pollen mitosis, pollen mother-cell mitosis). A Report of the US Environmental Protection Agency Gene-Tox Program. Mutat. Res. 99:293-302.

Tennant, R.W. and Ashby, J. (1991). Classification according to chemical structure, mutagenicity to Salmonella and level of carcinogenicity of a further 39 chemicals tested for carcinogenicity by the U.S. National Toxicology Program. Mutat. Res. 257:209-227.

Tennant, R.W., Elwell, M.R., Spalding, J.W. and Griesemer, R.A. (1991). Evidence that toxic injury is not always associated with induction of chemical carcinogenesis. Molec. Carcinogen 4:420-440.

Tennant, R.W., Margolin, B.H., Shelby, M.D., Zeiger, E., Haseman, J.K., Spalding, J., Caspary, W., Resnick, M., Stasiewicz, S., Anderson, B. and Minor, R. (1987). Prediction of chemical carcinogenicity in rodents from in vitro genetic toxicity assays. Science, NY 236:935-941.

Thilly, W.G., de Luca, J.G., Hoppe, I.V.H. and Penmann, B.W. (1976). Mutation of human lymphoblasts by methylnitrosourea. Chem. Biol. Interact. 99:293-302.

Tice, R.R., Erexson, G.L., Hilliard, C.J., Huston, J.L, Boehm, R.M., Gulati, D. and Shelby, M.D. (1990). Effect of treatment protocol and sample time on the frequencies of micronucleated polychromatic erythrocytes in mouse bone marrow and peripheral blood. Mutagenesis 5:313-321.

Tijo, J.W. and Whang, J. (1962). Chromosome preparation of bone marrow cells without prior in vitro culture of in vivo colchicin administrations. Stain Technol. 37:17-20.

Topham, J.C. (1980). Chemically-induced transmissible abnormalities in sperm head shape. Mutat. Res. 70:109-114.

Topham, J.C., Albanese, R., Bootman, J., Scott, D. and Tweats, D. (1983). In: Report of the UKEMS Sub-Committee on Guidelines of Mutagenicity Testing, Part 1, pp.119-142.

Tornqvist, M., Osterman-Golkar, S., Kautianen, A., Naslund, M., Calleman, C.J. and Ehrenberg, L. (1988). Methylations in human haemoglobin. Mutat. Res. 204:521-529.

Travis, C.C., Richter Pack, S.A., Saulsbury, A.W. and Yambert, M.W. (1990a). Prediction of carcinogenic potency from toxicological data. Mutat. Res. 241:21-36.

Travis, C.C., Saulsbury, A.W. and Richter Pack, S.A. (1990b). Prediction of cancer potency using a battery of mutation and toxicity data. Mutagenesis 5:213-219.

Travis, C.C., Wang, L.A. and Waehner, M.J. (1991). Quantitative correlation of carcinogenic potency with four different classes of short-term test data. Mutagenesis 6:353-360.

Trosko, J.E. (1988). A failed paradigm: carcinogenesis is more than mutagenesis. Mutagenesis 3:363-364.

Tu, A., Hallowell, W., Pallotta, S., Sivak, A., Lubet, R.A., Curren, R.D., Avery, M.D., Jones, C., Sedita, B.A., Huberman, E., Tennant, R., Spalding, J. and Kouri, R.E. (1986). An interlaboratory comparison of transformation in Syrian hamster embryo cells with model and coded chemicals. Environ. Mutagen 8:77-98.

Tweats, D.J., Bootman, J., Combes, R.D., Green, M.H.L. and Watkins, P. (1984). Assays for DNA repair in bacteria. In Report of the UKEMS sub-committee on Guidelines of Mutagenicity Testing, Part II, pp.5-27.

Valcovic, L.R. and Malling, H.V. (1973). An approach to measuring germinal mutations in the mouse. Environ. Hlth Perspect 6:201-206.

Valencia, R., Abrahamson, S., Lee, W.R., von Hallé, E.S., Woodruff, R.C., Wurgler, F.E. and Zimmering, S. (1984). Chromosome mutation tests for mutagenesis in *Drosophila melanogaster*. A report of the US Environmental Protection Agency Gene-Tox Program. Mutat. Res. 134:61-88.

Van't Hof, J. and Schairer, L.A. (1982). *Tradescantia* assay system for gaseous mutagens. A report of the US Environmental Protection Agency Gene-Tox Program. Mutat. Res. 99:303-315.

Venitt, S. and Crofton-Sleigh (1981). Mutagenicity of 42 coded compounds in a bacterial assay using *Escherichia coli* and *Salmonella typhimurium*. In: Evaluation of Short Term Test for Carcinogens, Progress in Mutation Research, Vol.II. Eds. F.J. de Serres and J. Ashby. Elsevier Biochemical Press, North-Holland, Amsterdam, pp.351-360.

Venitt, S., Forster, R.C. and Longstaff, E. (1983). Bacterial mutation assays. In: Report of the UKEMS Sub-Committee on Guidelines of Mutagenicity Testing, Part I, pp.5-40.

Vig, B.K. (1982). Soybean [*Glycine max* (L.) merrill] as a short term assay for the study of environmental mutagens. A report of the US Environmental Protection Agency Gene-Tox Program. Mutat. Res. 99:339-347.

Vogel, E. (1976). The function of *Drosophila* in genetic toxicology testing. In: Chemical Mutagens, Principles and Methods for their Detection. Ed. A. Hollaender. Plenum Press, New York, Vol.4, pp.93-132.

Vogel, E. (1977). Identification of carcinogens by mutagen testing in *Drosophila*. The relative reliability for the kinds of genetic damage measured. In: Origins of Human Cancer. Eds. H.H. Hiatt, J.D. Watson and J.A. Winsten. Cold Spring Harbor Laboratory, New York, Vol.C. pp.1483-1497.

Vogel, E. (1987). Evaluation of potential mammalian genotoxins using *Drosophilia*: the need for a change in a test strategy. Mutagenesis 2:161-171.

Wakabayashi, K., Ushiyama, H., Takahashi, M., Nukaya, H., Kim, S.B., Horse, M., Ochiai, M., Sugimura, T. and Nagao, M. (1993). Exposure to heterocyclic amines. Environ. Hlth. Perspect. 99:129-133.

Waters, M.D., Stack, H.F., Jackson, M.A. and Bridges, B.A. (in press). Hazard identification -efficiency of short-term tests in identifying germ cell mutagens and putative nongenotoxic carcinogens. Environ. Hlth Perspect.

Waters, R., Ashby, J., Barrett, R.H., Coombes, R.D., Green, M.H.L. and Watkins, P. (1984). Unscheduled DNA synthesis. In: Report of the UKEMS Sub-Committee on Guidelines of Mutagenicity Testing, Part 2. pp.63-89.

Weston, A. (1993). Physical methods for the detection of carcinogen-DNA adducts in humans. Mutat. Res. 288:19-29.

Whittaker, S.C., Zimmermann, F.K., Dicus, B., Piegorsch, W.W., Resnick, M.A. and Fogel, S. (1990). Detection of induced mitotic chromosome loss in *Saccharomyces cerevisiae* - interlaboratory assessment of 12 chemicals. Mutat. Res. 241:225-242.

Wild, D. (1973). Chemical induction of strptomycin-resistant mutations in *Escherichia coli*. Dose and mutagenic effect of dichlorvos and methyl methanesulphate. Mutat. Res. 19:33-41.

Williams, G.M., Mori, H. and McQueen, C.A. (1989). Structure-activity relationships in the rat hepatocyte DNA-repair test for 300 chemicals. Mutat. Res. 221:263-286.

Wurgler, F.E., Sobels, F.H. and Vogel, E. (1977). *Drosophila* as an assay system for detecting genetic changes. In: Handbook of Mutagenicity Test Procedures. Eds. B.J. Kilbey, M. Legator, W. Nichols and C. Ramel. Elsevier Biochemical Press, North-Holland, Amsterdam, pp.335-373.

Wyrobek, A.J. and Bruce, W.R. (1975). Chemical induction of sperm abnormalities in mice. Proc. Natl. Acad. Sci. (USA) 72:4425-4429.

Wyrobek, A.J., Gordon, L.A., Burkhart, J.G., Francis, M.W., Kapp, R.W. Jr., Letz, G., Malling, H.V., Topham, J.C. and Whorton, M.D. (1983a). An evaluation of the mouse sperm morphology test and other sperm tests in non-human mammals. A report of the US Environmental Protection Agency Gene-Tox Program. Mutat. Res. 115:1-72.

Wyrobek, A.J., Gordon, L.A., Burkhart, J.G., Francis, M.W., Kapp, R.W. Jr., Letz, G., Malling, H.V., Topham, J.C. and Whorton, M.D. (1983b). An evaluation of human sperm as indicators of chemically induced alterations of spermatogenic function. A report of the US Environmental Protection Agency Gene-Tox Program. Mutat. Res. 115:73-148.

Wyrobek, A.J., Watchmaker, G. and Gordon, L. (1984). An evaluation of sperm tests and indicators of germ cell damage in men exposed to chemical or physical agents. Teratogen, Carcinog. Mutagen 4:83-107.

Young, S.S. (1988). Do short-term tests predict rodent carcinogencity? Science 241:1232-1233.

Zeiger, E. (1987). Carcinogenicity of mutagens : Predictive capability of the *Salmonella* mutagenesis assay for rodent carcinogenicity. Cancer Res. 47:1287-1296.

Zeiger, E., Haseman J.K., Shelby M.D., Margolin B.H. and Tennant R.W. (1990). Evaluation of four *in vitro* genetic toxicity tests for predicting rodent carcinogenicity: confirmation of earlier results with 41 additional chemicals. Environ. Molec. Mutagen 16:(Suppl. 18), 1-14.

Zeiger, E. and Pagano D.A. (1989). Mutagenicity of the human carcinogen treosulphan in Salmonella. Environ. Molec. Mutagen 13:343-346.

Zeise, L., Wilson, R. and Crouch, E. (1984). Use of acute toxicity to estimate carcinogenic risk. Risk Analysis 4, 187-199 (cited in Travis *et al.* 1990a).

Zimmermann, F.K. (1975). Procedures used in the induction of mitotic recomination and mutation in the yeast *Saccharomyces cerevisiae*. Mutat. Res. 31:71-86

Zimmermann, F.K., Mayer, V.W., Scheel, I. and Resnick, M.A. (1985). Acetone, methyl ethyl ketone, ethyl acetate, acetonitrile and other polar aprotic solvents are strong inducers of aneuploidy in *Saccharomyces cerevisiae*. Mutat. Res. 149:339-351.

Zimmermann, F.K., von Borstel, R.C., von Halle, E.S., Parry, J.M., Siebert, D., Zetterberg, G., Barale, R. and Loprieno, N. (1984). Testing of chemicals for genetic activity with *Saccharomyces cerevisiae*: a report of the U.S. Environmental Protection Agency Gene-Tox Program. Mutat. Res. 133:199-214.

Use of Mechanistic Information for Adequate Metabolic Design of Genotoxicity Studies and Toxicological Interactions of Drugs and Environmental Chemicals

F. Oesch, F. Fähndrich[1], H.R. Glatt,
B. Oesch-Bartlomowicz, K.-L. Platt and D. Utesch
Institute of Toxicology
University of Mainz
Obere Zahlbacher Strasse 67
55131 Mainz
Germany

Introduction

Microorganisms as well as mammalian cells used for mutagenicity investigations have little or no activities for metabolism of premutagens and precarcinogens, i.e. of compounds ultimately leading to mutations and cancer but first requiring metabolic activation. Therefore, to such cells an exogenous activating system is added, generally the postmitochondrial supernatant fraction of the liver homogenate and a NADPH-generating system (Ames et al. 1976). In this situation enzymes requiring cofactors other than NADP(H) are unlikely to be active. Thus, this metabolic system is rather artificial. Monooxygenases are active in this system. They, for example, convert polycyclic aromatic hydrocarbons to epoxides. These epoxides may be substrates for epoxide hydrolase, an enzyme which is also active in the test system because it requires no cofactor. The resulting diols may be further metabolized by monooxygenases to diol epoxides that are potent mutagens and carcinogens (Conney 1982). On the other hand, enzymes for conjugation reactions by which precursors of the ultimate mutagens could be sequestered are inactive, due to the low concentrations of their cofactors. These include the glutathione transferases that conjugate epoxides with glutathione, or sulfotransferases and UDP-

---

[1]Department of Surgery, University of Kiel, 24105 Kiel, Germany

NATO ASI Series, Vol. H 90
Molecular Aspects of Oxidative Drug Metabolizing Enzymes
Edited by E. Arınç, J. B. Schenkman and E. Hodgson
© Springer-Verlag Berlin Heidelberg 1995

glucuronosyltransferases that couple phenols or diols to sulfate and UDP-glucuronic acid. Conjugation reactions are often detoxications but in some cases conjugations lead to reactive products (Rannug et al. 1978, Watabe et al. 1982, Weisburger et al. 1972). These conditions favoring NADPH-dependent reactions such as monooxygenase-mediated reactions and disfavoring reactions which depend on other cofactors such as most conjugating reactions lead to an increase in sensitivity in many cases but to severe distortion of the true control of genotoxic metabolites in most cases.

Distortion of the metabolic control of genotoxic species in *in vitro* systems and possibilities for correction

The mutagenicity of a series of structurally very similar but with respect to carcinogenic potency very dissimilar compounds were related to the known carcinogenicity of the same compounds (Table 1). Their carcinogenicity ranges from apparently inactive to very potent.

In the homogenate-mediated assays marked increases in the number of revertants were achieved with all compounds. Benz[a]anthracene (BA) and 11 methylbenz[a]anthracenes (11-MBA) produced the strongest absolute effects and also showed the dose-response curves with the steepest initial slope. 2-, 3-, 9-, and 12-MBA were the weakest mutagens in these tests. Obviously, the potency of these compounds as bacterial mutagens in the homogenate-mediated assays did not correlate with carcinogenicity observed *in vivo*. In the cell-mediated assays the effects were generally weaker. The exception was DMBA, which is the strongest carcinogen in this series. Its dose-response curve in the cell-mediated assay was similar to that in the homogenate-mediated assay and its potency and maximal effect were higher than those of the other BA derivatives. 7-MBA, the next strongest carcinogen, produced the next strongest maximal effect. However, its mutagenic potency was similar to those of the "moderately potent" and "weak carcinogens". The three

Table 1. *In vitro* mutagenicity and *in vivo* carcinogenicity of benz[*a*]anthracene (BA), the 12 isomeric monomethylbenz[*a*]anthracenes (MBA) and 7,12-dimethylbenz[*a*]anthracene (DMBA).

| Compound | In vivo Carcinogenicity[a] | In vitro Mutagenicity (*S. typhimurium* TA100) | | | |
| | | Homogenate[b] | | Hepatocytes[b] | |
		Efficacy[c]	Potency[c]	Efficacy[c]	Potency[c]
DMBA	++++	1130	35.6	1100	13.3
7-MBA	+++	1810	10.8	960	3.0
12-MBA	++	830	8.3	380	3.0
6-MBA	++	1560	11.4	210	1.2
8-MBA	++	1150	16.7	120	1.4
BA	+	3610	50.9	490	3.6
4-MBA	+	1230	10.5	160	1.7
5-MBA	+	2450	12.5	500	3.7
9-MBA	+	860	8.9	150	1.4
10-MBA	+	1970	28.8	210	2.0
11-MBA	+	3080	59.1	320	1.5
1-MBA	±	2480	23.8	240	0.9
2-MBA	±	450	10.9	58	(0.3)
3-MBA	±	430	6.8	35	(0.2)

a) In order of carcinogenic potencies (Glatt et al. 1981, Wislocki et al. 1982). 1-, 2- and 3-MBA produced a total of 3, 1 and 1 tumours in the combined carcinogenicity studies ("putative non-carcinogens").
b) Hepatocytes (10^6 cells per incubation) from Aroclor 1254 treated mature male Sprague-Dawley rats or an equivalent amount of hepatocyte homogenate with a NADPH-generating system.
c) Efficacy: Maximal increase in revertants above solvent control. Potency defined as the steepest initial slope of the dose-response curves expressed in revertants per nmol. If the increase in the number of revertants above solvent control is less than 2-fold, an accurate calculation of the potency is not possible (numbers in parentheses) (data from Utesch et al. 1987).

"putative non-carcinogens" 1-, 2- and 3-MBA were the weakest mutagens. Thus, a substantial improvement of the quantitative correlation of *in vitro* mutagenicity with *in vivo* carcinogenicity is observed when intact hepatocytes are used as metabolizing system instead of tissue homogenate fractions, but the sensivitiy is generally lower.Cell- and homogenate-mediated mutagenicity assays differ in that the cofactors necessary for conjugating reactions are strongly diluted in the homogenate-mediated assays but present in physiological concentrations in intact cells. This may be causally linked to the observed differences in the control of genotoxic metabolites. To test this, cofactors for conjugation reactions were added to homogenized hepatocytes. BA and DMBA were selected as a model compounds for a weak and a strong carcinogen. ATP, which is necessary for the formation of 3´-phosphoadenosine 5´-phosphosulfate, a cofactor for sulfotransferases, was added to some samples, to other incubations UDP-glucuronic acid and the UDP-glucuronosyltranserase activator UDP-N-acetylglucosamine were added to ensure glucuronosyltransferase activity. Furthermore, the homogenate was fortified with glutathione, to stimulate conjugation by glutathione transferases. These cofactors, added individually to the homogenates, reduced the mutagenicity of BA to 34, 35 and 56%, respectively. Given together, the addition of cofactors resulted in a further reduction of mutagenicity (2 % of standard homogenate). When BA was metabolized by intact hepatocytes without any additional exogenous cofactors, mutagenic effects were 15 % of the maximal effects with unsupplemented homogenate, i.e. weaker than those observed by metabolism with homogenate plus any single cofactor.

With DMBA, the reduction in mutagenicity was weaker when the cofactors were added individually to the homogenate (4, 16 and 32 %). However, when the cofactors were used together, they decreased the mutagenicity of DMBA similar to that of BA (93 % reduction). The maximal effects observed in the cell-mediated assays were similar or even higher (at high concentrations) than in the homogenate-mediated assays without additional cofactors.

Thus, the dilution of cofactors for conjugation reactions in subcellular metabolizing systems impaired the inactivation of reactive metabolites. They did this not equally for different compounds. Therefore, compounds that are weakly active in *in vivo* systems, such as BA, may be potent mutagens in homogenate-supplemented tests.

Genotoxic compounds and metabolites are almost always under the control of a complex set of activating, inactivating and precursor sequestering enzymes which differ greatly between test systems, animal species and man. An adequate metabolic design of genotoxicity studies requires attention to:

- Dilution of cofactors in *in vitro* tests which are present in the intact cell in much higher concentrations;
- Induction in high dose carcinogenicity bioassays of enzymes which arenot constitutively expressed and not induced at such doses of the compound which occur in the situations of its practical use;
- Control of expression or modification of enzymes which are effected by hormones or other endogenous factors which are differently influenced by high dose (bioassay) versus moderate dose (real exposure) or by conditions *in vivo* (endocrine regulation) versus *in vitro* (no endocrine regulation).

Toxicological interaction between foreign compounds

Foreign compounds may profoundly interact with each other based on enzyme inhibition. An example of inhibition of the major microsomal epoxide hydrolase mEHb[1] by a low concentration (0.3 µM) of 1,1,1-trichloropropene 2,3-oxide (Buecker et al. 1979a) is shown in figure 1. There are, at this concentration no influences known on any other enzyme contributing to the control of epoxides by either forming them or metabolizing them further. Phenanthrene does not induce mutations in *Salmonella typhimurium* TA 1537 in the absence of this modulator to a measurable extent in agreement with the known non-carcino-

---

[1]mEHb: The major microsomal epoxide hydolase possessing a broad substrate specificity diagnostically including benzo[a]pyrene 4,5-oxide (Timms et al. 1984).

genicity of the compound. 0.3 μM 1,1,1-trichloropropene oxide, however, leads to a pronounced mutagenicity. This concentration of the modulator leads to an inhibition of mEH$_b$ (measured with the standard substrate benzo[a]pyrene 4,5-oxide which has a similar affinity to the enzyme as phenanthrene 9,10-oxide) of approximately 95 %, that is to a lowering of the effective mEH$_b$ activity by a factor of roughly 20. Phylogenetic investigations had shown that the mEH$_b$ activities varied between individual vertebrate species by up to about 1000-fold (Walker et al. 1978). Thus, by sufficiently raising the steady state concentration of the responsible epoxide(s) inhibitor-mediated modulations of activities of mEH$_b$ much less pronounced than variations in mEH$_b$ activities naturally occuring between vertebrate species were sufficient to convert an apparent non-mutagen to a clearcut mutagen.

Benzo[a]pyrene is a well known mutagenic carcinogen but is virtually non-mutagenic when activated by liver microsomes from rats which had not been pretreated with enzyme inducers (Oesch and Glatt 1976). However, after induction by 3-methylcholanthrene high mutagenicity is observed. In Figure 2, the dramatic difference is shown. This is of high practical importance since in some cases, such enzyme inductions may be a prerequisite for the mutagenicity and carcinogenicity to occur. This may be caused by a dose of the compound itself which is sufficient for induction (potential threshold of genotoxicity and carcinogenicity due to induction of constitutively not expressed enzymes) or it may be caused by an enzyme inducer different from the putative carcinogen under investigation (interaction).

An especially intriguing and important case of interaction is that between xenobiotics based on shifts of routes of metabolism. A striking example of this is shown in table 2: [14C]-benzo[a]pyrene was incubated with control and with *trans*-stilbene oxide (TSO)-induced rat liver microsomes and the metabolites were separated by HPLC. Two different elution systems as well as standard compounds were used to characterize the

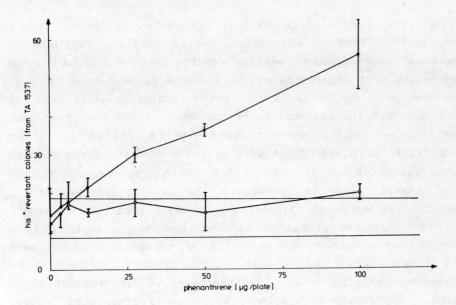

Fig. 1. Dose dependency of the mutagenic effect of phenanthrene after activation with liver microsomes from mice induced with Aroclor 1254 (500 mg/kg) for *Salmonella typhimurium* TA 1537. O, incubation without addition of an inhibitor for epoxide hydrolase. ●, 0.6 µl 1,1,1-trichloropropene 2,3-oxide in 10 µl dimethylsulfoxide was added to the incubation mixture. The horizontal lines indicate the range of numbers of colonies on plates without test compound (n = 8). From Buecker et al. 1979a, reproduced by permission of the publisher.

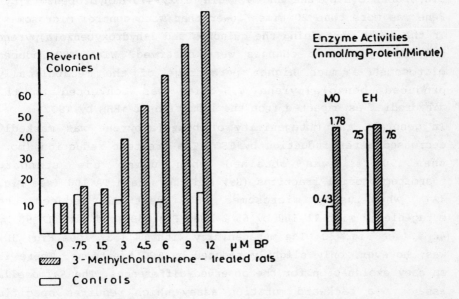

Fig. 2. Activation of benzo(a)pyrene to a mutagen: Potentiation by selective induction of monooxygenase. Benzo(a)pyrene and *Salmonella typhimurium* TA 1537 were incubated with hepatic microsomes from control male Sparague-Dawley rats or from rats that had been pretreated intraperitoneally with 10 mg/kg of 3-methylcholanthrene in sunflower oil and killed 3 days later. From: Oesch and Glatt 1976, reproduced by permission of the publisher.

metabolites. The results given in the table are derived from a separation using an acetonitrile-water gradient. By using a methanol-water gradient similar results were obtained with the exception that separation of the 4,5-epoxide peak from the 3,6-quinone peak was achieved with the acetonitrile-water gradient but not with the methanol-water gradient. It was confirmed that metabolites which were tentatively identified by their mobilities as dihydrodiols were markedly reduced or disappeared when the inhibitor of mEHb, 1,1,1,-trichloropropene 2,3-oxide was present. As shown in table 2, the total quantities of the metabolites were not significantly changed after TSO-treatment of the rats. Nevertheless, a most remarkable shift of the metabolism occurred. The quantity of metabolites which were oxidized at the benzo ring (7,8-dihydroxy-7,8-dihydrobenzo[a]pyrene, 9,10-dihydroxy-9,10-dihydrobenzo[a]pyrene, 9-hydroxybenzo[a]pyrene) was greatly diminished, whilst far more K-region metabolites (4,5-dihydroxy-4,5-dihydrobenzo[a]pyrene, benzo[a]pyrene 4,5-oxide) were formed. With TSO-induced microsomes, the ratio between 7,8-dihydroxy-7,8-dihydrobenzo[a]pyrene and 4,5-dihydroxy-4,5-dihydrobenzo[a]pyrene was more than 20 times lower than with control microsomes. In the peaks containing the quinones and 3-hydroxybenzo[a]pyrene only insignificant changes were observed. With TSO-induced microsomes a much higher percentage of the metabolically produced benzo[a]pyrene 4,5-oxide was converted to the dihydrodiol as expected from the induction of mEHb by TSO.

In general, the mutagenicity of benzo[a]pyrene was markedly decreased after induction by TSO. The greatest reduction (more than 90 %) was obtained when using the standard postmitochondrial fractions (S9) and the strain TA 100 (see Fig. 3A). When using microsomes TSO treatment reduced the mutagenicity with TA 100 by 65 % (see Fig. 3B) and with TA 98 by 40 % (see Fig. 3C) The mutagenicity with TA 1537 (see Fig. 3D) was, however, only slightly reduced by TSO treatment. There is an easy explanation for the observed differences. The Salmonella assay is a backward mutation assay which requires specific mutations to reconstruct a functional his gene. Different his⁻ strains vary in their susceptibility to reversion by different benzo[a]pyrene metabolites. TA 100 and TA 98 are easily reverted

Table 2. Effect of trans-stilbene oxide administration on the pattern of benzo[a]pyrene metabolism by rat liver microsomes[a])

Radioactivity with mobility of the following reference compounds	Control rats	Trans-stilbene oxide treated rats	Ratio[b]) TSO-treated/ control
9,10-Dihydroxy-9,10-dihydrobenzo[a]pyrene	1240 ± 90[c]	460 ± 10	0.37
4,5-Dihydroxy-4,5-dihydrobenzo[a]pyrene	430 ± 10	2670 ± 250	6.21
7,8-Dihydroxy-7,8-dihydrobenzo[a]pyrene	330 ± 10	90 ± 90	0.27
Benzo[a]pyrene 1,6-quinone and 3,6-quinone	3690 ± 20	3750 ± 480	1.01
Benzo[a]pyrene 4,5-oxide	430 ± 90	760 ± 70	1.76
9-Hydroxybenzo[a]pyrene	320 ± 10	120 ± 10	0.37
3-Hydroxybenzo[a]pyrene	2190 ± 170	1840 ± 320	0.84
Others	550 ± 40	1030 ± 80	1.87
Total	9180 ± 390	10700 ± 1250	1.16

a) From Buecker et al. 1979b, reproduced by permission of the publisher.

b) Ratio of metabolites present after 20 min incubation with microsomes from trans-stilbene oxide-treated and control rats

c) pmol per mg protein per 20 min

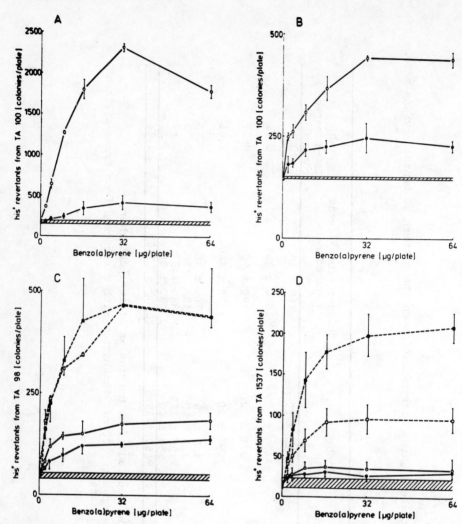

Fig. 3. Effect of *trans*-stilbene oxide treatment on the activation of benzo(a)pyrene to mutagens. The hatched zones show mean±SD of the number of revertant colonies in the absence of benzo(a)pyrene. O = control rats, ● = *trans*-stilbene oxide treated rats, □ = control rats with mEHb inhibitor 1,1,1-trichloropropene oxide, ■ = *trans*-stilbene-oxide treated rats, with mEHb inhibitor 1,1,1-trichloropropene oxide. A) *S. typhimurium* TA 100, 9000 *g* supernatant, B) *S. typhimurium* TA 100, microsomes, C) *S. typhimurium* TA 98, microsomes, D) *S. typhimurium* TA 1537, microsomes. From Buecker et al. 1979b, reproduced by permission of the publisher.

by benzo[a]pyrene 7,8-dihydrodiol 9,10-oxides and by benzo[a]pyrene 4,5-oxide. Towards the 7,8-dihydrodiol 9,10-oxides the strain TA 1537 is less sensitive, but this strain is highly sensitive to the 4,5-oxide (Bentley et al. 1977). When compared with a strain which is relatively insensitive towards the 7,8-dihydrodiol 9,10-oxides (TA 1537), the greater decrease of mutagenicity by TSO induction with the strains which are sensitive towards these dihydrodiol epoxides (TA 100, TA 98) suggests that the reduction of the mutagenicity is caused to a significant extent by the decreased oxidation of benzo[a]pyrene at the benzo ring. The decreased benzo ring metabolism is accompanied by increased K-region metabolism which leads to benzo[a]pyrene 4,5-oxide which is also mutagenic, but mEH$_b$, which is induced by TSO (Schmassmann and Oesch 1987), efficiently inactivates this monofunctional epoxide (Oesch and Glatt 1976, Wood et al. 1976).

The mutagenic effects with control and with TSO-induced microsomes (see Figs. 3C, 3D) were potentiated by the use of the mEH$_b$ inhibitor 1,1,1-trichloropropene oxide (Oesch et al. 1971). Since 1,1,1-trichloropropene oxide is too mutagenic for TA 100 (Glatt et al. 1979), this experiment can only be performed with TA 98 and TA 1537. Inhibition of mEH$_b$ increased the mutagenicity of benzo[a]pyrene activated with control microsomes for TA 98 about 3-fold, and with TSO-induced microsomes about 5-fold (cf. Fig. 3C). When TSO-induced microsomes were used for the activation, the inhibition of mEH$_b$ potentiated the mutagenicity for TA 1537 more strongly (Fig. 3D). This shows the increased importance of mEH$_b$ for the inactivation of mutagenic benzo[a]pyrene metabolites after a shift of the metabolism to the K-region, since the K-region epoxide benzo[a]pyrene 5,6-oxide is an excellent substrate of mEH$_b$ (used as its diagnostic substrate (Timms et al. 1984)). Thus, interactions between xenobiotics can lead to shifts in routes of metabolism which, in turn, can change their biological effects profoundly.

These examples show that interactions between foreign compounds may profoundly influence their toxic effects. It is expected that the permutations of possible interactions lead to an infinite number of conceivable situations which may never be amenable to individual investigation. Therefore, development of possiblities for reasonable approximations and predictions are desirable. They may be best possible if the underlying considerations are mechanism-based. Above a few important possiblities were outlined such as enzyme inhibition, enzyme induction and shift of routes of metabolism.

References

Ames BN, McCann J, Yamasaki E (1976) Methods for detecting carcinogens and mutagens with the *Salmonella*/mammalian-microsome mutagenicity test. Mutat Res 31:347-364

Bently P, Oesch F, Glatt HR (1977) Dual role of epoxide hydratase in both activation and inactivation. Arch Toxicol 39:65-75

Buecker M, Glatt HR, Platt KL, Avnir D, Ittah Y, Blum J, Oesch F (1979a) Mutagenicity of phenanthrene and phenanthrene K-region derivatives. Mutat Res 66:337-348

Buecker M, Golan M, Schmassmann HU, Glatt HR, Stasiecki P, Oesch F (1979b) The eypoxide hydratase inducer *trans*-stilbene oxide shifts the metabolic epoxidation of benzo[a]pyrene from the bay-to the K-region and reduces its mutagenicity. Mol Pharmacol 16:656-666

Conney AH (1982) Induction of microsomal enzymes by foreign chemicals and carcinogenesis by polycyclic aromatic hydrocarbons. Cancer Res 42:4875-4917

Glatt HR, Ohlsson A, Agurell S, Oesch F (1979) Delta[1]-tetrahydrocannabinol and 1α, 2α-epoxyhexahydrocannabinol: Mutagenicity investigation in the Ames test. Mutat Res 66:329-335

Glatt HR, Vogel K, Bentley P, Sims P, Oesch F (1981) Large differences in metabolic activation and inactivation of chemically closely related compounds: Effects of pure enymes and enzyme induction on the mutagenicity of the twelve monomethylated benz[a]anthracene in the Ames test. Carcinogenesis 2:813-821

Oesch F, Glatt HR (1976) Evaluation of the relative importance of various enzymes involved in the control of carcinogenesis. In: Montesano R, Bartsch H, Tomatis L (eds) Screening tests in chemical carcinogenesis. International Agency for Research on Cancer, Lyon, pp 255-274

Oesch F, Kaubisch N, Jerina DM, Daly J (1971) Hepatic epoxide hydrase: Structure-activity relationships for substrates and inhibitors. Biochemistry 10:4858-4866

Rannug U, Sundvall A, Ramel C (1978) The mutagenic effect of 1,2-dichloroethane on *Salmonella typhimurium*. I. Activation through conjugation with glutathione *in vitro*. Chem Biol Interactions 20:1-16

Schmassmann HU, Oesch F (1987) *Trans*-stilbene oxide: A selective inducer of rat liver epoxide hyratase. Mol Pharmacol 14:834-847

Timms C, Oesch F, Schladt L, Woerner W (1984) Multiple forms of epoxide hydrolase. In: Mitchell JF, Paton W, Turner P (eds) Proceedings of the 9th international congress of pharmacology. MacMillan Press, London, pp 321-237

Utesch D, Glatt HR, Oesch F (1987) Rat hepatocyte-mediated bacterial mutagenicity in relation to the carcinogenic potency of benzo[a]anthracene, benzo[a]pyrene, and twenty-five methylated derivatives. Cancer Res 47:1509-1515

Walker CH, Bentley P, Oesch F (1978) Phylogenetic distribution of epoxide hydratase in different vertebrate species, strains and tissues measured using three substrates. Biochim Biophys Acta 539:427-434

Watabe T, Ishizuka T, Isobe M, Ozawa N (1982) A 7-hydroxymethylsufate ester as an active metabolite of 7,12-dimethylbenz[a]anthracene. Science 215:403-405

Weisburger JH, Yamamoto RS, Williams GH, Grantham PH, Matsushima T, (1972) On the sulfate ester of N-hydroxy-2-fluorenylacetamide as a key ultimate hepatocarcinogen in the rat. Cancer Res 32:491-500

Wislocki PG, Fiorentini KM, Fu PP, Yang SK, Lu AYH (1982) Tumor-initiating ability of the twelve monomethyl-benz[a]anthracenes. Carcinogenesis 3:215-217

Wood AW, Levin W, Lu AYH, Yagi H, Hernandez O, Jerina DM, Conney AH (1976) Metabolism of benzo[a]pyrene and benzo[a]pyrene derivatives to mutagenic products by highly purified hepatic microsomal enzymes. J Biol Chem 251:4882-4890

Significance of Posttranslational Modification of Drug Metabolizing Enzymes by Phosphorylation for the Control of Carcinogenic Metabolites

F. Oesch, H.J. Arens, F. Fähndrich[1], T. Friedberg,
B. Richter, H. Yamazaki and B. Oesch-Bartlomowicz,
Institute of Toxicology
University of Mainz
Obere Zahlbacher Strasse 67
55131 Mainz
Germany

Introduction

The total activity of foreign compound metabolizing enzymes may change by altering the amount or the specific activity of the enzyme by induction or repression, or by activation or inhibition. The important contribution of enzyme induction is well known (Conney 1982, Oesch 1986, Nebert and Jones 1989). This is a relatively slow process which requires the biosynthesis of the enzyme protein. The possibility of a faster regulation of foreign compound metabolism by posttranslational modification by phosphorylation of an already preexisting protein molecule has only recently received attention. A central role in the metabolism of foreign compounds is played by the cytochrome P450-dependent monooxygenase. A few studies on phosphorylation of components of the cytochrome P450 dependent monooxygenases are available (Pyerin et al. 1983, Pyerin et al. 1987, Jansson et al. 1987, Epstein et al. 1989, Koch and Waxman 1989, Eliasson et al. 1990, Bartlomowicz et al. 1989a, 1989b, Oesch-Bartlomowicz et al. 1990), most of them in cell-free systems. The purpose of this chapter is to summarize the present information on the occurrence of phosphorylation of cytochromes P450 in intact cells, on its regulation by extracellular hormones and intracellular second messengers and on its impact

---

[1]Department of Surgery, University of Kiel, 24105 Kiel, Germany

NATO ASI Series, Vol. H 90
Molecular Aspects of Oxidative Drug Metabolizing Enzymes
Edited by E. Arınç, J. B. Schenkman and E. Hodgson
© Springer-Verlag Berlin Heidelberg 1995

in isoenzyme-selective metabolism of foreign compounds and on their genotoxic effects.

Donor and acceptor-selective phosphorylation of cytochrome p450 isoenzymes and its control by extracellular hormones and intracellular messengers

Investigations over the past 10 years by us and others (Pyerin et al. 1983, 1984, 1987, Pyerin and Taniguchi 1989, Epstein et al. 1989, Johansson et al. 1989) had shown that in cell-free systems purified cAMP-dependent protein kinase (PKA) or purified calcium/phospholipid-dependent protein kinase (PKC) catalyze donor and acceptor-selectively, the introduction of phosphate into some purified cytochrome P450 isoenzymes. None of the 13 cytochrome P450 isoenzymes were phosphorylated by the phosvitin/casein type II kinase. The question, therefore, arose whether these phosphorylation reactions also occurred in the compartmentalized situation of the intact cell. PKA and PKC transfer the γ-phosphoryl group from ATP to substrate proteins. However, when radioactive ATP is offered to cells in the medium, it does not enter the cells. Hepatocytes were, therefore, exposed to ^{32}P-orthophosphate in order to label the intracellular ATP pools. Hepatocytes were isolated from the liver of adult male Sprague-Dawley rats which had been pretreated with phenobarbital in order to induce the cytochrome P450 isoenzymes 2B1 and 2B2 (two major phenobarbital-inducible cytochromes P450). Amongst the 13 investigated isoenzymes these two had been the best substrates of the cAMP-dependent protein kinase in the cell free system (incubation of purified cytochromes P450 with purified protein kinases) (Pyerin et al. 1987). In absence of stimulation (see below) the incorporation of radioactive phosphate into those cytochrome P450 isoenzymes which serve as substrates (isolated and purified after the incubation with ^{32}P-orthophosphate) was very low. However, after stimulation of PKA either by extracellular hormones (glucagon) or by membrane permeating cAMP-derivatives (N^6, $O^{2'}$-dibutyryl-cAMP and 8-thiomethyl-cAMP) a marked incorporation of ^{32}P-phosphate into cytochrome P450 2B1 and related proteins took place (Table 1). Autoradiography of gel electrophoretically separated proteins from solubilized microsomes of these hepatocytes and of purified cytochrome P450 isoenzymes, combined with visualization on Western blots by specific antibodies (Fig. 1), showed that 4 cytochromes P450 were selectively phosphorylated, P4502B1, 2B2 and two P4502B1-related proteins. One of them was most closely related to P4502B2 (Oesch et al. 1989) and inducible by phenobarbital. The identity of the fourth phosphorylated protein (which was not induced by phenobarbital) remains to be established and probably represents a new cytochrome P450.

Table 1. The effects of glucagon and cAMP derivatives on the phosphorylation status of cytochrome P4502B1 related proteins.

Amount of protein applied onto gel	Treatment of hepatocytes		
	None	+Glucagon	+cAMP derivatives
25 µg	10646	18433[a]	75219[a]
		20729[a]	73416[a]
12.5 µg	6344	9978	39842
6.25 µg	-	-	15753

Partially purified cytochromes P450 (octylamino-Sepharose eluate) were obtained from hepatocytes which had been incubated in the prescence of ^{32}P-orthophosphate together with either glucagon (10^{-7}M) or N^6,O$^{2'}$-dibutyryl-cAMP (1mM) and 8-thiomethyl-cAMP (0.4 mM). The proteins were subjected to immunoblotting followed by autoradiography. The signals corresponding to the phosphorylated cytochromes P450 were integrated together. Values are expressed as arbitrary intensity units.

[a] Sample was applied in two lanes and each lane was separately scanned. (From: Bartlomowicz et al. 1989a, with permission by the publisher)

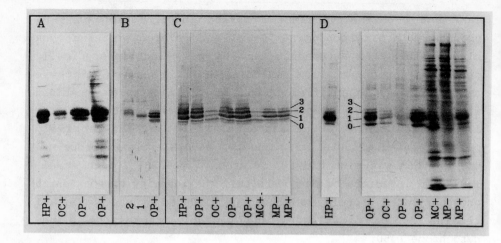

Fig. 1. Hepatocytes were isolated from either phenobarbital (P) treated or from untreated (C) rats. The hepatocytes were incubated in the presence (+) or in the absence (-) of cAMP derivatives. Microsomes (M) were prepared from these hepatocytes. Cytochromes P450 were partially purified from the microsomes by octylaminosepharose chromatography (O) followed by hydroxylapatite chromatography (H). The microsomes and the partially purified cytochromes P450 were analysed by SDS-PAGE. The proteins were then either stained by Coomassie blue (**A**) or the proteins were subjected to immunoblotting with anti-P4502B1 IgG as the first antibody (**B** and **C**). The immunoblot displayed in **C** was subjected to autoradiography (**D**). *MP*, microsomes from phenobarbital treated rats; *MC*, microsomes from untreated rats; *OP*, octylamino-Sepharose purified cytochromes from a phenobarbital treated rat; *OC* octylamino-Sepharose purified cytochromes P450 from a control rat; HP, octylamino-sepharose and hydroxylapatite purified cytochromes P450 from a phenobarbital treated rat. In **B** a partially purified cytochrome P450 fraction (*OP*) was analysed together with apparently homogenous cytochrome P4502B1 (lane 1) and 2B2 (lane 2). (From Bartlomowicz et al. 1989a, with permission by the publisher).

This phosphorylation was very selective in that none of the cytochromes P450 outside the range of the electrophoretic mobilities of the four bands were phosphorylated (Bartlomowicz et al. 1989a). Pyerin and Taniguchi (1989) and Koch and Waxman (1989) have also observed that the major phenobarbital-inducible cytochrome(s) P450 are phosphorylated in intact hepatocytes, Johansson et al. (1989) observed the cAMP-stimulated phosphorylation of the ethanol-inducible cytochrome P450 2E1 (P450j).

It is also known that the phosphorylation of rabbit cytochrome P4502B4 (closely related to rat cytochrome P4502B1) occurs on serine 128 within the cAMP dependent protein kinase recognition sequence (Müller et al. 1985). On the basis of other studies (Fujii-Kuriyama et al. 1982) one could predict that serine 128 is the amino acid which is phosphorylated in rat cytochrome P4502B1 and 2B2 as well. The presence of this recognition sequence is not always sufficient for phosphorylation to occur (absence of significant phosphorylation of cytochrome P4502C11 (P4502c) in isolated hepatocytes, Koch and Waxman 1989). This recognition sequence specific for PKA is present in many members of the cytochrome P450 gene family 2 of the rat liver (i.e. 2A1, 2B1, 2B2, 2C11, 2D1, 2D2 and 2E1).

Consequences of the phosphorylation of cytochromes P450 for the metabolism

The phosphorylation of cytochromes P4502B1 and 2B2 in intact hepatocytes led to a marked decrease in the microsomal monooxygenase activity towards substrates of these isoenzymes (Table 2; Bartlomowicz et al. 1989b): The O-dealkylation of 7-pentoxyresorufin which represents a highly selective substrate of cytochromes P4502B1 and 2B2 was markedly decreased; similarly, the metabolism of testosterone was markedly and highly regio- and stereo selectively changed in those positions of the steroid molecule which are attacked by cytochromes

Table 2. Isoenzyme-selective repression of monooxygenase activities in hepatocytes treated with cAMP derivatives

Treatment of hepatocytes	7-Pentoxyresorufin dealkylation (pmol resorufin per mg protein per min)	Testosterone metabolites a) (nmol hydroxytestosterone per mg protein per min)				
		16β-OH	16a-OH	7a-OH	2a-OH	A
None (control)	80.8±0.4	1.37±0.18	2.19±0.20	0.68±0.10	1.13±0.10	2.14±0.20
cAMP derivativesb)	53.6±1.67c) (66)	0.70±0.11c) (51)	1.52±0.14c) (69)	0.51±0.06 (75)	0.79±0.16 (70)	1.59±0.22 (74)

Microsomes were prepared from hepatocytes of phenobarbital-pretreated adult male Sprague-Dawley rats and incubated with 7-pentoxyresorufin or testosterone. The dealkylation of 7-pentoxyresorufin was monitored by spectrophotofluorimetry. Testosterone metabolites were resolved and quantitated by HPLC. Values represent means ± SEM, numbers in brackets represent percent of control.

a) The abbreviations denote the hydroxylated testosterone metabolites formed. "A" represents androst-4-ene-3,17-dione, the oxidation product of the 17β-hydroxy group of testosterone. The formation of 2β- and 6β-hydroxytestosterone was not decreased after treatment of the hepatocytes with cAMP derivatives (0.3 versus 0.3 and 1.39 versus 1.30 nmol hydroxytestosterone per mg protein per minute). No further testosterone metabolite peaks unequivocally above background noise were noted in the HPLC chromatogram.

b) 1 mM $N^6,O^{2'}$-dibutyryl-cAMP and 0.4 mM 8-thiomethyl-cAMP.

c) $p < 0.05$ compared with corresponding control (From: Bartlomowicz et al. 1989b, with permission by the publisher)

P4502B1 and 2B2, as opposed to other positions which are attacked by other isoenzymes (Bartlomowicz et al. 1989b). The hydroxylation in position 16ß is very selectively catalyzed by the two cytochrome P450 isoenzymes 2B1 and 2B2 (Levin et al. 1984; Waxman 1988). After treatment of the hepatocytes with cAMP derivatives leading to phosphorylation of cytochromes P4502B1 and 2B2 the hydroxylation of testosterone at this position was decreased by about 50 % (Bartlomowicz et al. 1989b). The influence of the phosphorylation on the hydroxylation at the 16α-position was "diluted" (decrease of about 30 %) (Bartlomowicz et al. 1989b) by those isoenzymes which, in addition to the phosphorylated cytochromes P4502B1 and 2B2, also catalyze this reaction. These are cytochromes P4502C7 (P450f, immunologically crossreactive with cytochrome P4502B1, constitutive isoenzyme), cytochrome P4502C11 (P450h, male specific testosterone 2α/16α-hydroxylase) and cytochrome P4502C13 (P450g, male specific, strain-dependent) (Levin et al. 1984; Waxman 1988). Cytochromes P4502B1 and 2B2 also catalyze the conversion of testosterone to androstenedione. This reaction is still less specific. Not only is it also catalyzed by a further cytochrome P450 isoenzyme (the male specific cytochrome P450 2C11 also called P450h), but also by a 17ß-hydroxysteroid dehydrogenase (Levin et al. 1984; Waxman 1988). In agreement with this, the decrease in activity due to phosphorylation was even more "diluted" at this position (it did not reach statistical significance at the $P < 0.05$ level). No differences of hydroxylation rates between hepatocytes treated with cAMP-derivatives and those which were not, were seen at other positions of the testosterone molecule (Bartlomowicz et al. 1989b). In a cell-free system phosphorylation of cytochrome P4502B4 (the major phenobarbital inducible cytochrome P450 in the rabbit liver) prior to its reconstitution with cytochrome b5 inhibited the interaction of the latter with the former (Jansson et al. 1987) leading to a loss of the stimulatory activity of cytochrome b5 on cytochrome P450-dependent monooxygenase activites. In a cell-free system the treatment of cytochrome P450 purified from the liver of a phenobarbital-treated rabbit with purified PKA led to a destruction of cytochrome P450 to cytochrome P420 (Taniguchi et al. 1985). Johansson et al. (1989)

reported that in hepatocytes the same agents which led to an increase of phosphorylation of the ethanol-inducible cytochrome P4502E1 (P450j) also led to an increase in its degradation rate. Moreover, the same agents which protected against the degradation of cytochrome P4502E1 (P4502E1 ligands such as imidazole) also protected against phosphorylation. Based on these observations the authors suggested that phosphorylation of cytochrome P4502E1 leads to a more rapid degradation of P4502E1. Thus, phosphorylation-induced changes in catalytic activity as well as phosphorylation-induced changes in the amount of enzyme protein will contribute to phosphorylation-induced changes of cytochrome P450-mediated metabolism.

Consequences of the phosphorylation of cytochromes P450 for the control of genotoxic metabolites

Hepatocytes were isolated from adult male Sprague-Dawley rats pretreated with phenobarbital in order to induce these cytochrome P450 isoenzymes which had been shown (Bartlomowicz et al. 1989a) to be phosphorylated in hepatocytes after their treatment with membrane-permeating cAMP derivatives. The hepatocytes were incubated with $N^6,O^{2'}$-dibutyryl-cAMP (dibt-cAMP) since such membrane-permeating cAMP derivatives had proven to lead to an isoenzyme-selective phosphorylation of cytochromes P4502B1 and 2B2 (Bartlomowicz et al. 1989a).

After treatment with dibt-cAMP and the phosphodiesterase inhibitor theophylline for 1 h the hepatocytes gave a markedly lower mutagenicity of cyclophosphamide in *Salmonella typhimurium* TA1535 compared with activation by hepatocytes that had not been stimulated by cAMP derivatives (Fig. 2, left part) (57 ± 3 %). The values given as percentages refer to the dibt-cAMP-induced difference of the metabolism-induced mutagenicity (i.e. they were calculated after subtracting the spontaneous mutations). In table 3 the pooled results from various experiments can be seen. The mutagenicity of cyclophosphamide (40-1280 μg) per plate was after stimulation with dibt-cAMP 51±18% of unstimulated

419

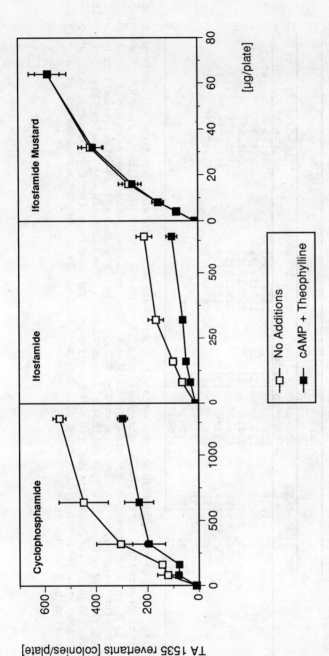

Fig. 2. Hepatocyte-mediated mutagenicity in *Salmonella typhimurium* TA1535 of cyclophosphamide, ifosfamide, and ifosfamide mustard. Hepatocytes were isolated from phenobarbital-pretreated rats and incubated with 1 mM N⁶,O²′-dibutyryl-cAMP in order to stimulate cAMP-dependent protein kinase and with 1 mM theophylline in order to inhibit phosphodiesterase activity. Values are means ± SD. (From: Oesch-Bartlomowicz et al. 1990, with permission by the publisher)

Table 3. Metabolic activation of cyclophosphamide by hepatocytes with and without stimulation by N6,O2'-dibutyryl-cAMP and theophylline

Revertant colonies per plate

Cyclophosphamide (µg/plate)	TA1535		TA100	
	No dibt-cAMP	Dibt-cAMP (1 h) + theophylline	No dibt-cAMP	Dibt-cAMP (1.5 h), no theophylline
0	13± 1 (10)	13± 2 (10)	167±12 (6)	186±15 (6)
10	18± 3 (5)	13± 5 (5)	197±19 (3)	203±12 (3)
20	31± 9 (5)	23±10 (5)	255±18 (3)	190±10* (3)
40	47± 8 (5)	31± 8* (5)	315±16 (3)	219± 7** (3)
80	93± 45 (5)	57±20 (5)	412±21 (3)	277±16** (3)
160	119± 23 (5)	65±13** (5)	549±16 (3)	354±21*** (3)
320	254± 16 (5)	135±59* (5)	nt	nt
640	375±107 (5)	193±41* (5)	nt	nt
1280	466±106 (5)	241±49** (5)	nt	nt

Mutagenicity of cyclophosphamide after metabolic activation by intact hepatocytes derived from phenobarbital-pretreated rats. The hepatocytes were incubated with N6,O2'-dibutyryl-cAMP (1 mM) for 1 h or 1.5 h in the presence or absence of 1 mM theophylline. Values are means ± SD from 2 experiments with Salmonella typhimurium strain TA1535 (duplicate and triplicate incubations) and 1 experiment with strain TA100 (triplicate incubations). The numbers of plates are given in parentheses. Significantly different from the corresponding sample without treatment with N6,O2'-dibutyryl-cAMP: *p<0.05; **p<0.01;***p<0.001. nt: Not tested (From: Oesch-Bartlomowicz et al. 1990, with permission by the publisher).

controls. Table 3 also presents results with an additional *Salmonella typhimurium* strain (TA100) and an additional time period of stimulation with dibt-cAMP (1.5h) in the absence of theophylline showing that also under these conditions the mutagenic cyclophosphamide metabolites were markedly reduced after pretreatment of the hepatocytes with dibt-cAMP (at concentrations of 40-160 µg cyclophosphamide per plate to 35±10% compared with unstimulated hepatocytes).

The accumulation of ifosfamide metabolites mutagenic for TA1535 was also markedly lower when bioactivation was mediated by hepatocytes that had been pretreated with dibt-cAMP in the presence of theophylline (47±10% of unstimulated controls, see Fig. 2, midle part). In repetition experiments the mutagenicity of pooled samples of 20-640 µg ifosfamide per plate was after pretreatment of hepatocytes with dibt-cAMP 38±11% of unstimulated controls, see Table 4).

Table 4. Metabolic activation of ifosfamide by hepatocytes with and without stimulation by $N^6,O^{2'}$-dibutyryl-cAMP and theophylline

Ifosfamide (µg/plate)	Revertant colonies per plate TA 1535	
	No dibt-cAMP	Dibt-cAMP +theophylline
0	13± 1(6)	12± 1 (6)
20	35± 6(6)	18± 4*** (6)
40	46±16(6)	24± 4* (6)
80	64± 9(6)	32± 8*** (6)
160	99±14(6)	48± 2****(6)
320	165±30(6)	61± 8****(6)
640	210±31(6)	103±21****(6)

Mutagenicty of ifosfamide in *Salmonella typhimurium* TA1535 after metabolic activation by intact hepatocytes derived from phenobarbital-pretreated rats. The hepatocytes were incubated with and without 1 mM $N^6,O^{2'}$-dibutyryl-cAMP and 1 mM theophylline for 1h. Values are means ± SD from 2 experiments (triplicate incubations). The numbers of plates are given in parentheses. Significantly different from the corresponding sample without treatment with cAMP derivatives: *p<0.05; ***p<0.001;****p<0.0001. (From Oesch-Bartlomowicz et al. 1990, with permission by the publisher).

Table 5. Metabolic activation of ifosfamide mustard by hepatocytes with and without stimulation by $N6,O2'$-dibutyryl-cAMP and theophylline

Ifosfamide mustard (μg/plate)	Revertant colonies per plate			
	TA1535		TA100	
	No dibt-cAMP	Dibt-cAMP (1 h) + theophylline	No dibt-cAMP	Dibt-cAMP (1.5 h), no theophylline
0	13± 1(9)	12± 2(9)	124±12(6)	95±12(6)
1	nt	nt	139±21(3)	94± 6(3)
2	nt	nt	145±11(3)	129±2(2)
4	84±14(9)	82± 9(9)	180± 6(3)	148±6(2)
8	155±23(9)	156±21(9)	228±18(3)	222±3(3)
16	271±38(9)	257±37(9)	347±19(3)	306±2(3)
32	418±47(9)	409±42(9)	nt	nt
64	584±74(9)	585±48(9)	nt	nt

Mutagenicity of ifosfamide mustard in *Salmonella typhimurium* TA1535 and TA100 after exposure to intact hepatocytes derived from phenobarbital-pretreated rats. The hepatocytes were incubated with $N6,O2'$-dibutyryl-cAMP (1 mM) 1 or 1.5 h in the presence or absence of 1 mM theophylline. Values are means ± SD from 3 experiments with *Salmonella typhimurium* strain TA1535 (triplicate incubations) and 1 experiment with strain TA100 (triplicate incubations). The number of plates are given in parentheses. nt: Not tested. (From: Oesch-Bartlomowicz et al. 1990, with permission by the publisher)

In contrast to cyclophosphamide and ifosfamide, which need metabolic activation, the mutagenicity of ifosfamide mustard, which does not need metabolic activation by cytochrome P450, was not different in the presence of hepatocytes that had been stimulated with dibt-cAMP and theophylline compared with unstimulated hepatocytes (Fig. 2, right part). Repetitions of the experiment gave almost identical results. The mutagenicity in the presence of dibt-cAMP-stimulated hepatocytes was 98±12% compared with those in the presence of unstimulated hepatocytes (Table 5). Table 5 also shows that prolongation of the time period of stimulation of the hepatocytes with dibt-cAMP to 1.5 h and use of a different detector strain (TA100 instead of TA1535) in the absence of theophylline also showed no dibt-cAMP-induced change in mutagenicity.

The absence of an effect of stimulation of the hepatocytes by dibt-cAMP on the mutagenicity of the active metabolite ifosfamide makes it also very unlikely, that changes in catalytic efficiency of inactivating enzymes contribute substantially to the dibt-cAMP-induced change in mutagenicity.

Table 6. Metabolic activation of cyclophosphamide by hepatocytes with and without stimulation by $N^6,O^{2'}$-dibutyryl-cAMP and theophylline in the presence of phenobarbital.

Cyclophosphamide (μg/plate)	TA1535 Revertant colonies per plate	
	No dibt-cAMP no theopylline	Dibt-cAMP +theophylline
0	15± 4(6)	13± 2 (6)
40	133±41(6)	44± 7***(6)
80	213±43(6)	77± 8****(6)
160	313±58(6)	116±15****(6)
320	500±46(6)	187±21****(6)
640	586±55(5)	254±34****(6)

Mutagenicity of cyclophosphamide after metabolic activation by intact hepatocytes derived from phenobarbital-pretreated rats. The hepatocytes were incubated for 1 h with 0.6 mM phenobarbital in presence or absence of $N^6,O^{2'}$-dibutyryl-cAMP (1 mM) and 1 mM theophylline. Values are means ± SD from 1 experiment with *Salmonella typhimurium* TA1535. The numbers of plates are given in parentheses. Significantly different from the corresponding sample without treatment with dibt-cAMP:***p<0.001;****p<0.0001.

The mutagenicity of cyclophosphamide was markedly increased by the presence of phenobarbital in the medium (compare Table 6 with Table 3) and the reduction of the mutagenicity by the treatment of the hepatocytes with dibt-cAMP was also more marked (down to 34±7% compared with the unstimulated controls). This indicates that continued induction of cytochrome P4502B1 and 2B2 leads to a further increase in the change of mutagenicity.

Table 7 Cytotoxicity of cyclophosphamide, ifosfamide and ifosfamide mustard metabolically activated by hepatocytes with and without stimulation by $N^6,O^{2'}$-dibutyryl-cAMP and theophylline.

Test compound (µg/plate)	Surviving fraction	
	No dibt-cAMP	Dibt-cAMP +theophylline
Cyclophosphamide		
40	0.90	0.97
80	0.91	0.95
160	0.92	0.97
320	0.72	0.87
640	0.59	0.70
Ifosfamide		
40	0.94	0.89
80	0.90	1.00
160	0.91	0.93
320	0.93	0.93
640	0.73	0.91
Ifosfamide mustard		
4	0.91	0.90
8	0.85	0.79
16	0.77	0.72
32	0.63	0.62
64	0.54	0.44

Hepatocytes isolated from phenobarbital-pretreated rats were incubated with $N^6,O^{2'}$-dibutyryl-cAMP for 1 h in the presence of 1 mM theophylline. Values represent means from 2 experiments.

Similar to the decrease in mutagenicity of metabolically activated cyclophosphamide and ifosfamide by pretreatment of the hepatocytes with dibt-cAMP, the cytotoxicity of metabolically activated cyclophosphamide and ifosfamide (but not that of the active metabolite ifosfamide mustard) was also markedly lower after pretreatment of the hepatocytes with dibt-cAMP (Table 7).

These data show that conditions which lead to the isoenzyme-selective phosphorylation of cytochromes P4502B1 and 2B2 in intact hepatocytes result in a marked change in the control of mutagenic and cytotoxic metabolites formed from cyclophosphamide and ifosfamide, without any change in the mutagenicity and cytotoxicity of the active metabolite ifosfamide mustard. Thus, the isoenzyme-selective phosphorylation of cytochromes P450 has important consequences for the control of genotoxic and cytotoxic metabolites.

References

Bartlomowicz B, Waxmann DJ, Utesch D, Oesch F, Friedberg T (1989a) Phosphorylation of carcinogen metabolizing enzymes: regulation of the phosphorylation status of the major phenobarbital-inducible cytochromes P450 in hepatocytes. Carcinogenesis 10:225-228

Bartlomowicz B, Friedberg T, Utesch D, Molitor E, Platt KL, Oesch F (1989b) Regio- and stereoselective regulation of monooxygenase activities by isoenzyme-selective phosphorylation of cytochrome P-450. Biochem Biophys Res Commun 160:46-52

Conney AH (1982) Induction of microsomal enzymes by foreign chemicals and carcinogenesis by polycyclic aromatic hydrocarbons: G.H.A. Clowes Memorial Lecture. Cancer Res 42:4875-4917

Eliasson E, Johansson I, Ingelman-Sundberg M (1990) Substrate-, hormone-, and cAMP-regulated cytochrome P450 degradation. Proc Natl Acad Sci USA 87:3225-3229

Epstein PM, Curti M, Jansson I, Huang CK, Schenkmann JB (1989) Phosphorylation of cytochrome P-450: Regulation by cytochrome b5. Arch Biochem Biophys 271:424-432

Fujii-Kuriyama Y, Mizukami Y, Kawajiri K, Sogawa K, Muramatsu M (1982) Primary structure of cytochrome P-450: coding nucleotide sequence of phenobarbital-inducible cytochrome P-450c cDNA from rat liver. Proc Natl Acad Sci (USA) 79:2793-2797

Jansson I, Epstein PM, Bains S, Schenkman JB (1987) Inverse relationship between cytochrome P-450 phosphorylation and complexation with cytochrome b5. Arch Biochem Biophys 259: 441-448

Johansson I, Eliasson E, Johansson A, Hagbjörk A-L, Lindros K, Ingelman-Sundberg M (1989) Mechanism of ethanol- and acetone-dependent induction of cytochromes P-450. In: Schuster I (ed) Cytochrome P-450: biochemistry and biophysics. Taylor and Francis, Vienna, pp 592-595

Koch JA, Waxman DJ (1989) Posttranslational modification of hepatic cytochrome P-450. Phosphorylation of phenobarbital-inducible P-450 forms PB-4 (IIB1) and PB-5 (IIB2) in isolated rat hepatocytes and in vivo. Biochemistry 28:3145-3152

Koch JA, Waxman DJ (1989) Posttranslational modification of hepatic cytochrome P-450. Phosphorylation of phenobarbital-inducible P-450 forms PB-4(IIB1) and PB-5(IIB2) in isolated rat hepatocytes and in vivo. Biochemistry 28:3145-3152

Levin W, Thomas PE, Reik LM, Wood AW, Ryan DE (1984) Multiplicity and functional diversity of rat hepatic microsomal cytochrome P-450 isozymes. In: Paton W, Mitchell J, Turnes P (eds) IUPHAR 9th international congress of pharmacology Vol 3. MacMillan Press, London, pp 203-209

Müller R, Schmidt WE, Stier A (1985) The site of cyclic AMP-dependent protein kinase catalyzed phosphorylation of P-450. FEBS Lett 187:21-24

Nebert DW, Jones JE (1989) Regulation of the mammalian cytochrome (CYP1A1) gene. Int J Biochem 21:243-252

Oesch F (1986) Short-term and long-term modulation of the enzymatic control of mutagenic and carcinogenic metabolites. In: Ramel C, Lambert B, Magnusson J (eds) Genetic toxicology of environmental chemicals,part A: Basic principles and mechanisms of action. Alan R. Liss, New York, pp 495-506

Oesch F, Waxman DJ, Morrisey JJ, Honscha W, Kissel W, Friedberg T (1989) Antibodies targeted against hypervariable and constant regions of cytochromes P450IIB1 and P450IIB2 Arch Biochem Biophys 270:23-32

Oesch-Bartlomowicz B, Oesch F (1990) Phosphorylation of cytochrome P450 isoenzymes in intact hepatocytes and its importance for their function in metabolic processes Arch Toxicol 64:257-261

Oesch-Bartlomowicz B, Vogel S, Arens HJ, Oesch F (1990) Modulation of the control of mutagenic metabolites derived from cyclophosphamide and ifosfamide by stimulation of protein kinase A. Mutat Res 232:305-312

Pyerin W, Taniguchi H (1989) Phosphorylation of hepatic phenobarbital-inducible cytochrome P-450. EMBO J 8:3003-3010

Pyerin W, Taniguchi H, Horn F, Oesch F, Amelizad Z, Friedberg T, Wolf CR (1987) Isoenzyme-specific phosphorylation of cytochromes P-450 and other drug metabolizing enzymes. Biochem Biophys Res Commun 142:885-892

Pyerin W, Taniguchi H, Stier A, Oesch F, Wolf CR (1984) Phosphorylation of rabbit liver cytochrome P-450LM2 and its effect on monooxygenase activity. Biochem Biophys Res Commun 122:620-626

Pyerin W, Wolf CR, Kinzel V, Kübler, D Oesch F (1983) Phosphorylation of cytochrome-P-450-dependent monooxygenase components. Carcinogenesis 5:573-576

Taniguchi H, Pyerin W, Stier A (1985) Conversion of hepatic
    microsomal cytochrome P-450 to P-420 upon phosphorylation by
    cyclic AMP dependent protein kinase. Biochem Pharmacol
    34:1835-1837
Watabe T, Ishizuka T, Isobe M, Ozawa N (1982) A 7-hydroxy-
    methyl sulfate ester as an active metabolite of 7,12-
    dimethyl-benz[a]anthracene. Science 215:403-405
Waxman DJ (1988) Interactions of hepatic cytochromes P-450 with
    steroid hormones. Regioselectivity and stereospecificity of
    steroid hydroxlation and hormonal regulation or rat P-450
    enzyme expresssion. Biochem Pharmacol 37:71-84

The Construction of Cell Cultures Genetically Engineered for Metabolic Competence towards Xenobiotics

Johannes Doehmer

Institut für Toxikologie und Umwelthygiene

Technische Universität München

Lazarettstrasse 62

80636 München

Germany

Introduction

Analytical tools have always played an important role for the study and understanding the complex biotransformation process. It needs technology to develop analytical tools. In this sense, gene technology is the most recent technology being applied in the construction of *in vitro* system useful for metabolism studies of xenobiotics. This includes cloning of genes encoding xenobiotics metabolizing enzymes, construction of recombinant expression vectors, and genetically engineering of cell cultures.

The first reports on the cloning of genes encoding xenobiotics metabolizing enzymes dates back to the beginning of the 80s and still continues (Adesnik et al. 1981; Fujii-Kuriyama

NATO ASI Series, Vol. H 90
Molecular Aspects of Oxidative Drug Metabolizing Enzymes
Edited by E. Arınç, J. B. Schenkman and E. Hodgson
© Springer-Verlag Berlin Heidelberg 1995

et al. 1981; Negishi et al. 1981; Gonzalez and Kasper 1982; Hardwick et al. 1983; Gonzalez et al. 1984).It was immediately evident, that this should be followed by expressing cloned genes in cultivated cells. The basic technology for having cloned genes expressed in cultivated cells was already established (Graham and van der Eb 1973, Pellicer et al. 1980; Doehmer et al. 1982).

The first expression of a cloned gene encoding a cytochrome P450 in a cultivated cell was reported for rat CYP1A1 in yeast cells (Oeda et al. 1985). Since then, a wide variety of cell cultures genetically engineered for cytochromes P450 have been reported. Expression systems are based on bacterial, yeast or mammalian cells, and express cytochrome P450 genes cloned from various species, including human, monkey, rat, mouse, rabbitt, probably soon fish, and many other interesting species. The choice of the cell system should depend on the question being asked. It is decisive to understand the technicality of these systems in order to choose the most appropiate system. The wrong choice might yield artefacts, false negative, or false positive results.

Genetically Engineering Cells

The technology for cloning genes and genetically engineering cells for heterologous expression was established during the 70s. The general strategy is depicted in Fig. 1. This technology is being constantly improved and advanced. Nevertheless, and even today, it is impossible to predict for sure which particular vector works best for a cloned gene in a chosen host cell. Some experience in handling and constructing recombinant expression vectors and host cells may help. But, there is no guarantee, if a vector works well in one particular system, it will also do perform well in another system. For this reason, most research groups stay with their vector, which has served them best in their system.

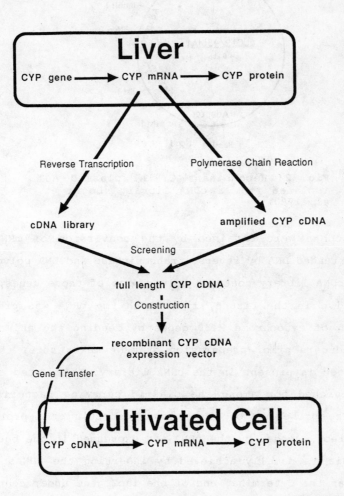

Fig. 1: Technical approach for heterologous expression in cells different from native tissue. Whatever genetic engineering procedures are to be applied, the resulting protein must remain authentic.

Cloning genes is routinely done by obtaining a full length cDNA from a gene library. A gene library consists of bacterial cells of which each cell contains a recombinant plasmid carrying a cDNA insert (Fig.2).

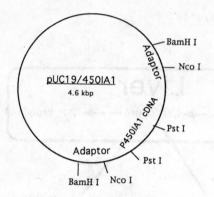

Fig. 2: Recombinant CYP1A1 plasmid as isolated from a cDNA library (Dogra et al. 1990)

The cDNAs were obtained by the conversion of mRNA into double stranded DNA by reverse transcriptase and DNA polymerase. Thus, a cDNA library contains only copies of those genes, which were active in the tissue, from which the mRNA was prepared. Induction of cytochrome P450 genes by feeding the animal with known inducers prior preparation of mRNA makes sure, that the cDNA wanted is present in the cDNA library. The wanted cDNA is being searched for among the plasmid carrying bacteria by a probe. The probe may be an antibody detecting the protein or part thereof encoded by the cDNA and expressed in the bacterial cell. This is usually achieved by inserting the cDNAs into a plasmid at the C-terminal end of the lacZ gene under control of the lac promoter resulting into a fusion protein consisting of galactosidase and cDNA encoded peptide. Due to cross hybrdization, antibodies raist against purified rodent cytochromes P450 usually detect also human cytochromes P450. Bacterial colonies expressing an immunoreactive fusion protein are detected by secondary binding to peroxidase coupled IgG.The wanted cDNA may also be searched for by a homologous nucleotide probe radioactivally labeled. Positive colonies are detected by X-ray autoradiographs. Either colonies, detected by immune-peroxidase staining or radiation, may be picked from the master plate for isolation of the cDNA recombinant plasmid. The cDNA insert is purified and verified for size, restriction sites, and sequence (Fig.3). The cDNA must be full length, i.e. containing the complete coding sequence starting with ATG as the first codon, and ending with TGA as the stop codon. Additional sequence from untranslated 5' and 3'-sequences help in constructing a functional recombinant expression vector. Full length cDNA may also be obtained by combining overlapping fragments.

In order to render a full length cDNA functional it needs to be connected to expression signals. Expression signals have to be compatible with the host cell. Prokaryotic promoters are to be chosen for expression in Escherichia coli, special yeast promoters for expression in yeast cells, and eukaryotic promoters for mammalian cells. This is achieved by inserting full length cDNA into the appropiate expression vectors. The cDNA has to be made compatible with the restriction sites in the expression vectors (Fig. 3). There are several ways to modify the ends of a cDNA. Small adaptor sequences may be hooked up which bridge between cDNA and vector DNA. The cDNA may also be inserted into an intermediary plasmid from which it may be isolated by making use of restriction sites in the intermediary plasmid adjacent to the 5'- and 3'-end of the cDNA. For a long time, expression of cytochrome P450 cDNA in E.coli was unsuccessful, although very potent prokaryotic promoters were used in the construction of expression vectors. A successful expression in E.coli became only possible by changing the cDNA coding sequence to the codon preference of E.coli

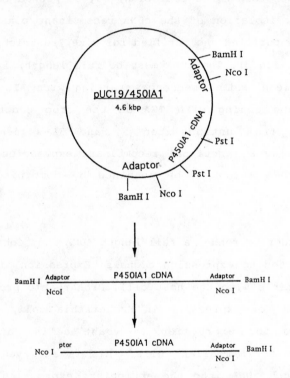

Fig. 3: Preparation of the CYP1A1 cDNA for insertion into the pSV expression vector is achievied by a complete digest with the restriction enzyme BamHI, followed by a partial digest with NcoI, resulting into a NcoI/BamHI fragment which its into NcoI/Bgl II of the pSV plasmid for expression.

The cDNA recombinant expression vector (Fig. 4) is being transferred into a cell for production of the cDNA encoded protein.

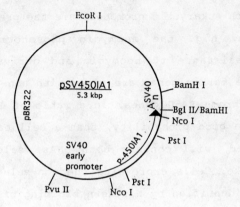

Fig. 4: Recombinant CYP1A1 expression vector as delivered into V79 Chinese hamster cells for stable expression (Dogra et al. 1990)

DNA mediated gene transfer may be achieved by various techniques, particularly developed for bacterial, yeast, and mammalian cells. A very common technique for transferring DNA into mammalian cells is the socalled Ca/P-coprecipitation technique, established for Adenovirus DNA (Graham and van der Eb 1973).A calcium and phosphate solution of certain concentration are mixed. Calcium and phosphate complexes and precipitates within 20 to 30 minutes. Any high molecular weight matter present, such as DNA, will then co-precipitate. In this state, DNA is added to the cultivated cells in the supernatant culture medium. Within hours the precipitated DNA will settle on the cell surface. Cells are treated with glycerol for a few minutes, which enhances uptake of DNA. Cells are kept in non-selective medium for another 2 days in order to allow stable integration of the transfected DNA into the genomic DNA of the cell. Cells are kept in selective medium from then on. Stably transfected cells show up as colonies after another 10 days in culture with selective medium. Bacterial antiobiotics resistance genes recombined with eukaryotic promoters are the preferred selective marker genes, e.g. the neomycin phosphotransferase gene conferring resistance to neomycin and derivatives, such as G418.Selective marker genes are mixed with non-selective genes, such as cytochrome P450 genes, in a ratio of 1:20 up to 1:50. This ensures a good probability, that a cell taking up DNA and being selected, will contain both DNAs, selective and non-selective alike. Cells expressing the non-selective cytochrome P450 DNA are identified by Northern blotting of total RNA, i.e. checking for authentic cytochrome P450 mRNA, by Western-blotting of total protein or immunofluorescence staining of the cells, i.e. checking for authentic protein, and finally checking for authentic enzyme activity by using the appropiate substrate.

Cell Systems for Cytochrome P450 cDNA directed Expression

Different types of cells have been genetically engineered for expression of cytochromes P450, including bacterial, yeast and mammalian cells. Some of those genetically engineered cell systems are listed in tables 1, 2, and 3. All these systems are unique in several ways. It is important to know their specific features in order to select the most appropiate system.

Escherichia coli has been shown over and over again to be very productive for a cDNA encoded protein, as shown for pharmaceuticals such as insulin. In the same way, Escherichia coli can be a rich source for large quantities of a specific cytochrome P450. As much as 15 mg of cytochrome P450 from 1 liter of bacterial culture has been reported. These are quantities no other genetically engineered system can deliver. Sometimes, it may be the only way to obtain sufficient quantities, as a particular cytochrome P450 may be present in native tissue at miniscule quantities. Large quantities are needed for some spectrophysical measurements or crystallization for X-ray structure analysis.

CYP Isoform expressed:	Reference:
17A (17α)	Barnes et al., 1991
2E1	Larson et al., 1991
2G1	Kempf et al., 1991
1A2	Fisher et al., 1991

Table 1: Genetically Engineered Bacterial Systems

Yeast cells are almost as easy to handle, to manipulate, and to maintain as bacterial cells. Yeast cells grow faster than mammalian cells with a doubling time of around 4 hours, and produce around 0.2 to 0.5 mg cytochrome P450 per liter of culture. Yet, yeast cells are eukaryotic cells, structurally organized like mammalian cells, containing a nucleus, endoplasmic reticulum, and a surface membrane. Therefore, yeast cells are somewhere between bacterial and mammalian cells. However, yeast cells naturally lack certain biological endpoints unique for mammalian cells. It is also known, that the yeast membrane is not as permeable like a mammalian cell membrane. For this reason, most studies on genetically engineered yeast cells have to be done in cell homogenate, and cannot be performed on live cells as needed for cytotoxicity studies.

CYP Isoform expressed:	Reference:
1A1 rat	Oeda et al., 1985
1A1 mouse	Kimura et al., 1987
1A1 human	Eugster et al., 1990
1A1 rabbit	Pompon, 1988
1A2 mouse	Cullin and Pompon, 1988
1A2 rat	Shimizu et al., 1986
1A2 rabbit	Pompon, 1988
2B1 rat	Black et al., 1989
2B2 rat	Zurbriggen et al., 1989
2E1 rabbit	Fujita et al., 1990
3A4 human	Renaud et al., 1990
17A1bovine	Sakaki et al., 1991
21B1bovine	Sakaki, et al., 1991

Table 2: Genetically engineered yeast cells

Although mammalian cells are not easy and as cheap to cultivate and to maintain, and in addition produce by far less cytochromes P450 than bacterial or yeast cells, there are several good reasons to genetically engineer mammalian cells. Most importantly, mammalian cells provide relevant and suitable biological endpoints not present in yeast and baterial cells. Combining these biological endpoints and metabolic competence in one and the same cell, makes genetically engineered mammalian cells powerful tools for the assessment of some cytochrome P450 mediated toxicological and pharmacological effects related to drug metabolism and activation of promutagen and procarcinogens.Various promoters and expression systems are routinely used for heterologous expression of xenobiotic metabolizing enzymes in various cells. Four of these systems will be dealt with in more detail.

CYP Isoform expressed	Reference
1A1 mouse	Battula et al., 1987
1A1 mouse	Battula, 1989
1A1 mouse	Aoyama et al., 1989
1A1 mouse	Trinidad et al., 1991
1A1 mouse	Puga et al., 1990
1A1 rat	Dogra et al., 1990
1A1 human	Schmalix et al., 1993
1A1 human	States et al., 1993
1A1 human	Crespi et al., 1989
1A2 mouse	Aoyama et al., 1989
1A2 mouse	Battula et al., 1991
1A2 mouse	Thompson et al., 1991
1A2 rat	Wölfel et al., 1991
2A2 human	Davies et al., 1989
2A3 human	Aoyama et al., 1990
2A6 human	Crespi et al., 1990
2B1 rat	Doehmer et al., 1988
2B7 human	Aoyama et al., 1990
2C8 human	Aoyama et al., 1990
2C9 human	Aoyama et al., 1990
2D6 human	Crespi et al., 1991
2E1 human	Nouso et al., 1992
2E1 human	Aoyama et al., 1990
2E1 human	Crespi 1991
2F1 human	Aoyama et al., 1990
3A3 human	Aoyama et al., 1990
3A4 human	Aoyama et al., 1990
3A4 human	Crespi 1991
3A5 human	Aoyama et al., 1990
17A human	Zuber et al., 1986

Table 3: Genetically engineered mammalian cells

The COS Cell Expression System

COS cells are African Green Monkey Kidney Cells. They are established as cell lines. They had been infected with the DNA tumor virus SV40. They do not produce infectious SV40 virus. But, they retained the expression of a viral gene, the "T-Antigen". The T-Antigen is responsible for maintaining the transformed status of this cell. It also serves as means to maintain genetically engineered SV40 recombinant plasmids episomally, which contain i.e. Cytochrome P450 encoding cDNA under control of the SV40 Early Promoter. Right after gene transfer SV40 recombinant plasmids are maintained in high copy number leading to a very high expression. However, over prolonged time of culturing the transfected cells, the plasmid is lost and expression slows down to zero. Maximum expression is observed 2 to 4 days after gene transfer. This expression is massive in most cases, and can be easily followed by immunofluorescence staining without a time consuming selection process. Therefore, this system is being frequently used for structure / funtion related studies on genes and proteins. It is noteworthy to mention, that the very first mammalian cell line genetically engineered for heterologous expression of a cytochrome P450 was the COS cell (Zuber et al. 1986).

The Vaccinia Virus Expression System

The cDNA encoding cytochrome P450 is being placed under transcriptional control of the vaccinia P7.5 promoter in the pSC11 plasmid first (Fig.5) (Gonzalez et al. 1990). This plasmid also contains the thymidine kinase gene, which is being used for spontaneous homologous recombination with thymidin kinase gene in the very large genomic DNA of vaccinia virus. The recombination leads to a disruption of the thymidine kinase gene in the vaccinia virus. Thus, recombinant vaccinia virus can be selected on TK$^-$ cells, that grow in the presence of 5-bromodeoxyuridine (BUdR). Only those cells survive BUdR that receive TK$^-$ recombinant vaccinia virus. The recombinant virus ist then plaque purified and used to infect HepG2 cells or other cells. The vaccinia virus is not restricted to a particular cell line. However, the recombinant vaccinia virus remains lytic by nature. Lysis of the infected cells occurs 3 to 4 days. This is enough time for production of sufficient quantities of cytochromes P450 encoded in a particular recombinant vaccinia virus strain. Usually, cells are harvested and lysed between one and two days after infection in order to yield a homogenate, which is being used for cytochrome P450 related metabolism studies. The vaccinia virus was established by expressing mouse cytochrome P450 1A1 and 1A2 (Battula et al. 1987; Faletto et al. 1988). Currently, this system is very efficiently and succesfully applied for human cytochromes P450. To date, 12 different human cytochromes P450 have been expressed in HepG2 cells (Aoyama et al. 1990) and have been applied in various studies, such as in a comparative study on steroid hydroxylase activities (Waxman et al. 1991). The HepG2 cells are perfectly suited in this case, as these cells have good levels of endogenous human NADPH-cytochrome P450 reductase. The extremely long doubling time of HepG2 cells is no drawback in the vaccinia expression system, due to the lytic natur of the system.

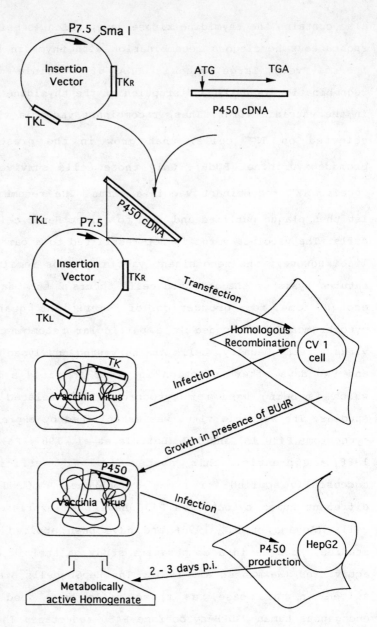

Fig. 5: The Vaccinia virus directed expression system

Certain health precautions are needed or advised, which make it obligatory or advisable to vaccinate the laboratory personell, depending on the law or guide lines that may apply in different countries.

The AHH-1 Cell Expression System

Human cytochrome P450 encoding cDNAs were placed under control of of the promoter/enhancer and polyadenylation signal derived from the Herpes Simplex virus thymdinine kinase gene (Fig. 6). The cDNA directed expression is constitutive and stable in mammalian cells under these conditions. The recombinant Herpes TK / human cDNA plasmids are introduced into the AHH1 cells. The AHH1 cell line is a human lymphoblastoic cell line derived from cell line RPMI 1788, and contains the Epstein-Barr virus EBNA-1 sequence which allows extrachromosomal replication of a recombinant plasmid containing sequences known as origin of replication ("OriP"). This is analogous to the COS cell system. However the Epstein-Barr virus directed maintenance of the plasmid appears to be more stable than the SV40 virus. AHH1 cells are TK +/-, and are therefore suitable not only for cytotoxicity but also for mutagenicity studies. The cells grow in suspension with a doubling time of less than 18 hours. Selection of cells for containing recombinant plasmids is achieved by resistance to hygromycin B. Due to high copy numbers of recombinant plasmids, levels of cytochromes P450 activities are very high. To date, 12 different human cytochromes P450 are expressed in AHH1-cells. There is one particular cell line "MCL-5", which expresses five different cytochromes P450 simultaneously at almost equal levels. The cell lines are commercially available from Gentest Corp., 6 Henshaw St., Woburn, MA 01801, U.S.A.

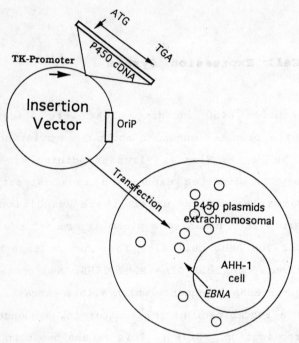

Fig. 6: The Epstein Barr virus directed expression system.

The Retroviral Vector Expression System

During the lifecycle of a retrovirus its genomic RNA is converted into doublestranded and circular DNA, which integrates in a very specific way into the genomic DNA of the infected host cell, where it remains as an additional entity to the genomic DNA of the host cell (Fig.6). These are the essential biological features for a retrovirus based expression systems: (1) gene transfer by infection; (2) integration of an intact recombinant viral genome; (3) efficient and stable expression by retroviral promoters; (4) no cell lysis.

The establishment of a retroviral based expression system starts with the construction of a recombinant retroviral genomic DNA. There are suitable sites in this genome for inserting additional DNA, such as cytochrome P450 cDNA, without disturbing the overall structural and functional organization of the viral genomic DNA. The recombinant viral genomic DNA is transferred by conventional DNA mediated gene transfer techniques into a special cell line, called packaging cell line. In case the recombinant viral genomic DNA is integrated intact into the genome of the host cell, recombinant viral genomic RNA is being expressed and viral proteins needed for an infectious virus particle is being produced. Recombinant viral genomic RNA and viral proteins assemble to recombinant infectious virus at the host cell membrane, and are released without lysis. The recombinant infectious virus may be used to infect other cells for expression of cytochromes P450.

For safety reasons, the retroviral expression vector system is designed to produce only defective viral particles, capable of only one infectious cycle. This is achieved by mixing and complementing two defective retrovirus. The genome of the first defective retrovirus is contained in the genomic DNA of the socalled packaging cell line. This retroviral genome is defective due to lack of its packaging signal. Therefore, retroviral genomic RNA of this virus is not packaged and is not contained in viral particles released from the packaging cell line. The recombinant retroviral genomic DNA lacks its viral gene reverse transcriptase, encoding the enzyme converting RNA into DNA for integration. Thus, both retroviral genomic DNAs in the packaging cell line complement each other by generating viral particles, which are infectious and contain the recombinant viral RNA only. Virus particles carry a few reverse transcriptase molecules to allow conversion of the RNA into doublestranded DNA for integration after infection. Due to lack of the viral gene encoding reverse transcriptase, the infectious cycle stops here.

So far, only a few cell cultures have been established, using the retroviral expression system. This is mostly due to the fact, that the construction of the packaging cell line is cumbersome and frustrating. Depending on the cell line chose for infection, the retrovirus has to be either ecotrophic or amphotrophic, i.e. retroviral particles are species and cell specific. In addition, retrovirus may raise safety concern in various countries.

The V79 Chinese Hamster Expression System

V79 Chinese hamster cells have been the preferred indicator cells in toxicological studies for several reasons. V79 cells grow extremely fast with a doubling time of 12 hours or less, and are able to start exponential growth from a small inoculum. Rapid growth allow experiments to be done quickly. V79 cells also have a high cloning efficiency of 90% and more. Therefore, results are not compromised by potential interaction between the low cloning efficiency of the cell culture itself and the toxicity of the test compounds. These cells have a stable karyotype with a modal chromosome number of 22 ±1 and are male, having only one X-chromosome. For these reasons, V79 cells are routinely used in mutagenicity, cytotoxicity, chromosomal abberration, and micronuclei formation studies. However, these cells lack cytochrome P450 activity. Therefore, these cells were always being used in conjunction with metabolically competent liver homogenate. In order to provide V79 cells with metabolic competence, these cells are being genetically engineered for stable expression of cytochromes P450 (Doehmer 1993) (Fig. 7).The recombinant expression vectors in this system are based on plasmids containing the SV40 Early Promoter and polyadenylation signal. The recombinant expression vectors are delivered to the V79 cells by the Ca/P-coprecipitation technique, and are selected for stable integration into the genomic DNA of the V79 cells. So far, six different rat and human cytochromes P450 have been successfully expressed in V79 cells. These recombinant V79 cell lines have been successfully applied in a variety of metabolism related problems in toxicology and pharmacology alike.

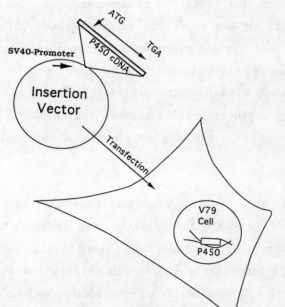

Fig. 7: The SV40 virus directed expression system.

Discussion

Gene technology is a perfect means to dissect the complex biotransformation by isolating a particular cytochrome P450 gene and making it to function again after transfer into another cell without confusing co-expression of other cytochrome P450 isoforms. The gene technological approach towards cytochrome P450 offers unique experimental possibilities and at the same time, can complement classical biochemical, toxicological, and pharmacological procedures.

Purification of cytochromes P450 starting from native tissues is always biased because of highly homologous cytochrome P450 species, which have similiar molecular weights and almost identical sequences. Together with overlapping substrate specificity there is no proof that a purified cytochrome P450 is of a single form. Genetically engineered cells provide a more defined source for isolating cytochromes P450. In some cases, it is difficult to get access to native tissue. This is certainly the case, when human cytochrome P450 needs to be purified. Some cytochromes P450 are impossible to isolate from native tissue due to low abundance in native tissues. In those cases, genetically engineered cells are not only a specific, but also a rich and probably the only source for a particular cytochrome P450.

Before gene technology, purification of a particular cytochrome P450 was the only direct way to study a cytochrome's P450 features, e.g. substrate specificity. This is being replaced by genetically engineered cells which are defined for the expression of one particular cytochrome P450.

Cells may have suitable and relevant biological endpoints, which interfere with cytochrome P450 activated xenobiotics. So far, biological endpoints and metabolism were contained in two separated components. For example, this is the case with the Ames-test, where the biological endpoint is contained in a bacterial cell, and the metabolism is given with a rat liver homogenate. Gene technology can be used to combine both, metabolism and biological endpoint, in one and the same cell. In this sense, gene technology is a means to improve classical test systems.

A variety of cell lines have been genetically engineered over the years using different expression systems. All of these cell lines have something unique that made it worthwhile to combine with cytochromes P450. Some of these approaches are listed in Table 1.

The above mentioned V79 cells are useful for a variety of studies, like mutagenicity, cytotoxicity or chromosomal abberration, which may be studied under the influence of cytochrome P450 mediated activation of xenobiotics.

NIH 3T3 and 10T1/2 cells are mouse fibroblast cell lines, which maintain a non-transformed status in cell culture for cell density depending growth. Cytochrome P450 expressing NIH 3T3 or 10T1/2 cells may be used in transformation studies that are related to reactive metabolites from pro-carcinogens.

There are CHO cell lines (Chinese hamster ovary) that are deficient and proficient for DNA repair may be used to study the effect on and of the repair system by reactive metabolites.

Thus, gene technology is not only a means to study gene sequences and evolutionary relationships, but has a growing impact in the testing of xenobiotics.

Acknowledgment

The generous support by the Bundesgesundheitsamt, Berlin - ZEBET - is gratefully acknowledged.

References

Adesnik M, Bar-Nun S, Maschio F, Zunich M, Lippman A, Bard E (1981) Mechanism of induction of cytochrome P-450 by phenobarbital. J Biol Chem 256: 10340-10345

Aoyama T, Gonzalez FJ, Gelboin HV (1989) Mutagen activation by cDNA-expressed P1-450, P3-450, and P450a. Mol Carcinog 1:253-259

Aoyama T, Yamano S, Guzelian PS, Gelboin HV, Gonzalez FJ (1990) Five of 12 forms of vaccinia virus-expressed human hepatic cytochrome P450 metabolically activate aflatoxin B1. Proc Natl Acad Sci USA 87:4790-4793

Barnes HJ, Arlotto MA, Waterman MR (1991) Expression and enzymatic activity of recombinant cytochrome P450 17α-hydroxylase in Escherichia coli. Proc Natl Acad Sci USA 88:5597-5601

Battula N. (1989) Transduction of cytochrome P3-450 by retroviruses: constitutive expression of enzymatically active microsomal hemoprotein in animal cells. J Biol Chem 264:2991-2996

Battula N, Sagara J, Gelboin HV (1987) Expression of P1-450 and P3-450 DNA coding sequences as enzymatically active cytochromes P-450 in mammalian cells. Proc Natl Acad Sci USA 84:4073-4077

Battula N, Schut HAJ, Thorgeirsson SS (1991) Cytochrome P450IA2 constitutively expressed from transduced DNA mediates metabolic activation and DNA-adduct formation of aromatic amine carcinogens in NIH3T3 cells. Mol Carcinog 4:407-414

Black SM, Ellard S, Meehan RR, Parry JM, Adesnik M, Beggs JD, Wolf CR (1989) Carcinogenesis 10:2139

Cullin C, Pompon D (1988) Functional expression of mouse cytochrome P-450 P1 and chimeric P-450 P3-1 in yeast saccharomyces cerevisiae. Gene 65:203-217

Crespi CL, Langenbach R, Rudo K, Chen YT, Davies RL (1989) Transfection of a human cytochrome P-450 gene into the human lymphoblastoid cell line AHH-1, and use of the recombinant cell line in gene mutation assays. Carcinogenesis 10:295-301

Crespi CL, Penman BW, Leakey JAE, Arlotto MP, Stark A, Parkinson A, Turner T, Steimel DT, Rudo K, Davies RL, Langenbach R (1990) Human cytochrome P450IIA3: cDNA sequence, role of the enzyme in the metabolic activation of promutagens, comparison to nitrosamine activation by human cytochrome P450IIE1. Carcinogenesis 11:1293-1300

Davies RL, Crespi CL, Rudo K, Turner TR, Langenbach R (1989) Development of a human cell line by selection and drug-metabolizing gene transfection with increased capacity to activate promutagens. Carcinogenesis 10:885-891

Dogra S, Doehmer J, Glatt HR, Mölders H, Siegert P, Friedberg T, Seidel A, Oesch F (1990) Stable expression of rat cytochrome P-450IA1 cDNA in V79 Chinese hamster cells and their use in mutagenicity testing. Mol Pharmacol 37:608-613

Doehmer J, Barinaga M, Vale W, Rosenfeld MG, Verma IM, Evans R (1982) Introduction of rat growth hormone gene into mouse fibroblasts via a retroviral DNA vector: expression and regulation. Proc Natl Acad Sci USA 85:5769-5773

Doehmer J, Dogra S, Friedberg T, Monier S, Adesnik M, Glatt HR, Oesch F (1988) Stable expression of rat cytochrome P-450IIB1 cDNA in Chinese hamster cells (V79) and metabolic activation of aflatoxin B1. Proc Natl Acad Sci USA 85:5769-5773

Doehmer J (1993) V79 Chinese hamster cells genetically engineered for cytochrome P450 and their use in mutagenicity and metabolism studies. Toxicol 82:105-118

Eugster HP, Sengstag C, Meyer UA, Hinnen A, Würgler FE (1990) Constitutive and inducible expression of human cytochrome P450IA1 in yeast Saccharomyces cerevisiae: An alternative enzyme source for in vitro studies. BBRC 172:737-744

Pompon D (1988) cDNA cloning and expression in yeast S.cerevisiae of beta-naphtoflavone induced rabbit liver P-450 LM4 and LM6. Eur J Biochem 177:285-293

Faletto MB, Koser PL, Battula N, Townsend GK, MacCubbin AE, Gelboin HV, Gurtoo HL (1988) Cytochrome P3-450 cDNA encodes aflatoxin B1-4-hydroxylase. J Biol Chem 263:12187-12189

Fisher CW, Caudle DL, Martin-Wixtrom CA, Quattrochi LC, Tukey RH, Waterman MR, Estabrook RW (1992) High-level expression of functional human cytochrome P450 1A2 in Escherichia coli. FASEB J 6:759-764

Fujii-Kuriyama Y, Taniguchi T, Mizukami Y, Saki M, Tashiro Y, Muramatsu M (1981) Construction and identification of a hybrid plasmid containing a DNA sequence complementary to phenobarbital-inducible cytochrome P-450 messenger RNA from rat liver. J Biochem 89:1869-1879

Fujita VS, Thiele DJ, Coon MJ (1990) Expression of alcohol-inducible rabbit liver cytochrome P-450 3a (P-450IIE1). DNA Cell Biol 9:111-118

Gonzalez FJ, Kasper CB (1982) Cloning of DNA complementary to rat liver NADPH-cytochrome c(P450) oxidoreductase and cytochrome P-450b mRNAs. J Biol Chem 257:5962-5968

Gonzalez FJ, Mackenzie PI, Kimura S, Nebert DW (1984) Isolation and characterization of full-length mouse cDNA and genomic clones of 3-methylcholanthrene-inducible cytochrome P1-450 and P3-450. Gene 29:281-292

Gonzalez FJ, Aoyama T, Gelboin HV (1990) Activation of promutagens by human cDNA-expressed cytochrome P450s. Mutation and the Environment, part B, 77-86

Graham FL, van der Eb A (1973) A new technique for the assay of infectivity of human adenovirus 5 DNA. J Virol 52:456-467

Hardwick JP, Gonzalez FJ, Kasper CB (1983) Cloning of DNA complementary to cytochrome P-450 induced pregnonolone 16a-carbonitrile: characterization of its mRNA, gene, and induction response. J Biol Chem 258:10182-10186

Kempf A, Nef P, Meyer UA (1991) Bacterial expression of spectrally active rat cytochrome P-450 2G1 (P-450 olf). 3rd International ISXX Meeting, Amsterdam

Kimura S, Smith HH, Hankinson N, Nebert DW (1987) Analysis of two benzo[a]pyrene-resistance mutants of the mouse hepatoma Hepa-1 P1-450 gene via cDNA expression in yeast. EMBO J 6:1929-1933

Larson JR, Coon MJ, Porter TD (1991) Alcohol-inducible cytochrome P-450IIE1 lacking the hydrophobic NH2-terminal segment retains catalytic activity and is membrane-bound when expressed in Escherichia coli. J Biol Chem 266:7321-7324

Negishi M, Swan DC, Enquist LW, Nebert DW (1981) Isolation and characterization of a cloned DNA sequence associated with the murine Ah locus and 3-methylcholanthrene-induced form of cytochrome P-450. Proc Natl Acad Sci USA 78:800-804

Nouso K, Thorgeirsson SS, Battula N (1992) Stable expression of cytochrome P450IIE1 in mammalian cells: metabolic activation of nitrosodimethylamine and formation of adducts with cellular DNA. Cancer Res 52:1796-1800

Oeda K, Sakaki T, Ohkawa H (1985) Expression of rat liver cytochrome P-450 MC cDNA in Saccharomyces cerevisiae. DNA 4:203-210

Pellicer A, Robins D, Wold B, Sweet R, Jackson J, Lowy I, Roberts JM, Sim GK, Silverstein S, Axel R (1980) Altering genotype and phenotype by DNA-mediated gene transfer. Science 209:1414-1422

Puga A, Raychaudhuri B, Kalman S, Zhang YH, Nebert DW (1990) Stable expression of mouse CYP1A1 and human CYP1A2 cDNAs transfected into mouse hepatoma cells lacking detectable P450 enzyme activity. DNA and Cell Biol 9:425-432

Renaud JP, Cullin C, Pompon D, Beaune P, Mansuy D (1990) Expression of human liver cytochrome P-450 IIIA4 in yeast: a functional model for the hepatic enzyme. Eur J Biochem 194:889-896

Sakaki T, Akiyoshi-Shibata M, Yabusaki Y, Manabe K, Murakami H, Ohkawa H (1991) Progesterone metabolism in recombinant yeast simultaneously expressing bovine cytochrome P450c17 (CYP17A1) and P450c21 (CYP21B1) and yeast NADPH-P450 oxidoreductase. Pharmacogenetics 1:86-93

Schmalix WA, Mäser H, Kiefer F, Reen R, Wiebel FJ, Gonzalez FJ, Seidel A, Glatt HR, Greim H, Doehmer J (1993) Stable expression of human cytochrome P450 1A1 cDNA in V79 Chinese hamster cells and metabolic activation of benzo[a]pyrene. Eur J Pharmacol Env Toxicol Pharmacol 248:251-261

Shimizu T, Sogawa K, Fujii-Kuriyama Y, Takahashi M, Ogoma Y, Hatano M (1986) Expression of cytochrome P-450d by saccharomyces cerevisiae. FEBS 207:217-221

States JC, Quan TH, Hines RN, Novak RF, Runge-Morris M (1993) Expression of human cytochrome P450 IA1 in DNA repair deficient and proficient human fibroblasts stably transformed with an inducible expression vector. Carcinogenesis, in press

Thompson LH, Wu RW, Felton JS (1991) Introduction of cytochrome P450IA2 metabolic capability into cell lines genetically matched for DNA repair proficiency/deficiency. Proc Natl Acad Sci USA 88:3827-3831

Trinidad AC, Wu RW, Thompson LH, Felton JS (1991) Expression of mouse cytochrome P450IA1 cDNA in repair-deficient and repair-proficient CHO cells. Mol Carcinog 4:510-518

Waxman DJ, Lapenson DP, Aoyama T, Gelboin HV, Gonzalez FJ, Korzekwa K (1991) Steroid hormone hydroxylase specificities of eleven cDNA-expressed human cytochrome P450s. Arch Biochem Biophysics 290:160-166

Wölfel C, Platt KL, Dogra S, Glatt HR, Wächter F, Doehmer J
(1991) Stable expression of rat cytochrome P450IA2 cDNA and
hydroxylation of 17b-estradiol and 2-aminofluorene in V79
Chinese hamster cells. Mol Carcinog 4:489-498

Zuber MX, Simpson ER, Waterman MR (1986) Expression of bovine
17a-hydroxylase cytochrome P-450 cDNA in nonsteroidogeni(COS
1) cells. Science 234:1258-1261

Zurbriggen B, Boehlen E, Sanglard D, Kappeli O, Fiechter A
(1989) Controlled expression of heterologous cytochrome P-
450c cDNA in Saccharomyces cerevisiae I. Construction and
expression of a complete rat cytochrome P-450c cDNA. Int
Biotechnol 9:255-272

In Vitro Systems as Analytical Tools for Understanding and Predicting Drug Metabolism In Vivo

Johannes Doehmer
Institut für Toxikologie und Umwelthygiene
Technische Universität München
Lazarettstrasse 62
80636 München
Germany

Introduction

One of the most classical and well known *in vitro* system is the Ames test, making use of a liver homogenate providing the metabolic competence needed for the detection of mutagenic reactive metabolites in Salmonella bacterium. Liver cells were isolated and have been tried over many years to maintain their differentiation status. The application of established cell lines for metabolism related studies was developed. The importance of cytochromes P450 became evident in the meantime, resulting into studies based on individual purified and reconstituted cytochromes P450. Finally, genetically engineered cell systems for metabolism studies were created (Doehmer 1993).

By now, there is wide variety of *in vitro* systems for metabolism studies. The crucial criteria for selecting the right system depends on its properties needed to answer an experimental question. It is therefore important to know the details of the different *in vitro* system, in order to understand, why one system may serve you well and the other not.

NATO ASI Series, Vol. H 90
Molecular Aspects of Oxidative Drug Metabolizing Enzymes
Edited by E. Arınç, J. B. Schenkman and E. Hodgson
© Springer-Verlag Berlin Heidelberg 1995

The wrong choice of an *in vitro* system may be the reason for artifacts, false negative or false positive results. In Fig. 2, several in vitro systems are listed and discussed for their advantages and disadvantages.

Comparison between primary hepatocytes and genetically engineered V79 cells

An example is given for phenacetin-*O*-deethylation in rat primary hepatocytes and V79 cells genetically engineered for rat CYP1A2 (Jensen et al. 1993). Table 1 shows enzyme kinetic data as obtained from rat hepatocytes and genetically engineered cells. The difference between both systems becomes obvious, when these data are presented as an Eadie-Hofstee plot (Fig. 1). A biphasic curve results from hepatocytes due to low affinity sites, whereby a linear relation is found in genetically engineered V79 cells. Consequently, for enzyme kinetic studies, genetically engineered cells defined for the particular cytochrome to be tested has to be preferred.

In contrast to genetically engineered cells, primary hepatocytes are expected to exhibit the whole in vivo spectrum of cytochromes P450. However, the application of primary hepatocytes is restricted to more qualitative studies - as shown by the above mentioned example - due to an immediate loss of a large number of cytochromes P450 and a constantly changing cytochrome P450 enzyme profile upon cultivation (Rogiers et al. 1990). Beside these problems, reasonable quality human primary hepatocytes are not readily available.

Table 1

Comparative study on phenacetin O-deethylation in genetically engineered V79MZr1A2 cells and freshly prepared hepatocytes from rats (Jensen et al. 1993)

	V_{max_1} (pmol/min/10^6 cells)	K_{m_1} (µM)
Hepatocytes		
Rat 1	13.46 ±1.66	0.36 ±0.14
Rat 2	18.30 ±1.71	0.20 ±0.10
Rat 3	7.43 ±0.63	0.17 ±0.04
Rat 4	33.30 ±2.60	0.20 ±0.06
V79MZr1A2 cells		
Assay 1	20.27 ±0.62	0.84 ±0.06
Assay 2	11.21 ±0.49	0.90 ±0.09
Assay 3	15.24 ±0.48	1.27 ±0.09
Assay 4	12.88 ±0.51	0.96 ±0.08

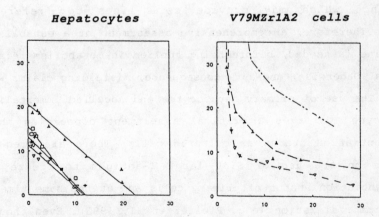

Fig. 1: Eadie-Hofstee plots of the kinetics of phenacetin-O-deethylation (Jensen et al. 1993).

Discussion

There is no such thing like an ultimate or best *in vitro* system for studying xenobiotic metabolizing enzymes. Each of these systems has its advantages and limitations (Fig. 2). It needs a full understanding of the biology of these systems and an awareness for system intrinsic advantages and limitations. Only then, the proper choice for the most appropriate *in vitro* system can be made. There will never be any *in vitro* system available, which is a substitute for the *in vivo* situation. Already by definition, *in vitro* systems are made by technical means. There is no technology for developing *in vitro* systems, which display all features given in an *in vivo* system, including interindividual variations. Established cell lines, like V79 or HepG2 may vary extensively in their metabolic make up and physiological features, due to continued subcloning. There are probably as many HepG2 cell lines around as there are laboratories working with this cell line. Variant HepG2 cell lines exist with inducible and constitutive CYP expression, with fast and slow growth. So far, there is no central unit in existence which may provide cell lines for reference reasons.Therefore, a comprehensive assessment of a established cell line is needed, before being applied in metabolism related studies. Otherwise, and as a consequence, misleading claims will result.The use of primary hepatocytes and socalled immortalised hepatocytes deserves a critical assessment concerning their differentiation status as cultured cells. There are frequent reports on the use of SV40 large T-antigen to freeze the differentiation status of primary cells and at the same time to cause immortalization (e.g. Pfeifer et al. 1993). Even though,

SV40 T-antigen transformed hepatocytes continue to produce albumin and hepatocyte specific cytokeratin type 18, which changes to the expression of cytokeratin type 19 in late passages. However, type 19 is bile duct specific marker (Moll et al. 1982). These cells have also lost cytochrome P450 activities to the extent, that no CYP mRNAs nor CYP protein can be detected (Pfeifer et al. 1993). Amazingly, these cells were claimed to contain CYPs, because they were found to be sensitive in cytotoxicity studies when exposed to chemicals known to be metabolically activated by cytochromes P450. However, toxicity data were not shown (Pfeifer et al. 1993). This is just one example, which demonstrates, that the area of *in vitro* systems appears to be prone to false claims and overstatements.

**IN VITRO SYSTEMS
AS TOOLS FOR METABOLISM STUDIES**

ADVANTAGES / LIMITATIONS

Primary Hepatocytes

Close but not identical to in vivo
Rapid loss of CYP activities
Uncontrolled loss of CYP activities
Inducible CYP activities
Limited growth capacity
Limited availability of human cells

Established Cell Lines

Unlimited growth capacity
Biological endpoints
CYP activities restricted or not present

Genetically Engineered Cell Lines

Unlimited growth capacity
Biological endpoints
Highly defined CYP activity

Fig. 2:   Advantages and limitations of *in vitro* systems.

Unfortunately, a discussion on ethical issues and alternatives to animal experimentation linked to *in vitro* systems has clouded a critical and scientific judgment on the possibilities given by *in vitro* systems.

An intelligent combined application of all available *in vitro* systems should form the basis for a strategy on conducting any relevant metabolism related study. This approach allows to take advantages and to overcome shortcomings and disadvantages of the various *in vitro* system.

References

Doehmer J (1993) V79 Chinese hamster cells genetically engineered for cytochrome P450 and their use in mutagenicity and metabolism studies. Toxicol 82:105-118

Jensen KG, Loft S, Doehmer J, Poulsen HE (1993) Metabolism of phenacetin in V79 Chinese hamster cell cultures expressing rat liver cytochrome P450 1A2 compared to isolated rat hepatocytes. Biochem Pharmacol 45:1171-1173

Moll R, Franke WW, Schiller DL (1982) The catalog of human cytokeratins: Patterns of expression in normal epithelia, tumors and cultured cells. Cell 31:11-24

Pfeifer AMA, Cole KE, Smoot DT, Weston A, Groopman JD, Shields PG, Vignaud JM, Juillerat M, Lipsky MM, Trump BF, Lechner JF, Harris CC (1993) Simian virus 40 large tumor antigen-immortalized normal human liver epithelial cells express hepatocyte characteristics and metabolize chemical carcinogens. Proc Natl Acad Sci USA 90:5123-5127

Rogiers V, Vandenberghe Y, Callaerts A, Verleye G, Cornet M, Mertens K, Sonck W, Vercruysse A (1990) Phase I and phase II xenobiotic biotransformation in cultures and co-cultures of addult rat hepatocytes. Biochem Pharmacol 40:1701-1708

The Role of Oxidative Enzymes in the Metabolism and Toxicity of Pesticides

Ernest Hodgson, Randy L. Rose, Nancy H. Adams,
Nora J. Deamer, Mary Beth Genter,
Krishnappa Venkatesh, and Patricia E. Levi
Department of Toxicology,
Box 7633, North Carolina State University,
Raleigh NC 26795 USA

Introduction

Xenobiotics, including pesticides (Hodgson & Levi,1992, Levi & Hodgson, 1991), are metabolized by many enzymes including: cytochrome P450s (P450); flavin-containing monooxygenases (FMO); prostaglandin synthetase; molybdenum hydroxylases; alcohol dehydrogenase; aldehyde dehydrogenase; esterases; and a variety of transferases, particularly the glutathione transferases. Of these P450 appears to be the most important, followed by the FMO. It should also be remembered that pesticides may serve not only as substrates for these enzymes but, particularly in the case of P450, may serve also as inhibitors and/or inducers. The toxicological implications of these multiple roles are important and are illustrated by our studies of methylenedioxyphenyl (MDP) compounds, of the herbicide synergist tridiphane and the herbicide, dichlobenil, of metabolism in target tissues and portals of entry, and of insect resistance to insecticides.

Methylenedioxyphenyl (MDP) Compounds

Piperonyl butoxide (PBO) and sesamex (SES) have been used as synergists with pyrethroid and carbamate pesticides, and isosafrole (ISO) and safrole (SAF) are found in many common foods of plant origin. ISO, SAF, and other plant MDP compounds may act as natural synergists for insecticidal chemicals produced by plants (Berenbaum and Neal, 1985).

MDP compounds exert several different biological effects. SAF has been shown to be a liver carcinogen in rodents at high doses (Miller and Miller, 1983, Iannides et al.,

NATO ASI Series, Vol. H 90
Molecular Aspects of Oxidative Drug Metabolizing Enzymes
Edited by E. Arınç, J. B. Schenkman and E. Hodgson
© Springer-Verlag Berlin Heidelberg 1995

1981, Homberger et al., 1961, Long et al., 1963). MDP compounds affect multiple enzyme pathways (Hodgson and Philpot, 1974, Goldstein et al., 1973), including the cytochrome P450 dependent monooxygenase system. The effect of MDP compounds on P450s is biphasic, with an initial inhibition of activity followed by an increase above control levels (Philpot and Hodgson, 1971/72, Kamienski and Murphy, 1972). The inhibitory effect of MDP compounds has been attributed to the formation of a stable metabolite complex between the heme iron of the P450 and the carbene species formed when water is cleaved from the hydroxylated methylene carbon of the MDP (Dahl and Hodgson, 1979). MDP exposure induces several P450 isozymes not found in detectable quantities in unexposed animals (Lewandowski et al., 1990).

Several studies have been published regarding the effects of MDP compounds on mammalian liver enzymes (Lewandowski et al., 1990, Fujii et al., 1970, Yeowell et al., 1985, Thomas et al., 1983, Wagstaff and Short, 1971, Cook and Hodgson, 1985, 1986). Cook and Hodgson (1985) showed that ISO increased the level of Ah receptor in mice but did not displace receptor-bound 3-methylcholanthrene (3MC) or 2,3,7,8-tetrachlorodibenzo-p-dioxin (TCDD), both of which are known to interact with the Ah receptor to induce P450s. Cook and Hodgson (1986) also demonstrated that there was comparable induction of P450 in a congenic strain of C57 mice which lacked a functional Ah receptor and in Ah receptor proficient C57 mice. MDP compounds have been reported to induce P450 isozymes 1a-2 and 2b-10 in the mouse (Lewandowski et al., 1990). Some investigators have reported that P450 1a-1 is also induced in mice by MDP compounds (Marcus et al., 1990). In induction studies in rats using 4-n-alkyl MDPs, the length of the alkyl side chain affected which P450 isozymes were preferentially induced, with the six-carbon side chains favoring the maximal induction of 1A1 enzyme activity and the two carbon side chains favoring 2B1, the rat gene most similar to mouse 2b-10 (Marcus et al., 1990). In another study, MDP compounds with electron-donating side chains were reported to be P450 inducers, while MDP compounds with electron-withdrawing groups were not (Murray et al., 1985).

Regulation of cytochrome P450 isozymes 1a-1, 1a-2, and 2b-10 by MDP compounds was studied in our laboratory by measuring levels of mRNA, protein and enzyme activity in hepatic tissue from C57BL/6 (Ah$^+$) and DBA/2 (Ah-) mice dosed with ISO or PBO (Adams et al., 1993, 1993a). Increases in 1a-2 and 2b-10 were observed for ISO and PBO in both strains of mice, suggesting an Ah receptor-independent mechanism for induction of these isozymes; 1a-1 induction, however, was seen only in C57 mice. Piperonyl butoxide was the more potent inducing agent in both strains. In C57 mice treated with five dose levels of PBO, induction of 1a-1 mRNA, protein, and enzyme activity were seen at doses equal to or greater than 104 mg/kg, but were not detected at lower doses. With ISO, however, induction of 1a-1 mRNA was observed only at the highest dose tested (400 mg/kg); moreover, neither 1a-1 protein nor increased 1a-1 associated enzymatic activity was seen at this dose. Dose-response studies showed maximum inducible levels for 1a-2 and 2b-10 protein, beyond which the mRNAs continued to increase while the protein levels remained constant.

Further studies of the induction of the P450 isozymes 1a-1, 1a-2, and 2b-10 were carried out (Adams 1993, 1993a) using four MDP compounds, SAF, ISO, PBO, and SES, and the non-methylenedioxyphenyl analog of SAF, allyl benzene (AB), in male C57BL/6N mice. P450 1a-1 was not detected in control animals, and was induced by SES and PBO, with SES inducing higher levels of 1a-1 protein than PBO. P450 1a-2 mRNA was detected in the livers of control animals and was also increased by all MDP compounds (SES > PBO ≅ ISO > SAF), although SAF treatment did not increase 1a-2 protein. P450 2b-10 mRNA and protein, not detected in untreated animals, were also increased by all MDP compounds (PBO > SES ≅ ISO > SAF). AB treatment did not induce detectable levels of 1a-1, 1a-2, or 2b-10, suggesting that, in this series of compounds, the methylenedioxyphenyl moiety is important in induction.

Tridiphane

The herbicide synergist, tridiphane, (2-(3,5-dichlorophenyl)-2-(2,2,2-trichloroethyl) oxirane) is registered as a postemergent herbicide and is used as a herbicide synergist in conjugation with atrazine. The ability of tridiphane to synergize s-triazine, α-chloroacetamide and thiocarbamate herbicides in plants (Lamoureux and Rusness, 1986, and Dionigi and Dekker, 1990) and the insecticide diazinon in insects (Lamoureux and Rusness, 1987) has been attributed to its ability to inhibit glutathione S-transferases. Tridiphane is also known to be a peroxisome proliferator and to induce epoxide hydrolase in rodents (Moody and Hammock, 1987). It is also an inhibitor of P450 (Moreland et al., 1989), appearing to be relatively specific for P450 2b10.

With hepatic microsomes isolated from mice pretreated with phenobarbitol (PB) as well as with purified P450 isozymes, tridiphane efficiently inhibits demethylase and deethylase activities catalyzed by PB-induced P450 isozymes, i.e., I_{50} values of approximately 4.0 μM (Moreland et al., 1989). However, tridiphane is a weak inhibitor of enzyme activities catalyzed by hepatic microsomes isolated from 3-methylcholanthrene (3-MC)-pretreated mice. Type I binding spectra were obtained with hepatic microsomes isolated from uninduced mice (K_s = 12.6 μM) and PB-induced mice (K_s 1.0 μM), and with a purified enzyme isolated from PB-induced mice (K_s = 1.26 μM). However, tridiphane does not form difference spectra with microsomes isolated from 3-MC-treated mice or with a purified P450 enzyme isolated from 3-MC-treated mice. (Table 1) Inhibition of p-nitroanisole O-demethylase suggested that tridiphane acted as a competitive inhibitor of enzyme activity and produced a K_m of 440 μM with a K_i between 0.72 and 1.04 μM.

TABLE 1

Effects of Tridiphane on the Activity of Purified Mouse Hepatic Cytochrome P450s[a]

Source of P450	Substrate		
	p-Nitroanisole[b]	Benzphetamine[c]	Ethoxyresorufin[d]
Phenobarbital-treated			
Control	9.3 ± 0.4	4.1 ± 0.1	0.22 ± 0.02
10 μM tridiphane	3.4 ± 0.4 (63)	1.5 ± 0.1 (63)	0.12 ± 0.03 (45)
40 μM tridiphane	2.7 ± 0.6 (71)	0.8 ± 0.1 (80)	0.03 ± 0.02 (86)
3-Methylcholanthrene-			
treated control	6.0 ± 0.3	1.2 ± 0.1	2.97 ± 0.08
10 μM tridiphane	4.9 ± 0.1 (18)	1.0 ± 0.1 (17)	2.87 ± 0.02 (3)
40 μM tridiphane	4.0 ± 0.3 (33)	0.7 ± 0.1 (42)	2.55 ± 0.22 (14)

[a]Data are presented as arithmetic average of activity ± SD obtained from three replications. Percentage inhibition is shown parenthetically.
[b]nmol p-nitrophenol produced/nmol P450/min.
[c]nmol formaldehyde produced/nmol P450/min.
[d]nmol resorufin produced/nmol P450/5min.

Studies in our laboratory (Levi et al., 1992), using isozyme specific activities and antibodies to P450 isozymes have shown that tridiphane is also an inducer of P450, although the isozyme specificity for induction is quite different to that for inhibition. Like other peroxisome proliferators, tridiphane induces P450 4a-1. Male C57BL/6N mice were given tridiphane, 250 mg/kg ip, for 3 days. Liver weight and P450 content were increased after tridiphane treatment, and SDS-PAGE showed an increase in microsomal proteins in the 50-60 kDa range. No significant increases in enzymatic activities were observed with the substrates benzphetamine, p-nitroanisole, benzo(a)pyrene, or ethoxyresorufin. There was, however, a 10-fold elevation in microsomal hydroxylation of lauric acid, an activity specifically associated with induction of P450 4A1 (Table 2) Western Blot analysis using an antibody specific for P450 4A1 showed a dramatic increase in P450 4A1.

TABLE 2

Effect of *In Vivo* Tridiphane Treatment on Selected Microsomal P450 Enzyme Activities

	Activities			
	per mg protein		per nmol P450	
Substrate	Control	Tridiphane	Control	Tridiphane
p-Nitroanisole	2.22 ± 0.37	2.21 ± 0.67	4.18	2.41
Benzphetamine	8.18 ± 0.36	12.3 ± 0.45	15.4	13.3
Benzo(a)pyrene	1.03 ± 0.15	2.03 ± 0.25	1.94	2.21
Ethoxyresorufin	0.09 ± 0.12	0.27 ± 0.05	0.18	0.29
Lauric Acid	1.85 ± 0.57	18.4 ± 4.08	3.49	20.0

Dichlobenil

2,6-Dichlorobenzonitrile (Dichlobenil, Casoron) is an herbicide with numerous terrestrial and aquatic uses (USEPA, 1987). This compound is among those comprising a small but growing list of chemicals which cause degeneration of the olfactory mucosa of the nasal cavity following systemic (i.e. non-inhalation) routes of exposure (Table 3). Intraperitoneal administration of a single 12 mg/kg dose of dichlobenil results in degeneration of the olfactory mucosa of the dorsal medial nasal cavity in the mouse (Brandt et al., 1990), sparing the epithelium of the more ventral and lateral airways. The olfactory toxicity of dichlobenil is dependent upon cytochrome P450 activity (Brandt et al., 1990; Eriksson and Brittebo, 1991), can be enhanced by administering a glutathione-depleating agent, and is mitigated by administration of a N-acetylcysteine (Brittebo et al., 1992). Preliminary data from this lab suggest that cytochrome P450 2E1 may be the isozyme involved in the bioactivation of dichlobenil. The rationale for this observation is two-fold. First, the distribution of the dichlobenil-induced olfactory epithelial lesion coincides with the distribution of P450 2E1 in the nasal mucosa (Deamer et al., 1993a). Secondly, inhibitors of P450 2E1, namely diethyldithiocarbamate (Deamer et al., 1993a) and diallyl sulfide (unpublished observation), also inhibit the olfactory toxic effects of dichlobenil. Interestingly, we have also demonstrated that microsomal epoxide hydrolase is markedly reduced in the regions of the nasal cavity affected by dichlobenil; therefore we hypothesize that localized high levels of a particular cytochrome P450 isozyme (e.g. 2E1) coupled with a paucity of detoxication capability results in the site-specific generation of a reactive epoxide intermediate which causes tissue damage in the dorsal medial regions of the nasal cavity (Genter et al., submitted).

TABLE 3

Compounds Toxic to the Olfactory Mucosa Following Non-Inhalation Exposure

Compound	Route	Reference
ß,ß'-iminodipropionitrile	i.p.	Genter et al., 1992
2,6-dichlorothiobenzamide	i.p.	Brittebo et al., 1991
2,6-dichlorobenzamide	i.p.	Brittebo et al., 1991
2,6-dichlorobenzonitrile	i.p., i.v	Brandt et al., 1990
bromobenzene	i.v., i.p.	Brittebo et al., 1990
phenacetin	diet	Bogdanffy et al., 1989
dihydropyridines	i.p.	Reed et al., 1989
3-methylindole	i.p.	Turk et al., 1986, 1987
N-nitrosodiethylamine	i.v.	Jensen and Sleight, 1987
	i.p.	Brittebo et al., 1981
diethyldithiocarbamate	i.p.	Genter et al., unpublished
disulfiram	i.p.	Genter et al., unpublished
methimazole	i.p., gavage	Genter et al., unpublished

Given the potential for human exposure to an agricultural chemical such as dichlobenil (which is applied at rates ranging from 1.4-20.0 lb of active ingredient/acre (USEPA, 1987)), the effects on the olfactory system resulting from a more relevant route of exposure, namely dermal administration were examined. Olfactory system damage resulting from a single or repeated (designed to mimic an acute exposure or that which might be expected following a five day work week for humans) dermal exposure to dichlobenil was assessed using nasal cavity histopathology and a quantitative analysis of the astroglial response to injury using an ELISA (O'Callaghan, 1991) for glial fibrillary acidic protein (GFAP). Dichlobenil caused olfactory mucosal damage following a single or multiple dermal applications of dichlobenil in acetone at treatment doses of 50 mg/kg and greater. Mice receiving 12.5 mg/kg dichlobenil i.p. had extensive olfactory mucosal damage, involving primarily the mucosa adjacent to the dorsal medial airways, sparing the more ventral and lateral areas (Deamer et al., 1993b), as has previously been described for dichlobenil (Brandt et al., 1990) and IDPN (Genter et al., 1992). The most severe damage in dermally-treated mice was observed following a single 200 mg/kg application. The toxicity did not appear to be cumulative, as doses which did not cause olfactory mucosal damage following a single application were also not toxic following 5 consecutive daily applications. GFAP levels in the olfactory bulb were significantly elevated following administration of 150 and 200 mg/kg dichlobenil dermally as well as 12.5 mg/kg i.p. (Table 4). The increase in olfactory bulb GFAP was greatest in the mice receiving dichlobenil by ip injection, correlating with the increased severity of olfactory epithelial damage in these animals compared to those exposed by dermal application (Deamer et al., 1993b).

These studies confirm previous findings (Brandt et al., 1990; Eriksson and Brittebo, 1991; Brittebo et al., 1992) indicating that exposure to dichlobenil is associated with olfactory system damage in the C57Bl mouse. These studies extend previous findings in that we demonstrate that administration of dichlobenil by a more relevant route of exposure for an agricultural chemical also results in olfactory epithelial damage. The relevance of these findings to human health remains to be established. Whether dichlobenil is readily absorbed across human skin is currently unknown, and data are also sparse as to the exact mechanism of olfactory system damage. Available data suggest that dichlobenil is activated within the nasal mucosa to its olfactory toxic metabolite. The degree of olfactory function impairment and the effects of dichlobenil-induced olfactory system damage on other nervous system functions have not yet been evaluated.

TABLE 4

Olfactory Olfactory Bulb GFAP Levels Following Acute Exposure to Dichlobenil[1,2]

Dichlobenil dose	Route*	GFAP (SEM)
(mg/kg)		mg/mg protein
0	D	0.65 (0.03)
10	D	0.66 (0.05)
25	D	0.74 (0.01)
50	D	0.82 (0.03)
100	D	0.57 (0.09)
150	D	1.04 (0.09)**
200	D	1.21 (0.11)**
12.5	i.p.	2.30 (0.31)**

[1]Each value represents the mean of two studies in which three mice were treated per group;
[2]From Deamer et al., 1993b.
*D = dermal; i.p. = intraperitoneal
**Significantly greater than controls, Students t-test (2-tailed), p < 0.05

Consistent with the hypothesis that damage to the olfactory mucosa can have effects within the central nervous system (CNS), it is important to note that relatively minor damage to the olfactory mucosa is associated with significant increases in glial fibrillary acidic protein in the olfactory bulb of the brain. For example, the acute olfactory epithlelial damage following dermal application of dichlobenil results in epithelial folding and sensory cell loss in the dorsal medial meatus of the mouse nasal cavity. Though the damage appears histologically relatively minor, this lesion is not completely repaired (i.e. the epithelium which replaces most of the damaged olfactory epithelium is a respiratory epithelium) and significant increases in olfactory bulb GFAP occur (Deamer et al., 1993b). Based on similar data following i.p. administration of IDPN, the CNS effects are likely permanent (Genter et al., 1992). Future studies to establish the functional changes associated with the olfactory system damage resulting from dichlobenil exposure are needed.

Metabolism in Target Tissues and Portals of Entry

The general methods for examination of the importance of different oxidative pathways and the metabolism of the insecticide phorate have been discussed in a preceding paper (Hodgson and Levi, 1992) along with the expression of the flavin-containing monooxygenase in target tissues and portals of entry. We have also studied additional pesticide substrates in mouse skin (Venkatesh et al., 1992). In these studies the P450 content, the cytochrome c reductase activity, the metabolism of a variety of P450 pesticide substrates, and the presence and role of FMO in xenobiotic metabolism were studied in skin microsomes and compared to those of liver. The P450 content of skin as determined by CO-dithionite-reduced minus CO-oxidized spectra was approximately 6.8% of the liver P450 content. By comparison, cytochrome c reductase activity in skin microsomes was high, being equivalent to approximately one-third of the liver microsomal enzyme activity. Skin microsomes metabolized several known P450 substrates and, depending upon the substrated used, the specific activity ranged from 2.5 to 13.4% of the corresponding rates seen in liver microsomes. Skin microsomes exhibited the highest enzymatic activity with benzo(a)pyrene and ethoxyresorufin, moderate activity with parathion and aldrin (Table 5) and low activity with benzphetamine and ethoxycoumarin. Skin microsomes also metabolized the triazine herbicides atrazine, simazine, and terbutryn, with the activity being 2 to 5% of the liver microsomal activity. (Table 6) FMO activity in skin microsomes with thiobenzamide and methimazole as substrates ranged from 10 to 20% of the liver FMO activity. Immunohistochemical studies using antibodies to mouse liver FMO showed localization primarily in the epidermis. Additional studies using pig skin showed a similar distribution pattern. Antibodies developed to mouse liver FMO and the constitutive liver P450 isozyme, 1A2, showed cross-reactivity on Western blots with proteins in skin microsomes that appeared identical to the cross-reacting proteins present in liver microsomes. The relative contribution of P450 and FMO in mouse skin to the sulfoxidation of phorate was investigated and compared to that of liver microsomes.

Several procedures were employed to selectively inhibit either P450 or FMO so that the role of each monooxygenase system in the absence of the other system could be determined. In liver microsomes, P450 was responsible for 68 to 85% of the phorate sulfoxidation activity. In contrast, in skin microsomes 66 to 69% of the phorate sulfoxidation activity was due to FMO, while P450 was responsible for the remainder of the activity. Thus, although the overall phorate sulfoxidation rate in mouse skin microsomes was only 3 to 4% of the rate seen in liver, FMO appears to assume a greater relative role to P450 in the metabolic processes in skin.

TABLE 5

Cytochrome P450 Content, Cytochrome c Reductase Activity and Cytochrome
P450-Dependent Monooxygenation of Pesticides in
Skin Microsomes and Liver Microsomes

Assay	Activity in		
	Liver	Skin	Ratio (%) skin/liver
P450 content (pmol/mg)	758 ± 97	52 ± 11	7
Cytochrome c reductase (nmol/min/mg)	62 ± 9	23 ± 4	37
Parathion (pmol paraoxon/min/mg)	680 ± 240	48 ± 24	7
Aldrin epoxidation (nmol/min/mg)	2.5 ± 0.3	0.14 ± 0.02	6

[a]Values are means ± standard deviation of 9 to 12 determinations. Enzyme activities are expressed per milligram of microsomal protein.

TABLE 6

Metabolism of Triazine Herbicides by Mouse Liver and Skin Microsomes[a]

Product formed (pmol/min/mg protein)

Tissue	Atrazine		Terbutryn		Simazine
	-ipr[a]	-et[b]	-t bu[c]	-et[d]	-et[e]
Liver	249 ± 15	269 ± 36	217 ± 31	31	252 ± 18
Skin	ND	15 ± 1.1	7.5 ± 1.7	1.7	5.2 ± 0.7

[a]Values are means ± standard deviation of three independent determinations. ND, not detected.
[a]2-Chloro-4-amino-6-(isopropylamino)-s-triazine.
[b]2-Chloro-4-(ethylamino)-6-amino-s-triazine.
[c]2-(Methylthio)-4-(ethylamino)-6-amino-s-triazine.
[d]2-(Methylthio)-4-amino-6-(tert-butylamino)-s-triazine.
[e]2-Chloro-4-(ethylamino)-6-amino-s-triazine.

Insect Resistance to Insecticides

Resistance of insects to insecticides is a worldwide problem and an increasing threat to the production of food and fiber. While many different mechanisms of resistance have been demonstrated, increased metabolism of the insecticide is the most common. Metabolic enzymes of primary importance include P450, glutathione S-transferases and esterases.

The genus *Heliothis* is arguably the most important insect pest in agriculture, many of its members can feed on a wide variety of food plants and resistance to both plant allelochemicals and sythetic organic chemicals is common within the group (Rose et al., 1991; Sparks, 1981). Most recently, the advent of pyrethroid resistant populations has been an area of increasing concern in Australia and the southeastern United States (Plapp and Campanhola, 1986; Leonard et al., 1988; Gunning et al., 1984).

Our most recent studies have involved the biochemistry and molecular biology of resistance in a field-collected resistant population of *Heliothis virescens*. This population, collected from an area where multiple pesticides had been in use, was subjected to additional selection pressure with the carbamate insecticide, thiodicarb. Bioassays conducted in the ninth generation demonstrated resistance levels to cypermethrin and thiodicarb of 90 and greater than 150-fold, respectively. Comparative biochemical assays conducted upon microsomal and cytosolic fractions from various tissues demonstrated increased levels of P450, esterases and glutathione S-transferases as compared to a susceptible laboratory population (Tables 7 & 8).

TABLE 7

P450 Content and Monooxygenase Activity in Resistant (R) and Susceptible (S) *Heliothis virescens* Midgut Tissue

Population	P450 (nmole/mg pr.)	Methoxyresorufin (nmole/min mg pr.)	Benzo(a)pyrene (f.u/min mg pr.)
Hebert (R)	1.07 ± .02	77.7 ± 12.6	1.84 ± 0.24
Wake (S)	0.29 ± .003	2.6 ± 0.3	0.07 ± 0.01

TABLE 8

Esterase[1] and Glutathione S-Transferase[2] Activity in Resistant (R) and Susceptible(s) *Heliothis virescens* Midgut Tissue

Population	α-Naphthyl acetate (μmole/min. mg pr.)	p-Nitrophenyl acetate (nmole/min. mg pr.)	CDNB (nmole/min. mg pr.)
Herbert (R)	362.3 ± 19.3	138.2 ± 0.7	1.01 ± 0.07
Wake (S)	105.4 ± 3.7	39.6 ± 1.1	0.62 ± 0.09

[1]Esterase substrates are: α-naphthyl acetate and p-Nitrophenyl acetate
[2]Glutathione S-transferase substate is CDNB (1-chloro-2,4 dinitrobenzene)

Modest increases in P450 content were accompanied by particularly high levels of activity for certain monooxygenase substrates, particularly for methoxyresorufin and benzo(a)pyrene. The former substrate had previously been shown to be associated with high monooxygenase activity in pyrethroid resistant strains of housefly (Lee and Scott, 1989) and diamondback moth (Hung and Sun, 1989).

The high metabolic activity observed in the resistant population of *Heliothis virescens* for both monooxygenase and esterase substrates suggested that biochemical assays might be utilized to effectively monitor resistance development in field populations. A microtiter plate assay, utilizing *p*-nitroanisole (a monooxygenase substrate) or *p*-nitophenyl acetate (an esterase substrate) has been developed which can effectively discriminate between resistant and susceptible individuals, thus enabling rapid detection of resistance based on these mechanisms.

In insects, recent progress has been made in the characterization of P450's at the molecular level. The first P450 gene sequence was identified in an insecticide resistant strain of housefly which had been induced with phenobarbital (Feyereisen et al., 1989). Two other P450 families have recently been isolated from *Drosophila*, one belonging to family 6 (Waters et al., 1992) and the other to family 4 (Ghandi et al., 1991). Comparisons of gene sequences isolated from resistant and susceptible *Drosophila* strains revealed the presence of a long terminal repeat of a transposable element in the susceptible strain. Features of this long terminal repeating element suggest that a posttranscriptional mechanism involving mRNA stability are involved in the regulation of this P450 gene (Waters et al., 1992). Other P450 seqeunces known to have been sequenced recently include those from the cockroach (family 4) (Bradfield et al., 1991), one from *Papilio polyxenes* (family 6) (Cohen et al., 1992) and five from a susceptible population of houseflies (Feyereisen, personal communication).

A cDNA library constructed in our laboratory from the resistant population was screened for P450 using an antibody derived from an insecticide resistant strain of *Drosophila* (courtesy of Larry Waters, Oak Ridge National Laboratory). The cDNA for an isozyme that is associated with this oxidation has been cloned and sequenced and its characteristics are currently under study.

Conclusions

Although the interaction of a number of pesticides with P450 and other oxidative and phase 2 enzymes is widely known, neither the toxicological implications nor many of the interactions are well understood. In order to advance the science of toxicology in the service of agriculture and public health, a more holistic approach will be necessary. It is not sufficient to know that a particular pesticide is a substrate for P450. Certain other questions must be answered: is the reaction a detoxication or an activation reaction; what is the mechanism of action and all of the products; what is the isozyme specificity; what is the organ specificity relative to portals of entry and target tissues; what is the relationship to other phase 1 enzymes; what is the relationship to phase 2 enzymes; how are these relationships changed by other xenobiotics. Moreover, it must also be determined whether the pesticide can act as an inducer or an inhibitor and, if so, what are the toxicological implications.

Acknowledgements:

Investigation carried out at North Carolina State University were supported in part by NIH grants number ES-00044 and ES-07046.

References

Adams NH, Levi PE, Hodgson E (1993) Regulation of Cytochrome P450 isozymes by methylenedioxyphenyl compounds. Chem-Biol Interactions 86:255-274

Adams NH, Levi PE, Hodgson E (1993a) Differences in induction of three P450 isozymes by piperonyl butoxide, sesamex, safrole and isosafrole. Pesticde Biochem and Physiol 46: (in press)

Berenbaum M, Neal JJ (1985) Synergism between myristicin and xanthotoxin, a naturally occuring plant toxicant. J Chem Ecology 11:1349-1358

Bogdanffy MS, Mazaika TJ, Fasano WJ (1989) Early cell proliferative effects of phenacetin on rat nasal mucosa. Toxicol Appl Pharmacol 1989; 98:100-112

Bradfield JY, Lee YH, Keeley LL (1991) Cytochrome P450 4 in a cockroach: molecular cloning and regulation by hypertrehalosemic hormone. Proc Natl Acad Sci USA 88:4558-4562

Brandt I, Brittebo EB, Feil VJ, Bakke JE (1990) Irreversible binding and toxicity of the herbicide dichlobenil (2,6-dichlorobenzonitrile) in the olfactory mucosa of mice. Toxicol Appl Pharmacol 103:491-501

Brittebo EB, Lofberg B, Tjalve H (1981) Sites of metabolism of N-nitrosodiethylamine in mice. Chem-Biol Interact 34:209-221

Brittebo EB, Eriksson C, Brandt I (1990) Activation and toxicity of bromobenzene in nasal tissues in mice. Arch Toxicol 64:54-60

Brittebo EB, Eriksson C, Feil VJ, Bakke J, Brandt I (1991) Toxicity of 2,6-dichlorothiobenzamide (Chlorthiamid) and 2,6-dichlorobenzamide in the olfactory nasal mucosa of mice. Fundam Appl Toxicol 17:92-102

Brittebo EB, Eriksson C, Brandt I (1992) Effects of glutathione-modulating agents on the covalent binding and toxicity of dichlobenil in the mouse olfactory mucosa. Toxicol Appl Pharmacol 114:31-40

Cohen MB, Schuler MA, Berenbaum MR (1992) A host-inducible cytochrome P450 from a host-specific caterpillar: molecular cloning and evolution. Proc Natl Acad Sci USA 89:10920-10924

Cook JC, Hodgson E (1985) The induction of cytochrome P-450 by isosafrole and related methylenedioxyphenyl compounds. Chem Biol Interact 54:299-315

Cook JC, Hodgson E (1986) Induction of cytochrome P-450 in congenic C57BL/6J mice by isosafrole: lack of correlation with the Ah locus. Chem Biol Interact 58:233-240

Dahl AR, Hodgson E (1979) The interaction of aliphatic analogs of methylenedioxyphenyl compounds with cytochromes P-450 and P-420. Chem Biol Interact 27:163-175

Deamer NJ, O'Callaghan JP, St Clair MB (1993) Olfactory system damage following dermal exposure to 2,6-dichlorobenzonitrile. Toxicologist 13:261

Deamer NJ, O'Callaghan JP, Genter MB (1993) Olfactory toxicity resulting from dermal application of 2,6-dichlorobenzonitrile (dichlobenil) in the C57Bl mouse. NeuroToxicology in press

Eriksson C, Brittebo EB (1992) Metabolic activation of the herbicide dichlobenil in the olfactory mucosa of mice and rats. Chem-Biol Interact 79:165-177

Fennel TR, Sweatman BC, Bridges JW (1980) The induction of hepatic cytochrome P-450 in C57 BL/10 and DBA/2 by isosafrole and piperonyl butoxide. A comparative study with other inducing agents. Chem Biol Interact 31:189-201

Feyereisen R, Koener JF, Farnsworth DE, Nebert DW (1989) Isolation and sequence of cDNA coding a cytochrome P450 from an insecticide-resistant strain of the house fly, Musca domestica. Proc. Natl Acad Sci USA 86:1465-1469.

Fujii K, Jaffe H, Bishop Y, Arnold E, Mackintosh D, Epstein SS (1970) Structure-activity relations for methylenedioxyphenyl and related compounds on hepatic microsomal enzyme function, as measured by prolongation of hexobarbital narcosis and zoxazolamine paralysis in mice. Toxicol Appl Pharmacol 16:482-494

Genter MB, Llorens J, O'Callaghan JP, Peele DB, Morgan KT, Crofton KM (1992) Olfactory toxicity of ß,ß'-iminodipropionitrile in the rat. J Pharmacol Exp Therap 263:1432-1439

Ghandi R, Varak E, Goldberg ML (1992) Molecular analysis of a cytochrome P450 gene of family 4 on the Drosophila X chromosome. DNA Cell Biol 11:397-404.

Goldstein JA, Hickman P, Kimbrough RD (1973) Effects of purified and technical piperonyl butoxide on drug-metabolizing enzymes and ultrastructure of rat liver. Toxicol Appl Pharmacol 26:444-453

Gunning RV, Easton CS, Greenup LR, Edge VE (1984) Pyretheroidresistance in *Heliothis armigera* (Hubner) (Lepidoptera: Noctuidae) in Australia. J.Econ Entomol 77:1283-1287.

Hodgson E, Levi PE (1992) The role of the flavin-containing monooxygenase (EC 1.14.13.8) in the metabolism and mode of action of agricultural chemicals. Xenobiotica 22:1175-1183

Hodgson E, Philpot RM (1974) Interactions of methylenedioxyphenyl (1,3-benzodioxole) compounds with enzymes and their effects on mammals. Drug Metabol Rev 3:231-301

Homberger F, Kelley T, Friedler G, Russfield AB (1961) Toxic and possible carcinogenic effects of 4-allyl-1,2-methylenedioxybenzene (safrole) in rats on deficient diets. Med Exptl 4:1-9

Hung CF, Sun CN (1989) Microsomal monooxygenases in diamondback moth larvae resistant to fenvalerate and piperonyl butoxide. Pestic Biochem Physiol 33:168-175.

Iannides C, Delaforge M, Parke DV (1981) Safrole: its metabolism, carcinogenicity and interactions with cytochrome P-450. Food Cosmet Toxicol 9:657-669

Jensen RK, Sleight SD (1987) Toxic effects of N-nitrosodiethylamine on nasal tissues of Sprague-Dawley rats and golden Syrian hamsters. Fundam Appl Toxicol 8:217-229

Kamienski FX, Murphy SD (1971) Biphasic effects of methylenedioxyphenyl synergists on the action of hexobarbital and organophosphate insecticides in mice. Toxicol Appl. Pharmacol. 18:883-894

Rose RL, Gould F, Levi PE, Hodgson E (1991) Differences in cytochrome P450 activities in tobacco budworm larvae as influenced by resistance to host plant allelochemicals and induction. Pestic Biochem Physiol 99:535-540

Rose RL, Gould F, Levi PL, Konno T, Hodgson E (1992) Resistance to plant allelochemicals in Heliothis virescens (Fabricius). Chapter 11 in Molecular Mechanisms of Insecticide (Mullin CA, Scott JG, eds) ACS Symposium Series No 505, Washington DC.

Sparks TC (1981) Development of insecticide resistance in Heliothis zea and Heliothis virescens in North America. Bull Entomol Soc 27:186-192

Thomas PE, Reik LM, Ryan DE, Levin W (1983) Induction of two immunochemically related rat liver cytochrome P-450 isozymes, cytochrome P-450c and P-450d, by structurally diverse xenobiotics. J Biol Chem 258:4590-4598

Turk MAM, Flory W, Henk W (1986) Chemical modulation of 3-methylindole toxicosis in mice; Effects on bronchiolar and olfactory mucosal injury. Vet Pathol 23:563-570

Turk MAM, Henk W, Flory W (1987) 3-Methylindole-induced nasal mucosal damage in mice. Vet Pathol 24:400-403

US Environmental Protection Agency (1987) Pesticide Fact Sheet (Dichlobenil). Office of Pesticide Programs. Washington DC 8 pp

Venkatesh K, Levi PE, Inman AO, Monteiro-Riviere NA, Misra R, Hodgson E (1992) Enzymatic and immunohistochemical studies on the role of cytochrome P450 and the flavin-containing monooxygenase of mouse skin in the metabolism of pesticides and other xenobiotics. Pestic Biochem Physiol 43:53-66

Wagstaff DJ, Short CR (1971) Induction of hepatic microsomal hydroxylating enzymes by technical piperonyl butoxide and some of its analogs. Toxicol Appl Pharmacol 19:54-61

Waters LC, Zelhof AC, Shaw BJ, Chang LY (1992) Possible involvement of the long terminal repeat of transposable element 17.6 in regulating expression of an insecticide resistance-associated P450 gene in Drosophila. Proc Natl Acad Sci USA 89:4855-4859

Whitten CJ, Bull DL (1970) Resistance to organophosphorus insecticides in tobacco budworms. J Econ Entomol 60:1492-1495

Yeowell HN, Linko P, Hodgson E, and Goldstein JA (1985) Induction of specific cytochrome P-450 isozymes by methylenedioxyphenyl compounds and antagonism by 3-methylcholanthrene. Arch Biochem Biophys 243:408-419

Lee SST, Scott, JG (1987) Microsomal cytochrome P450 monooxygenases in the housefly (Musca domestica L.): biochemical changes associated with pyrethroid resistance and phenobarbital induction. Pestic Biochem Physiol 35:1-10.

Leonard BR, Graves JB, Sparks TC, Pavloff AM (1988) Variation in resistance of field populations of tobacco budworm and bollworm (Lepidoptera: Noctuidae) to selected insecticides. J Econ Entomol 81:1521-1528.

Levi PE, Hodgson E (1990) Monooxygenations: interactions and expression of toxicity. pp 233-244 in Insecticide Action, From Molecule to Organism (Narahashi T, Chambers JE, eds) Plenum Press New York

Levi PE, Rose RL, Adams NH, Hodgson E (1992) Induction of cytochrome P450 4A1 in mouse liver by the herbicide synergist tridiphane. Pestic Biochem Physiol 44:1-19

Lewandowski M, Chui YC, Levi PE, Hodgson E (1990) Differences in induction of hepatic cytochrome P450 isozymes by mice in eight methylenedioxyphenyl compounds. J Biochem Toxicol 5:47-55

Long EL, Nelson AA, Fitzhugh OG, Hansen WH (1963) Liver tumors produced in rats by feeding safrole. Arch Pathol 75:595-607

Marcus CB, Wilson NM, Jefcoate CR, Wilkinson CF (1990) Selective induction of cytochrome P450 isozymes in rat liver by 4-n-alkyl methylenedioxybenzenes. Arch Biochem Biophys 277:8-16

Miller JA, Miller EC (1983) The metabolic activation and nucleic acid adducts of naturally occuring carcinogens: recent results with ethyl carbamate and the spice flavors safrole and estragole. Br J Cancer 48:1-13

Moreland DE, Novitzky WP, Levi PE (1989) Selective inhibition of cytochrome P450 isozymes by the herbicide synergist tridiphane. Pestic Biochem Physiol 35:42-49

Murray M, Wilkinson CF, Dube CE (1985) Induction of rat hepatic microsomal cytochrome P-450 and aryl hydrocarbon hydroxylase by 1,3-benzodioxole derivatives. Xenobiotica 15:361-369

O'Callaghan, JP (1991) Quantification of glial fibrillary acidic protein: Comparison of slot-immunobinding assays with a novel sandwich ELISA. Neurotoxicol Teratol 13:275-281

Philpot RM, Hodgson E (1971/72) The production and modification of cytochrome P-450 difference spectra by in vivo administration of methylenedioxyphenyl compounds. Chem Biol Interact 4:185-194

Plapp FW Jr, Campanhola C (1986) Synergism of pyrethroids by chlordimeform against susceptible and resistant Heliothis. Proc Beltwide Cotton Prod and Res Conf Natl Cotton Council 167-169

Metabolism of Xenobiotic Proestrogens and Estrogens by Cytochrome P450[1]

David Kupfer
Worcester Foundation for
Experimental Biology
Shrewsbury, MA 01545
USA

Introduction

This presentation examines the metabolism of chlorinated hydrocarbon pesticides (CHPs) and their interactions with cytochrome P450. Additionally, the non-pesticidal biological activity (i.e., estrogenicity) of CHPs and their metabolites that might affect non-target organisms yielding potential toxic effects are explored. The CHPs to be discussed have been selected from those which are currently widely used and from those which are essentially banned in the industrially developed countries, but are still being used in certain developing countries. In this regard, the recent finding that women with high exposure to the pesticide DDT, appear to have a greater risk of contracting breast cancer, is of concern (Wolff et al. 1993).

Are CHPs estrogens or proestrogens? are their contaminants the responsible for their estrogen action?

DDT(Fig. 1), represents a CHP that has been banned in the industrially developed countries, but is manufactured and used in certain developing countries. Also even in the developed countries, where the use of DDT has been discontinued, DDT is still present in the environment because of its relatively long half-life. The pesticidal (technical) grade preparation of DDT is composed primarily of the p,p' isomer; however, in addition to several contaminants, it contains approximately 15%

[1]This lecture is dedicated to the memory of Dr. Anthony D. Theoharides, who has died prematurely. Anthony was my first Ph.D. student and over the years has become a great friend. Anthony collaborated with us and made a major contribution to the identification of certain methoxychlor metabolites.

NATO ASI Series, Vol. H 90
Molecular Aspects of Oxidative Drug Metabolizing Enzymes
Edited by E. Arınç, J. B. Schenkman and E. Hodgson
© Springer-Verlag Berlin Heidelberg 1995

of the **o,p'DDT** isomer (Fig. 1), as a major contaminant. Earlier findings demonstrated that o,p'DDT exhibits pronounce estrogenic activity (Welch et al. 1969; Bitman and Cecil 1968). Subsequently, we observed that the estrogenic activity of o,p'DDT mimics that of classical estrogens with respect to the various **estrogenic markers:** uterotropic activity (increased uterine weight), binding to the uterine estrogen receptor (**ER**), increased localization of the ER in the nuclei and diminished cytosolic ER and eliciting a dramatic induction of uterine ornithine decarboxylase (**ODC**) (Kupfer and Bulger 1976; Bulger and Kupfer 1978; Kupfer and Bulger 1980).

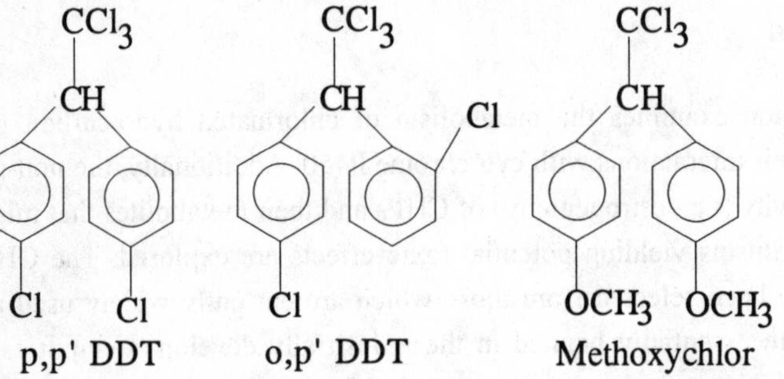

Figure 1

It is difficult to determine conclusively *in vivo* whether a given compound is a *pro*estrogen or an estrogen *per se*. The difficulty rests in the inability to design experiments so that the metabolism of a given compound *in vivo* is totally inhibited, so that it would be possible to determine whether the estrogenic activity is actually due to the parent compound. Such determination of proestrogenic activity, would require highly time consuming and extensive investigations. With that problem in mind, we developed a relatively simple *in vitro* assay that determines whether a compound is an estrogen or a proestrogen (Kupfer and Bulger 1979; Bulger et al. 1985). Indeed, the assay was employed to resolve the question of whether o,p'DDT is an estrogen *per se* or a proestrogen (see below).

Assay to determine whether a compound is an estrogen or a proestrogen (referred to as PPAA): In this assay the compound in question is co-incubated with immature rat uterine strips (**estrogen detecting system**) and male rat liver microsomes (**metabolizing system**) in the same vessel. The compound is considered an estrogen *per se*, if it does not require an active metabolizing system for eliciting an elevation of the nuclear content of the ER and is considered a proestrogen, if it requires an active metabolizing system (presence of NADPH) for that activity to be manifested (Kupfer and Bulger 1979 ; Bulger et al. 1985).

o,p'DDT. It was previously suggested that o,p'DDT is a proestrogen, i.e., that the estrogenic activity resides in the o,p'DDT metabolite(s) (Welch et al. 1969). This conclusion was based on the observation that carbon tetrachloride, a potent inh ibitor of hepatic P450 reactions, inhibited the uterotropic activity of o,p'DDT. By contrast, the finding of others including our own indicate that o,p'DDT is active *per se* (Nelson et al. 1978; Kupfer and Bulger 1979, 1980; Stancel et al 1980).

Chlordecone (Kepone). (Fig. 2). Of interest is the observation that Kepone is an estrogen (Eroschenko and Palmiter 1978; Bulger et al. 1979; Hammond et al. 1979). Kepone, exhibits no structural resemblance to either estradiol or diethylstilbestrol. Nevertheless, kepone binds to the ER and exhibits estrogenic activity, apparently acting as an estrogen *per se*. However, the variety of structurally diverse compounds that are estrogenic (Katzenellenbogen et al. 1980), indicates that the estrogen receptor is highly promiscuous. Surprisingly, Mirex, a structurally related Kepone analog, exhibits no estrogenic activity (Hammond et al. 1979). Thus, it seems unlikely that it will be possible in the immediate future, to p redict *priori*, based solely on structure, whether a given compound would be estrogenic and exhibit estrogenic activity in vivo.

Interestingly, both Kepone and Mirex are inducers of P450, apparently both belonging to the PB-class of inducers. In this regard, it is noteworthy that the endogenous factors involved in induction of P450s 2B and 3A are not highly selective with respect to structures of the "PB-type" inducers. This has led to speculations that the PB-type inducers do not involve a receptor mediated mechanisms. However, by analogy to the variety of compounds binding to the estrogen receptor, it would be risky to conclude that because of the diversity of structures of these PB-type of P450 inducers, a receptor-like mechanism for the PB type of P450 induction, is untenable

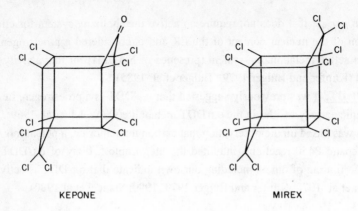

KEPONE   MIREX

Figure 2

Methoxychlor (see Figure 1).

Methoxychlor is a widely employed CHP, serving as a substitute for certain pesticidal activity of DDT. The selection of methoxychlor as a pesticide to replace DDT was based on its relatively low acute toxicity in animals and on its rapid biodegradability in the environment, classifying methoxychlor as a biodegradable pesticide (Metcalf 1976).

     Methoxychlor metabolism.   Earlier studies showed that methoxychlor is metabolized by hepatic microsomal enzymes into the mono- and bis-demethylated derivatives (Kapoor et al. 1970). However, these studies did not explore in detail the enzymatic activities catalyzing those reactions. Subsequently we observed that liver microsomes from untreated rats (**control-microsomes**), supplemented with NADPH, sequentially demethylate methoxychlor into mono-OH-M and bis-OH-M, and that liver microsomes from phenobarbital-treated rats (**PB-microsomes**) catalyze, in addition to demethylation, the ring-hydroxylation of bis-OH-M, yielding the catechol tris-OH-M (Kupfer et al. 1990). Most recently, we observed that PB-microsomes catalyze the ring-hydroxylation of methoxychlor proper, forming the ring-OH-M, which is then demethylated into tris-OH-M (Dehal and Kupfer 1992). Based on such observations, we concluded that methoxychlor metabolism involves two metabolic pathways, yielding the same product (Fig 3) and that these reactions are catalyzed by a variety of P450 isoforms (see below).

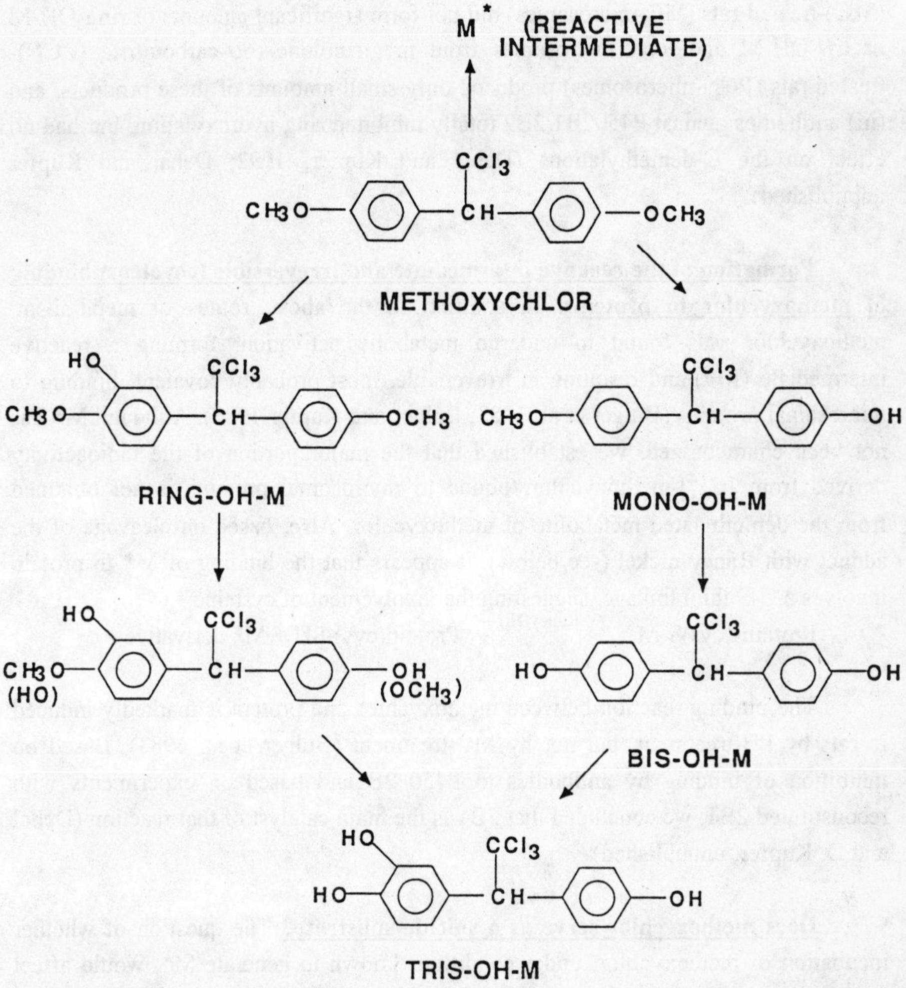

Fig. 3. Proposed pathways of methoxychlor metabolism

P450s catalyzing methoxychlor metabolism. It appears that the O-demethylation reactions of methoxychlor are catalyzed by both 'constitutive' (i.e., present in untreated rats) and by induced P450s. By contrast, ring-hydroxylation is catalyzed solely by the induced P4502B1(P450b)/2B2(P450e). These conclusions are based on our observations that: (i) control- and PB-microsomes catalyze the demethylation reactions, (ii) PB-microsomes catalyze effectively the ring-hydroxylation, however, liver microsomes from <u>control</u> and methylcholanthrene

(**MC**)-treated rats (MC-microsomes) did not form significant amounts of ring-OH-M or tris-OH-M and liver microsomes from pregnenolone-16α-carbonitrile (**PCN**)-treated rats (PCN-microsomes) produced only small amounts of these products, and (iii) antibodies against P4502B1/2B2 totally inhibited ring-hydroxylation, but had no effect on the O-demethylations (Dehal and Kupfer, 1992; Dehal and Kupfer unpublished).

Formation of the reactive intermediate and irreversible (covalent) binding of methoxychlor to proteins. In addition to the above routes of metabolism, methoxychlor was found to undergo metabolic activation, forming a reactive intermediate (**M***) and resulting in irreversible, most probably covalent, binding to microsomal proteins (Bulger et al. 1983; Bulger and Kupfer 1989). Although M* has not been characterized, we established that the major portion of the radioactivity derived from $[C^{14}]$-methoxychlor, bound to microsomal proteins is not obtained from the demethylated metabolite of methoxychlor. Also, based on cleavage of the adduct with Raney nickel (see below), it appears that the binding of M* to protein involves a C-S thiol linkage, suggesting the involvement of cysteine.

Protein-(cys)S-M* $\xrightarrow{\text{Raney Ni}}$ Protein(cys)-SH + M* derivative

The binding reaction between methoxychlor and protein is markedly induced in rats by PB-treatment, but not by MC-treatment (Bulger et al. 1983). Based on inhibition of binding by antibodies to P450 2B and based on experiments with reconstituted 2B1, we concluded that 2B1 is the main catalyst of that reaction (Dehal and D. Kupfer, unpublished).

Does methoxychlor serve as a suicide substrate? The question of whether incubation of methoxychlor, under conditions known to generate M*, would affect P450- monooxygenases, was addressed. Indeed, methoxychlor elicited a time-dependent inhibition of microsomal 6β-hydroxylation of testosterone, but there was no inactivation of other P450-mediated enzymatic activities, indicating inactivation of P450 3A (Li et al. 1993). Furthermore, the addition of glutathione, which we previously found to inhibit the covalent binding of methoxychlor, inhibited the time-dependent inactivation of the 6β-hydroxylation. This provided supportive evidence that the inactivation of P4503A was associated with the generation of M*. However, it was not established whether the inactivation of P4503A was actually due to the covalent linkage of M* to P4503A or possibly was because of inactivation by a reactive oxygen species formed during the generation of M*.

Estrogenic activity of Methoxychlor. Pesticide grade methoxychlor contains numerous contaminants (Lamoureux and Feil 1980). In earlier studies we demonstrated *in vitro* that purified methoxychlor does not bind to the cytosolic uterine estrogen receptor (**ER**). Additionally, we observed that the impurities (probably mostly phenolic) present in small quantities even in the 'laboratory grade' methoxychlor, bind to the ER. Furthermore, using the PPAA assay, we demonstrated that methoxychlor is strictly a proestrogen (Kupfer and Bulger 1979; Bulger et al 1985). Namely, we showed that ultra pure methoxychlor does not bind to the ER, but requires metabolism for estrogenic activity. Additionally, we observed that the metabolites mono-OH-M and bis-OH-M are estrogenic; the bis-OH-M being the more potent estrogen of the two metabolites. *In vivo* studies confirmed these findings (Bulger et al. 1985). Preliminary evidence suggests that tris-OH-M is weakly estrogenic, however because of lability of ring-OH-M, the question of whether ring-OH-M is estrogenic, has not been addressed.

Is methoxychlor a toxin of reproductive processes? Studies in several laboratories demonstrated that methoxychlor exhibits certain toxic manifestations on reproductive functions in mammals (Gray et al. 1989; Cummings and Gray 1989; Cummings and Laskey 1993). It has not been established, however, whether these toxic effects are related to the estrogenic activity of methoxychlor metabolites. Also, it is not known whether some of these toxic effects could be caused by the formation of the reactive intermediate of methoxychlor.

Induction of P450s in rats by CHPs. Numerous earlier studies demonstrated that treatment of rats with various CHPs (e.g., DDT, chlordane) induce the hepatic P450-monooxygenases. Based on the induction of certain enzymatic activities and on the findings that there was an increase in levels of RNA which hybridize to specific oligonucleotide probes, it appears that p,p'DDT behaves like a PB-type inducer (Cambell et al. 1983; Lubet et al. 1992). By contrast earlier attempts, including our own, to induce P450 with methoxychlor in rats, have not been successful. This led us to the erroneous assumption that methoxychlor was not P450 inducer. Most recently we conducted experiments to examine the possibility that the lack of P450 induction by methoxychlor was not due to its being intrinsically inactive, but rather that it may have been due to its extremely brief half-life in the animal. In those studies, we utilized a regimen of methoxychlor that would maintain methoxychlor at levels that hopefully will induce P450. Indeed, in these preliminary experiments, we observed that methoxychlor moderately enhanced the hepatic microsomal 16α- and 6β-hydroxylation of testosterone. However, methoxychlor treatment did not alter the total level of microsomal cytochrome P450 nor did it affect the levels of the P450 reductase activity, indicating that the increased enzymatic activity was due to induction of certain specific P450 isoforms. This conclusion was further substantiated by the marked increase in P450 2B and 3A proteins on western immunoblots. (Li, Dehal and Kupfer, unpublished).

Conclusion

Several chlorinated hydrocarbon pesticides, among these o,p'DDT, chlordecone (Kepone) and methoxychlor exhibit estrogenic activity. Whereas o,p'DDT and kepone appear to be estrogens *per se*, methoxychlor is a proestrogen. Thus, the advantages of the rapid 'biodegradability' of methoxychlor might be outweighed by its conversion into estrogenic metabolites. These estrogenic metabolites (mono- and bis-demethylated methoxychlor) are formed by the hepatic cytochrome P450 monooxygenases. Untreated rat and human liver microsomes catalyze solely the demethylation of methoxychlor, indicating catalysis by constitutive enzymes. Thus, exposure to methoxychlor, shown to elicit estrogenic effects in the rat, will be expected to produce estrogenic activity in other species including humans. Interestingly, PB-treated rat liver microsomes catalyze both methoxychlor demethylation and its ring hydroxylation, yielding the tris-OH-M (catechol) metabolite. Whether tris-OH-M is estrogenic has not been yet established. Thus the metabolism of methoxychlor in PB-treated rat involves two pathways, both yielding tris-hydroxy derivative as the major final product (Dehal and Kupfer, unpublished). Our evidence indicates that the ring hydroxylated products are catalyzed solely by the PB-inducible 2B1/2B2. Additionally, methoxychlor metabolism yields a reactive intermediate **M***, that binds covalently to proteins. Also, methoxychlor metabolism by liver microsomes is accompanied by inactivation of cytochrome P4503A, observed by diminished 6β-hydroxylation of testosterone. It is not known, however, whether the covalent binding of methoxychlor to proteins and/or the inactivation of P450 3A would result in a mild or severe toxic effects. Indeed, several studies in animals indicate that methoxychlor could elicit considerable reproductive toxicity. Whether this toxicity is due to methoxychlor *per se* or due to its metabolites or due to the formation of the reactive intermediate, has not been established. Most recently, we demonstrated that methoxychlor treatment could induce several P450s. Further studies are needed to establish whether that induction will affect the metabolism and biological activity of methoxychlor.

The described studies, supported by a USPHS grant ES00834 from the National Institute of Environmental Health Sciences, NIH, were conducted in our laboratory by W.H. Bulger, R.M. Muccitelli, S.S. Dehal, H-C. Li & D. Kupfer and carried out in part through collaboration with V.J. Feil (USDA, Fargo,ND), A.D. Theoharides (Walter Reed Army Res, Washington, DC), H.V. Gelboin and S.S. Park (NCI, Bethesda, MD).

References

Bitman J, Cecil HC, Harris SJ, Fries GF (1968) Estrogenic activity of o,p'-DDT in the mammalian uterus and avian oviduct. Science 162:371-372

Bulger WH, Kupfer D (1978) Studies on the induction of rat uterine ornithine decarboxylase by DDT analogs. I. Comparison with estradiol 17β activity. Pest Biochem Physiol 8:253-262?

Bulger WH, Kupfer D (1989) Characteristics of monooxygenase-mediated covalent binding of methoxychlor in human and rat liver microsomes. Drug Metab Dispos 17:487-494

Bulger WH, Feil VJ, Kupfer D (1985) Role of hepatic monooxygenases in generating estrogenic metabolites from methoxychlor and from its identified contaminants. Mol Pharmacol 27:115-124

Bulger WH, Muccitelli RM, Kupfer D (1978) Interactions of methoxychlor, methoxychlor base-soluble contaminant, and 2,2-bis(p-hydroxyphenyl)-1,1,1 trichloroethane with rat uterine estrogen receptor. J Toxicol Environ Health 4:881-893

Bulger WH, Muccitelli RM, Kupfer D (1979) Studies on the estrogenic activity of chlordecone (Kepone) in the rat: Effects on uterine estrogen receptor. Mol Pharmacol 15:515-524

Bulger WH, Temple JE, Kupfer D (1983) Covalent binding of ^{14}C-methoxychlor metabolite(s) to rat liver microsomal components. Toxicol Appl Pharmacol 68:367-374

Campbell MA, Gyorkos J, Leece B, Homonko K, Safe S (1983) The effects of 22 organochlorine pesticides as inducers of the .,hepatic drug-metabolizing enzymes. Gen Pharmacol 14:445-454

Cummings AM, Gray LE Jr. (1989) Antifertility effect of methoxychlor in female rats: Dose- and time-dependent blockade of pregnancy. Toxicol Appl Pharmacol 97:454-462

Cummings AM, Laskey J (1993) Effect of methoxychlor on ovarian steroidogenesis: Role in early pregnancy loss. Reprod Tox 7:17-23

Dehal SS, Kupfer D (1992) Formation of a novel ring-hydroxylate is product of the proestrogenic pesticide methoxychlor by phenobarbital treated rat /-liver microsomes. ISSX Proc. 2:76

Eroschenko, VP, Palmiter RD (1980) Estrogenicity of Kepone in birds and mammals. In: McLachlan JA (ed) Estrogens in the environment. Elsevier, North-Holland, pp 305-325

Gray LE Jr, Ostby J, Ferreii J, Rehnberg G, Linder R, Cooper R, Goldman J, Slott V, Laskey J (1989) A dose-response analysis of methoxychlor-induced alterations of reproductive development and function in the rat. Fundam Appl Toxicol 12:92-108

Hammond B, Katzenellenbogen BS, Krauthammer N, McConnell J (1979) Estrogenic activity of the insecticine cnlordecone (Kepone) and interaction with uterine estrogen receptors. Proc Natl Acad Sci USA 76:6641-6645

Kapoor IP, Metcalf RL, Nystrom RF, Sangha GK (1970) Comparative metabolism of methoxychlor, methiochlor and DDT in mouse, insects and in a model ecosystem. J Agric Food Chem 18:1145-1152

Katzenellenbogen JA, Katzenellenbogen BS, Tatee T, Robertson DW, Landvatter SW (1980) The chemistry of estrogens and antiestrogens. In McLachlan J (ed) Estrogens in the environment. Elsevier, North-Holland, pp 33-51

Kupfer D, Bulger WH (1976) Studies on the mechanism of estrogenic actions of o,p'DDT: Interactions with the estrogen receptor. Pestic Biochem Physiol 6:561-570

Kupfer D, Bulger WH (1977) Interaction of o,p'-DDT with the estrogen-binding protein (EBP) in human mammary and uterine tumors. Res Comm Chem Pathol Pharmacol 16:451-452

Kupfer D, Bulger WH (1980) Estrogenic properties of DDT and its analogs. In: McLachlan J (ed) Estrogens in the environment. Elsevier, North-Holland, pp 239-263

Kupfer D, Bulger WH (1979) A nov3lel in vitro method for demonstrating proestrogens. Metabolism of methoxychlor and o,p'-DDT by liver microsomes in the presence of uteri and effects on intracellular distribution of estrogen receptors. Life Sci 25:975-987

Kupfer D, Bulger WH, Theoharides AD (1990) Metabolism of methoxychlor by hepatic P-450 monooxygenases in rat and human. Characterization of a novel catechol metabolite. Chem Res Toxicol 3:8-16

Lamoureux CH, Feil VJ (1980) Gas chromatographic and mass spectrometric characterization o42f impurities in technical methoxychlor. J Assoc Off Anal Chem 63:1007-1037

Li H-C, Mani C, Kupfer D (1993) Reversible and time-dependent inhibition of the hepatic cytochrome P450 steroidal hydroxylases by the proestrogenic pesticide methoxychlor in rat and human. J Biochem Toxicol 21: (in press)

Lubet RA, Dragnev KH, Chauhan DP, Nims RW, Diwan BA, Ward JM, Jones CR, Rice JM, Miller MS (1992) A pleiotropic response to phenobarbital-type enzyme inducers in the F344/NCr rat. Effect of chemicals of vari53ed structure. Biochem Pharmacol 43:1067-1070

Metcalf RL (1976) Organochlorine insecticides, survey and prospects. In: Metcalf RL, McKelvey JJ Jr (eds) The future for insecticides. Needs and prospects. John Wiley and Sins, Inc, pp 223-285

Nelson JA, Struck RF, James R (1978) Estrogenic activities of chlorinated hydrocarbons. J Toxicol Environ Health 4:325-340

Stancel GM, Ireland JS, Mukku VR, Robison AK (1980) The estrogenic activity of DDT: in vivo and in vitro induction of a specific estrogen inducible uterine protein by o,p'-DDT. Life Sci 27:1111-1117

Welch RM, Levin W, Conney AH (1969) Estrogenic action of DDT and its analogs. Toxicol Appl Pharmacol 14:358-367

Wolff MS, Toniolo PG, Lee EW, Rivera M, Dubin N (1993) Blood levels of organochlorine residues and risk of breast cancer. J Natl Cancer Inst 85:648-652

Metabolism of Antiestrogenic Anti-Cancer Agents by Cytochromne P450 and Flavin-Containing Monooxygenases. Mechanism of Action Involving Metabolites

David Kupfer
Worcester Foundation for
Experimental Biology
Shrewsbury, MA 01545
USA

Introduction

This presentation describes the metabolism of triphenylethylene derivatives, chlorotrianisene (TACE) and tamoxifen by hepatic microsomal enzymes in rat and human. Both compounds appear to exhibit partial agonist/antagonist estrogen activity in the rat. The antiestrogenic activity of triphenylethylenes is thought to involve the interactions of these compounds and/or of their respective metabolites with the estrogen receptor (ER). However, the exact mechanism of their antiestrogenic action, has not been fully elucidated.

TACE has been used as a therapeutic agent for postpartum breast engorgement and for prostatic cancer. Those activities of TACE were originally

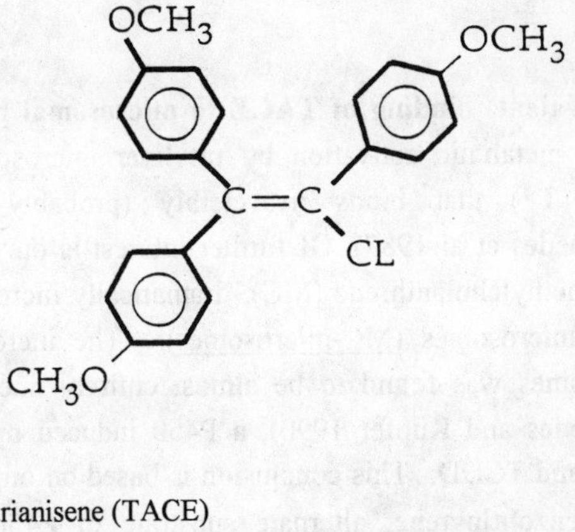

Figure 1. Chlorotrianisene (TACE)

NATO ASI Series, Vol. H 90
Molecular Aspects of Oxidative Drug Metabolizing Enzymes
Edited by E. Arınç, J. B. Schenkman and E. Hodgson
© Springer-Verlag Berlin Heidelberg 1995

attributed to be functioning via its being a long acting estrogen. However, recent studies demonstrated that TACE is not a pure estrogen agonist, but is a partial estrogen agonist/antagonist (Powers et al. 1989). Interestingly, TACE and several other estrogenic contaminants are present in the technical grade methoxychlor (Lamoureux and Feil 1980; Bulger et al 1985). Hence, TACE, most probably contributes to the higher estrogenic activity exhibited by the pesticidal grade than by the more purified preparations of methoxychlor (e.g., laboratory grade).

Based on the extremely low binding affinity of TACE to the estrogen receptor (ER) and on its low estrogenic activity *in vitro* (Jordan and Lieberman 1984; Kupfer and Bulger unpublished) and on its relatively high estrogenic activity **in vivo** (TABLE 1), TACE could be classified as a proestrogen or proantiestrogen; i.e., being converted into active metabolites. Indeed, liver microsomes metabolize TACE via a sequential demethylation, apparently to the mono-, bis- and tris-demethylated products (Reunitz and Toledo 1981). Furthermore, we observed an additional route of TACE metabolism, involving the formation of a reactive intermediate and irreversible binding to proteins (see below).

Table 1. Estrogenic activity of TACE (chlorotrianisene) in ovariectomized rats

Treatment	Ornithine Decarboxylase (ODC) pmol/uterus/hr	Uterine/BW (x 100)
none	17	38
TACE	2089	64

TACE (8 mg/kg BW) was given i.p. Uteri were removed 6 hr later and ODC activity was determined for each uterus. Values for treated groups, represent the mean of 6 animals (Kupfer & Bulger unpublished).

Irreversible (covalent) binding of TACE to microsomal proteins. TACE was found to undergo metabolic activation by rat liver microsomes, yielding a reactive intermediate (T*) that binds irreversibly (probably covalently) to microsomal proteins (Juedes et al 1987). Of further interest is our observation that treatment of rats with methylcholanthrene (MC), dramatically increases the binding of TACE to rat liver microsomes (MC-microsomes). The increase in covalent binding in MC-microsomes was found to be almost entirely due to catalysis by P4501A1 (P450c)[1] (Juedes and Kupfer 1990), a P450 induced by treatment with MC, β-naphthoflavone and TCDD. This conclusion is based on our findings that in MC-microsomes: (i) benzo[a]pyrene, alternate substrate of P450 1A1, strongly inhibited TACE binding, (ii) antibodies to P4501A1/1A2, but not antibodies to P4501A2 (P450d), strongly inhibited the binding of TACE and (iii) reconstituted P4501A1 effectively catalyzed that reaction.

In MC-microsomes, the binding sites for T* appear to be limiting, since the addition of albumin to the incubation, markedly increased the binding of TACE, mostly to albumin (Juedes and Kupfer 1990). There was little or no increase in binding of TACE by addition of albumin to liver microsomes from control rats (control-microsomes). These findings suggested that T* generated in MC-microsomes is sufficiently stable to migrate from its site of formation and to react withot her proteins. However, the possibility that T* is not released with ease from the enzyme, but merely reacts while bound at the enzyme site with the exogenously added albumin, was not ruled out[2].

Interaction of TACE with the estrogen receptor (ER). The question of whether the metabolism of TACE could affect the estrogen receptor, was addressed. Using the PPAA assay (see previous manuscript), we observed that the incubation of TACE under conditions known to form T*, caused "inactivation" of the estradiol(E_2)-binding sites in the ER (TABLE 2); i.e., lowering of the amount of E_2 bound to the ER (Kupfer and Bulger 1990). The question of whether inhibition of the E_2 binding to the ER is entirely the result of covalent binding of T* to the estrogen binding site, has not been resolved. There is the possibility that the "inactivation" of the binding of E_2 to the ER occurs by a different mechanism; i.e., due to the reaction of the ER with some form of a reactive oxygen species, such as superoxide anions or H_2O_2. Indeed, the incubation of another triphenylethylene, tamoxifen, with hepatocytes was found to yield superoxide anions (Turner et al. 1991). Additionally, it is not known whether incubation of TACE actually yields a biologically inactive ER or whether the binding of T* the ER produces a 'permanently' activated receptor. Further studies are needed to resolve these important questions.

---

[1]Occasionally, the classical nomenclature of Levin for P450s in rat liver was used: e.g., P450c, d, b, e
[2]. Sincere thanks are due to Dr. R. Philpott (NIEHS), for suggesting that possibility. However, assuming such a mechanism, it appears unlikely that albumin will selectively 'pick off' the reactive intermediate (RI) of TACE off the enzyme site in MC-microsomes, but not in control-microsomes. Also, it seems unlikely that albumin will be inert towards the RIs of methoxychlor and tamoxifen; albumin did not increase covalent binding of these two compounds. Nevertheless, the possibility raised by Dr. Philpot is requires further testing.

Table 2. Evidence for the involvement of the flavin-containing monooxygnase (FMO) in rat hepatic tamoxifen N-oxidation

Treatment of microsomes	N-Oxide (% of control)
None	100
Heat (50°, 90 seconds)	9
Methimazole, 50 μM^-	59
Methimazole, 200 μM^-	38

Modified from Mani et al. 1993a

Tamoxifen is currently used as the sole endocrine therapeutic agent for breast cancer, primarily in post-menopausal women. Recently, large clinical trials were initiated in the USA, to determine whether the prophylactic treatment of tamoxifen in pre- and post-menopausal women considered at risk of breast cancer, will prevent breast cancer (Fisher and Redmond 1991).

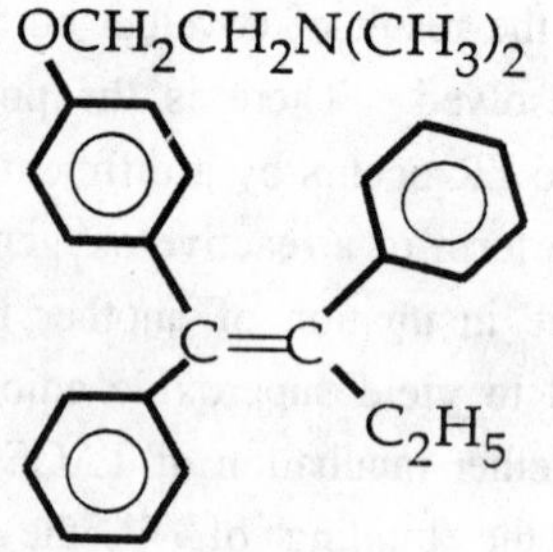

Figure 2. Tamoxifen

Tamoxifen metabolism. Tamoxifen is metabolized by mammalian liver microsomes into three primary major products: tam-N-oxide, N-desmethyl-tam and 4-hydroxy-tam (Fig. 3) (Foster et al 1980, Reunitz et al 1984, McCague and Seago 1986). It has been suggested that the antiestrogenic activity of tamoxifen is primarily due to its 4-hydroxy metabolite (4-OH-Tam) (Lyman and Jordan 1986). This tempted a speculation that 4-OH-tam could be a more proximate anti- cancer

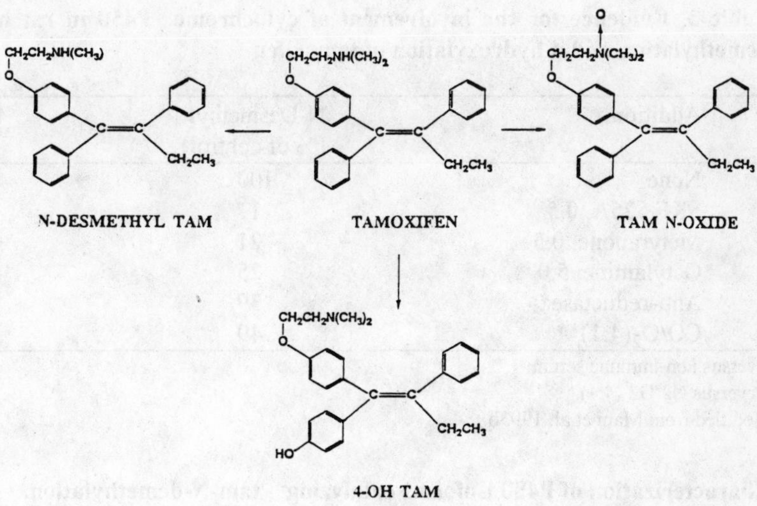

Figure 3. Tamoxifen metabolism

agent than the parent drug. Thus, we undertook a study to determine which enzymes catalyze the formation of tamoxifen metabolites and whether induction of these enzymes will alter quantitatively the pattern of tamoxifen metabolism:

Flavin-containing monooxygenase (FMO). We demonstrated that the N-oxidation is catalyzed by the hepatic microsomal FMO (Mani et al.1993a). This observation was based on our findings that heat and methimazole, known to inhibit hepatic FMO, inhibited primarily the N-oxide formation (TABLE 3), but had little or no effect on N-demethylation and 4-hydroxylation. Additionally, antibodies to the NADPH-reductase did not inhibit N-oxidation. Lastly, purified microsomal FMO catalyzed effectively the N-oxidation of tamoxifen, but did not form any of the other products (Mani et al. 1993a).

P450. We observed that the N-demethylation and 4-hydroxylation are catalyzed by P450s (Mani et al. 1993b). The evidence rested on the following: Inhibitors of P450 and carbon monoxide inhibited both reactions. Additionally, antibodies to NADPH-P450 reductase inhibited both reactions. (TABLE 3).

Table 3. Evidence for the involvement of cytochrome P450 in rat hepatic N-demethylation and 4-hydroxylation of tamoxifen

Addition,	N-Desmethyl (% of control)	4-Hydroxy
None	100	100
SKF 525A, 0.5	17	8
Metyrapone, 0.5	21	23
Octylamine, 5.0	25	18
Anti-reductase*	39	50
$CO/O_2(4:1)$**	49	32

* versus non-immune serum
** versus N2/O2 (4:1).
Modified from Mani et al. 1993b.

Characterization of P450 isoforms catalyzing tam-N-demethylation.

P4503A: a) Treatment of rats with phenobarbital (PB) or pregnenolone-16α-carbonitrile (PCN), increased the rate of N-demethylation, but not 4-hydroxylation or N-oxidation[3], (b) alternate substrates (cortisol and erythromycin) and inactivator (TAO) of P4503A inhibited N-demethylation and (c) antibodies against P4503A strongly inhibited N-demethylation in microsomes from PB and PCN-treated rats. Hitherto, we have not been able to catalyze the N-demethylation with reconstituted 3A1 or by engineered 3A4 in a human cell line. That is not surprising, since several investigators have experienced limitted success with reconstitution of the enzymatic activity of 3A isozymes. Our indirect findings, based on inhibition by alternate substrates of P4503A, suggested that N-demethylation in human liver is also catalyzed by 3A isozymes. However, much stronger support for 3A involvement was obtained by others (Jacolot et al 1991), who showed that there was a high correlation between levels of 3A in individual human livers and tam-N-demethylation. The same investigators demonstrated that anti-3A IgG immunoinhibited N-demethylation in human liver microsomes. Unfortunately, these investigators encountered difficulties in reconstituting P4503A, and hence direct evidence for the involvement of 3A in N-demethylation, has not been achieved.

P4502B: Antibodies against P4502B1/2B2 did not inhibit N-demethylation and reconstituted 2B1 did not catalyze that reaction, indicating that 2B1/2B2 do not catalyze the N-demethylation (Mani et al. 1993b).

---

[3]. PCN was previously shown to induce primarily P450 3A1 and to a lesser extent 2B1. Others using purified PCN did not observe induction of 2B1. Similarly, no induction of 2B1 by PCN was found in Parkinson's laboratory (Parkinson, personal communication)

P501A1/1A2: MC treatment increased N-demethylation, suggesting that P4501A1/P4501A2 catalyzed the N-demethylation. Indeed, reconstituted 1A1 catalyzed effectively the N-demethylation. However, monoclonal antibodies (MAb) against 1A1/1A2 and polyclonal antibodies (PAb) against 1A1 produced little and no inhibition in MC-microsomes (Mani et al. 1993b). This indicated that although reconstituted 1A1 catalyzes tam-N-demethylation, it does not catalyze significantly this reaction in the intact microsomes.

Characterization of P450s catalyzing tam-4-hydroxylation. P450 inducers (PB, PCN or MC) in the rat, did not elevate tam-4-hydroxylation. Furthermore, the antibodies to P450 2B and 3A did not inhibit the 4-hydroxylation and reconstituted 2B1 and 1A1 did not catalyze that reaction. This indicated that P450 2B, 3A and 1A1 do not catalyze that activity and suggested that 4-hydroxylation is catalyzed primarily by constitutive P450s. However, monoclonal anti 1A1/1A2 and estradiol (alternate substrate of 1A2 and 3A) elicited some inhibition of 4-hydroxylation in MC-microsomes, suggesting the possibility that 1A2 catalyzes in part the 4-hydroxylation (Mani et al. 1993b). Due to the low level of the 4-hydroxylation in rats and man, it was anticipated that it will be difficult to identify the P450 catalyzing that reaction. Thus, we examined liver microsomes from several species to identify those with higher rates of 4-hydroxylation. Indeed, we and others found a much higher rate of tam-4-hydroxylation in chicken liver microsomes. Also, in collaboration with Dr. Arleen Rifkind's laboratory (Cornell U., NY), it was demonstrated that chick embryo liver microsomes exhibit similar rate of tam-4-hydroxylase activity to that of adult chicken liver (Kupfer et al. 1993). Furthermore, βnaphthoflavone (BNF) or 2,3,7,8-tetrachlorodibenzo-p-dioxin (TCDD) treatment of chick embryos markedly enhanced that catalytic activity. This enzymatic activity in the TCDD- and BNF-treated chick embryos was attributed to catalysis by a TCDD and BNF inducible P450 $TCDD_{AA}$ (Kupfer et al., 1993 & **unpublished results**). This P450, which exhibits higher immunochemical resemblance to rat P4501A2 than to 1A1, catalyzes effectively arachidonate metabolism and estradiol-2-hydroxylation (A. Rifkind personal comm., Nakai et al 1992). If indeed 4-hydroxy-tam is the proximate anticancer metabolite of tamoxifen in the human, then it is tempting to speculate that alteration in the levels of P4501A2 could have pronounced effect on tamoxifen metabolism as well as on its therapeutic activity.

Metabolic activation of tamoxifen and irreversible binding to microsomal proteins. In the course of examining the metabolism of ^{14}C-labeled tamoxifen by liver microsomes, we observed that tamoxifen undergoes an NADPH dependent activation and irreversible binding to microsomal proteins (Mani and Kupfer 1991).

This binding appears to be covalent, since numerous washings with a variety of organic solvents do not dissociate the radioactivity. Also, the radioactivity remains associated with the microsomal proteins on SDS-polyacrylamide gel electrophoresis (SDS-PAGE). The major -radioactive zone on SDS-PAGE exhibits an Mr of 54 KDa. The identity of the macromolecule associated with the radioactivity, has not been uncovered.

The formation of the reactive intermediate of tamoxifen (TAM*) resulting in covalent binding was found to involve cytochrome P450 and possibly FMO (Mani and Kupfer 1991). Inhibitors of P450 and carbon monoxide markedly inhibited covalent binding. PB and PCN treatment of rats elevated the rate of covalent binding. The P450 isoforms catalyzing that reaction have not been yet been identified. The possible involvement of FMO in the covalent binding was suggested by the observation that heating liver microsomes at 50 °C for 90 sec or addition of methimazole (alternate substrate of FMO) diminished binding. Also, incubations at pH 8.6, optimal for FMO, exhibited higher binding than at pH 7.4. The possibility was considered that the reactive intermediate TAM* was generated by a series of reactions involving both P450 and FMO: a) involving a P450 catalyzed N-demethylation forming N-desmethyl-tam, followed by an FMO-mediated nitrone formation and b) the subsequent hydrolysis of the nitrone into an aldehyde, with the aldehyde serving as the proximate reactive intermediate. To examine that possibility, incubations were conducted with [^{3}H]- and [^{14}C]-labeled tamoxifen (Mani and Kupfer 1991). Based on the observation that tamoxifen covalent binding was quantitatively similar with either of the two isotopes and that the tritium derived from [^{3}H]-tamoxifen remained associated with the TAM*-protein complex, aldehyde formation was ruled out[4]. The possibility that tam-N-oxide was a more proximate reactive intermediate than tamoxifen was also ruled out, since the rate of tam-N-oxide covalent binding was at a much lower rate than that obtain with tamoxifen (Mani and Kupfer, unpublished).

Since tamoxifen metabolism *in vitro* appears to yield tamoxifen epoxide (Reunitz et al 1984), future studies will examine the possibility that tamoxifen epoxide is a proximate reactive intermediate.

It is not known whether tamoxifen activation, like that of TACE, affects the binding of E$_2$ to the estrogen receptor. Also, the findings that rats treated with tamoxifen elicit hepatic DNA-adducts (Han and Liehr 1992, White et al 1992) raises the question of whether the reactive intermediate of tamoxifen may be

---

[4]The possibility, that the tritium in the adduct was derived from the [□3□H]-HCHO, released from the nitrone was unlikely. In a separate study we demonstrated that the incubation of radiolabled formaldehyde with liver microsomes with or without NADPH, did not yield radiolabeled proteins (Bulger and Kupfer, unpublished)

involved in certain toxic manifestations. Indeed, it is conceivable that the hepatocellular carcinoma in rats and the increase in human endometrial cancer after tamoxifen treatment, may be due to the formation of TAM*-protein and/or of Tam-DNA adducts.

Conclusion

The triphenylethylenes, TACE and tamoxifen, are structurally similar, but nevertheless are metabolized by different subsets of P450 isoforms. The formation of the reactive intermediate of TACE (T*) is dramatically enhanced by treatment of rats with MC, but not with PB. Our studies demonstrated that P4501A1 is the prime P450 catalyzing the formation of T* in MC-microsomes. By contrast, the formation of the reactive intermediate of tamoxifen (**TAM***) is substantially increased by PB- or PCN-treatment of rats. Though we have not characterized the P450 isoform(s) catalyzing the formation of **TAM***, it appears that P4502B is probably involved. Additionally, whereas T* appears to be released from the enzyme site and reacts with exogenously added albumin, suggesting that T* is relatively stable, **TAM*** does not react with the added albumin. These studies need further exploration (see footnote 2).

The mechanism of the antiestrogenic activity of triphenylethylenes is not understood. Our studies demonstrated that during co-incubation of TACE with liver microsomes and uteri, the uterine estrogen receptor (ER) becomes inactivated towards binding of estradiol. The possibility that the mechanism of the antiestrogenic activity of triphenylethylenes involved the inactivation of the ER, needs further exploration.

The P450s catalyzing the demethylation reactions of TACE have not been hitherto characterized. By contrast, N-demethylation of tamoxifen in rat and human was found to be catalyzed by the P4503A isozymes. The P450 catalyzing the 4-hydroxylation of tamoxifen in rat and human, has not been identified. However, in the chick embryo, the 4-hydroxylation of tamoxifen is catalyzed by a 450 isoform present in livers of TCDD and BNF induced embryos. This P450 appears to resemble the rat 1A2 closer than that of rat 1A1 (A. Rifkind, personal commun.).

Since, 4-hydroxy-tamoxifen appears to be a more proximate therapeutic agent than tamoxifen, it would be of importance to determine whether human 1A2 catalyzes that reaction and whether alteration in levels of 1A2 would affect the biological activity of administered tamoxifen.

The above studies, supported by a USPHS grant ES00834 from the National Institute of Environmental Health Sciences, were conducted in our laboratory by M.J. Juedes, C. Mani & D. Kupfer in collaboration with:

H.V. Gelboin (NCI, Bethesda, MD)
S.S. Park (NCI, Frederick, MD)
A. Parkinson, R. Pearce (U. Kansas)
E. Hodgson (N. Carolina State U.)
A.B. Rifkind, C. Lee (Cornell U.Med Coll.)

References

Kupfer D (1985) Role of hepatic monooxygenases in generating estrogenic metabolites from methoxychlor and from its identified contaminants. Mol Pharmacol 27:115-124

Fisher B, Redmond C (1991) New perspective on cancer of the contralateral breast: A marker for assessing tamoxifen as a preventive agent. J Nat Cancer Inst 83:1278-1280

Foster AB, Griggs LJ, Jarman M, vanMaanen JMS, Schulten HR (1980) Metabolism of tamoxifen by rat liver microsomes: formation of the oxide, a new metabolite. Biochem Pharmacol 29:1977-1979

Han XL, Liehr JG (1992) Induction of covalent DNA adducts in rodents by tamoxifen. Cancer Res 52:1360-1363

Jacolot F, Simon I, Dreano Y, Beaune P, Riche C, Berthou F (1991) Identification of the cytochrome P-450 IIIA family as the enzymes involved in the N-demethylation of tamoxifen in human liver microsomes. Biochem Pharmacol 41:1911-1919

Jordan VC, Lieberman ME (1984) Estrogen-stimulated prolacten synthesis in vitro. Mol Pharmacol 26:279-285

Juedes MJ, Bulger WH, Kupfer D (1987) Monooxygenase-mediated activation of chlorotrianisene (TACE) in covalent binding to rat hepatic microsomal proteins. Drug Metab Disp 15:786-793

Juedes MJ, Kupfer D (1990) Role of P-450c in the formation of a reactive intermediate of chlorotrianisene (TACE) by hepatic microsomes from methylcholanthrene-treated rats. Drug Metab Dispos 18:131-137

Kupfer D, Mani C, Lee CA, Rifkind AB (1993) Identification of P450 TCDD$_{AA}$ as the catalyst of tamoxifen 4-hydroxylation (TAM-4-OH) induced by 2,3,7,8-tetrachlorodibenzo-p-dioxin and β-naphthoflavone (BNF) in avian liver. FASEB J (7): Abstract 878, pA1203.

Kupfer D, Bulger WH (1990) Inactivation of the uterine estrogen receptor binding of estradiol during P-450 catalyzed metabolism of chlorotrianisene (TACE). Speculation that TACE antiestrogenic activity involves covalent binding to the estrogen receptor. FEBS Lett 261:59-62

Lyman SD, Jordan VC (1986) Metabolism of nonsteroidal antiestrogens in estrogen/antiestrogen action and breast cancer therapy. In: Jordan VC (ed) Estrogen/antiestrogen Action and Breast Cancer. The University of Wisconsi42n Press, Madison, WI. pp 191-219

Lamoureux CH, Feil VJ (1980) Gas chromatographic and mass spectrometric characterization of impurities in technical methoxychlor. J Assoc Off Anal Chem 63:1007-1037

Mani C, Kupfer D (1991) Cytochrome P-450-mediated activation and irreversible binding of the antiestrogen tamoxifen to proteins in rat and human liver: Possible involvement of flavin-containing monooxygenases in tamoxifen activation. Cancer Res 51:6052-6058

Mani C, Hodgson E, Kupfer D (1993a) Metabolism of 53the antimammary cancer antiestrogenic agent tamoxifen. II. Flavin-containing monooxygenase-mediated N-oxidation. Drug Metab Disp 21:657-661

Mani C, Gelboin HV, Park SS, Pearce R, Parkinson A, Kupfer D (1993b) Metabolism of the antimammary cancer antiestrogenic agent tamoxifen. I. Cytochrome P-450 -catalyzed N-demethylation and 4-hydroxylation. Drug Metab Disp 21:645-656

McCague R, Seago A (1986) Aspects of metabolism of tamoxifen by rat liver microsomes. Biochem Pharmacol 35:827-345

Nakai K, Ward AM, Gannon M, Rifkind AB (1992) β-naphthaflavone induction of a cytochrome P-450 arachidonic acid epoxygenase in chick embryo liver distinct from the aryl hydrocarbon hydroxylase and from phenobarbital induced arachidonate epoxygenase. J Biol Chem 267:19503-19512

Powers CA, Hatala MA, Pagano PJ (1989) Differential responses of pituitary kallikrein and prolactin to tamoxifen and chlorotrianisene. Mol Cell Endocrinol 66:93-100

Reunitz PC, Toledo MM (1981) Chemical and biochemical characteristics of O-demethylation of chlorotrianisene in the rat. Biochem Pharmacol 30:2203-2207

Reunitz PC, Bagley JR, Pape CW (1984) Some chemical and biochemical aspects of liver microsomal metabolism of tamoxifen. Drug Metab Disp 12:478-483

Turner MJ III, Fields CE, Everman DB (1991) Evidence for superoxide formation during hepatic metabolism of tamoxifen. Biochem Pharmacol 41:1701-1705

White, IN, deMatteis F, Davies A, Smith LL, Crofton-Sleigh C, Venitt S, Hewer A, Phillips DH (1992) Genotoxic potential of tamoxifen and analogues in female Fischer F344/2 rats, DBA/2 and C57BL/6 mice in human MCL-5 cells. Carcinogenesis 13:2197-2203

Therapeutic Agents and Cytochrome P450

Anthony Y. H. Lu
Department of Drug Metabolism
Merck Research Laboratories
Rahway, NJ 07065 USA

Introduction

The discovery and development of therapeutic agents for the treatment of human diseases involve a long and complex process (Table I). Through coordinated efforts, researchers with diverse scientific disciplines work together as a team to solve many of the issues illustrated in Table 2. In the past, drug metabolism studies were only conducted to support safety assessment in animals and clinical pharmacology studies in humans in the development process. However, scientists in drug metabolism are now playing an increasingly important role in the design, discovery, and development of new therapeutic agents. The evolution of this approach is in part due to the tremendous progress in cytochrome P450 research and the better understanding of the role of each human cytochrome P450 in drug metabolism and drug-induced toxicity. In fact, much of the knowledge regarding human and animal cytochrome P450s can now be either directly or indirectly applied to many aspects of the drug discovery and development process.

Therapeutic Targets

Several cytochrome P450 enzymes are associated with the pathogenesis of a number of diseases. Therefore, specific inhibitors of these cytochrome P450s are of potential importance as therapeutic agents (Correia and Ortiz de Montellano, 1992; VandenBossche, 1992). For example, aromatase plays a pivotal role in the synthesis of estrogens and catalyzes the conversion of androstenedione to estrone, and testosterone to estradiol. This unique cytochrome P450 reaction involves three consecutive oxygenation steps resulting in the removal of the 19-methyl group and the aromatization of the A-ring. Aromatase inhibitors such as aminoglutethimide, fadrozole and other compounds are potentially useful agents for the control of estrogen-dependent mammary tumors (Table 3).

Fungi such as *candida tropicalis* require ergosterol, a vital plasma membrane component, for their normal aerobic proliferation and growth. Lanosterol 14α-

NATO ASI Series, Vol. H 90
Molecular Aspects of Oxidative Drug Metabolizing Enzymes
Edited by E. Arınç, J. B. Schenkman and E. Hodgson
© Springer-Verlag Berlin Heidelberg 1995

Table 1

DISCOVERY AND DEVELOPMENT PROCESS OF THERAPEUTIC AGENTS

Discovery Phase	Identify therapeutic target. Identify active compounds through in vitro and in vivo testings.
Preclinical Phase	Establish mechanism of action. Select candidates based on potency, toxicity, metabolism and disposition in animals; numerous modifications to optimize molecular structures. Identify compound for development and generate detailed toxicity, metabolism and disposition data in animals to support human study; select formulation.
Clinical Phase	Establish efficacy and safety in human clinical trials (Phase I, II, and III).
Regulatory Phase	Submit documents for regulatory approval.

demethylation, a cytochrome P450-catalyzed reaction, is a key step in the ergosterol synthesis. This enzyme catalyzes three successive two-electron oxidation steps converting lanosterol to 14,15-desaturated sterol. Some lanosterol 14α-demethylase inhibitors (Table 4) are potent antifungal agents (Correia and Ortiz de Montellano, 1992; VandenBossche, 1992).

Another interesting enzyme is the recently discovered nitric oxide synthase (Nathan, 1992; Marletta, 1993). This is a catalytically self-sufficient mammalian P450 enzyme (containing FMN, FAD, heme, tetrahydrobiopterin), catalyzing the conversion

Table 2

THERAPEUTIC AGENTS: DISCOVERY AND DEVELOPMENT ISSUES

1. Therapeutic target: May take years of basic research to define the target.

2. Compounds available for screening: Need capacity to screen huge numbers of compounds.

3. In vitro activity screening: Specificity of screening method critical.

4. In vivo animal model screening for activity: If a compound is active in vitro but not in vivo, should find out why. (Poor absorption? Extensive metabolism? Very high plasma binding? Wrong animal model?)

5. In vitro and in vivo toxicity testing in animals: If toxic, should find out why. (Significance? Mechanism? Animal model relevant? Species differences in pharmacokinetics and toxic metabolic pathways?)

6. Human trial:

    •   No efficacy: Therapeutic hypothesis incorrect?
                                   Experimental models wrong?

    •   Serious side effects: Unique metabolic pathways in humans?
                                   Different pharmacokinetics?

    •   Idiosyncratic reactions: Critical to understand the mechanism.

---

of L-arginine to citrulline and nitric oxide, a messenger molecule. In different tissues, nitric oxide has different biological functions. It exerts cytotoxic effects in circulating macrophages against tumor cells and microorganisms. In blood vessels, it is identified as the endothelium-derived relaxing factor. In brain cells, it functions as a neurotransmitter. Thus, nitric oxide has been implicated in cytokine-induced hypotension and neurotoxicity associated with stroke and neurodegenerative diseases. Inhibitors of nitric oxide synthase may be used as antiinflammatory agents or other therapeutic purposes.

Table 3

HUMAN AROMATASE INHIBITORS

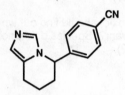

Aminoglutethimide

Used clinically for estrogen-responsive breast tumors in postmenopausal women; Ki = 0.54 µM for human placental aromatase; not a specific aromatase inhibitor; 30-40% patients with side effects.

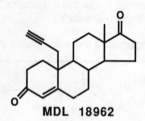

Fadrozole

Ki = 0.17 nM for human placental aromatase; more specific to aromatase; potent inhibitor (2 mg/day) of estrogen biosynthesis in postmenopausal women; well-tolerated in patients.

MDL 18962

Potent suicide inhibitor, Ki = 3 nM, t$_{1/2}$ = 9 min.; orally active to lower estrogen levels.

Mercaptoandrostenedione

Suicide inhibitor, Ki = 30 nM, t$_{1/2}$ = 8 min.

Table 4

LANOSTEROL 14α-DEMETHYLASE INHIBITORS

Effective topical antifungal ag
rapidly metabolized.

Miconazole

Orally active, broad spectrum
effects at high dose.

Ketoconazole

Water soluble, long-acting, o
selective for fungal P450; littl

Fluconazole

Metabolism of Drug Candidates

Because of the major advances in biochemistry and molecular biology, many therapeutic targets (such as receptors and enzymes) are now available in pure forms for structural-activity relationship analysis. While the *in vitro* system is convenient for large scale screening of compounds and the available structural information of biological targets is vital for rational design of potent compounds, it is not unusual to find that compounds with potent *in vitro* activity show little or no activity (or plasma level) in the animal models. Various *in vitro* and *in vivo* studies can be carried out to determine if the lack of *in vivo* activity is due to a lack of absorption or extensive metabolism by drug-metabolizing enzymes, especially cytochrome P450. Structure-metabolism relationships can often help guide synthetic efforts to design more stable compounds, and active metabolites can serve as new leads for further synthetic work. Thus, cytochrome P450 can influence the biological activity of therapeutic agents by either inactivating the compounds through biotransformation or producing active metabolites with various degrees of potency.

The impact which pharmacokinetics and drug metabolism studies have made to the drug discovery programs has recently been reviewed by Humphrey and Smith (1992). For example, by incorporating pharmacokinetic and metabolism studies at the discovery stage, fluconazole was designed to provide a more metabolically stable, lower plasma binding and longer plasma half-life antifungal agent.

Enzyme Induction in Animals

For chronically used therapeutic agents, long-term carcinogenicity tests in rodents are generally required by Regulatory Agencies in order to assess the long-term risk in human use. Since there are some known "associations" between cytochrome P450 inducers and rodent tumorigenesis, potential drug development candidates are often screened for cytochrome P450 induction in rats and mice. Although the relevance of cytochrome P450 induction in human safety is controversial, the selection of biologically active drug candidates without enzyme induction properties can be highly desirable.

Enzyme Inhibition in Animals

Many classes of compounds are known inhibitors of cytochrome P450. Reversible inhibitors (such as pyridine and imidazole derivatives, *e.g.*, metyrapone, cimetidine) bind to the hydrophobic domain, heme iron or the active site residues of cytochrome P450. Mechanism-based inhibitors (such as parathion, chloramphenicol, terminal olefins, and acetylenes) covalently bind to the protein or heme. Understanding the inhibition mechanism and the structural features required for inhibition would minimize the selection of drug candidates which could either inhibit or destroy cytochrome P450s in animals or in humans. Decreased cytochrome P450 levels could result in unacceptably high plasma drug level and long drug half-life.

Cytochrome P450 Mediated Toxicity

Various *in vitro* systems containing purified or stably expressed cytochrome P450s from animals and humans are now available for promutagen testing and other toxicity evaluation of potential drug candidates (Gonzalez *et al*, 1991). These systems not only allow the elimination of potentially toxic drug candidates but also the identification of specific cytochrome P450 isoforms responsible for the activation process. Furthermore, through mechanistic studies and the understanding of specific metabolic pathways for the generation of reactive species, safer compounds can be designed. For example, metabolic activation of ronidazole (as measured by covalent binding and mutagenicity) involves reduction of the nitro group, the removal of the carbamate side chain, and requires an unsubstituted 4 position (Lu *et al*, 1988). Thus, substituent at the 4 position is the key strategy to modify 5-nitroimidazoles in order to decrease or eliminate mutagenicity but retain good antiprotozoal activity (Walsh *et al*, 1987). Indeed, 1,2-dimethyl-4-(2-hydroxyethyl)-5-nitroimidazole shows good antitrichomonal activity *in vitro* but possessing only 0.28% of the mutagenicity of ronidazole.

Selection of Safety Species

In principle, animal species selected for safety evaluation of drug candidates should generate similar metabolic profiles as those from humans. *In vitro* and *in vivo* studies can now be conducted to establish the metabolic profiles of drug candidates in animal species and human, providing scientific basis for the selection of animal species for safety evaluation.

Clinical Studies in Humans

A. Identification of cytochrome P450 -- Significant progress in basic research on human cytochrome P450 has provided valuable tools for the identification of specific cytochrome P450(s) responsible for the metabolism of therapeutic agents in man (Gonzalez *et al*, 1991; Guengerich, 1989; Wrighton and Stevens, 1992; Table 5). Using antibodies, specific inhibitors and stably expressed human cytochrome P450, it is now possible to identify the major human cytochrome P450(s) responsible for the metabolism of drugs before clinical study. For example, lovastatin, a potent cholesterol-lowering agent, is metabolized by human liver microsomes primarily at the 6'-position to give 6'ß-hydroxy- and 6'-exomethylene-lovastatin. Antibody inhibition studies indicate that P450 3A is the major enzyme

Table 5

HUMAN CYTOCHROME P450s

Enzyme	Substrates	Inducers	Inhibitors
1A1	benzo(a)pyrene	smoking	7,8-benzoflavone
1A2	phenacetin, caffeine, arylamines	smoking, cruciferous vegetables, charcoal broiled meats	7,8-benzoflavone
2A6	coumarin		
2C	tolbutamide, S-mephenytoin, S-warfarin	phenobarbital	sulfaphenazole
2D6	debrisoquine, imipramine, bufuralol, propanolol, dextromethorphan		quinidine
2E1	chlorzoxazone, acetaminophen	ethanol	4-methylpyrazole
3A4/5	nifedipine, quinidine, cyclosporin, lidocaine, lovastatin, testosterone (6β), dapsone	dexamethasone, rifampicin, phenobarbital	gestodene

involved in lovastatin metabolism at the 6'-position as well as the side chain hydroxylation (Wang *et al*, 1991). In other instances, more than one cytochrome P450 may be responsible for the metabolism of therapeutic agents in humans. For example, propafenone, an antiarrhythmic agent, is metabolized by P450 2D6 to 5-hydroxypropagenone and by 3A4 and 1A2 to the dealkylated metabolite in human liver microsomes (Botsch *et al*, 1993).

B.   Drug-drug interaction -- Since most therapeutic agents examined so far are metabolized by a limited number of human cytochrome P450s, namely, CYP 3A4, CYP 2D6, CYP 2C8/9/18, CYP 1A2, and CYP 2E1, knowledge of the involvement of specific cytochrome P450 for the metabolism of new therapeutic agents provides a rational approach to design drug-drug interaction studies in man. In principle, it is increasingly possible to predict whether a new drug might inhibit the metabolism of known therapeutic agents, or vice versa. It should be noted that *in vitro* inhibition studies can only address the drug-drug interaction potential, *in vivo* studies would ultimately establish whether such interactions are clinically relevant.

Recently, the Food and Drug Administration has expressed concern on reports of serious ventricular arrhythmias associated with the coadministration of antihistamine terfenadine and antifungal agent ketoconazole. A healthy young woman experienced life-threatening ventricular arrhythmia after taking both medications. The patient had high blood levels of terfenadine, a compound usually undetectable due to rapid metabolism, suggesting ketoconazole inhibits terfenadine metabolism (Honig *et al*, 1993). Yun *et al* (1993) have shown that cytochrome P450 3A4 is responsible for the N-dealkylation and C-hydroxylation of terfenadine in human liver microsomes. Since ketoconazole is a potent inhibitor of 3A4 (Maurice *et al*, 1992), it could alter the metabolism of terfenadine resulting in accumulation of the drug and a potential increased risk of cardiotoxicity (Woosley *et al*, 1993). Thus, understanding the metabolic pathways of drugs and the enzymes involved in metabolism could provide valuable information on the potential of drug-drug interactions and human safety of drug use.

C.   Genetic polymorphism -- Enzyme identification allows one to determine whether a new therapeutic agent is metabolized by one of the cytochrome P450s exhibiting polymorphism, *e.g.,* CYP 2D6. In addition, probes are now available for genotyping as well as phenotyping of CYP 2D6 in individuals. In cases where a major metabolic route or the formation of an active metabolite are polymorphically controlled, knowledge about a patient's cytochrome P450 status might be of practical value for dose adjustments so that therapeutically adequate dose levels can be maintained and potentially adverse side effects can be avoided.

D.   Enzyme induction -- Induction of specific cytochrome P450 isoforms is known to be dependent on species and other factors. Thus, compounds which induce cytochrome P450 in animals may or may not be enzyme inducers in man. Human hepatocyte culture systems have been used to examine the induction potential of therapeutic agents in man for certain cytochrome P450 species. It would be desirable to have other *in vitro* systems to screen drug candidates for the potential of human cytochrome P450 induction.

E.   Idiosyncratic reactions -- Idiosyncratic adverse drug reactions are rare and unexpected responses to therapeutic agents and at times may be fatal. Since these reactions are not seen in toxicity studies in animals, mechanisms regarding idiosyncratic reactions in human are generally not known. Recent studies on neoantigens, population genetics, and the application of new biochemical tools such as the lymphocyte toxicity assay have suggested that cytochrome P450 may play some central role in some of the observed idiosyncratic reactions (Spielberg, 1984).

Acknowledgments

I would like to thank Terry Rafferty for her assistance in preparing this manuscript.

References

Botsch SB, Gautier JC, Beaune P, Eichelbaum M, Kroemer HK (1993) Identification and characterization of the cytochrome P450 enzymes involved in N-dealkylation of propafenone:molecular base for interaction potential and variable disposition of active metabolites. Mol. Pharmacol. 43:120-126.

Correia MA, Ortiz de Montellano PR (1992) Inhibitors of cytochrome P450 and possibilities for their therapeutic application, in Frontiers in Biotransformation, Volume 8, Ruckpaul K, Rein H (eds). Akademie Verlag Berlin, p 74-144.

Gonzalez FJ, Crespi CL, Gelboin HV (1991) cDNA-expressed human cytochrome P450s: A new age of molecular toxicology and human risk assessment. Mut. Res. 247:113-127.

Guengerich FP (1989) Characterization of human microsomal cytochrome P450 enzymes. Annu. Rev. Pharmacol. Toxicol. 29:241-264.

Honig PK, Wortham DC, Zamani K, Conner DP, Mullin JC, Cantilena LR (1993) Terfenadine-ketoconazole interaction:pharmacokinetic and electrocardiographic consequences. JAMA 269:1513-1518.

Humphrey MJ, Smith DA (1992) Role of metabolism and pharmacokinetic studies in the discovery of new drugs -- present and future perspectives. Xenobiotica 22:743-755.

Lu AYH, Miwa GT, Wislocki PG (1988) Toxicological significance of covalently bound drug residues. Rev. Biochem. Toxicol. 9:1-27.

Marletta MA (1993) Nitric oxide synthase structure and mechanism. J. Biol. Chem. 268:12231-12234.

Maurice M, Pichard L, Daujat M, Fabre I, Joyeux H, Domergue J, Maurel P (1992) Effects of imidazole derivatives on cytochromes P450 from human hepatocytes in primary culture. FASEB J. 6:752-758.

Nathan C (1992) Nitric oxide as a secretory product of mammalian cells. FASEB J. 6:3051-3064.

Spielberg SP (1984) In vitro assessment of pharmacogenetic susceptibility to toxic drug metabolites in humans. Federation Proc. 43:2308-2313.

VandenBossche H (1992) Inhibitors of P450-dependent steroid biosynthesis: From research to medical treatment. J. Steroid Biochem. Molec. Biol. 43:1003-1021.

Walsh JS, Wang R, Bagan E, Wang CC, Wislocki PG, Miwa GT (1987) Structural alterations that differentially affect the mutagenic and antitrichomonal activities of 5-nitroimidazoles. J. Med. Chem. 30:150-156.

Wang RW, Kari PH, Lu AYH, Thomas PE, Guengerich FP, Vyas KP (1991) Biotransformation of lovastatin. IV. Identification of cytochrome P450 3A proteins as the major enzymes responsible for the oxidative metabolism of lovastatin in rat and human liver microsomes. Arch. Biochem. Biophys. 290:355-361.

Woosley RL, Chen Y, Freiman JP, Gillis RA (1993) Mechanism of the cardiotoxic actions of terfenadine. JAMA 269:1532-1536.

Wrighton SA, Stevens JC (1992) The human hepatic cytochromes P450 involved in drug metabolism. Critical Reviews in Toxicol. 22:1-21.

Yun CH, Okerholm RA, Guengerich FP (1993), Oxidation of the antihistaminic drug terfenadine in human liver microsomes--role of cytochrome P450 3A4 in N-dealkylation and C-hydroxylation. Drug Metab. Disp. 21:403-409.

Drug Metabolism, Lipid Peroxidation and Glutathione

Wolfgang Klinger
Institute of Pharmacology and Toxicology
Friedrich-Schiller-University of Jena
Loebderstr. 1 D-07743 Jena
Germany

Introduction

Cytochrome P-450 (P450) mediated drug metabolism can be connected with the creation of free oxygen radicals such as superoxide radical anion (O_2^{-}) and hydroxyl radical ($OH^{\cdot}$), and hydrogen peroxide (H_2O_2), so-called reactive oxygen species (ROS), by different ways and mechanisms. This is the beginning of a multistep chain mechanism (initiation, propagation and termination) in the further production of radicals and of peroxidation of lipids, proteins and DNA, leading finally to cell damage and cell death. The tripeptide glutathione (GSH) serves both as coupler in drug phase II metabolism to form glutathione conjugates and later on evtl. mercapturic acid derivatives, but also as radical scavenger, thus protecting liver parenchymal and other cells throughout the whole organism from damage. Thus there are different connections between the microsomal electron transport chain with P450 (with its functions as monooxygenase, oxidase and peroxidase) and GSH.

Reactive oxygen species (ROS) such as singlet oxygen, superoxide radical anion (O_2^{-}) or at acid pH superoxide radical (hydroperoxyradical)($HOO^{\cdot}$), hydrogen peroxide (H_2O_2), which is not a radical, and finally hydroxyl radical (OH^{-}) are constantly produced in dependence on the oxygen concentration. ROS comprise the reactive singlet forms and the four reduction steps of molecular (atmospheric) oxygen (dioxygen) to the final product water.

The four reduction steps and the intermediates in different protonated forms (in dependence on pH) are given in the following scheme (Fig. 1):

Abbreviations

b5: cytochrome b5, b5 red: NADH-cytochrome b5 reductase, DMSO: dimethylsulfoxide, EH: epoxidehydrolase, GGT: gamma-glutamyltranspeptidase, GP: glutathione peroxidase, GR: glutathione reductase, GT: glutathione S-transferase(s), luc: lucigenin, lum: luminol, MDA: malondialdehyde, PUFA: polyunsaturated fatty acid, TBARS: thiobarbituric acid reactive substances (mainly MDA)

NATO ASI Series, Vol. H 90
Molecular Aspects of Oxidative Drug Metabolizing Enzymes
Edited by E. Arınç, J. B. Schenkman and E. Hodgson
© Springer-Verlag Berlin Heidelberg 1995

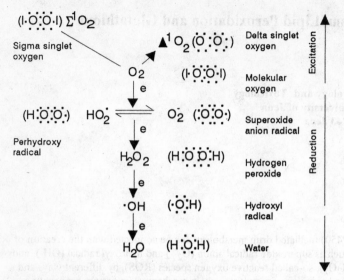

Fig. 1: Reactive oxygen species (ROS) produced during reduction of dioxygen to form water in dependence on pH. e = negatively charged electrons.

Superoxide radical anions have a halflife time of about 9 sec and are relatively inert. H_2O_2, which is not a radical, can readily permeate membranes, is relatively stable, and is highly reactive, therefore it is of general biological and especially toxicological importance and must be detoxicated by the antioxidative system (enzymes catalase and peroxidases, hydrophilic antioxidants such as ascorbic and uric acid and GSH, and lipophilic antioxidants such as vit E) immediately after formation to avoid injury.

Hydrogen peroxide is formed not only by direct divalent reduction of dioxygen, but also by slow nonenzymatic or fast enzymatic, superoxide dismutase (SOD) catalyzed dismutation of O_2^{-}:

$$O_2^{-} + O_2^{-} + 2H^+ = O_2 + H_2O_2$$

Of highest reactivity is the hydroxyl radical with a half life of only 10^{-9} sec. Hydroxyl radical can be formed from superoxide radical anion and H_2O_2 in the so-called Haber-Weiss reaction:

$$O_2^{-} + H_2O_2 = O_2 + OH^- + OH^{\cdot}$$

This slow reaction is accelerated by metal catalysis in living organisms by the so-called Fenton reaction:

$$O_2^{-} + Fe^{3+} = Fe^{2+} + O_2$$
$$Fe^{2+} + H_2O_2 = Fe^{3+} + OH^- + OH^{\cdot}$$
In summary:
$$O_2^{-} + H_2O_2 = O_2 + OH^- + OH^{\cdot}$$

Instead of H_2O_2 also an organic hydroperoxide may participate in the second step of this reaction:

$$Fe^{2+} + ROOH = Fe^{3+} + RO^{\cdot} + OH^-$$

The alkoxy radical RO˙ may react with the biological systems itself and create a chain reaction, cp. lipid peroxidation (LPO).

ROS are produced in all cells investigated so far and by a variety of cell organelles (mitochondrial and endoplasmic electron transport chains, peroxisomes, plasma membranes, envelop of the nuclei), by various enzymes and macromolecules under physiological conditions. Physiological free radical reactions are under strict control by the different antioxidant systems. ROS are produced at an increased rate in pathological states, e.g. inflammation, cell injury by irradiation or toxicants, in postischemic states, carcinogenesis, by different xenobiotics, among them drugs and environmental pollutants, antibiotics, cytostatics etc. (Fehér et al. 1987, Cotgreave et al. 1988, Southeron and Powis 1988, Csomós and Fehér 1992, Spatz and Bloom 1992, Halliwell and Atuoma 1993, Poli et al. 1993, Slater and Cheeseman 1993, Tarr and Samson 1993). Also physiological and pathological aging is attributed to the constant production of ROS (Harman 1984, Emerit and Chance 1992). The generation of ROS by phagocytes involved in host defence and inflammation with both beneficial and destructive consequences is well established (Dahlgren and Stendahl 1983, Elsbach and Weiss 1983, Briheim et al. 1984, Allen 1986, Gyllenhammar 1987, Williams and Cole 1991). Only phagocytes appear to possess the membrane-bound NAD(P)H oxidase which reduces molecular oxygen to the superoxide radical anion (Allen 1986).

The P450-GSH-System

In the hepatocyte superoxide radical anion is also formed in different organelles, predominantly in the endoplasmic reticulum by the so-called oxidase function of P450 (Kuthan and Ullrich 1982). Substrates for the monooxygenase activity of P450 can be "true substrates": the cofactors NADPH and oxygen are completely consumed to form the hydroxylated product and water. Many substrates have an "uncoupling activity": part of NADPH and oxygen (21 - 83%) are used to form either superoxide radical anions or hydrogen peroxide besides the hydroxylated product. Some substrates are "pseudosubstrates", "complete uncouplers", mainly fluorinated compounds (Archakov and Zhukov 1989).

All possible aspects of this oxidase activity are discussed by Archakov and Zhukov (1989): The stochiometry, the superoxide anion and hydrogen peroxide formation and finally direct water formation. Either after the release of superoxide anions with consecutive formation of hydrogen peroxide, spontaneously or enzymatically via the superoxide dismutase (SOD) (Coon and Vaz 1987) or after direct formation of hydrogen peroxide predominantly at low pH (Zhukov and Archakov 1985, Klinger et al. 1986, Blanck et al. 1991) the highly reactive hydroxyl radicals can be formed, which may extract a hydrogen from the lipid, especially from fatty acids with at least two double bonds (polyunsaturated fatty acids, PUFA), a free lipid radical (R.) will be formed, which reacts readily with molecular oxygen to form a peroxy free radical (ROO.), which is the beginning of a chain reaction called lipid peroxidation (LPO), in which PUFA peroxidation is only one step, cp. Fig. 2.

I. Initiation

$$X^\bullet + -CH_2-CH=CH-CH_2-CH=CH-CH_2-$$

$\downarrow$ -H

Alkyl radical

$$XH + -CH_2-CH=CH-CH^\bullet-CH=CH-CH_2-$$

Alkyl radical, isomer   Alkyl radical, isomer

$$-CH_2-CH=CH-CH=CH-CH-CH_2- \quad \text{and} \quad -CH_2-CH-CH=CH-CH=CH-CH_2-$$

$\downarrow \cdot O_2$ $\downarrow \cdot O_2$

$$-CH_2-CH=CH-CH=CH-CH \quad \text{and} \quad -CH_2-CH-CH=CH-CH=CH-CH_2-$$
$$\begin{array}{c} | \\ O-O^\bullet \end{array} \qquad\qquad \begin{array}{c} | \\ O-O^\bullet \end{array}$$

$\downarrow$

$$R^\bullet +-CH_2-CH=CH-CH=CH-CH-CH_2- \text{ and } -CH_2-CH-CH=CH-CH=CH-CH_2-+R^\bullet$$
$$\begin{array}{c} | \\ O-OH \end{array} \qquad\qquad \begin{array}{c} | \\ O-OH \end{array}$$

II. Catalysis and propagation

$$-CH_2-CH=CH-CH=CH-CH-CH_2-$$
$$\begin{array}{c} | \\ O-OH \end{array}$$
Alkoxy radical

$\downarrow$ + Spontaneously or metal catalyzed

$$-CH_2-CH=CH-CH=CH-CH-CH_2 \quad + \quad {}^\bullet OH$$
$$\begin{array}{c} | \\ O^\bullet \end{array} \qquad\qquad \downarrow +RH$$

$\downarrow$ +RH $HOH+R^\bullet$

$$-CH_2-CH=CH-CH=CH-CH_2 + R^\bullet$$
$$\begin{array}{c} | \\ OH \end{array}$$

III. Fragmentation and termination

Alkoxy radical

$$-CH_2-CH=CH-CH-CH=CH- \qquad\qquad + {}^\bullet OH$$
$$\begin{array}{c} | \\ O^\bullet \end{array}$$

$\downarrow$ $\downarrow$

Aldehyde   Alkyl radical

$$-CH_2-CH=C-CH \qquad\qquad + \quad {}^\bullet CH=CH-$$
$$\begin{array}{c} \| \\ O \end{array}$$
$$-CH_2-CH=CH-CH-CH=CH-$$
$$\begin{array}{c} | \\ O-CH=CH- \end{array}$$

Fig 2: In the first step: Initiation alkyl radical isomers are formed, while the double bonds are arranging in dien- conjugations in a progressive manner. Simultaneously molecular oxygen binds to the C with radical character and alkoxy radicals are formed in the second step: Catalysis and Propagation. In the following third step: Fragmentation and termination the fatty acid radicals are fragmented to form different smaller molecules, among them malondialdehyde (MDA) which represents more than 60% of the original lipid. (Slightly modified from Fehér et al. 1987).

LPO causes membrane damage, but this is not the only and main toxicological event: target molecules are also DNA, causing cytotoxicity, mutagenicity and carcinogenicity (Halliwell and Aruoma 1993), and proteins, causing enzyme and membrane damage. Due to the easily detectable TBARS most investigations have been performed on LPO as one consequence of ROS formation, but the influence of ROS on membrane structure is a much more complex one, s. Fig. 3.

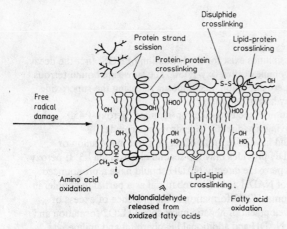

Fig. 3: ROS induced membrane damage. (From Fehér et al. 1987).

Hydroxylated and reactive metabolites (epoxides) of xenobiotics are formed by the monooxygenase function of P450 which can be detoxified by hydrolysis, catalyzed by epoxide hydrolase (EH), cp. Thomas et al. 1990, and conjugation with GSH, catalyzed by a family of glutathione-S-transferases (GT), cp. Ketterer and Taylor (1990).

GSH is also the main antioxidant to detoxify ROS and lipidperoxides as a scavenger. Thus GSH is the complementary or counterpart of both the monooxygenase and oxidase functions of P450. Therefore it seems to be necessary to discuss P450 activities and their general biological im-portance only in connection with the reactions of the GSH system as it is pointed out in Fig. 4.

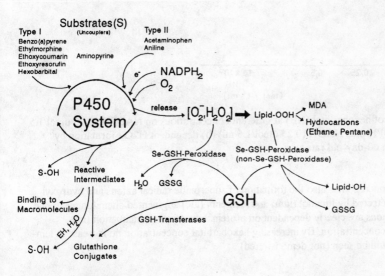

Fig. 4: The P450-GSH-System

The Oxidative System

<u>Formation of Reactive Oxygen Species (ROS)</u>
According to the P450 cycle two possibilities exist for the formation of ROS: first the decay of the one-electron-reduced ternary complex called oxy-P450. The substrate bound ferrous hemoprotein, complexed with the superoxide anion, decomposes releasing the superoxide anion which can dismutate to form hydrogen peroxide.
The second alternative is the protonation of the two-electron-reduced peroxy-P450-complex, the dissociation of which giving rise directly to hydrogen peroxide.
If oxy-P450 decays then NADH should have little or no effect on superoxide anion or hydrogen peroxide formation, as NADH provides only the second electron via b5. If peroxy-P450 decomposes to form hydrogen peroxide directly NADH should have a pronounced effect. We investigated the influence of NADH and of hexobarbital as a partial uncoupler in various concentrations on the microsomal H2O2 formation in the presence of excess of NADPH. Moreover the postnatal development of hepatic microsomal H2O2 formation and the influence of NADPH, NADPH + NADH and additional hexobarbital and aniline was investigated. The results are given in Figs. 5-7.

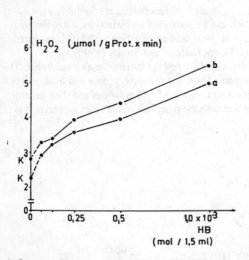

Fig. 5: Influence of hexobarbital (HB) at different concentrations on NADPH (0.75 umol/1.5 ml) (a) and NADPH plus NADH (0.25 umol/1.5 ml) (b) dependent H2O2 formation by microsomes from 60-day-old rats.

With rat hepatic microsomes also the formation of superoxide radical anions and hydroxyl radicals can be detected by luminol (lum) and lucigenin (luc) augmented chemiluminescence (CL). Both reactions are clearly dependent on proteinconcentration, incubation time, NADPH and Fe concentration. By increasing hexobarbital concentration no increase in lum-CL or luc-CL could be seen (not demonstrated).

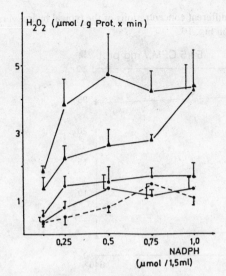

Fig. 6: Influence of hexobarbital at different concentrations (10^{-2}-10^{-5}, upper to lower lines) on NADPH dependent H2O2 formation by microsomes from 10-day-old rats. Controls = dashed line. (From Klinger et al. 1986).

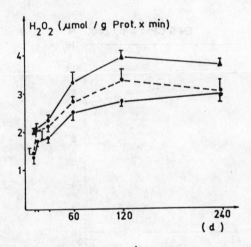

Fig. 7: Influence of hexobarbital (10^{-4} M) (upper line) and aniline (10^{-4} M) (lower line) compared with controls (dashed line) on NADPH (0.75 umol/1.5 ml) dependent H2O2 formation by microsomes from rats of different age. It can be seen that hexobarbital increases H2O2 formation both in the presence of excess NADPH and with additional NADH. Thus we concluded that H2O2 is released after the reduction by a second electron via b5 by dissociation of the peroxy-P450-complex. And this holds true in all age groups.

Both CL reactions could be totally abolished by increasing concentrations of Cu, ascorbic acid or GSH. The influence of increasing activities of catalase and superoxide dismutase is demonstrated in Figs. 8 and 9, that of different concentrations of the typical hydroxyl radical scavenger dimethylsulfoxide (DMSO) in Fig. 10.

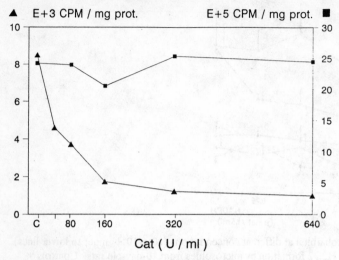

Fig. 8: Influence of catalase (Cat) with increasing activities on luminol- (▲) and lucigenin-amplified (■) chemiluminescence.

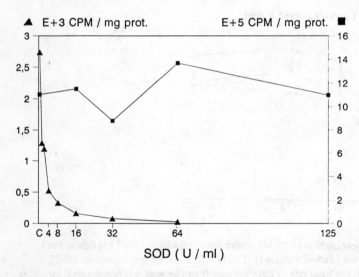

Fig. 9: Influence of superoxide dismutase (SOD) with increasing activities on luminol- (▲) and lucigenin-amplified (■) chemiluminescence.

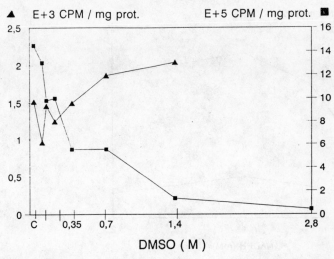

Fig. 10: Influence of dimethylsulfoxide (DMSO) on luminol- (▲) and lucigenin-amplified (■) chemiluminescence.

From these results we conclude, that lum-CL specifically indicates $O_2.'$, whereas luc-CL measures $OH.'$.

Calibration curves for both CL reactions can be performed with H_2O_2. As the different ROS can change from one form to another (cp. Fig. 1) these measurements do not permit conclusions which ROS is originally released from P450. Lum- and luc-CL have been widely used to determine e.g. phagocyte activity (Müller-Peddinghaus 1984), to test oxygen radical scavengers (Müller-Peddingshaus and Wurl 1987) etc., but a more direct proof for the specificity of CL has not been given to our best knowledge. According to our former experiments (Klinger et al. 1986) we are convinced that the primary form is H_2O_2. H_2O_2 production has also been shown in vivo (Premereur et al. 1985).

Lipidperoxidation (LPO) initiated by ROS

For decades the work in our institute is dedicated to problems of developmental pharmacology (cp. Klinger et al. 1987).

Among the practical problems and tasks of this field of investigation pathophysiology and therapy of hypoxia in the neonate have highest importance, the central nervous system evidently is most endangered by hypoxia and ischemia and posthypoxic and postischemic states.

We were unable so far to detect lum-CL with brain microsomes, but luc-CL can be demonstrated. Also the formation of TBARS can be shown, demonstrated is the dependence of both reactions on NADPH, cp. Figs. 11 and 12.

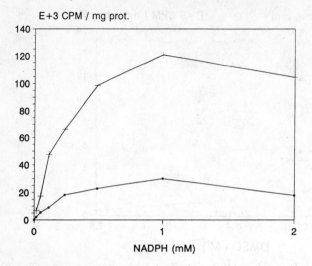

Fig. 11: Influence of NADPH on luc-CL by brain microsomes of newborn (■) and 60-day-old (+) rats.

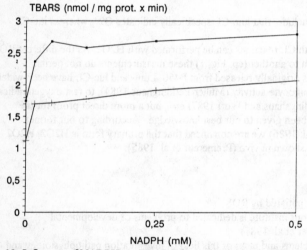

Fig. 12: Influence of NADPH on TBARS formation by brain microsomes of newborn (■) and 60-day-old (+) rats.

Though the ROS formation is much higher by brain microsomes of young adult rats, TBARS production is higher in newborn rats. This may indicate a much higher antioxidative capacity in adult compared with newborn rats. This shows a high vulnerability of the newborn brain, which could be demonstrated by an enhanced TBARS and GSSG formation in newborn brain by hypoxia, whereas liver is much more stable, cp. Fig. 13.

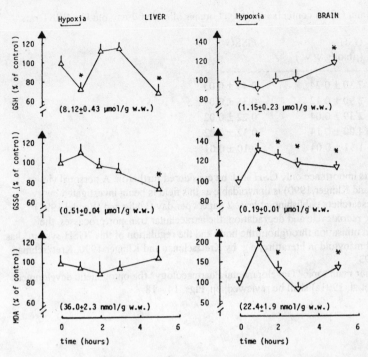

Fig. 13: The influence of reversible hypoxia and reoxygenation on GSH, GSSG and TBARS concentration in liver and brain of newborn rats. (From Kretzschmar et al. 1990).

This example shows that ROS and TBARS formation are not closely linked, distinctly enhanced TBARS production can be expected only after an exhaustion of the antioxidative capacity.

The Antioxidative System

The GSH-System
This system comprises hydrophilic (e.g. ascorbic acid, uric acid, GSH, but also proteins) and lipophilic (e.g. vit E) substances and enzymes such as catalase and superoxide dismutase. By extraction of a proton these substances can form instable reactive radicals and can be oxidized. Membrane bound lipophilic vit E forms a radical after oxidative attack by ROS and can be restored by hydrophilic cytosolic GSH, the GSH radical can be restored e.g. by ascorbic acid, after oxydation to GSSH it can be reproduced by GSH reductase (GR) with NADPH as cofactor. Thus the tripeptide GSH acts like the center of different hydrophilic and lipophilic, cytosolic and membrane bound redox pairs, having itself the highest antioxidative capacity, due to its high concentration mainly in liver, but also in other organs, cp. Tabl. 1.

Table 1: GSH and GSSG contents in different organs of male 60-day-old Uje:WIST rats
(N = 10).

	GSH (µmol/g w.w.)	GSSG
liver	7.30 ± 0.24	0.41 ± 0.03
lung	2.90 ± 0.12	0.38 ± 0.04
heart	2.39 ± 0.04	0.22 ± 0.02
kidney	4.00 ± 0.14	0.32 ± 0.02
brain	1.51 ± 0.04	0.10 ± 0.01

According to its importance only GSH shall be considered furtheron. A personal view
(Kretzschmar and Klinger 1990) is unavoidable as this field is being investigated by
uncountable researchers publishing at least 2 papers per day (Uhlig and Wendel 1992).
GSH synthesis, redox cycle and degradation, the intracellular transportprocesses, the
distribution and utilisation throughout the body and the regulation of this "GSH-system" has
been presented manyfold in literature, e.g. by Kretzschmar and Klinger 1990, Kretzschmar
and Müller 1993.
According to our main topic "Developmental Pharmacology" the ontogenetic development
(Kretzschmar et al. 1991a) shall be reviewed, cp. Figs. 14 - 18.

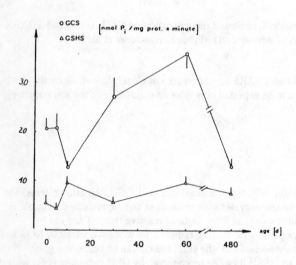

Fig. 14: Age course of hepatic activities of gamma- glutamycysteinyl synthetase (o) and GSH
synthetase (Δ).

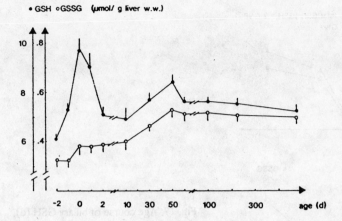

Fig. 15: Age course of hepatic GSH (●) and GSSG (o).

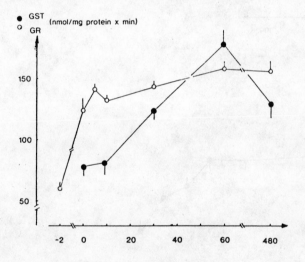

Fig. 16: Age course of hepatic GT (o) and GR (●).

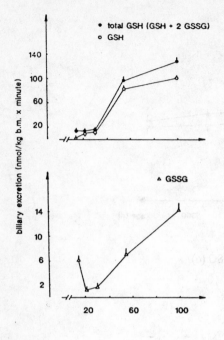

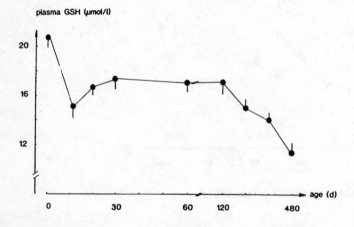

Fig. 17: Age course of biliary GSH (o), GSSG ( ) and total GSH (GSH + 2 GSSG)(●).

Fig. 18: Age course of plasma GSH concentration.

Remarkable are the activities of the first step of GSH synthesis around birth and the high GSH concentrations in liver and in plasma (highest in the whole life span). This can be explained as a rebound phenomenon due to hypoxia/oxidative stress in the perinatal period. Also the unexpected and extremely high activities of gamma- glutamyltranspeptidase (GGT) in the perinatal period (Kretzschmar and Klinger 1989) indicate the high activity of the GSH-system in the perinatal period, as GGT may serve the cystein transport for GSH synthesis in this period.

The capacity of the GSH-system can be exhausted under conditions of hypoxia in the neonatal period: in brain and lung of newborn rats as well as in plasma the GSSG concentrations increased, whereas in liver, considered to be the main organ for GSH synthesis and export at least in the neonatal period, the GSH concentration fell (Reuter and Klinger 1992).

Evidently the GSH status is well reflected by the GSH and GSSG concentrations in plasma. Investigations on GSH and GSSG concentrations as well on TBARS in plasma in different hypoxic, ischemic, infectious and traumatic diseases are in progress with highly interesting preliminary results, e.g. that a tremendous decline of plasma GSH within the first 3o min after a severe accident is frequently followed by SIRS (systemic inflammatory reaction syndrome). Final fatal multiorgan failure is frequently preceded by an almost total loss of plasma GSH. On the other hand cases were observed with prefinal extremely high plasma levels of GSH (Kretzschmar 1994).

The Influence of Aging, Training and Exercise on the P450 - GSH - System

Until recently plasma levels of GSH, GSSG and TBRS were believed to reflect the GSH status of the liver in normal and most pathological states, as liver was considered to be the only organ for GSH export. The active GSH transport through the sinusoidal membranes of the hepatocytes into blood was determined to range in between 12 to 18 nmol/min/g liver. This assumption was only partly confirmed for the age courses: in rats the GSH concentrations in liver as well as in plasma decreased immediately after the neonatal period, but with increasing age the plasma levels continued to fall whereas the liver concentration remained almost stable (cp. Kretzschmar et al 1991a, Kretzschmar and Müller 1993). Decreasing values with age have also been observed in human beings (Kretzschmar et al. 1991b). Experiments in hepatectomized rats led to the conclusion that organs or tissues other than liver contribute considerably to the total influx into circulation (Kretzschmar et al. 1992). Training of rats, dogs and human beings and comparison of the influence of acute exercise on oxidative and antioxidative parameters in trained and untrained individuals confirmed specified the results gained in hepatectomized rats: striated muscles play an important role in the GSH homoeostasis of the whole body. The GSH export rate of striated muscle was estimated at nearly 2 nmol/min/g, thus being the second important source for plasma GSH in untrained individuals. In trained muscles GSH synthetase activities as well as GSH concentrations were distinctly higher. The age dependent decrease of plasma GSH concentrations can be compensated by physical training. Decreased plasma GSH concentrations following physical exercise demonstrate increased GSH consumption by heart and skeletal muscle accompanied by a reduced export rate, which cannot be compensated by an exercise-induced stimulation of the GSH export from the liver by increased vasopressin levels (cp. Kretzschmar and Müller 1993).

Thus not only the monooxygenase function of P450 is enhanced by physical training (Frenkl et al. 1980a,b), but also the capacity of the GSH system. An overview on synthesis, export, transport and utilisation of GSH is given in Fig. 19.

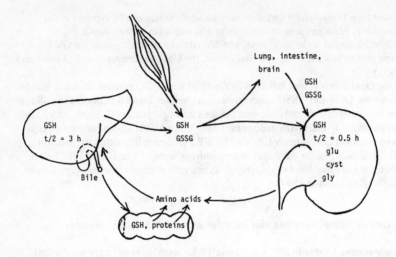

Fig. 19: Highest GSH synthesis activities have been shown in liver (export rate into blood 12 - 18 nmol/min/g), heart and skeletal muscle (export rate about 2 nmol/min/g) in rats, minor activities e.g. in kidney and brain, no activity in lung. The highest GSH utilisation is observed in kidney.

References

Allen R C (1986) Phagocytic leukocyte oxygenation activities and chemiluminescence: a kinetic approach to analysis. Methods in Enzymology 133:449-493

Archakov A I, Zhukov A A (1989) Multiple activities of cytochrome P-450. In: Ruckpaul K, Rein H (eds): Frontiers in biotransformation. Basis and mechanisms of regulation of cytochrome P-450. Akademie-Verlag Berlin, 1:151-175

Blanck J, Ristau O, Zhukov A A, Archakov A I, Rein H, Ruckpaul K (1991) Cytochrome P-450 spin state and leakiness of the monooxygenase pathway. Xenobiotica 21:121-135

Briheim G, Stendahl O, Dahlgren C (1984) Intra- and extracellular events in luminol-dependent chemiluminescence of polymorphonuclear leukocytes. Infect Immun 45:1-5

Coon M J, Vaz A D N (1987) Radical intermediates in peroxide-dependent reactions catalyzed by cytochrome P-450. J Biosciences 11:35-40

Cotgreave I A, Moldeus P, Orrenius S (1988) Host biochemical defense mechanisms against prooxidants. Ann Rev Pharmacol Toxicol 28:189-212

Csomós G, Fehér J (eds)(1992) Free radicals and the liver. Springer Berlin

Dahlgren C, Stendahl O (1983) Role of myeloperoxidase in luminol-dependent chemiluminescence of polymorphonuclear leukocytes. Infect Immun 39:736-741

Elsbach P, Weiss J (1983) A reevaluation of the roles of the O_2-dependent and

O₂--independent microbial systems of phagocytes. Rev Infect Dis 5:843-853

Emerit I, Chance B (eds)(1992) Free radicals and aging. Birkhäuser Basel

Fehér J, Csomós G, Vereckei A (eds) (1987) Free radical reactions in medicine. Springer Berlin

Frenkl R, Gyre A, Szeberenyi S (1980a) The effects of muscular exercise on the microsomal enzyme system of the rat liver. Europ J Appl Physiol 44:135-140

Frenkl R, Gyre A, Meszaros J, Szeberenyi S (1980b) A study of the enzyme inducing effect of physical exercise in man. J Sports Med 20:371-376

Gyllenhammar H (1987) Lucigenin chemiluminescence in the assessment of neutrophil superoxide production. J Immun Meth 97:209-213

Halliwell B, Aruoma O I (eds) (1993) DNA and free radicals.Ellis Horwood Chichester

Harman D (1984) Free radicals and the origination, evolution, and present states of the free radical theory of aging. In: Armstrong D, Sohal R S, Cutler R G, Slater T F (eds) Free radicals in molecular biology, aging, and disease, Raven Press New York, pp 1-12

Kahl R, Weimann A, Weinke S, Hildebrandt A G (1987) Detection of oxygen activation and determination of the activity of antioxidants towards reactive oxygen species by use of the chemiluminogenic probes luminol and lucigen. Arch Toxicol 60:158-162.

Ketterer B, Taylor J B (1990) Glutathione Transferases. In: Ruckpaul K, Rein H (eds): Frontiers in biotransformation. Principles, mechanisms and biological consequences of induction. Akademie-Verlag Berlin, 2:244-277

Klinger W, Freytag A, Schmitt W (1986) Influence of age, hexobarbital, and aniline on NADPH/NADH dependent hydrogen peroxide production in rat hepatic microsomes. Arch Toxicol, Suppl 9:382-385

Klinger W, Müller D, Kleeberg U, Jahn F, Glöckner R (1987) Developmental pharmacology. In: Rand M J, Raper C (eds) Pharmacology. Proceedings of the Xth International Congress of Pharmacology (IUPHAR), Sydney, 23-28 August 1987, pp 753-763

Klinger W (1990) Biotransformation of xenobiotics during ontogenetic development. In: Ruckpaul K, Rein H (eds) Frontiers in Biotransformation. Principles, mechanisms and biological consequences of induction. Akademie-Verlag Berlin, 2:113 - 149

Kretzschmar M, Klinger W (1989) Gamma-glutamytranspeptidase in liver homogenates of rats of different ages: enzyme kinetics and age course of Km and Vmax. Z.Versuchstierk. 32:41-47

Kretzschmar M, Glöckner R, Klinger W (1990) Glutathione levels in liver and brain of newborn rats: investigations on the influence of hypoxia and reoxidation on lipid peroxidation. Physiol Bohemoslov 39:257-260

Kretzschmar M, Klinger W (1990) The hepatic glutathione system - influences of xenobiotics. Exp Pathol 38:145-164

Kretzschmar M, Bach G, Ratzmann I, Fleck C, Klinger W (1991a) Ontogenetic changes in hepatic glutathione system (synthesis, catabolism, export) of male Uje:Wist rats. J Exp Anim Sci 34:132-139

Kretzschmar M, Müller D, Hübscher J, Marin E, Klinger W (1991b) Influence of aging, training and acute physical exercise on plasma glutathione and lipid peroxides in man. Int J sports Med 12:218-222

Kretzschmar M, Pfeifer U, Machnik G, Klinger W (1992) Glutathione homeostasis and turnover in the totally hepatectomized rat: evidence for a high glutathione export capacity of extrahepatic tissues. Exp Toxic Pathol 44:273-281

Kretzschmar M, Müller D (1993) Aging, trainig and exercise. Sports Med 15:196-209

Kretzschmar M (1994) The role of free radicals in the pathophysiology. In: Vincent J L, Sprung C, Eyrich K, Reinhart K (eds) Update in intense care and emergency medicine. Springer Berlin (in press)

Kuthan H, Ullrich V (1982) Oxidase and oxigenase function of the microsomal cytochrome P-450 monooxygenase system. Eur J Biochem 126:583-588

Müller-Peddinghaus R (1984) In vitro determination of phagocyte activity by luminol- and lucigenin-amplified chemiluminescence. Int J Immunpharmac 6:455-466

Müller-Peddinghaus R, Wurl M (1987) The amplified chemiluminescence test to characterize antirheumatic drugs as oxygen radical scavengers. Biochem Pharmacol 36:1125-1132.

Poli G, Albano E, Dianzani M U (eds)(1993) Free radicals: from basic science to medicine. Birkhäuser Basel

Premereur N, Van den Branden C, Roels F (1986) Cytochrome P-450-dependent H_2O_2 production demonstrated in vivo. FEBS 199:19-22.

Rao, P S, Luber J M, Milinowicz J, Lalezari P, Mueller H S (1988) Specificity of oxygen radical scavengers and assessment of free radical scavenger efficiency using luminol enhanced chemiluminescence. Biochem Biophys Res Commun 150:39-44

Reuter A, Klinger W (1992) The influence of systemic hypoxia and reoxygenation on the glutathione redox system of brain, liver, lung and plasma in newborn rats. Exp Toxic Pathol 44:339-343

Sies H, Murphy M E, Di Mascio P, Stahl W (1992) Tocopherols, carotenoids and the glutathione system. In: Ong A S H, Packer L (eds) Lipid-soluble antioxidants: biochemistry and clinical applications. Birkhäuser Basel, pp 160-165

Slater T F, Cheeseman K H (eds)(1993) Free radicals in medicine. Brit Med Bull 49, No 3

Southeron P A, Powis G (1988) Free radicals in medicine. I. Chemical nature and biologic reactions. II. Involvement in human disease. Mayo Clin Proc 63:381-408

Spatz L, Bloom A D (eds)(1992) Biological consequences of oxidative stress. Implications for cardiovascular disease and carcinogenesis. Oxford University Press New York/Oxford

Tarr M, Samson F (eds)(1993) Oxygen free radicals in tissue damage. Birkhäuser Basel

Thomas H, Timms C W, Oesch F (1990) Epoxide hydrdolase: molecular properties, induction, polymorphisms and function. In: Ruckpaul K, Rein H (eds) Frontiers in Biotransformation. Principles, mechanisms and biological consequences of induction. Akademie-Verlag Berlin, 2:278-337

Uhlig S, Wendel A (1992) Minireview. The physiological consequences of glutathione variations. Life Sci 51:1083-1094

Vereckei A, Fehér E, Blázovics A, György J, Toncser H, Fehér J (1992) Free radical reactions in the pathomechanism of amiodarone liver toxicity.In: Csomós G, Fehér J (Eds.): Free radicals and the liver, Springer Berlin etc. pp.124-157.

Weimann A, Hildebrandt A G, Kahl R (1984) Different efficiency of various synthetic antioxidants towards NADPH induced chemiluminescence in rat liver microsomes. Biochem Biophys Res Commun 25:1033-1038.

Williams A J, Cole P J (1991) In vitro stimulation of alveolar macrophage metabolic activity by polystyrene in the absence of phagocytosis. Br J exp Path 62:1-7.

Health Environment of Altai Territory and Monooxygenase and Conjugating Activities in Placentae of Newborn

Sergey V. Kotelevtsev, Victor V. Obrazsov,
Cyril N. Novicov and Ludmila I. Stepanova
Laboratory of Physical-Chemistry Biomembranes,
Department of Biology,
M.V. Lomonosov Moscow State University,
119899 Moscow, Russia.

Introduction

Intensification of the industry and agriculture in Russia, despite a number of measures on the environmental protection, is accompanied by the exhausts in the air, in water and ground ecosystems of constantly increasing amount of xenobiotics, which possess the ability not only to accumulate in animals and human tissues but causing development of pathological processes of various etiology. Another important factor that initiates different pathological states, is accumulation of radionuclides in the environment, increasing to a great extent after the Chernobyl tragedy. Long-time consequences of minor dosages of the ionized radiation with the chronic character of this influence on the molecular level at present are not thoroughly studied. However, it is known, that the genetic apparatus is the first to suffer, irreversible disturbances in the system of haemogenesis, in the membrane structures of the cells of the blood take place. For the last time in connection with appearance of new data about the character of the radioactive pollution caused by some accidents at the plants of the military industrial complex and consequences after the nuclear arms testing, there occurred the question concerning the associated action of radionuclides and environmental pollution by xenobiotics, that reveal the mutagenous, carcinogenic and teratogenic action.

Accumulation of radionuclides in the environment, increasing to a great extent in Altay region after surface

NATO ASI Series, Vol. H 90
Molecular Aspects of Oxidative Drug Metabolizing Enzymes
Edited by E. Arınç, J. B. Schenkman and E. Hodgson
© Springer-Verlag Berlin Heidelberg 1995

nuclear tests had been carried out in 1949 (29 August), 1955 (6 November), 1962 (7 August and 25 September). The radioactive trashy cover the territory more then 500 km from Semipalatinsk's polygon and the effective irradiation doses obtained by same grope of people was more then 35 cSv.

The Altay region is a place with intensive industry and agricultural, with produce strong chemical pollution. The majority of these chemical compounds possess hydrophobic characteristics, they are capable of accumulating in the biological membranes, and changing their physical-chemical properties and causing modification activity of membrane-related enzymes. A considerable amount of hydrophobic xenobiotics (in the first place polycyclic aromatic hydrocarbons and polychlorinated biphenyls) are subjected to detoxication and metabolic activation in the membrane monooxygenases - the cytochrome P450 system and are able of inducting to a great extent the activity of these enzymes. Haem destruction precedes bilirubin formation and its evacuation as a result of the conjugation with the glucuronic acid. Disturbances of different types in the membrane system functioning of the endoplasmatic reticulum (in the first place, liver, where membrane-related enzymes of detoxication possess the maximum activity, and bear the main loading on fatty acid hormones metabolism, cholesterol, haemoglobin and xenobiotics) can serve one of the reasons of haemolytic jaundice development. The cytochrome P450 system and the enzymes conjugation system composing these enzyme complexes, are liable and are subjected to modification by a number of factors. Disturbance in microsomal monooxygenases functioning is caused by radioactive radionuclides actions, that was demonstrated not long ago on the liver microsomal fraction in rats under chronic radiation in the Chernobyl electric power station region (Akhalaya et al. 1990).

The amount of newborns with the general pathology and stillborns in some regions of the Altai territory is considerable and continues to increase as well as mortality of grown-up population from different oncological diseases. More then 25% newborns at same region suffer from haemolytic jaundice (Kolyadko et al. 1991). According to the character and the severity degree jaundice of this etiology considerably differs from physiological jaundice and can be corrected only in intensive medicamentous therapy. There are available some convincing indications of an uninfectious character of this pathology, one of the reasons of its occurrence is the unfavorable influence of the environment factors both at present and in the period of the development of parents whose children are subjected to this form of jaundice.

Besides, the activity of the membrane-related enzymes is under multistage genetic control of special genetic families and according to our minds, can be modified as a result of the mutation, including the radiation activity. On the other hand, enzymes of the microsomal system, the so called enzymes of the phase II detoxication, performing the conjugation reactions, take place in the detoxication of xenobiotics (Pasanen and Pelkonen 1989/90). Thus, study of the placental glutathione S-transferase (GST) activity of a human, especially in combination with other placental detoxication enzymes (monooxygenase enzymes, epoxyhydrolase etc.) is the necessary step in understanding the nature of different pathologies in newborns, including jaundice (Pacifici and Rane 1981).

Among other detoxication enzymes in placentae the important role belongs to UDP-glucuronosyltransferase (UDP-GT), performing the conjugation of xenobiotics with UDP-glucuronic acid (Tephly and Burchell 1990). The importance of this enzyme is stressed by the fact, that it takes part in the endogenic compounds metabolism: bilirubin, thyroxin, tetrahydrocortizol, steroid hormones. It is known, that tenfold differences in rates of glucuronization of different compounds are the usual picture, observed in healthy human population. Moreover, at present, it is un-known, whether these variations in activity of placental UDP-GT are the consequences of different age, pathological states, presence of xenobiotics in environment or have some genetic basis (Tephly and Burchell 1990).

We conducted investigations of the activity both as isoforms of cytochrome P450 and its content in the placental tissues of women in labor of the different regions of the Altai territory, and the activity of enzymes of the phase II xenobiotic detoxication which performs the conjugation with glutathione and glucuronic acid. Parallel we studied the content in the placental tissues of mutagenious and carcinogenic compounds with Ames's test.

Materials and Methods

Placentae was isolated at once after childbirth, perfused by the isolating medium contained: 1,15% KCl, 15 mg/l BSA (V fraction), 1 mg/l phenylmethylsulfonylftoride, and was froze in liquid nitrogen for the transport in laboratory. Placentae was kept at -70 C^0 no more than three weeks. The placental microsomal fraction was isolated by the method of the differential centrifugation (Pelkonen and Saarni 1980) with some modifications and was used for the determination of cytochrome P450 concentration and monooxygenase and conjugating activities.

Cytochrome P450 was measured as described by Johannesen and De Pierre (Johanessen and De Pierre 1978). The activities of 7-ethoxycoumarin (ECOD) and 7-ethoxyresorufin O-deethylase (EROD) were measured by direct fluorometric method (Ullrich and Weber 1972; Prough et al. 1978).

The activities of microsomal and cytoplasm GST were determined by Habig's method (Habig et al. 1974). The enzyme activity was monitored by spectrophotometrically by means the detection of glutathione conjugates with 1-Cl-2,4-dinitrobenzene at 340 nm. The incubation medium contains 2 ml of 0.1 M phosphate buffer solution with 1 mM EDTA, pH 6.5, 30 mkl of microsomal suspension or post microsomal supernatant and 30 mkl of 20 mM 1-Cl-2,4-dinitrobenzene solution. The reaction was initiated by addition of 30 mkl of 20 mM glutathione solution. The molar extinction coefficient of the forming product-conjugate was 9.6 mM^{-1} cm_{-1}.

The activity of UDP-GT was determined by a method of Isselbacher (Isselbacher 1956) by means a detection of the loss of p-nitrophenol at 400 nm. The enzyme activity was measured in

a medium of following composition: 200 mkl of microsomal
suspension, 200 mkl of 1% digitonin solution, 50 mkl of 75 mM
$MgCl_2$ and 30 mkl of 2.5 mM p-nitrophenol solution. The reaction
was started by 150 mkl of 1.24% UDP-glucuronic acid solution.
The reaction was carried out at 37^0 C during 20 min under a
continual agitation and was stopped by 3 ml of cold glicine
buffer (pH 7,4).

The mutagenicity of placentae tissue extracts were
analyzed by the modified semiquantitative Salmonella/microsomes
Ames's test (Ames et al. 1975). As standard promutagenes for
the control of metabolic activity, we used 2-aminoanthracene
and benz(a)pyrene, whose methabolites effectively induce gene
mutation of both types. For Ames's tests the placental tissue
was extracted using chloroform/acetone/hexane mixture (1:1:1)
as previously described (Kotelevtsev et al. 1986).

Results and Discussion

No noticeable distinctions in the cytochrome P450 content
of placental microsomes isolated from different placentae have
been found. The experiments fulfilled with various methods of
placentae perfusion show their importance and a necessary of
repeated perfusion. Pollution of the placental samples by
haemoglobin has marked influence on the cytochrome P450 level.
Apparently, the Johannesen's method of measurement of the low
cytochrome concentrations is more convenient then Omura and
Sato (Omura and Sato 1964) one inasmuch as it allows to avoid
of haemoglobin influence under the spectrophotometric
determination of cytochrome P450 although in this case it is
not possible to measure the cytochrome P420 - inactivated form
of cytochrome P450.

The cytochrome P450 concentration in the investigated
samples of placentae does not exceed 0.05 nm per 1 mg of
microsomal protein that is corresponded to earlier published
data (Pelkonen et al. 1980a).

Microsomal placental ECOD demonstrates low activity. In
some samples the activities of ECOD was not detected at all

that corresponded to earlier published data. It was shown in works of Jacobson and Pelkonen (Jacobson et al. 1974; Pelkonen and Moilanen 1979; Pelkonen et al. 1986) that in some cases there was an absence of the ECOD activity in placentae of non-smoking women. The results of an analysis of monooxygenase activities are presented in Fig. 1 and 2.

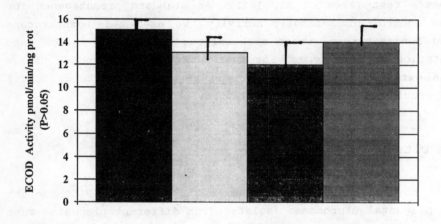

■ Series 1 □ Series 2 ▨ Series 3 ▤ Series 4

    1 - Healthy newborn,
    2 - haemolitic jaundice,
    3 - physiological jaundice,
    4 - central nerve system diseases.

Fig. 1. ECOD activities in placentae (newborn with different diseases).

The ECOD activity does not strong correlate with that of EROD. The activity of EROD also is detected not at all samples and varies rather widely (results are not presented). The highest values of the EROD activity was detected in the samples of placentae of the women which smoked during pregnancy.

The level of monooxygenase activities in placentae of some mothers is significantly lower than the normal one. In Fig. 2 the results are presented as groups joining the mothers living in the territories having suffered from radioactive deposits during surface nuclear weapon tests in Semipalatinsk's polygon. The groups are conventionally formed taking into account the radioactive traces of four explosions with maximal effect to investigated regions.

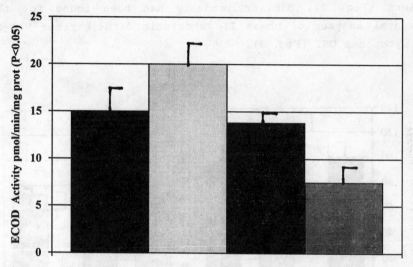

■ Series 1 ▢ Series 2 ■ Series 3 ▨ Series 4

1 - Healthy newborn from all region,

2 - healthy newborn of mothers from safe region,

3 - unsafe region (near 1 cSv),

4 - unsafe region (near 10 cSv).

Fig. 2. ECOD activities in placentae (mothers of newborn from different region).

The radioactive pollution in some regions was rather large. The irradiation doses obtained by thyroid glands reached 3000 cGr in the first group and 1000 cGr in the second one. As it is seen from the Fig. 1 the groups are jointed taking into account the affection of newborns by etiology of jaundice.

Children living in conventionally safe territories and all
healthy children have been taken as the control group. The data
obtained do not allow to find statistic significant
dependencies in the levels of activities of monooxygenase
enzymes in placentae in the groups of healthy and jaundice-
affecting children.

However, there is a tendency to decrease of monooxygenase
activities in the groups of mothers living in the polluted
regions (Fig. 2). Similar tendency has been found for the
placental enzymes of phase II xenobiotic detoxication: UDP-GT
and cytoplasm GST (Fig. 3).

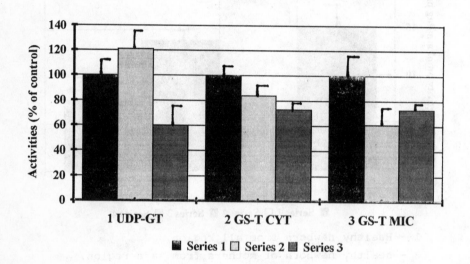

1 - Healthy newborn from all region,
2 - healthy newborn of mothers from safe region,
3 - unsafe region (> 1 cSv),
4 - unsafe region (> 10 cSv).

Fig. 3. The activities of detoxication enzymes (II Phases) in
placentae of mothers from different region of Altay territory.

The GST activity both in placental microsomes and in supernatant was few changeable for various samples. This results are corresponded to the data concerning that placental GST is uninducible enzyme which activity has low inter individual variability (Pacifici et al. 1981). Therefore the activity of this enzyme is either little decreased or kept unchangeable in the polluted groups. Another picture has been observed for UDP-GT. The enzyme activation in the second group is replaced by its inhibition in the third one. It should be note the activities of microsomal UDP-GT and GST in the second group are changed rather contrariwise. Inasmuch as the above enzymes have some identical substrates, one can consider this opposite changes of activities as compensate reaction of placentae. Naturally, the detected pathologies can be connected with alterations microsomal membrane not only placentae but liver and some other tissues, lymphocytaric for example. It is proved by our parallel experiments deal with studying of ions transport in red blood cell membranes and "oxygen burst" in the blood of mothers and newborns (Orlov et al. 1993).

At present there is not convincing proofs that induction or other alteration of monooxygenase activities in liver can be reflected by a character of activities of corresponding cytochrome P450 isoforms in human placentae. It should be note, that we didn't investigate the cytochrome P450 isoform composition in placentae at this stage of work.

On the other hand some toxic xenobiotics can appear to inhibit the monooxygenase activities in placentae altering microsomal membrane structure and initiating lipid peroxidation (Kulkarni 1987). By the way, it is seems for us also logically enough, that low activities of placental detoxication enzymes result in accumulation of mutagenic and cancerogenic compounds in such placentae.

At present the obtained results with the rear exception show an absence of any monooxygenases induction in placentae (Fig. I, 2). This fact allows to do a preliminary conclusion in respect of a lack in mother's organism of xenobiotics-inducers such as polychlorinated biphenyls and polycyclic aromatic hydrocarbons. Nevertheless the above compounds have been detected in the milk of some nurse mothers (Koliado et al. 1991).

However, as it is seen from the Fig. 1 and 2, the variety of monooxygenase activity values as well the cytochrome P450 content in placental microsomes are rather large. This fact is corresponded with published data. The variety is determined not only by an influence of environment but also by genetic factors. Indeed, a lot of works shows that monooxygenase activities can be markedly induce in placentae of smoking mothers (Pasanen and Pelkonen 1989/90).

Today it is not yet known is this induction a result of an influence of environmental conditions or genetic factors play the main role in a regulation of monooxygenase activities. Apparently, both liver and placental monooxygenase activities are associated with Ah locus (Shum et al. 1979; Juchau 1980; Manchester et al. 1987). The animal experiments (Pelkonen et al. 1980) show that placental AHH activities are well induced by 3-methylcholanthren in guinea pigs, rats, mice of C57B1/6 strain and (B6D2)xD2 backcross. On the other hand there is not induction in rabbits placentae or in placentae of insensitive mice line DBA/2(D2) and hybrid (B6Ds)xD2. As induction of placental AHH activity connects with induction of liver one in animals it can be suggested that genetic background is also important under induction of placental AHH activity (Pelkonen et al. 1980).

The investigation of the AHH activity induction being under genetic control is rather hard in human but indirect studies concerned with the dibenzoanthracene induction in cord blood and mother lymphocytes have showed that there is statistically significant correlation ($r = 0.75$, $p < 0.01$) between placental AHH activity and the inducibility of cord blood lymphocytes. The correlation was poor with lymphocytes of mother (Pelkonen et al. 1981). The genetic factors appear to do the main contribution to intra individual variety of AHH activity among smokers (Gurtoo et al. 1983). Thus, it is impossible to separate an influence of environment and genetic background during the induction of AHH activities in human placentae (Pelkonen et al. 1981; Fujino et al. 1984).

On the other hand the induction can be connected with affinity and quantity of Ah receptors in placentae (Shum et al. 1979; Pelkonen and Karki 1982; Harris et al. 1984; Manchester et al. 1987). High concentration of Ah receptors in cytoplasm has been detected by means of 2,3,7,8-tetrachlorinedibenzodioxine as ligand. The Ah concentration in human placentae is similar that in liver of rats and mice but the affinity of Ah receptor in human placentae is lower (Manchester et al. 1987). Intra individual variety of Ah receptor properties has not been demonstrated yet but it would be of interest to know could human population to be class according to individual affinity of Ah receptor or to another genetic characteristics.

The radiation effect on the monooxygenase and conjugation activities in placentae of mothers who were born more later then 1962 and were not exposed to direct irradiation can be explained as an alteration of genetic apparatus of their parents living in the polluted territories. At present such hypothesis have been suggested although it seems poor convincing for us.

Ames's test is the one of common accepted and sensitive techniques using to an analysis of some genotoxic compounds. The special Salmonella strains allow to detect various gene mutations caused by the mutagens. The complete Ames's test (Salmonella + microsomes) allows to detect all types of gene mutations caused by both direct mutagenes and promutagene compounds. More than 90% of mutagene compounds detected in Ames's test show the cancerogenic properties in animal organism (Ames et al. 1975).

Increased human placental microsomal monooxygenase actvities are produced active metabolites from polycyclic aromatic hydrocarbons, which are mutagenic when tested with Salmonella typhimurium (Berry et al. 1977; Vaught et al. 1979).

The main troubles under Ames's testing are the work connected with the tissue extraction and the concentrating of mutagene compounds. The processes are very complex because it is necessary to extract compounds with unknown chemical nature. The rather simple technique of mutagene extraction from animal tissues has been developed in our laboratory (Kotelevtsev et al. 1986).

The attempts have been undertaken to adapt the above technique to an analysis of mutagenic compounds from placentae of the mothers which children are sicked by unknown etiology jaundice. Such kinds of experiments could be prove in a smuch as the supposed reason of the jaundice was an attack of mother's organism by organochlorinate compounds. It is known this substances have a mutagenic activity detected in Ames's test (Fonshtein et al. 1977). On the other hand if the effect of chemical compounds is not a base of the investigated pathology that the presence of such xenobiotics in placentae can result in combined influence of the xenobiotics and some other harmful factors (irradiation for example). An appearance of this influence can to have a prolong effects and it is not revealed in first period of the child development.

The placental extracts have been made by means the above technique and have been tested with two Salmonella strains TA 98 and TA 100 with and without microsomal metabolic activation system. Mutagenic and cancerogenic compounds have been detected only in some samples. Apparently, there are 3 reasons result in the negative results. Firstly, the mutagenic xenobiotics is lacked in placental tissue. Secondly, the harmful xenobiotics are presented in placentae but they do not cause the gene mutation and therefore are not detected in Ames's test. At last thirdly, our extraction technique of xenobiotics is not sufficiently effective.

The one of reasons of the xenobiotics penetration in the mother organism is smoking resulting in the formation of some benz(a)pyrene (BP) metabolites: epoxides, quinones, phenols and diols (Wang et al. 1977; Berry et al. 1977; Vaught et al. 1979; Pelkonen and Saarni 1980; Blanck et al 1980). Among BP metabolites the BP-7,8-diol is more toxic as it is metabolized by placental cytochrome P450-dependent system in vivo to mutagenic and cancerogenic BP-7,8-diol-9,10-oxide which is bound with DNA (Berry et al. 1977; Namkung and Juchau 1980; Wang et al. 1977). This binding is correlated with the AHH activity.

The formation of the BP-diol-epoxide-DNA adducts in human placentae have been confirmed by synchronous fluorescence spectrophotometric measurements (Gelboin 1980;). The defective induction of human placental monooxygenase activities has been suggested to be associated with birth defects (Lambert and Nedert 1977) and, as it seems more probable for us, with an influence of environmental factors. The placental monooxygenase system in itself can to form mutagenic intermediates with potential hazard for placentae and child (Weistler et al. 1984; Namkung and Juchau 1980).

The our experimental data do not allow to make reliable conclusions about an accumulation of mutagenes in placentae as a result of smoking. In one case the high content of direct mutagenes in placentae could be explained that mother had a constant professional contact with organochlorinate compounds.

The placental extracts from the mothers living in the unsafe territories more often demonstrate mutagenic activity in the Ames's test. It can be explained that the low activity of monooxygenase and conjugation enzymes results in a decrease of the rate of mutagene metabolism. As the result an increase of the stationary level of mutagenes and in placentae of the mothers living in the unsafe territories appear to observe.

The investigations of the bilirubin level in blood of newborns, the monooxygenase and conjugation enzyme activities as well as of a character of mutagen accumulation in placentae have not find a correlation between the above parameters and an appearance of newborn jaundice. However the experimental data allow to suggest that there are membrane alterations in placental tissue from the mothers living in unsafe regions and to refer this women as a group of increased risk. The obtained results do not reply to a question about the mechanism and the reasons of the investigated pathology. Nevertheless the alteration of detoxication enzymes in placentae because of mother irradiation as a result of nuclear weapon tests in Semipalatinsk's polygon appears to seem a probable cause. It should be also examined a possibility of a joint influence of irradiation and xenobiotics on membrane systems of placentae.

Additional investigations need to final conclusions about reasons of children pathologies in unsafe regions of Altai.

References

Akhalaya MJ, Deev LI, Platonov AG, Khakimov A (1990) Influence of rentgen irradiation on cytochrome P-450-dependent monooxygenase hepatic system of different animal speacies. Biological Sci (Biologicheskie nauki) 11:47-52 (in Russian)

Ames BN, McCaun J, Jamasaki E (1975) Methods for detecting carcinogens and mutagens with the sallmonella/mammalian-microsome mutagenicity test. Mutat Res 31(3):347-364

Berry DL, Zacharaih PK, Slaga TJ, Juchau MR (1977) Analysis of the biotransformation of benzo(a)pyrene in human fetal and placental tissues with high-pressure liquid chromatography. Eur J Cancer 13(7):667-675

Blanck A, Rane A, Toftgard R, Gustafsson JA, (1980) Cytochrome P-450 in human fetal liver and placentae. Metabolism of benzo(a)pyrene and 7-ethoxyresorufin. SDS-polyacrilamide gel electrophoresis. In: Gustafsson JA, Carlstedt-Duke J, Mode A, Rafter J (eds) Dev Biochem vol 13. (Biochemistry, Biophysics and Regulation of Cytochrome P-450) Elsevier. Amsterdam, pp 93-96

Fonshtein LM, Kalinina LM, Poluhina GN, Abilev SK, Shapiro AA (1977) Test-system for the estimation of mutagenic activity of pollutants in environmental. Moscow, 107 pp (in Russian)

Fujino T, Gottlieb K, Manchester DK, Park SS, West D, Gurtoo HL, Tarone RE, Gelboin HV (1984) Monoclonal antibody phenotyping of inter individual differences in cytochrome P-450-dependent reactions of single and twin human placentae. Cancer Res 44(9):3916-3923

Gelboin HA (1980) Benzo(a)pyrene metabolism, activation, and carcinogenesis: role and regulation of mixed-function oxidases and related enzymes. Physiol Rev 60(4):1107-1166

Gornall AG, Bardawill CJ, David MM (1949) Determination of serum proteins by means of the biuret reaction. J Biol Chem 177(2):751-766

Gurtoo HL, Williams CJ, Gottlieb K, Mulbern AI, Caballes L, Vaught JB, Marinello AJ, Bansal SK (1983) Population distribution of placental benzo(a)pyrene metabolism in smokers. Int J Cancer 31(1):29-37

Habig W, Pabst M, Jacoby WB (1974) Glutathione S-transferases. First enzymic step in mercapturic acid formation. J Biol Chem 249(22):7130-7139

Harris CC, Autrup HN, Vahakangas K, Trump BF (1984). Interindividual variation in carcinogen activation and DNA repair. In: Omenn GS and Gelboin HV (eds) Genetic Variability in Responses to Chemical Exposure Banbury Report. vol 16. Gold Spring Harbor Laboratory, pp 145-154

Isselbacher KJ (1956) Enzymic mechanisms of hormone metabolism. II. Mechanism of hormonal glucuronide formation. In: Pincus G (ed) Recent progress in hormone research, vol 12. Academic Press Inc. New York, pp 134-151

Jacobson M, Levin W, Poppers PJ, Wood AW, Conney AH (1974) Comparasion of the O-dealkylation of 7-ethoxycoumarin and the hydroxylation of benzo(a)pyrene in human placentae. Effect of cigaret smoking. Clin Pharmacol Ther 16(4):701-710

Johannesen KAM de Pierre JW (1978) Measurement of cytochrome P-450 in the presence of large amounts of contaminating haemoglobin and methemoglobin. Analyt Biochem 86(2): 725-732

Juchau MR (1980) Drug biotransformation in the placentae. Pharmacol Ther 8(3):501-524

Koliado VV, Ulanov AN, Lotosh EA, Karpov AV, Martynenko AI, Golovin NM, Plugin SV, Kotovschikova EF, Yakovleva EL, Zheleznikova LI, Sviridov SV, Gafarov NI, Luzina FA (1991) The analysis of dynamics of the state of health of Altai territory population and five model the country-side regions (Talmensky, Tret'yakovsky, Uglovsky, Loktevsky, Altaisky) in with increasing of ecological loading on these regions. Novosibirsk. pp 131 (in Russian)

Kctelevtsev SV, Stvolinskyi SL, Beim AM (1986) Ecotoxicological analysis on base the biological membrane. Moscow University. Moscow, pp 120

Kulkarni AP, Kenel MF (1987) Human placental lipid peroxidation. Some characteristics of the NADPH-supported microsomal reaction. Gen Pharmacol 18(5):491-496

Lambert GH, Nebert DW (1977) Genetically mediated induction of drug-metabolizing enzymes associated with congenital defects in the mouse. Teratology 16(2):147-153

Manchester DK, Gordon SK, Golas CL, Roberts EA Okey AB (1987) Ah receptor in human placentae: stabilization by molybdate and characterization of bonding of 2,3,7,8-tetrachlorodibenzo-p-dioxin, 3-methylcholanthrene, and benzo(a)pyrene. Cancer Res 47(18):4861-4868

Namkung MJ, Juchau MR (1980) On the capacity of human placental enzymes to catalyze the formation of diols from benzo(a)pyrene. Toxicol Appl Pharmacol 55(2):253-259

Omura T, Sato R (1964) Carbon monooxide-binding pigment of liver microsomes. I. Evidence for its hemoprotein nature. J Biol Chem 239(7):2370-2378

Orlov SN, Kotelevtsev SV (1993) The pathology of biological membrane in the blood forms and placentae tissue of the mothers and newborn from Altay region. In: Shoyhed J.N. (ed) Nuclear test and environmental health in Altay region. Barnaul medicine institute, pp 120 (in Russian)

Pacifici GM, Rane A (1981) Glutathione S-epoxidetransferase in the human placentae at different stage of pregnancy. Drug Metab Dispos 9(5):427-475

Pasanen M, Pelkonen O (1989/1990) Human placental xenobiotic and steroid biotransformation catalyzed by cytochrome P-450, epoxide hydrolase, and glutathione S-transferase activities and and their relationships to maternal cigarette smoking. Drug Metabol Rev 21(3):427-461

Pelkonen O, Karki NT (1983) Is aryl hydrocarbon hydroxylase induction systematically regulated in man? In: Armstrong B and Bartsch H (eds) Host Factors in Human Carcinogenesis. vol 39. IARC Scientific Publications. Lyon, pp 421-426

Pelkonen O, Karki NT, Tuimala R (1981) A relationship between cord blood and maternal blood lymphocytes and term placentae in the induction of aryl hydrocarbon hydroxylase activity. Cancer Lett 13(2):103-110

Pelkonen O, Moilanen M-L (1979) The specifity and multiplicity of human placental xenobiotic-metabolizing monooxygenase system studied by potential substrates, inhibitors and gel electrophoresis. Med Biol 57(5):306-312

Pelkonen O, Pasanen M, Kuha H, Gachalyi B, Kairaluoma M, Sotaniemi EA, Park SS, Friedman FK, Gelboin HV (1986) The effect of cigarette smoking on 7-ethoxyresorufin O-deethylase and other monooxygenase activities in human liver, analysis with monoclonal antibodies. Br J Pharmacol 22(2):125-134

Pelkonen O, Saarni H (1980a) Unusual patterns of benzo(a)pyrene metabolities and DNA-benzo(a)pyrene adducts produced by human placental microsomes in vitro. Chem Biol Interact 30(3):287-296

Pelkonen O, Vahakangas K, Karki NT (1980b) Genetic and environmental regulation of aryl hydrocarbon hydroxylase in the placentae. In: Gustafsson JA, Carlstedt-Duke J, Mode A, Rafter J (eds) Dev Biochem vol 13. (Biochemistry, Biophysics and Regulation of Cytochrome p-450) Elsevier. Amsterdam, pp 163-170

Prough RA, Burke MD, Mayer RT (1978) Direct fluorometric methods for measuring mixed-function oxidase activity. In: Colowick SP, Kaplas NO (eds) Methods Enzymol. vol 52c. Academic Press. New York, pp 372-377

Shum S, Jensen NM, Nebert DW (1979) The murine Ah locus: In utero toxicity and teratogenesis associated with genetic differences in the benzo(a)pyrene metabolism. Teratology 20(3):365-376

Tephly TR, Burchell B (1990) UDP-glucuronosyl transferases: a family of detoxifying enzymes. Trends Pharmacol Sci 11(7): 276-279

Ullrich V, Weber P (1972) O-deethylation of 7-ethoxycoumarin by liver microsomes. Hopper-Seyler's Z Physiol Chem 353(7): 1171-1177

Vaught JB, Gurtoo HL, Parker NB, LeBoeuf R Doctor G (1979) Effects of smoking on benzo(a)pyrene metabolism by human placental microsomes. Cancer Res 39(8):3177-3183

Wang IY, Rasmussen RE, Creasey R, Crocker TT (1977) Metabolites of benzo(a)pyrene produced by placental microsomes from cigarette smokers and nonsmokers. Life Sci 20(7):1265-1272

Weistler OD, Kleihues P, Rice JM, Ivankovic S (1984) DNA methylation in maternal administration of procarbazine. J Cancer Res Clin Oncol 108(1):56-59

Monooxygenase Measurements as Indicators of Pollution in the Field

Richard F. Addison
Department of Fisheries and Oceans
Ocean Chemistry Division, Institute of Ocean Sciences,
P.O. Box 6000 SIDNEY BC Canada V8L 4B2

Introduction

Since the mid-1970's, measurements of monooxygenase induction have been used to show the direct, sub-lethal impact of chemicals on organisms. In this lecture, I review the application of these measurements --- which, in practice, are almost always of the induction of CYP 1A1, or its analogue, by a limited suite of organic compounds --- to assessing the impact of water pollution on fish.

Interest in applying monooxygenase measurements to water pollution developed during the mid-1970's, partly in response to the growing literature on the pharmacology of monooxygenase induction at that time. There was also a growing need to demonstrate (for regulatory purposes) the "effects" of chemical residues, rather than simply their presence, and monooxygenase induction was viewed as a possible measure of impact, i.e., as a sub-lethal bioassay. The early measurements of monooxygenase induction in fish suffered from the fact that the basic biochemistry of the system being studied was not well defined at that time. For example, it had previously been established (Buhler and Rasmussen, 1968) that the insecticide DDT and the drug phenobarbitone did not induce monooxygenase systems in fish, although they did so in rats. At that time, however, these observations could not be explained in terms of failure to express (for unknown reasons) the gene for CYP 2B1 in fish (as we would probably now

NATO ASI Series, Vol. H 90
Molecular Aspects of Oxidative Drug Metabolizing Enzymes
Edited by E. Arınç, J. B. Schenkman and E. Hodgson
© Springer-Verlag Berlin Heidelberg 1995

explain it: cf. Kleinow et al., 1990). Furthermore, methods to detect induction at that time were limited to measurements of catalytic activity or of total P450, and the specificity of substrates for isozymes in P450-catalysed reactions was not fully appreciated. Finally, monooxygenase induction was often measured in response to an environmental emergency, without any "calibration" of the response in the test organism measured under controlled conditions of exposure. The early field studies of monooxygenase induction were therefore not well interpreted or understood by today's standards.

By the mid-1980's, that situation had changed significantly. Antibodies to specific P450's were becoming available, and this led to a better description of the inducible forms of P450 in fish, and to a clearer understanding of the specificity of substrates for P450 isozymes. Laboratory studies also defined the major "nuisance variables" which could confound the interpretation of monooxygenase induction: these included sex, reproductive status, age, and (possibly) nutritional status or condition of the test organism. As a result of this work, by the late 1980's the specificity of induction in fish had been well characterized, methods to detect induction (either by catalytic measurements of monooxygenase activity or by immunochemical measurements of isozyme content) had been established, and the role of "nuisance variables" had been assessed, at least for a few species. Also, time course and dose-response relationships between contaminant exposure and induction had been established for a few species. At this stage, then, monooxygenase induction in fish had been "calibrated" as a fairly specific sub-lethal bioassay of the presence or effects of certain compounds.

The most recent phase of this work has been the convincing demonstrations of hepatic monooxygenase induction in response to the presence of contaminants, often at the relatively low concentrations of "chronic" pollution. These studies have reflected the development of methods to measure simultaneously several different aspects of induction. It is now possible, for example, to measure induction via catalytic activity (such as that of EROD:

ethoxyresorufin O-de-ethylase), by concentrations of CYP 1A1 (which catalyses EROD) and by concentrations of the mRNA which codes for CYP 1A1 --- and to show that these are all inter-related.

Early studies (mid-1970's to mid-1980's)

This phase of the development of monooxygenase induction as an environmental monitoring technique has been reviewed by Addison and Payne (1986). Although Ahokas et al. (1976) had found that fish from a lake polluted with pulp mill effluents (among other wastes) had reduced MFO activity compared to those from a reference site, there was considerable interest during the mid-1970's in showing that contamination of habitat or environment led to hepatic monooxygenase induction in fish. During this phase, field studies dealt almost exclusively with the impact of hydrocarbon contamination, either from chronic or sudden oil spills. Benthic fish were usually chosen as the indicator species, presumably because they would experience the greatest exposure to hydrocarbons bound to particles or sediments. The commonest response measured was the induction of aryl hydrocarbon hydroxylase (AHH) using benzo[a]pyrene as a substrate, usually based on the fluorescence method of Nebert and Gelboin (1968) (e.g., Payne and Penrose, 1975; Payne, 1976; Kurelec et al., 1977; Spies et al., 1982; Kezic et al., 1983; Binder and Lech, 1984; Davies et al., 1984; Payne et al., 1984; Lindstrom-Seppä et al., 1985). The cause-effect relationship inferred from the results was usually that of environmental contamination by hydrocarbons inducing hepatic AHH activity. Many of these studies were based on comparative analyses of AHH activity in "reference" fish with those in "impacted" fish (e.g., Payne et al., 1984) and as the hydrocarbon stress was often unexpected (as in the case of an accidental spill) it was usually not possible to "calibrate" the response of the indicator species in advance, in terms of the time-course of induction, the specificity of the response or a dose-response relationship. There were, however, later exceptions to this rule (e.g., Spies et al., 1988a, b) in which laboratory studies to define these factors supported the correlations observed in the field.

Although most indices of induction at this time were based on AHH measurements in response to hydrocarbon exposure, there were occasional exceptions. Pequegnat and Wastler (1980) used total P450 (and P420) concentrations to indicate induction. In retrospect, the success of this approach is somewhat surprising, as CYP 1A1 (the main inducible form in fish) may not account for much of the total P450. Thus, Stegeman et al. (1988) calculated that P450E (the analogue of CYP 1A1 found in scup, Stenotomus chrysops) represented up to about 20% of total P450 in environmentally stressed European flounder, Platichthys flesus. If this were a fairly typical proportion of CYP 1A1, measurement of total P450 would not be a very sensitive index of CYP 1A1 induction in fish. Similarly, Burns (1976) and Ridlington et al. (1982) used aldrin epoxidase to indicate hepatic monooxygenase induction by exposure to hydrocarbons. This enzyme is usually considered to be catalysed by CYP 2B1, at least in mammals, and although it is inducible by experimental exposure of fish to PAH (polynuclear aromatic hydrocarbons) (Addison et al., 1978; Ridlington et al., 1982) it is a relatively insensitive indicator of CYP 1A1 induction. The fact that these relatively insensitive measurements did indeed demonstrate induction in the field indicates the overall sensitivity of the approach, and by the early 1980's, fish hepatic monooxygenase induction was becoming established as a moderately reliable indicator of the sub-lethal effects of some organic environmental contaminants.

Studies from the mid-1980's to the present.

This period coincided with the development of specific antibodies which allowed the detection by immunochemical methods of specific forms of P450 (especially CYP 1A1); this in turn led to the identification of the P450 isozymes which catalysed specific reactions. Also about this time, a number of substrates were developed which were selectively metabolised by specific P450's, e.g., ethoxyresorufin (by CYP 1A1) and pentoxy- or benzyloxy-resorufin (by CYP 2B1).

One aspect of water pollution to which monooxygenase measurements have been applied has been the impact of pulp and paper processing effluents (Förlin et al., 1985; Andersson et al., 1987; Lindström-Seppä et al., 1989; Lindström-Seppä and Oikari, 1990; Oikari and Lindström-Seppä, 1990; Lehtinen et al., 1990; McMaster et al., 1991; Mather Mihaich and di Giulio, 1991; Munkittrick et al., 1991; 1992). These effluents have been shown repeatedly to induce EROD (an indicator of CYP 1A1 induction). In contrast to the earlier studies of induction by hydrocarbons, the pulp mill studies have often used pelagic indicator species which are more likely to accumulate their contaminant burden from the water column or from pelagic food sources. This choice is probably based on practical considerations of availability of fish, as the inducers in pulp mill effluents should be mainly particle-bound. CYP 1A1 (and EROD) induction has usually been attributed, tacitly or otherwise, to the presence of chlorinated dibenzodioxins or dibenzofurans (produced during chlorine bleaching) in pulp mill effluents. (A dose-response relationship between EROD induction in trout and exposure to 2,3,7,8 TCDD has been established: van der Weiden et al., 1990.) However, recent results (K. Munkittrick, pers. comm.) suggest that since EROD induction is also observed in fish exposed to pulp mill effluents which do not arise from chlorine bleaching processes, chlorinated dioxins and/or furans may not be the sole cause. This aspect of monooxygenase induction by pulp mill effluents remains to be clarified.

The development of immunochemical probes for CYP 1A1 has greatly increased the reliability of monooxygenase induction measurements: not only can the catalytic activity of the enzyme be measured, but the concentration of the enzyme itself can be determined. These approaches have now been applied to several field studies of the impact of contamination on monooxygenase induction; results from some of these are summarised in Table 1. Usually, though not always, measurement of CYP 1A1 itself (or its analogue) seem to be less sensitive than measurements of its catalytic activity, and it is not clear why this should be the case.

Stegeman et al. (1986) have shown that CYP 1A1 (or its analogues) and EROD activity are both induced in a deep sea fish (rat-tail) which had accumulated large concentrations of chlorobiphenyls. Chlorobiphenyl concentrations in fish collected from the (contaminated) Hudson Canyon were about eight times higher than those in fish from a relatively clean area off Newfoundland; EROD and AHH activities were (respectively) about seven and nine times higher in the contaminated fish than in the "clean" fish. There was clearly a higher concentration of CYP 1A1 in the Hudson Canyon fish, though this could not be quantified due to lack of standards. Winter flounder (Pseudopleuronectes americanus, a benthic flatfish) from Boston Harbour, which were probably exposed to high levels of PAH, contained a P450 isozyme analogous to that induced in "control" fish by treatment with ß-naphthoflavone (ß-NF) and also had higher EROD activity than fish from less contaminated sites (Stegeman et al., 1987). In Fundulus heteroclitus taken from the Seekonk River, which is fairly heavily contaminated with PAH and some chlorinated compounds, EROD activity was about four times higher than that in a control group, and CYP 1A1 concentrations almost twice as high (Elskus and Stegeman, 1989).

Other recent field studies of monooxygenase induction in fish have dealt with other forms of pollution. In an industrialised fjord in Norway which was contaminated with PAH, polychlorinated biphenyls (PCB) and other chlorinated compounds, EROD and (less convincingly) AHH activity in flounder (Platichthys flesus) liver increased with increasing contaminant concentrations in sediments and invertebrates (Addison and Edwards, 1988). In the same fish, concentrations of the analogue of CYP 1A1 in liver microsomes increased along the same gradient

Table 1

Summary of selected field studies of mono-oxygenase induction in fish involving CYP 1A1 (or equivalent) measurements
(Note that contaminant listed may not be agent responsible for mono-oxygenase induction.)

Site	Contaminant (possible inducer)	Response (extent of induction in liver unless otherwise stated)	Reference
NW Atlantic	Chlorobiphenyls 8X difference, contam. v. reference fish	EROD and AHH 7-9X in contam. v. reference fish P450E (CYP 1A1) increased	Stegeman et al., 1986
MA, USA. (several sites)	PAH in sediment (12,000X range in biota)	EROD and AHH 2X over sites P450E(CYP 1A1) increased	Stegeman et al., 1987
Langesundfjord Norway (four sites)	Chlorobiphenyls 7X in fish over all sites	EROD 13X, liver AHH 3X over sites. P450E (CYP 1A1) 14X over sites	Addison and Edwards 1988 Stegeman et al., 1988
RI, USA. (two sites)	PCB, 3X difference in fish between sites	EROD 4X difference between sites P450E (CYP 1A1) 2X difference	Elskus and Stegeman 1989
Elizabeth R., Va., USA	PAH (from 9 - 96000 µg/kg dry wt. sediment	Gut EROD 33X; CYP 1A1 130X; Liver EROD 8X; CYP 1A1 23X	Van Veld et al., 1991
Glomma Estuary, Norway (several sites)	PAH 4X, PCB 6X in fish over several sites	EROD 50X between sites P450c (CYP 1A1) 2X	Goksøyr et al., 1991b

(Stegeman et al., 1988); EROD, CYP 1A1 and organochlorine residues in the fish were all well correlated with each other. However, although monooxygenase activity generally increased with increasing contamination, it was not possible to attribute induction to any single chlorinated residue, or even to any major group of residues. In another study of the European flounder, EROD activity increased appreciably in fish caught downstream from a major petrochemical complex in Scotland compared to those caught upstream, or in controls from an apparently clean site (Sulaiman et al., 1991).

Using spot (Leiostomus xanthurus) as an indicator species, Van Veld et al. (1991) showed that fish from the Elizabeth River, Virginia, whose sediments are heavily contaminated with PAH, contained elevated hepatic and intestinal EROD activity and CYP 1A1 concentrations. Sediment PAH concentrations ranged over about 10,000-fold between the reference and most contaminated sites; hepatic EROD activity ranged over 8-fold, and CYP 1A1 about 23-fold. In gut tissue, EROD activity increased about 33-fold and CYP 1A1 more than 130-fold. However, although monooxygenase components of intestinal tissue appeared more sensitive to induction than those in liver, the latter were much better correlated with environmental PAH concentrations.

Goksøyr et al. (1991b) have studied a variety of flatfish species living at the Hvaler Archipelago at the mouth of the industrialised Glomma River in southeastern Norway. A consistent response to PAH and various chlorinated hydrocarbon residues was the induction of EROD activity, often up to 20-fold, and of CYP 1A1 measured indirectly by an ELISA method. This is one of the few studies in which EROD activity in most of the species studied seemed well correlated with the general pattern of residue accumulation, but CYP 1A1 showed a rather different pattern. The reason for this is not clear, but the variation in response shows the importance of making more than one measurement of monooxygenase induction.

In Sydney Harbour, N.S., there is a steep gradient of PAH contamination in sediments resulting from the activities of a local steel mill. PAH concentrations in sediments range over 1000-fold from the vicinity of the mill to the open sea. Winter flounder collected at various points along the harbour two years in succession showed elevated levels of EROD activity (up to about 5-fold); the substrate 3-cyano-7-ethoxycoumarin (which allows a kinetic determination of ethoxycoumarin de-ethylation: White, 1988) increased about 50-fold. Benzo[a]pyrene hydroxylase increased up to 10-fold over the range of sites, and CYP 1A1 increased 5-fold. Rather surprisingly, total P450 increased up to 3-fold (Addison et al., 1993).

The most recent field studies include those carried out during an international collaborative study at Bremerhaven, Germany. The southern North Sea has been generally recognised as being appreciably polluted particularly by contaminants from the rivers Elbe, Weser and Rhine. A study was recently undertaken to attempt to relate chemical distribution to a range of biological responses, including monooxygenase induction, in a local benthic flatfish, the dab (Limanda limanda). Briefly, monooxygenase activity (EROD, CN-ECOD) declined with increasing distance from the mouths of the Elbe and Weser rivers, and activities were well correlated with concentrations of chlorobiphenyls in the fish (Renton and Addison, 1992). CYP 1A1 in other samples of dab from the same sites showed similar trends to those of EROD and CN-ECOD (Goksøyr et al., 1992). Finally, CYP 1A1 mRNA measured by a synthetic cDNA oligonucleotide probe was present at higher concentrations at inshore than at offshore sites, although the correlations with chemical distribution were poorer than was the case with EROD (Renton and Addison, 1992). The variability in the indices of induction measured probably reflects the time elapsed between exposure to the causative agent and sampling: CYP 1A1 mRNA, for example, would be expected to reach a maximum relatively soon after exposure, whereas CYP 1A1 itself and its catalytic activity might turn over more slowly than the mRNA. One point worth emphasising about this study was that, in general, induction responded to "background" concentrations of organic pollutants at relatively distant points far out in the North Sea in the combined "plume" of the Elbe and Weser Rivers; this indicates the sensitivity of the induction response.

Discussion

There is now a large body of field evidence which shows reasonably convincingly that monooxygenase induction is well correlated either with organic contaminant burdens in the fish or with contaminant concentrations in their habitat. Since there is increasing interest in using monooxygenase induction to complement contaminant distribution data obtained by conventional chemical analyses, it is worth asking <u>why</u> these observations are consistent. The answer probably lies in the results of experimental studies which have complemented the field studies.

We can summarise the sequence of events relating the presence of contaminants in the environment to monooxygenase induction (in fish) as follows:

(1)   Accumulation of inducer by the fish, either by direct uptake from water or sediment or from food;

(2)   Transfer through the circulatory system to the liver;

(3)   Binding of inducer to cytosolic Ah receptor;

(4)   Inducer-receptor complex "switches on" the gene for CYP 1A1;

(5)   Transcription to CYP 1A1 mRNA;

(6)   Translation to CYP 1A1;

(7)   Catalytic activity of CYP 1A1 indicated by EROD, AHH,   &

Describing the sequence of events in this way emphasises the point that almost all these steps can be measured. Thus, at steps (1) and (2), the concentrations of contaminants in the fish's environment (water or sediment), in its circulatory system and in its liver can all be measured (though, admittedly, this is rarely done). Furthermore, fairly good compartmental models exist to relate all these measurements (e.g., Gibaldi and Perrier, 1982). At step (5) the concentration of the CYP 1A1 mRNA can be measured using either synthetic cDNA probes (Renton and Addison, 1992) or cloned cDNA probes (Haasch et al., 1988; Kreamer et al., 1991). At step (6) the concentration of CYP 1A1 can be measured using probes developed for analogues of CYP 1A1 in several species of fish (Kloepper-Sams et al., 1987; Goksøyr et al., 1991a, c). Finally, the catalytic activity of the CYP 1A1 can be measured with various selective or specific substrates. Only steps (3) and (4) are not routinely measurable, at least in fish. Perhaps most importantly, most of these lab studies establish the mechanistic relationship underlying the correlation between the presence of a contaminant and its induction of CYP 1A1, and it is our understanding of the induction process that gives us confidence in its application.

It is worth emphasising, however, that although monooxygenase induction in fish from the wild is a reliable measure of the general impact of some organic contaminants, it rarely identifies the specific cause. This is probably because most environmental contaminants co-vary: it is difficult to find an example of environmental contamination by a single organic contaminant. Thus, in the example of monooxygenase induction by environmental contaminants in Langesundfjord, Norway, (Addison and Edwards, 1988; Stegeman et al., 1988) the distribution of chlorobiphenyls, PAHs, and industrial chlorinated by-products such as octachlorostyrene all followed the same pattern (Bayne et al., 1988). In such cases, it is impossible to ascribe monooxygenase induction to any single chemical cause. The problem is compounded by the fact that some inducers (such as PAH) may be extensively degraded by the induced monooxygenases with the result that tissue concentrations of the "parent" compounds may be undetectable; in this case, it might be useful to analyse concentrations of (e.g.,) biliary metabolites. A further complication is that in the case of PCBs, many of the major congeners may not be effective CYP 1A1 inducers as they have a non-co-planar structure; the effective CYP 1A1 inducers, however, are usually present at such low concentrations that they are not yet routinely analysed. Taking all these points together, it seems that CYP 1A1 induction may be a good general indicator of the presence and effects of a group of organic contaminants all of which interact with the cytosolic Ah receptor. All of these probably co-vary in the environment because they are released from generally similar sources, and all have roughly similar physico-chemical properties which control their environmental behaviour. However, monooxygenase induction is not yet a specific enough measurement to identify its own cause, at least in terms of individual chemical species.

Most field studies of the environmental impact of contamination on fish hepatic monooxygenases have been carried out without appropriate calibrations having been made of induction in response to contaminant exposure. For example, time course or dose-response relationships have been established in only a few fish species, and even then by experimental exposure to contaminants which may not be the major inducers in the field (cf. Addison and Payne, 1986; Addison et al., 1993). In one sense, the usefulness of field monooxygenase induction measurements which are not supported by such calibrations shows the robustness of the approach. On the other hand, to measure a response in the field without having a "calibration curve" established by controlled studies with which to compare it is intellectually unsatisfactory, and ultimately may miss some potentially useful information.

What is the future of "effects monitoring" based on monooxygenase induction? The trend over the last twenty years or so has been to move from demonstrating the effects of rather massive incidents of pollution (e.g., major oil spills) to detecting subtle effects , e.g., of the "plume" of industrial wastes in the open sea. These field studies have been complemented by laboratory work which has led to the development of more sensitive methods to measure induction (and which have made the more recent field work possible). It seems fairly clear now that we understand the main steps in the induction process, and we can measure most of these routinely in fish (except for the interaction of the inducer with the Ah receptor); as noted above, it is the fact that we can show that these steps are all inter-related, and can explain why this is so, that gives us our confidence in using monooxygenase induction as a field measurement. It seems that the obvious areas for further work are:

(i)   definition of the relationship between the inducer and Ah receptor in fish, and development of methods for measuring it;

(ii) increased application of monooxygenase induction measurements to complement conventional analytical chemistry measurements, using indicator species in which the induction process has been well-calibrated, and focussing on the linkage between the presence of a chemical and several components of the induction process;

(iii) definition of the significance of monooxygenase induction in terms of "higher-order"effects, i.e., in terms of whole organism, population and/or community responses.

References

Addison RF, Edwards AJ (1988) Hepatic microsomal monooxygenase activity in flounder Platichthys flesus from polluted sites in Langesundfjord and from mesocosms experimentally dosed with diesel oil and copper. Mar Ecol Prog Ser 46: 51 - 54

Addison RF, Payne JF (1986) Assessment of hepatic mixed function oxidase induction in winter flounder (Pseudopleuronectes americanus) as a marine petroleum pollution monitoring technique, with an Appendix describing practical field measurements of MFO activity. Can Tech Rept Fish Aquat Sci no. 1505.

Addison RF, Willis DE, Zinck ME (1993) Liver microsomal monooxygenase induction in winter flounder (Pseudopleuronectes americanus) from a sediment PAH gradient in Sydney Harbour, N.S. Mar Env Res, In press.

Addison RF, Zinck ME, Willis DE (1978) Induction of hepatic mixed function oxidase (MFO) enzymes in trout (Salvelinus fontinalis) by feeding Aroclor[R] 1254 or 3-methylcholanthrene. Comp Biochem Physiol 61C: 323 - 325

Ahokas JT, Karki, NT, Oikari A and Soivio A (1976) Mixed function mono-oxygenase of fish as an indicator of pollution of aquatic environment by industrial effluent. Bull Env Contam Toxicol 16: 270 - 274.

Andersson T, Bengtsson B-E, Förlin L, Haerdig J, Larsson A (1987) Long-term effects of bleached kraft mill effluents on carbohydrate metabolism and hepatic xenobiotic biotransformation enzymes in fish. Ecotoxicol Environ Saf 13: 53 - 60

Bayne BL, Clarke KR, Gray JS (Eds.) (1988) Biological effects of pollutants: results of a practical workshop. Mar Ecol Prog Ser 46: nos. 1 - 3

Binder RL, Lech JJ (1984) Xenobiotics in gametes of Lake Michigan lake trout (Salvelinus namaycush) induce hepatic monooxygenase activity in their offspring. Fund Appl Toxicol 4: 1042 - 1054

Buhler DR, Rasmusson ME (1968) The oxidation of drugs by fishes. Comp Biochem Physiol 25: 223 - 239

Burns KA (1976) Microsomal mixed function oxidases in an estuarine fish, Fundulus heteroclitus, and their induction as a result of environmental contamination. Comp Biochem Physiol 53B: 443 - 446

Davies JM, Bell JS, Houghton C (1984) A comparison of the levels of hepatic aryl hydrocarbon hydroxylase in fish caught close to and distant from North Sea oil fields. Mar Env Res 14: 23 - 45

Elskus AA, Stegeman JJ (1989) Induced cytochrome P-450 in Fundulus heteroclitus associated with environmental contamination by polychlorinated biphenyls and polynuclear aromatic hydrocarbons. Mar Env Res 27: 31 - 50.

Förlin L, Andersson, T, Bengtsson, B, Harding, J, Larsson, A (1985) Effects of pulp bleach plant effluents on hepatic xenobiotic biotransformation enzymes: laboratory and field studies. Mar Env Res 17: 109 - 112

Gibaldi M, Perrier D (Eds.) (1982). "Pharmacokinetics", 2nd Ed., Dekker, N.Y.

Goksøyr A, Andersson T, Buhler DR, Stegeman JJ, Williams DE, Förlin L. (1991a) Immunochemical cross-reactivity of beta-naphthoflavone-inducible cytochrome P450 (P450IA) in liver microsomes from different fish species and rat. Fish Physiol Biochem 9: 1 - 13

Goksøyr A, Husøy A-M, Larsen H, Klungsøyr J, Wihelmsen S, Maage A, Brevik E, Andersson T, Celander M, Pesonen P, Förlin L. (1991b) Environmental contaminants and biochemical responses in flatfish from the Hvaler Archipelago in Norway. Arch Env Contam Toxicol 21: 486 - 496

Goksøyr A, Larsen HE, Husøy AM (1991c) Application of a cytochrome P-450 IA1-ELISA in environmental monitoring and toxicological testing of fish. Comp Biochem Physiol 100C: 157 - 160

Goksøyr A, Larsen HE, Blom S, Förlin L (1992) Detection of cytochrome P4501A1 in North Sea dab liver and kidney. Mar Ecol Prog Ser 91: 83 - 88.

Haasch ML, Kleinow KM, Lech JJ (1988) Induction of cytochrome P-450 mRNA in rainbow trout: in vitro translation and immunodetection. Tox Appl Pharmacol 94: 246 - 253

Kezic N, Britvic S, Protic M, Simmons JE, Rijavec M, Zahn RK, Kurelec B. (1983) Activity of benzo(a)pyrene mono-oxygenase in fish from the Sava River, Yugoslavia: correlation with pollution. Sci Tot Env 27: 59 - 69

Kleinow KM, Haasch ML, Williams DE, Lech JJ (1990) A comparison of hepatic P450 induction in rat and trout (Oncorhynchus mykiss): delineation of the site of resistance of fish to phenobarbital-type inducers. Comp Biochem Physiol 96C: 259 - 270

Kloepper-Sams PJ, Park SS, Gelboin HV, Stegeman JJ (1987) Specificity and cross-reactivity of monoclonal and polyclonal antibodies against cytochrome P-450E of the marine fish scup. Arch Biochem Biophys 253: 268 - 278

Kreamer GL, Squibb K, Gioeli D, Garte SJ, Wirgin I. (1991). Cytochrome P4501A mRNA expression in feral Hudson River tomcod. Environ Res 55: 64 - 78

Kurelec B, Britvic S, Rijavec M, Muller WEG, Zahn RK (1977) Benzo(a)pyrene monooxygenase induction in marine fish --- molecular response to oil pollution. Mar Biol 44: 211 - 216

Lehtinen KJ, Kierkegaard A, Jakobsson E, Waendell A (1990) Physiological effects in fish exposed to effluents from mills with six different bleaching processes. Ecotoxicol Environ Saf 19: 33 - 46

Lindström-Seppä P, Oikari A (1990) Biotransformation activities of feral fish in waters receiving bleached pulp mill effluents. Environ Toxicol Chem 9: 1415 - 1424

Lindström-Seppä P, Hanninen O, Hudd R (1985) Biomonitoring of the petroleum spill in Vaasa Archipelago, Finland, by biotransformation activities in fish. In: Cytochrome P-450: Biochemistry, biophysics and induction, Eds. L. Vereczkey and K. Magyar, pp. 357 - 360, Akademiai, Kiado, Budapest

Lindström-Seppä P, Vuorinen PJ, Vuorinen M, Haenninen O (1989) Effect of bleached kraft pulp mill effluent on hepatic biotransformation reaction in vendace (Coregonus albula L.). Comp Biochem Physiol 92C: 51 - 54

Mather Mihaich E, di Giulio RT (1991) Oxidant, mixed-function oxidase and peroxisomal responses in channel catfish exposed to a bleached kraft mill effluent. Arch Environ Contam Toxicol 20: 391 - 397

McMaster ME, van der Kraak GJ, Portt CB, Munkittrick KR, Sibley PK, Smith IR, Dixon DG (1991) Changes in hepatic mixed-function oxygenase (MFO) activity, plasma steroid levels and age at maturity of a white sucker (Catostomus commersoni) population exposed to bleached kraft pulp mill effluent. Aquat Toxicol 21: 199 - 218

Munkittrick KR, Portt CB, van der Kraak GJ, Smith IR, Rokosh DA (1991) Impact of bleached kraft mill effluent on population characteristics, liver MFO activity, and serum steroid levels of a Lake Superior white sucker (Catostomus commersoni) population. Can J Fish Aquat Sci 48: 1371 - 1380

Munkittrick KR, van der Kraak GJ, McMaster ME, Portt CB (1992) Response of hepatic MFO activity and plasma sex steroids to secondary treatment of bleached kraft pulp mill effluent and mill shutdown. Environ Toxicol Chem 11: 1427 - 1439

Nebert DW, Gelboin HV (1968) Substrate inducible microsomal aryl hydroxylase in mammalian cell culture. J Biol Chem 243: 6242 - 6249

Oikari A, Lindström-Seppä P (1990) Responses of biotransformation enzymes in fish liver: Experiments with pulpmill effluents and their components. Chemosphere 20: 1079 - 1085

Payne JF (1976) Field evaluation of benzo[a]pyrene hydroxylase induction as a monitor for marine pollution. Science 191: 945 - 946

Payne JF (1984) Mixed function oxygenases in biological monitoring programs: review of potential usage in different phyla of aquatic animals. In: "Ecotoxicological testing for the marine environment" (G. Persoone, E. Jaspers and C. Claus, eds.) Vol. 1: 625 - 655. State University of Ghent and Institute for Marine Scientific Research, Bredene, Belgium.

Payne JF, Penrose WR (1975) Induction of aryl hydrocarbon benzo[a]pyrene hydroxylase in fish by petroleum. Bull Environ Contam Toxicol 14: 112 - 116

Payne JF, Bauld C, Dey AC, Kiceniuk JW and Williams U (1984) Selectivity of mixed function oxygenase induction in flounder (Pseudopleuronectes americanus) collected at the site of the Baie Verte, Newfoundland, oil spill. Comp Biochem Physiol 79: 15 - 19

Pequegnat WE, Wastler TA (1980) Field bioassays for early detection of chronic impacts of chemical wastes upon marine organisms. Helg Meeres 33: 531 - 545

Renton KW, Addison RF (1992) Hepatic microsomal monooxygenase activity and P450IA mRNA in North Sea dab (Limanda limanda) from contaminated sites. Mar Ecol Prog Ser 91: 65 - 69

Ridlington JW, Chapman DE, Boese BL, Johnson VG, Randall R (1982) Petroleum refinery wastewater induction of the hepatic mixed-function oxidase system in Pacific Staghorn Sculpin. Arch Environ Contam Toxicol 11: 123 - 127

Spies RB, Felton JS, Dillard L (1982) Hepatic mixed-function oxidases in California flatfishes are increased in contaminated environments and by oil and PCB ingestion. Mar Biol 70: 117 - 127

Spies RB, Rice DW Jr, Felton J (1988a) Effects of organic contaminants on reproduction of the starry flounder Platichthys stellatus in San Francisco Bay. 1. Hepatic contamination and mixed-function oxidase (MFO) activity during the reproductive season. Mar Biol 98: 181 - 189

Spies RB, Rice DW Jr, Felton J (1988b) Effects of organic contaminants on reproduction of the starry flounder Platichthys stellatus in San Francisco Bay. 2. Reproductive success of fish captured in San Francisco Bay and spawned in the laboratory. Mar Biol 98: 191 - 200

Stegeman JJ, Kloepper-Sams, PJ, Farrington JW (1986) Mono-oxygenase induction and chlorobiphenyls in the deep-sea fish Coryphaenoides armatus. Science 231: 1287 - 1289

Stegeman JJ Teng FY, Snowberger EA (1987) Induced cytochrome P450 in winter flounder (Pseudopleuronectes americanus) from coastal Massachusetts evaluated by catalytic assay and monoclonal antibody probes. Can J Fish Aquat Sci 44: 1270 - 1277

Stegeman JJ, Woodin BR, Goksøyr A (1988) Apparent cytochrome P-450 induction as an indication of exposure to environmental chemicals in the flounder Platichthys flesus. Mar Ecol Prog Ser 46: 55 - 60

Sulaiman N, George S, Burke MD (1991) Assessment of sublethal pollutant impact on flounders in an industrialised estuary using hepatic biochemical indices. Mar Ecol Prog Ser 68: 207 - 212

van der Weiden MEJ, van der Kolk J, Penninks AH, Seinen W, van den Berg M (1990) A dose/response study with 2,3,7,8-TCDD in the rainbow trout (Oncorhynchus mykiss). Chemosphere 20: 1053 - 1058

Van Veld PA, Westbrook DJ, Woodin BR, Hale RC, Smith CL, Hugget RJ, Stegeman JJ (1990) Induced cytochrome P450 in intestine and liver of spot (Leiostomus xanthurus) from a polycyclic aromatic hydrocarbon contaminated environment. Aquat Toxicol 17: 119 - 132

White INH (1988) A continuous fluorimetric assay for cytochrome P-450-dependent mixed function oxidase activity using 3-cyano-7-ethoxycoumarin. Anal Biochem 172: 304 -310

Biochemical and Genotoxicological Monitoring of Ecosystems with Special Reference to Lake Baikal and Northern Black Sea

Sergey V. Kotelevtsev and Ludmila I. Stepanova
Laboratory Physical-Chemistry Biomembrane,
Department of Biology,
M.V. Lomonosov Moscow State University,
119899, Moscow, Russia

Introduction

The test-systems for biological indication and biological monitoring of carcinogenic and mutagenic compounds in water ecosystems are demonstrated the increasing content of mutagenic compounds in the tissues of the fresh water and marine animal. We use the Ames test for assay of tissue extracts from the water animals. The mutagen content in tissue of fish and other water animals was found in different region of Caspian and Black Sea, Rybinsk Reservoir and Baikal Lake.

It is customary to divide anthropogenic effects on ecosystems into two groups:

1 - acute stress, which is characterized by a sudden start, a rapid increase of intensity and a short duration of disturbances;

2 - chronic stress, where disturbances of low intensity are prolonged or repeat frequently - these are "constantly disturbing" effects (Odum, 1954; Phiodorov and Gilmanov, 1980).

As a rule, ecosystems possess resistance and resiliency that help them to be a periodic severe or acute effects. Moreover, a number of populations need seven stochastic disturbing effects for their development, such as fires or sharp changes in climatic conditions. Water ecosystems restore successfully after effects given on one occasion only. These effects may not be natural like storms, volcano eruptions or sudden sharp changes in water temperature. Exposure, only once

NATO ASI Series, Vol. H 90
Molecular Aspects of Oxidative Drug Metabolizing Enzymes
Edited by E. Arınç, J. B. Schenkman and E. Hodgson
© Springer-Verlag Berlin Heidelberg 1995

to even a considerable amount of toxic substances may not be especially dangerous if these compounds are not stable and are quickly removed from the ecosystem. However, chronic disturbances can result in marked, stable and irreversible consequences, especially in the case of pollution by waste chemicals that did not exist in the environment before (Lyakhovich and Tsyrlov, 1981). As a rule these compounds, alien to organisms, xenobiotics, do not take part neither in energetic nor in structure metabolism, do not accumulate in tissues and undergo (with this or that speed) chemical transformations while reacting with vitally important biological molecules, such as DNA, RNA, protein and lipids.

Carcinogenic and mutagenic compounds are the most dangerous among them; according to some estimates they make up not less than 5% of the total amount of anthropogenic pollutants of ecosystems (Dubinin, 1977).

These compounds may not possess acute toxicity but when accumulating in the organism they display prolonged action with especially dangerous consequences. Among the direct genotoxic substances a discrimination should be made between chemicals that are intrinsically reactive and thus may form DNA adducts directly, and those compounds which require a metabolic transformation or activation before they become genetically active.

Xenobiotics are distributed world-wide. Nowadays nobody is surprised by the presence of mutagenic pesticides in fatty tissues of whales or by the presence of polychlorinated biphenyls (PCBs) in breast milk of women from Pacific islands, locations where these chemicals had never been used. The increase of frequency of some diseases, disturbances of the immune system maid some hereditary pathologies among human population is attributed to the presence of xenobiotics in the environment. With that it is noted that from 5 to 25 % of human cancer diseases is connected with the pollution of air, food products and drinking water by carcinogenic and mutagenic substances (Poroschenko and Abilev, 1988).

Mutagenic and carcinogenic compounds produced as a result of industrial and agricultural activity, sooner or later will reach the water ecosystem, transported by rivers or by atmospheric deposition. A considerable part of these substances is introduced into the seas directly as spills from ships carrying oil, oil products or other toxic chemicals, sometimes from calamities. Substantial amounts of mutagenic compounds get into the marine ecosystem in the course of oil and gas extraction in shelf areas.

During the last ten years there was a sharp increase in the number of tumors recorded in tissues of marine fish, including fish for human consumption (De Flora et al., 1991; Moore et al., 1989). Some of these diseases have a virus nature but the majority is connected with the pollution of seas and oceans by mutagenic and carcinogenic xenobiotics (Chemical contaminants and fish tumors, 1990). One of the features of mutagenic and carcinogenic compounds is that they can display biological effects in very low concentrations. It hinders their chemical-analytical assay in particular in biological tissues (Hmelnitsky and Brodsky, 1990). On the other hand, it is not possible to determine by means of chemical techniques whether a given substance has carcinogenic and mutagenic properties. That is the reason why the biological testing and the biological indication of carcinogenic and mutagenic compounds obtains the growing significance. The appearance of the increasing content of mutagenic compounds in water animals used in human nutrition accounts for the necessity of monitoring of genotoxic effects in water ecosystems are as and especially in those areas where the commercial fisheries takes place.

Indeed, mutagenic and carcinogenic substances accumulating in water animals can not only modify existing ecosystems but represent a direct danger for man as well. Mutations are hereditary changes resulting in the increase or decrease of the amount of genetic material or in changes of chromosome structure or DNA nucleotide sequence. Mutations can arise spontaneously and under the action of various factors: chemical substances, and physical ultraviolet or ionizing radiation.

The different mutation types are:

1 - changes of ploidy, i.e. chromosome number in the cell;

2 - chromosome mutations - changes of gene location in the chromosome;

3 - point mutations - changes of sequence in DNA at the level of single nucleotides.

Test-Systems for Analysis of Mutagenic and Carcinogenic Compounds

Various test-systems are employed for the investigation whether environmental factors alter the genetic material, ranging from chemical detection of adduct formation, assays for DNA effects for the determination of specific mutations. The complexity and duration of a test is highly varying from one test system to another. Moreover, it is impossible to obtain information about all aspects of genotoxicity in one single test. For a description of all mutagenic and carcinogenic effects in an aquatic ecosystem one requires the use of different test-systems allowing to reveal most types of mutations. Such research is time-consuming and expensive, and is thus seldom carried out and then usually only while performing model compound experiments.

Genotoxicity tests can be classified according to the information they provides follows:

1 - assays for gene mutation; they commonly make use of mutants of wild types. These mutants lack a characteristic function which is used to test a back-mutation reaction as e.g. in the Ames test (Ames et al., 1975; Maron and Ames, 1983). Other tests include in vivo and in vitro eucariotic systems;

2 - assays for chromosomal aberrations (CA); eucariotic in vivo and in vitro tests are known, which can often be detected using a light microscope. An example is the in vivo micronuleus test (Schmid, 1975; Mackey and McGregor, 1979; Iliinskich et al., 1992);

3 - assays for DNA effects; not the specific mutations are detected, but information is based on the induction of the

DNA-repair system. Examples are bacteria tests such as the *Bacillus subtilis* rec-assay (Matsui, 1980), or the *in vivo* and *in vitro* sister chromatid exchange (SCE) in eucariotic systems.

As a rule, the first two types of mutations, are studied by means of cytogenetic techniques when animal organisms and higher plants are used as test-objects for genetic monitoring. One of the conditions limiting the use of this method is the number and the size of chromosomes in the cells of the test-object. Today flow laser fluorimeters are used to investigate direct lesions of the DNA structure by a chemical mutagens. They permit to determine fluorescent characteristics of the DNA of separate cell nuclei in special capillaries in the flow of the medium containing the cells. In the case of fish, both immunocompetent blood cells and erythrocytes can be used.

The analysis of the results is carried out by means of specialized computer programs. Mutagen-induced lesions in the DNA structure change the nature of fluorescence and allow to judge possible gene mutations (Jenner et al., 1990). Besides, the analysis of DNA adducts by chromatographic and other chemical-analytical techniques can be carried out (McCarthy et al., 1989; Peakal, 1992; Stein et al., 1989).

To investigate point mutations in mammals or fish, observing their origin in a series of generations by morphological characteristics, is a long and expensive process. Nevertheless, studies on mice, rats and aquarium fish are carried out while analyzing carcinogenic features of a number of chemical compounds.

For a quick estimation of the mutagenic effects of chemical substances microorganisms are used. The advantages of using bacteria (or yeasts) as test-objects are evident. First they produce a great number of generations in only a few hours. Second, the use of genetic engineering techniques allows to construct special strains, highly sensitive to chemical mutagens that makes it possible to detect mutagenic properties already at low concentrations of the investigated substances. A significant short coming of these test-systems is that microorganisms almost completely lack enzymes accomplishing metabolic activation of xenobiotics. The following diagram shows the two routes of the induction of genotoxic effects:

```
        --------- Direct action ------------->
Compound                                          Genotoxic
                                                   effect
        ----- Metabolic activation --------->
```

For this reason it is possible to register in microorganism cultures the effect of direct mutagens only, i.e. only the effect of compounds interacting directly with the genetic material and displaying mutagenic effect themselves. However, the great number of xenobiotics acquire mutagenic properties only after metabolic activation by the monooxygenase system.

To overcome this problem systems were developed where the investigated substance was introduced into the abdominal cavity of animals (mice, rats) together with tester bacterial cells. After a few hours, the animals were killed and the tester microorganisms from the abdominal cavity plated on Petri dishes where the mutant colonies were registered. This method ("host-mediated assay") is still used in some laboratories (Frezza et al., 1982; Abilev, 1986).

Ames et al. (1975) proposed a test where enzymes from the rat liver cytochrome P450 system were used for *in vitro* activation of the investigated substances. For a special analysis of effects of mutagenic and carcinogenic compounds on aquatic organisms it is possible to use the microsomal fraction from the liver of fish (including marine fish) which were preliminary exposed to injection of 20-methylcholanthrene (Kotelevtsev et al. 1986a; Kotelevtsev et al., 1987a) and even monooxigenase system from the mollusc tissue (Marsh et al., 1991). The Ames test *Salmonella*/microsomes is based on the use of the set of tester strains specially constructed by Ames and his collaborators (Ames et al., 1975). The strains bear mutations of histidine deficiency of different types.

Various mutagenic agents cause reversions to histidine prototrophy (*his*-phenotype) in tester strains, of which the frequency can be estimated by a number of colonies grown on the plates with minimal selective medium without histidine. The use of different tester salmonella strains in the assay makes it possible to distinguish betwen a mutagen causing a mutation of the frame-shift or of the base-substitution type. To improve sensitivity, in addition to one of the *his*-mutations, all tester strains bear *rfa* and *uvrB*.

The test consists of the addition to half of the probes a suspension of the tester bacteria, the studied substance (pollutant), rat liver postmitochondrial fraction S9 and cofactors of monooxygenases ("+MA") into 2 ml of melted soft agar (45 °C). The sample is quickly mixed and covered into selected minimal agar. The other half of probes ("-MA") does not contain the metabolic activation components. The plates are incubated at 37 °C for 48 h.. If metabolic activation takes place with the formation of mutagenic substances, it results in the increase of His^+-revertant frequency in a positive metabolic activation: "+MA" samples, in contrast to the negative metabolic activation: "-MA" ones. Thus one may draw a conclusion about the possible promutagenic activity of the assayed substance. If mutations are induced in both "+MA" and "-MA" samples it indicates a direct mutagenic activity of the tested substance.

This test is used in a great number of laboratories and is convenient for analysis of not only chemical compounds but of waste waters, food products etc. as well. By statistical analysis one may distinguish between the various degrees of induced effect. If the number of colonies in the experiment is more than two fold higher compared to the background level, the result is considered to be statistically significant but weak (it is marked as "+"), more than tenfold higher - a strong result (++), more than hundred fold higher - a very strong effect (+++).

In studies of several hundreds of chemical compounds it was shown that about 90% of substances causing carcinogenesis in mammals proved to be mutagenic in the Ames test. On the other hand, more than 90% of compounds that didn't cause mutations don't display carcinogenic effect in this test although some of such "non-carcinogens" were active in the Ames test (so-called "false-positive results"). So if the substance displays mutagenic properties in the Ames test it will have both a mutagenic and carcinogenic effect upon man and animals with high probability (Ames et al., 1975).

Principles of Biological Testing of Mutagenic and Carcinogenic Compounds in Water Ecosystems

For the biological (toxicological) testing of a chemical substance or a water mixture of substances, the aquatic organisms are exposed under controlled laboratory conditions to defined concentrations of the chemical(s) and the possible effects are recorded as function of time. That concentration of the tested substance upon which animals or plants do not die, and show no other biological effects such as marked changes in growth rate or their development and functional state, even after chronic exposure, is considered to be harmless.

This principle is used for the analysis of the potential toxicity of individual chemical compounds or their mixtures and of industrial waste materials they are or may be discharged into the aquatic ecosystem. The method is, however, of limited use for biological (genotoxic) testing of mutagenic and carcinogenic compounds. Nevertheless some methods have been worked out for the detection of mutagenic effects in higher organisms, e.g. with daphnia or with small marine pelagic crustacean.

They are able to produce several generations in 1-2 months allowing to register the appearance of abnormalities and some other hereditary changes over the generations. As they do not have a sufficiently developed system of metabolic activation of xenobiotics and are only accumulators of promutagenic compounds, one will observe the action of direct mutagenic substances only (Kotelevtsev et al., 1987a).

For the biological testing of mutagenic compounds it is necessary to use special genetic test-systems. Firstly, it is possible to study chromosome aberrations or DNA damages with aquatic animals exposed to injections of the tested substances or caught in the areas polluted by mutagenic and carcinogenic xenobiotics (fish, molluscs) (Smith, 1990; Iliinskich, 1992).

Secondly, it is possible to place selected test-organisms (e.g. molluscs, fish) in special cages directly into the to be tested aquatic system. After a certain time period the organisms may be used for determination of chromosome aberrations, conducting micronuclear test or studying damages in DNA by the above techniques (Waranazi et al., 1989).

Thirdly, it is possible to study (by means of genetic a test-system) the accumulation of mutagenic and carcinogenic substances in tissues of marine animals, algae and sediments. In this case it is necessary to extract and isolate the mutagenic compounds from water animals tissues (Scarpato et al., 1989).

The latter method seems to be the most valuable for carrying out biological monitoring in the field, since it allows, on the one hand, to use various genetic test-systems for analysis of extracts, such as the SOS-chromotest, a test -systems that lends itself to automation (Kotelevtsev et al., 1986b). On the other hand, it makes possible to judge the extent of metabolic activation of xenobiotics in tissues of various organisms, using the monooxygenase systems from liver of warm-blooded animals and fish for metabolic activation.

Extraction of Mutagenic Compounds from Tissues of Marine Organisms, Water and Sediments

Separation and concentration of xenobiotics is accomplished by various chromatographic techniques. Two types of extraction are performed for preparing samples for analysis with the use of biological tissues: aqueous extraction and extraction by organic solvents. From water xenobiotics are usually extracted using one or more hydrophobic solvents The different methods of xenobiotic extraction from tissues of marine and fresh water organisms and from water have no fundamental differences. The choice of the method depends upon what kind of compounds one intends to isolate from the to be tested sample. For extraction of mutagenic compounds that are the most widespread in the water ecosystems (including oil products, pesticides and herbicides) various organic solvents are more often used (Parkinson, 1974; Perry et al., 1976).

The preparation of tissue extracts for mutagen activity assay is beginning from homogenization of the tissue, then followed I or II procedure.

I. Extraction aqueous, aqua-acidic or aqua-alkaline; sedimentation of proteins; pH adjusting; concentration of mutagens.

II. Extraction by organic solvents; concentration of mutagens (removal of solvent by a rotor evaporator or concentration on resins of a XAD type).

The mechanical destruction of tissues is performed with the help of homogenizers of a "POTTER" or "POLTROON" type with addition of wateror organic solvent. Aqueous extracts of animal tissues contain a relatively large amount of proteins that can hamper isolation of mutagenic components in the subsequent procedures. So such coagulants as water free sodium sulphate (Parkinson, 1974; Osborne et al., 1982; Krone and Iwaoka, 1983), ammonium sulphate (Commoner et al., 1978), sulphosalycilic acid (Belisario et al., 1984) are added to the homogenate. The method of hydrolysis of proteins by hydrochloric and nitric acids is widespread (Commoner et al., 1978; Bjeldanes et al., 1982; Kadhim and Perry, 1984). The use of polar organic solvents (e.g. acetone, methanol, ethanol) for the extraction of tissues of aquatic organisms is more common (Felton et al., 1981; Krone and Iwaoka, 1983). These solvents allows not only the proteinsto coagulate but they will also transfer (as much as possible) mostly hydrophobic mutagenic and promutagenic compounds to a soluble phase. The method of the acetone extraction is the easiest and the most convenientfor isolation of mutagens from organic samples (Felton et al., 1981). The yield of mutagens from the investigated material is about two-fold higher by this method compared to the aqueous extraction method (Commoner et al.,

1978). Acetone in mixture (in equal ratios) with other organic solvents: benzene, water, hexane (Vian et al., 1982), methanol, ethanol (Krone and Iwaoka, 1983) is widely used. Parallel to these methods nonpolar solvents are also widely used forextraction of mutagens from the various compartments: hexane (Parkinson, 1974; Kurelec et al., 1981), chloroform (Loper et al., 1981), dichloromethane (Grabow et al., 1981), benzene (Smith, 1982), cyclohexane, benzene (Osborne et al., 1982). They are often used together with protein coagulants (Krone and Iwaoka, 1983). Taking into account the different chemical nature of the various mutagens and promutagens, sometimes mixtures of polar and nonpolar solvents are used for their extraction (Smith, 1982).

However in each concrete case, taking into account chemical featuresof mutagens and the nature of the studied material, it is necessary to selectthe corresponding method of extraction. With that one should take into consideration that some procedures of extraction as well as the presence of ammonium ions in solvents, can induce formation of artificial mutagenssince they have potential ability to form amides and amines in the course ofammonium ions reactions with organic solvents (Felton et al., 1981; Iwaoka et al., 1981). It was also shown that in the course of extraction many solvents can alter the tested substances via oxidation (Jori et al., 1969; Gennari and Jori, 1970; Holt et al., 1977). The use of chlorinated solventscan also result in artifacts in testing of mutagens since the latter contains tabilizers. Besides, chlorinated solvents are not stable and can form hydrochloric acid with subsequent formation of radicals that easily polymerize non-volatile toxic substances in solution (Anthony, 1979). But there is a wide range of organic solvents (alcohols, ethers, alphabetichydrocarbons) that change the tested material via reaction of ether ification without affecting the mutagenic potential. Therefore from this point of view they are inert and are the most suitable for extraction. Different chemical nature of mutagens and promutagens supposes their different solubility in water and aqueous solutions. That's why for the most complete extraction of the tested compound use the cascade method of extraction by organic solvents with preliminary consecutive change of pH of the investigated samples is used (Commoner et al., 1978; Grabow et al., 1981; Bjeldanes et al., 1982). However this method also has shortcoming since the possibility exists to obtain artifact results while studying the extracted mutagens in the Ames test (Grabow et al., 1981; Krone and Iwaoka,1983). The use of protein coagulants in this method leads to the binding of mutagens by precipitated proteins owing to electrostatic, ionic and hydrophobic interactions (Krone and Iwaoka, 1983). Taking into account hydrophobic nature of mutagens it is possible to say that the use of organic solvents for isolation of mutagens is more preferable compared to aqueous extraction. In comparison with other methods of isolation of mutagen compounds (aqua-acidic extraction by Commoner et al. (1978), acetone extraction by Felton et al. (1981), method of prolonged extraction ofacidic-neutral-alkaline fractions by a system of solvents methanol /dichloromethane / water) (Krone and Iwaoka, 1983) the use of synthetic resinsof a XAD type does not cause the appearance of artifact mutagens. Secondly, it separates quantitatively much yield of mutagens, not changing their chemical structure and, as a result, increases in many times the mutagenic response in the Salmonella/microsomes Ames assay (Jamasaki and Ames, 1977; Krone and Iwaoka, 1983). However, despite of the relative universality of the enumerated methods, ineach concrete case, taking into account the investigation aims, one shouldchoose the corresponding techniques or combine various methods.

Mutagens after extraction are dissolved in the medium of assay and analyzed in biological test-systems.

Genotoxicity of Tissue Extracts of the Water Organisms

In our laboratory the biological monitoring of genotoxicity of water ecosystems of the southern part of Lake Baikal, the Rybinsk Reservoir, the Black and the Caspian Seas has been carried out for a long time by means of the Ames test and the SOS - chromotest (Kotelevtsev et al., 1986a,b; Kotelevtsev and Hanninen, 1992). In some cases we used the metabolic activation system not only from liver of rats but from liver of fish inhabiting the investigated areas as well (Kotelevtsev et al., 1986a, 1987a). It turned out that the activity of the metabolic activation system of standard carcinogens from the fish liver in the Ames test depended considerably upon both their species and the method of induction of monooxygenases (Kotelevtsev et al., 1987a). For instance, benzo(a)pyrene and 2-aminoanthracene are metabolized more actively in the liver of the Black Sea scad compared to the liver of bullheads (Kotelevtsev et al., 1986b, 1987b). Therefore in the scad tissues xenobiotics would be oxidized and released from tissues, but the population of bullheads pays for it by an elevated risk of rise of malignant tumors. In the bullheads tissues the metabolism of xenobiotics is slower so in this fish they would accumulate in significantly greater amounts (Kotelevtsev et al., 1987b). However it is necessary to take into account the possibility of concentration of xenobiotics in food chains as for instance: molluscs - bullheads - scad - man.

In the case of the Caspian Sea the mutagenic compounds in the extracts from water animal tissues are detected even in relatively pure, remote from settlements areas. Mutagenic compounds are present in tissues of almost all organisms inhabiting polluted areas of the Black Sea. The character of accumulation of mutagenic compounds in marine animals tissues depends upon not only the type of pollutant but the activity of metabolizing enzymes in tissues as well. In this section we will examine as an example the results of the analysis of mutagenic compounds in tissues of the Black Sea animals.

The expedition work had been carried out in June-July 1989-1991 on the Black Sea on board the Moscow University's scientific-research ship "Experiment". The material was collected by trawling (beam-trawl) of the coastal areas at 7-12 meters depth, while individual species were collected by divers. Immediately after collection the species was identified and the material was fixed in twofold volume of acetone and stored at about 4 °C.

In total six groups of species collected at 32 Black Sea stations were investigated. Unfortunately, at some stations we failed to catch representatives of the main classes of these organisms. Nevertheless it is possible to consider species described in this work to be typical for each of the explored areas.

Isolation of xenobiotics from the organisms' tissues and from the bottom sediments was carried out by means of the repeated extractions by the mixture of organic solvents (acetone:hexane:chloroform = 1:1:1). The obtained xenobiotics were separated from the extraction mixture in a rotor evaporator, dried by lyophilization, dissolved and diluted in dimethylsulphoxide (DMSO) proportionally to the dry weight of a sample at a rate of 1 ml of DMSO per 100 mg of tissue dry weight. Thus the samples were tested for genotoxicity.

Analysis of mutagenic properties of each sample was performed in the Ames test in four variants. The first two of them identify the presence of direct mutagens ("-MA") in the extracts of xenobiotics causing the both types of point gene mutations in the test-organism *Salmonella typhimurium*: frameshift of the genetic code (TA 98 strain) and base substitution (TA 100 strain). In the two other variants the presence of indirect mutagens ("+MA") were determined causing a drastic increase of the frequency of point gene mutations in the conditions of the Arochlor 1254 - induced metabolic activation system from the rat liver (S9 fraction).

The results of typical experiments on assay of extracts from water animal tissues in the Ames test are shown in Table 1. They are represented as a number of colonies of His^+-revertants of salmonella per plate. The number of spontaneous mutations in the control experiment was determined in the presence of DMSO in the incubation medium. As a control promutagen causing mutagenic effects after metabolic activation in the microsomal monooxygenase system, 2-aminoanthracene and benzo(a)pyrene were used.

Table 1. Mutagenic activity of different tissue extracts (Black Sea, Pomorie) in tester strains TA 98, TA 100 of *S. typhimurium* (MA = metabolic activity)

Sample	Dose ml per plate	TA 98		TA 100	
		+MA	−MA	+MA	−MA
Dimethylsulphoxide	0.1	38	20	112	108
2-Aminoanthracene	0.5 mkg	4420	18	2150	104
2-Nitrofluorene	10.0 mkg	843	34	682	107
Gobius sp.(liver)	0.1	40	21	154	115
Rapana thomasiana (digestion gland)	0.1	58	23	141	103
Chamella gallina	0.1	60	114	110	101
Clibanarius erythropus	0.1	61	54	190	165
Polychaete sp.	0.1	76	18	121	111

The main results of the investigation of extracts of tissues of the Black Sea organisms are represented in Tables 2 and 3.

The analysis of the data showed that in the majority of the investigated areas including those relatively remote from direct sources of pollution, a significant amount of mutagenic compounds was detected. They displayed both direct mutagenic and promutagenic effects.

The TA 98 strain is the most informative among the two test-strains (Tables 2, 3). The results indicate that the investigated Black Sea coastal organisms accumulate mainly direct mutagens causing mutations of a frame-shift type.

It is possible to arrange the investigated animals in the decreasing line by the ability to accumulate direct mutagens.

For the TA 98 strain (upon the mentioned species): polychaetes = shrimps > bullheads > sea snails = bivalves = hermit-crab;

For the TA 100 strain: shrimps > bullheads > polychaetes = hermit crab = sea snails > bivalves.

Table 2. Genotoxicity of tissue extracts from Black Sea organisms in the Ames test with/without (=/=) metabolic activation from liver of rats, inducted with Arochlor 1254 (tester strain *S. typhimurium* TA 100).

Extracts of marine organisms *)

Station name	1	2	3	4	5	6
Poty	0	0	0	-	0	-/+
Ochamchira	0	0	-/+	0	0	-
Sukchumi (town)	+/-	-	-	0	0	0
Sukchumi (port)	0	-	-	0	0	+/-
Gagra	0	-	-	0	0	0
Sochi	0	-	+/+	-	0	0
Tuapse	0	-	+++/+	-	0	-/+
Gelendgick	0	-/+	+/+	-	0	-/+
Anapa	0	+/	-/+	-	0	-
Kerch	0	-/+	-	0	0	-/+
Caradag	0	+/+	+++/+	0	0	-
Phaodosia	0	+/+	0	-/+	0	+/+
Sudack	+/-	-	-	-	0	-
Alushta	-	-/+	-	-	0	0
Ialta	0	+/+	-	-	0	0
Varna	0	0	0	-	-/+	0
r. Camchiah	-	0	+/-	0	0	-/+
Necebr (port)	-/+	-	0	0	0	-
Necebr (town)	+/+	+/+	0	-	0	-/+
Pomoria	-	-	0	-/++	+/-	-/+
Burgas	0	0	0	0	+/+	0
r. Velena	0	0	-	-	0	-
Achtopol	-/+	0	-	-/+	-	+/+
Michurin	-/+	0	-	-/+	0	0
Pomorsk	-	-	-	+/+	0	+/+
Sozopol	-/+	-	-	-	0	0
Albena	-/+	+/-	0	-	0	-/+
Balchick	0	-/+	-	-/+	-/+	-/+
c. Cavarna	-	-	-/+	0	-/+	-
c. Calavcar	+/+	-/+	-	-/+	0	-/+
Zlatni Piasci	-	-/+	-	0	0	0

*) Species tested:
1. - Bullhead (muscles)   - *Gobius sp.*
2. - Sea snail (digestion gland) - *Rapana thomasiana*
3. - Hermit crab   - *Clibanarius erythropus*
4. - Bivalves   - *Chamella gallina*
5. - Polychaete   - *Nereis sp.*
6. - Shrimps   - *Crandon crandon*

0: Animals was not collected; -: mutagen effect not found; +: weak mutagen effect; ++:medium mutagen effect; +++: strong mutagen effect.

Table 3. Genotoxicity of tissue extracts from Black Sea water animals in the Ames test with/without (=/=) metabolic activation from liver of rats, inducted with Arochlor 1254 (tester strain *S. typhimurium* TA 98).

Station names	Extracts of marine organisms *)					
	1	2	3	4	5	6
Poty	+/-	0	0	-	0	-
Ochamchira	+/-	0	+/-	0	0	-
Sukchumi (town)	+/-	-	+/-	0	0	0
Sukchumi (port)	0	-	-	0	0	0
Gagra	0	-	-	-	0	0
Sochi	0	-	-	-	0	0
Tuapse	0	-	-	-	0	-
Gelendgick	0	-	+/-	-	0	-
Anapa	0	-	-	-	0	-
Kerch	0	-	-	0	0	+/+
Caradag	0	+/+	-	0	0	-
Phaodosia	0	-	0	-	0	+/-
Sudack	-	-	-	-	0	-
Alushta	-	-	-	-	0	0
Ialta	0	-	-	-	0	0
Varna	0	0	0	-	-	0
r. Camchiah	+/-	0	-	-	-	+/+
Necebr (port)	+/+	-	0	0	0	+/-
Necebr (town)	+/+	-	0	-	0	+/+
Pomoria	-	-	0	-	-	+/-
Burgas	0	0	0	0	+/-	0
r. Velena	0	0	-	-	0	-
Achtopol	+/+	0	-	-	0	+/+
Michurin	-	0	-	-	0	0
Pomorsk	-	-	-	-	-	+/+
Sozopol	+/+	-	-	-	-	0
Albena	+/-	-	0	-	0	-
Balchick	0	-	-/+	-	-/+	-
c. Cavarna	-	+/-	-	-	+/-	-
c. Calavcar	+/+	-	-	-	0	+/+
Zlatni Piasci	-	-	-	0	0	0

*) Species tested:
1. Bullhead (muscles)   - *Gobius sp.*
2. Sea snail (digestion gland) - *Rapana thomasiana*
3. Hermit crab   - *Clibanarius erythropus*
4. Bivalves   - *Chamella gallina*
5. Polychaete   - *Nereis sp.*
6. Shrimps   - *Crandon crandon*

0: Animals was not collected; -: mutagen effect not found; +: weak mutagen effect; ++: medium mutagen effect; +++: strong mutagen effect.

This approach allows to choose the most active accumulators of the mutagens of the Black Sea such as polychaetes, shrimps, bullheads (Table 4). These animals can thus be used as biomonitors for genotoxicity in Black Sea coastal waters.

Table 4. Species for monitoring of genotoxicity in the Black Sea.

Species	% samples content mutagenic compound
Bullhead (muscles) - *Gobius sp.*	53 ± 7.0
Sea snail (digestion gland)- *Rapana thomasiana*	27 ± 3.1
Hermit crab - *Clibanarius erythropus*	46 ± 4.2
Bivalves - *Chamella gallina*	27 ± 4.5
Polychaetes - *Nereis sp.*	77 ± 6.0
Shrimps - *Crandon crandon*	51 ± 3.2

In extracts from tissues compounds are also found that have promutagenic action which is revealed in the Ames test after the metabolic activation by the monooxygenase system from the liver of Arochlor 1254-induced rats.

The Black Sea water organisms accumulate these compounds in different extent: *Nereis*, *C. erythropus* and *R. thomasiana* should be selected as typical accumulators of these xenobiotics. It indicates that the system of metabolic activation of xenobiotics in tissues of these animals is not sufficiently active in comparison with other species and warm-blooded animals and can not hydroxylize effectively mutagenic substances getting into the organism. Rarely promutagens are accumulated in organisms of the Black Sea bivalve molluscs (*C. gallina*). It is necessary to mention that *R.thomasiana* and *C. erythrophus* display the same tendencies in the frequency of accumulation direct mutagenic and promutagenic substances.

The Black Sea polychaetes and bullheads are the most active accumulators of both types of mutagenic compounds.

We should realize, however, that it is natural that in each specific case the type of compounds accumulating in the tissues depends upon the character of pollutants entering the ecosystem, the features of nutrition and the activity of the organism's detoxication system.

We found both promutagens and direct mutagens in muscle and liver of bream from different sites of the Rybinsk reservoir. A numerous mutagens were detected with tester strain *S. typhimurium* TA 100).

In the majority of cases both promutagens and direct mutagens more often were detected in muscles then in liver Of fish from Rybinsk reservoir. The frequency of detecting of frame shift promutagens was in muscles similar with liver. But direct frame shift mutagens were more often detected in liver.

Also we studied the changes of mutagen contents in the tissues of breams collected in different seasons. In autumn genotoxity of bream tissue extracts (the frequency of mutagen detection) considerable increased. However both promutagens detecting with TA 98 and TA 100, and direct mutagens detecting with TA 100 in liver were discovered in spring more often then in autumn.

Probably, the hier level of genotoxity of tissue extracts may be connected with storing of fatty tissue by fish where xenobiotics can accumulate.

Therefore to study migrations of mutagens along the trofic chains in both certain populations and ecosystems is so interesting for interpretation of obtaining results.

Comparing these data with results of assessing of 7-ERODE activity we can observe distinction between the level of monooxigenase activity and content of cytochrome P450 isozymes in liver. Possibly, xenobiotics can also reduce induction of cytochrome P450 system that results in fall of monooxygenase activity in fish capturing near by throw off point.

Thus, it is evident, that both evaluation of mutagen accumulation in tissues and study the functions of monooxygenase systems taking part in their detoxication are to use for ecotoxicological monitoring of xenobiotics in environment.

Induction of the monooxygenases activities in fish liver often but not always correlated with the amounts of mutagenic substances in their tissues (Kotelevtsev and Hanninen, 1992). These investigations have to involve the detecting both of isozymes concentration, their functional activity and xenobiotics content in fish tissue.

We have studied the effects of waste water of Baikalsk Bleached Craft Pulp Mill, as an example, on the isozymes cytochrome P450 in the fish liver microsomes. The tissue extracts of fish caught in polluted areas or exposed in waste water were also examined by the Ames test (Kotelevtsev and Hanninen, 1992).

The Lake Baikal fish species used in this study were as follows: omul (*Coregonus migratorius*), Baikal grayling (*Thymallus arcticus baikalensis* D.), lenok (*Brachymystax lenok*), bullherd (*Cottocomephorus greminski*).

We used Ames test to study mutagenic activity waste water of Baikal Bleach Craft Mill and tissue extracts water animals (fish, mollusc, crayfish) in polluted and control regions Baikal Lake.

Keeping Baikal grayling and bullhead in waste water of BBKM (diluted 1:5) didn't induce accumulation of mutagenic substances in tissues of skin, muscle or liver, but extracts of plankton caught in polluted region near Baikal Bleach Craft Mill had undirected mutagenic activity in TA 98 salmonella strain.

Mutagenic and carcinogenic compounds are now more often registered in the water animals tissues. The mutual action of these factors may be especially dangerous for water animals. On the other hand, accumulation of mutagenic and carcinogenic compounds in water animals tissues, as it was mentioned above, represents a direct danger for man as a consumer of sea products.

It is necessary to carry out a regular monitoring of genotoxicity of sea ecosystems, to reveal and, if possible, to liquidate the sources of pollution of water ecosystems. These problem can be solved only by means of international cooperation.

References

Abilev SK 1986. Metabolic activation of mutagens. In: Poroshenko GG (ed) Sum of sciences technique. Genetic. vol 9. VINT. Moscow, pp 5-96 (in Russian)

Ames BN, McCann J, Yamasaki E (1975) Method for detecting carcinogens and mutagens with the *Salmonella*/mammalian microsomes mutagenicity test. Mutat Res 31:347-364

Anthony T (1979) Methylene chloride. In: Bushey GJ, Campbell L, Eastman C, Klingsberg A, Van Nes L (eds) Kirk-Othmer encyclopedia of chemical technology. vol 5. Wiley. New York, pp 686-693

Belisario MA, Buonocore V, De Marinis E, De Lorenco F (1984) Biological availability of mutagenic compounds absorbed into diesel exhaust particulate. Mutat Res 135:1-9

Bjeldanes LF, Grose KR, Davis PH, Stuermer DH, Healy SK, Felton JS (1982) An XAD-2 resin method for efficient extraction of mutagens from fried ground beef. Mutat Res 105:43-49

Chemical contaminants and fish tumors (1990) Sci Total Environ 94:168

De Flora S, Bagnasco M, Manacchi P (1991) Genotoxic carcinogenic, and teratogenic hazards in marine environment with special reference to the Mediterranean Sea. Mutat Res Rev Genet Toxicol 258:285-320

Dubinin NP (1977) The genetic results of the environmental pollution. Nauka, Moscow, pp 3-20 (in Russian)

Felton JS, Healy S, Stuermer D, Berry C, Timourian H, Hatch FT (1981) Mutagens from the cooking of food, improved extraction and characterization of mutagenic fraction from cooked ground beef. Mutat Res 88:33-34

Frezza D, Pegoraro B, Presciuttini S (1982) A marine host-mediatedassay for the detection of mutagenic compounds in polluted seawaters. Mutation Res 104:215-223

Gennari G, Jori G (1970) Acetone-sensitized anaerobic photooxidation of methionine. FEBS Lett 10:129-131

Grabow WOK, Burger JS & Hilner CA (1981) Comparison of liquidextraction and resin adsorption for concentrating mutagens in Ames Salmonella/microsomal assays on water. Bull Environ Contam Toxicol 27:442-449

Hmelnitsky RA, Brodsky ES (1990) Mass Spectrometric of Environmental Pollution. Chimia. Moscow, pp 184 (in Russian)

Holt LA, Milligan B, Rivett DF, Stewart FHC, (1977) The photodecomposition of tryptophan peptides. Biochem Biophys Acta 499:131-138

Iliinskich NN, Novitskyi VV, Vanchugova NN, Iliinskich IN (1992) The micronucleus analysis and cytogenetic unstability. Tomsk University. Tomsk, pp 272 (in Russian)

Iwaoka WT, Krone CA, Sullivan JJ, Meaker EH, Johnson CA, Miyasato LS (1981) A source of error in mutagen testing of foods. Cancer Lett 11:225-230

Jamasaki E, Ames BN (1977) Concentrations of mutagens from urineby adsorption with the nonpolar resin XAD-2: Cigarette smokers havemutagenic urine. Proc Natl Acad Sci (U.S.A.) 74: 3555-3559

Jenner NK, Ostrander GK, Kavanagh TJ, Livesey IS, Shen MW, Kim SC, Holmes EH (1990) A flow cytometric comparison of DNA content and glutathione levels in hepatocytes of English sole (Parophrys vetulus) from areas of differing water quality. Environ Contam Toxicol 19:807-815

Jori G, Galiazzo G, Scoffone E (1969) Photodynamic action ofporphyrins on amino acids and proteins. 1. Selective photooxidation ofmethionine in aqueous solution. Biochem 8: 2868-2875

Kotelevtsev SV, Stvolinskyi SL, Beim AM (1986a) Ecotoxicological Analysis on Base of the Biological Membrane. Moscow State University. Moscow, pp 120 (in Russian)

Kotelevtsev SV, Kozlov YP, Stepanova LI (1986b) Ecotoxicological biomonitoring by the methods of physico-chemical biology Biol Sci 1:19-30 (in Russian)

Kotelevtsev SV, Stepanova LI, Ponomariova LV, Buevitch GV, Limarenko IM, Beim AM, Glazer VM, Kozlov YP (1987a) Polycyclic aromatic hydrocarbon induction of the monooxygenase activity in fish tissues to use for biomonitoring of the pollution. Exper Oncology 9:46-49 (in Russian)

Kotelevtsev SV, Stepanova LI, Limarenco IM, Kozlov YP, 1987b. Different metabolic pathways of polycyclic xenobiotics in fish tissues. In: Lyakhovich VV, Tsyrlov IB (eds) Cytochrome P-450 and environmental health. CO AMH. Novosibirsk, pp 21-22

Kotelevtsev SV, Hanninen O (1992) Biotechnology in analysis water ecosystems: Hydrobionts tissue isoforms of cytochrome P-450. In: Archakov AI, Bachmanova GV (eds) Cytochrome P-450: Biochemistry and Biophysics. Proc 7th Int Conf INCO-TNC. Joint Stock Company. Moscow, pp 475-480

Krone CA, Iwaoka WT (1983) Differences in observed mutagenicity associated with the extraction of mutagens from cooked fish. J Agr Food Chem 31:428-431

Kurelec B, Protic M, Britvic S, Kesic N, Rijavec M, Zahn RK (1981) Toxic effects in fish and the mutagenic capacity of water from the Sava river. Bull Environ Contam Toxicol 26:179-187

Loper JC, Tabor MV, Distlerath L(1981) A new microsomal activation dependent mutagen isolated from old residues of drinking water. Environ Mutagens 3:306-310

Lyakhovich VV, Tsyrlov IB (1981) The induction of the metabolic xenobiotics enzyme. Nauka. Novosibirsk, pp 242 (in Russian)

Mackey BE, McGregor JT (1979) The micronucleus test: Statistical design and analysis. Mutation Res 64:195-204

Maron DM, Ames BN (1983) Revised methods for the Salmonella mutagenicity test. Mutation Res 113:173-215

Marsh S, Chipman JK, Livingstone D (1991) Activation of carcinogens to mutagenic products by the mussel, Mytilus edulis. Mutat Res Environ Mutagens and Related Subj 252:191

Matsui S (1980) Evaluation of a Bacillus subtilis rec-assay for the detectionof mutagens which may occur in water environments. Wat Res 14:1613-1619

McCarthy JF, Jacobson DN, Shugart LR, Jimenez BD (1989) Preexposure to 3-methylcholanthrene increases benzo(a)pyrene adducts on DNA of bluegill sunfish. Mar Environ Res 28:323-32

Moore MS, Smolowitz R, Stegeman JJ (1989) Cellular alterna-
tions preceding neoplasia in *Pseudopleuronectes americanus*
from Boston Harbor. Mar Environ Res 28:425-429

Odum EP (1954) Fundamentals of ecology. Saunders. Philadelphia, pp 384

Osborne LL, Davies RW, Dixon KR, Moore RL (1982) Mutagenic
activity of fish and sediments in the sheep river Alberta.
Wat Res 16:899-902

Parkinson JA (1974) Organic pesticides. In: Allen SF (ed)
Chemical analysis of ecological materials. Wiley. New York,
pp 332-356

Peakal DB (1992) Studies on genetic material. In: Peakall DB,
Depledge MH, Sanders B (eds) Animal biomarkers as pollution
indicators. Chapman and Hall. London, pp 290

Perry JM, Tweats DJ, Al-Mossawi MAJ (1976) Monitoring of
marine environment for mutagens. Nature 264:538-540

Phiodorov VD, Gilmanov TG (1980) Ecology. Moscow State University. Moscow, Leninskie Gori, pp 464 (in Russian)

Poroschenko GG, Abilev SK (1988) Anthropogenic mutagens and
natural antimutagens, In: Shevchenko VA (ed) Sum of science
and technique. Genetic. vol 12. VINT. Moscow, pp 207 (in
Rassian)

Scarpato R, Cegnetti AG, DiMarino F, Migliore L (1989) Mutagenic monitoring of marine environments. Mutat Res Environ
Mutatgenes and Related Subj 216:315

Schmid W (1975) The micronucleus test. Mutation Res 31:9-15

Smith IR (1990) Erythrocytic micronuclei in wild fish from
lakes Superior and Ontario that have pollution-associated
neoplasia. J Great Lakes Res 16:139-142

Smith JW (1982) Mutagenicity of extracts from agricultural
soil in *Salmonella*/microsome test. Environ Mutagens 4:369-
370

Stein JE, Reichert WL, Varanasi U (1989) Covalent binding of
environmental contaminants to hepatic DNA of marine
flatfishes: laboratory and field studies with English sole
and winter flounder. Mar Environ Res 28:345-346

Van der Gaag MA, Gauthier L, Noordsij A, Levi Y, Wrisberg MN
(1990) Methods to measure genotoxins in waste water:
evaluation with *in vivo* and *in vitro* tests. In: Waters MD
et al.(eds) Genetic toxicology of complex mixtures. Plenum
Press. New York, pp 219-236

Varanasi U, Reichert WL, Le Eberhart BT, Stein JE (1989)
Formation and persistence of benzo(a)pyrene diolepoxide-DNA
adducts in liver of English sole *(Parophrus wentulus)*. Chem
Biol Interact 69:203-216

Relating Biochemical Responses to "Higher Order" Effects

Richard F. Addison
Department of Fisheries and Oceans
Ocean Chemistry Division, Institute of Ocean Sciences,
P.O. Box 6000 SIDNEY BC Canada V8L 4B2

Introduction

In my previous paper, I described how measurements of monooxygenase induction, usually in fish, are being used increasingly to complement chemical analyses of the presence of contaminants. "Monooxygenase induction" in this context refers to induction of CYP 1A1 (or its analogue), usually in liver, and usually by a limited range of organic environmental contaminants. Usually the monooxygenase measurements are undertaken to answer the question: "Does the presence of a chemical in the environment or in the organism have any measurable effect on the organism?" It has been suggested that the simple observation of <u>any</u> biological change resulting from the presence of a contaminant should be sufficient reason to implement appropriate regulations (Gray, 1992). However, the demonstration that subtle biochemical or enzymatic changes occur in biota exposed to sub-lethal levels of contamination often elicits the response "so what?". This could be phrased more elegantly as a question as to whether hepatic mono-oxygenase induction in an individual has any effect at the whole organism, the population or the community level. That is (as may be expected) a difficult question to answer. In this paper, I will approach it in the specific context of water pollution and its effects on fish, and I will discuss four aspects of the question: the contribution of monooxygenase induction to reducing body burdens of unwanted chemicals; the increase in pre-carcinogen production arising from induced

NATO ASI Series, Vol. H 90
Molecular Aspects of Oxidative Drug Metabolizing Enzymes
Edited by E. Arınç, J. B. Schenkman and E. Hodgson
© Springer-Verlag Berlin Heidelberg 1995

monooxygenase activity; the relevance of monooxygenase induction to steroid hormone metabolism, and lastly, correlations between monooxygenase induction in individual fish and whole organism or community responses in which no causal chain between the responses should be inferred.

1. Does hepatic monooxygenase induction reduce body burdens of contaminants in individual fish?

Although monooxygenase induction is usually assayed by measuring the catalytic activity of enzymes in vitro, a more relevant measure to the health of the organism or to the ecosystem is the capacity of an organism to degrade and/or excrete the contaminants it has accumulated, i.e., its metabolic capacity in vivo. So far, relatively few studies have addressed this question, perhaps because the approach is technically fairly demanding: monooxygenase activity must first be induced, and then a study of foreign compound metabolism and/or clearance must be undertaken in both induced and control organisms. Those studies which have been carried out show that monooxygenase induction indeed contributes to reducing body burdens of contaminants; some of the evidence is discussed below.

Treatment of trout with piperonyl butoxide --- a mono-oxygenase inhibitor --- changed the tissue disposition, and reduced the biliary excretion of di-2-ethylhexyl phthalate (DEHP) (Melancon et al., 1977). Furthermore, hydrolysis to mono-ethylhexyl phthalate in vivo was reduced. Although this work did not strictly address the question of the role of monooxygenase induction, the data are consistent with mono-oxygenases playing a role in vivo in the metabolism of DEHP. Statham et al. (1978) showed that pre-treatment of trout with the CYP 1A1 inducer 2,3-benzanthracene increased by about 4-fold the biliary excretion of a subsequent dose of methylnaphthalene; concentrations of methylnaphthalene in other tissues were lower in induced than in control fish, and the concentrations of polar methylnaphthalene metabolites in bile of treated fish was higher. Later, Melancon and Lech (1979) found similar results using various other species of fish treated with the CYP 1A1 inducer ß-naphthoflavone (ß-NF). In a similar study of coho salmon, Collier et al. (1985) found that pre-treatment of fish with Aroclor 1254, a polychlorinated biphenyl (PCB) mixture and an inducer of CYP 1A1 and other P450s in mammals, also changed the tissue distribution in vivo and the excretion of a dose of 2,6-dimethylnaphthalene. Finally, the benthic flatfish English sole (Parophrys vetulus) exposed to PCBs had enhanced biliary excretion of benzo[a]pyrene metabolites (Stein et al., 1984). (In this last study, however, sole exposed to benzo[a]pyrene, itself a CYP 1A1 inducer, did not show any very clear enhancement of PCB degradation or clearance, which suggests that CYP 1A1 induction may not play a major role in this process.)

Taken together, these data provide convincing evidence that induction of the CYP 1A1 system may contribute to reducing total body burdens of unwanted contaminants, particularly polynuclear aromatic hydrocarbons (PAH). In other words, and addressing the "so what" question, monooxygenase induction may represent a real advantage to the organism by enhancing its ability to degrade and eventually excrete certain contaminants. However, as Okey (1990) has pointed out, the enhanced metabolism of PAH (in particular) may not be completely beneficial, as some of the metabolic intermediates may lead eventually to mutagenesis or carcinogenesis (discussed further below). Nevertheless, many of the compounds which are effective inducers of the CYP 1A1 system also induce the phase II (conjugation) enzyme systems (at least in mammals) and this latter process may contribute to increased rates of clearance of some metabolites.

2.   **Does monooxygenase induction lead to DNA damage and (ultimately) to carcinogenesis?**

As noted above, CYP 1A1 induction enhances foreign compound

metabolism not only in vitro but also in vivo. In the case of PAH degradation, hydroxylated products are formed, which may be further transformed to metabolites which can form DNA adducts. Most of the experimental and field studies in this area (relating to marine pollution and its effects on fish) have emerged from studies in Puget Sound, parts of which are heavily contaminated with PAH. This work has been the subject of some recent reviews (Varanasi et al., 1987; Stein et al., 1990).

In the particular case of the metabolism of benzo[a]pyrene (B[a]P) by fish, the putative sequence of events is as follows: B[a]P is metabolised (by CYP 1A1) to an arene oxide (B[a]P -7,8-epoxide); this is probably the rate limiting step in the chain of reactions. The epoxide is hydrolysed to B[a]P-7,8-diol which is a major (or the major) B[a]P metabolite in fish (Varanasi et al., 1986; Stegeman and Lech, 1991). This diol is epoxidised further to B[a]P-7,8-diol-9,10-epoxide (B[a]PDE), which is considered to be the ultimate carcinogen. This compound (or one isomer of it) can bind covalently to DNA. Formation of a covalent adduct is considered to be a key step in the action of most carcinogens (Dunn, 1991). However, it is worth emphasising that at the arene oxide, 7,8-diol and diol-epoxide stages, the opportunity exists for detoxification by conjugation with glutathione or via glucuronide or sulphate conjugates (Varanasi et al., 1987).

There is strong evidence, particularly from laboratory and field studies in Puget Sound, that PAH accumulated by fish exposed to contaminated sediments indeed follow the metabolic pathways outlined above. For example, fish exposed to PAH in sediments produced much more fluorescent aromatic compounds in bile (indicating the formation of hydroxylated or other polar metabolites) than those produced by fish from uncontaminated environments; fish experimentally exposed to B[a]P also contained DNA adducts chromatographically identical to those formed synthetically with B[a[PDE; finally, fish from contaminated field sites apparently contained these or closely-related DNA adducts (Stein et al., 1990). The evidence which most convincingly relates environmental contamination to neoplasm development comes from the work of Schiewe et al. (1991): solvent extracts from contaminated sediments injected at monthly intervals into sole from a clean site produced (after eighteen months) a range of histopathological lesions including pre-neoplastic foci. Control animals, receiving solvent injections or no injections at all, did not show such lesions. There is therefore convincing evidence that exposure to environmental PAH can lead to neoplasms, even if the detailed sequence of events between the demonstration of the presence of DNA-adducts and neoplastic formation are not yet clear. However, the preliminary steps, involving conversion of the "parent" hydrocarbon to a metabolite capable of forming DNA-adducts, has been well worked out --- and involves at one stage, monooxygenase enzymes.

In summary, then, induction of hepatic monooxygenase activity may imply that an organism could have increased susceptibility to carcinogenesis.

3. Does monooxygenase induction by contaminants affect steroid hormone metabolism or distribution, and hence affect reproduction?

There is now a fairly extensive corpus of literature dating from the late 1970's which describes the relationship between sex and maturity in fish and the sensitivity of the fish hepatic monooxygenase system to induction by contaminants. Most of this work has been undertaken with the aim of defining how sex or maturity affects basal or induced monooxygenase activity, so that they can be eliminated as "nuisance variables" in interpreting monooxygenase measurements. It is now clear that age and sex affect both "basal" monooxygenase activity (Stegeman and Chevion, 1980) and the extent to which monooxygenase systems are inducible (e.g., Förlin, 1980). This effect is probably mediated to some extent through 17ß-estradiol (Vodicnik and Lech, 1983; Förlin et al., 1984); testosterone appears to have a less significant role. These laboratory demonstrations of the depression of monooxygenase activity by hormones, particularly estradiol, are consistent with seasonal changes in "basal" monooxygenase activity from uncontaminated (and therefore putatively un-induced) populations of fish, which declines around spawning time (e.g., Edwards et al., 1988; Spies et al., 1988a).

There is therefore good evidence that an interaction exists between hormonal activity and monooxygenase activity, particularly CYP 1A1 as indicated by EROD and/or AHH activity. But is the reverse true? --- that is, does MFO induction bring about changes in normal hormone distribution? This is less easy to answer. Some reports in the earlier literature inferred a direct effect, on the grounds that monooxygenases were involved in steroid hormone metabolism. However, we now know that the P450 systems involved in steroid hormone metabolism (i) differ from those involved in (e.g., PAH metabolism) and (ii) are found in different tissues (e.g., Waterman et al., 1986). On the other hand, CYP 1A2 (which is inducible by ß-NF) is known to catalyse the hydroxylation of estradiol (e.g., Graham et al., 1988). Taking all these observations together, it seems possible that CYP 1A1 induction could affect steroid hormone metabolism indirectly, through mechanisms which remain to be clarified. However, it is a little too simple to suggest that an alteration in hepatic CYP 1A1 concentrations (or in its catalytic activity) would directly affect hormone metabolism or distribution in tissues other than liver, or would affect the circulatory levels of hormones other than estradiol.

Having said that, some laboratory studies show that hormonal metabolism is disrupted by CYP 1A1 inducers. Thus, administration of PCBs or 3-methyl-cholanthrene affected hormonal metabolism in trout directly (Hansson et al., 1980). Administration of the CYP 1A1 inducer ß-NF to rainbow trout approximately doubled the rate of biliary excretion of 17ß-estradiol (Förlin and Haux, 1985). These observations are supported by field studies which show that high concentrations of contaminants either in the environment or in fish tissues are associated both with monooxygenase (CYP 1A1) induction and with reduced reproductive success (e.g., Spies et al., 1988b). More recent studies have shown connections between contaminant exposure, CYP 1A1 induction and circulatory steroid levels: post-spawning white suckers (<u>Catostomus commersoni</u>) exposed to bleached kraft mill effluent (BKME) had significantly reduced testosterone (but not estradiol) concentrations and reduced fecundity (Munkittrick et al., 1991). In a second sampling a year later a more complex pattern emerged: pre- and post-spawning males and females usually showed some reduction in testosterone following exposure to BKME but BKME appeared to have no effect during spawning. Estradiol was reduced by BKME in females only in one post-spawning sample. Both 11-keto testosterone and 17α,20β-dihydroxyprogesterone showed complex and variable responses to BKME (McMaster et al., 1991). Furthermore, Stein et al., (1991) have found that English sole treated with a CYP 1A1 inducing sediment extract have reduced plasma 17β-estradiol, and changed <u>in vitro</u> steroid metabolising activities, but a distribution of estradiol metabolites <u>in vivo</u> no different from controls.

The connection between contaminant exposure, CYP 1A1 induction, disruption of steroid metabolism and reproductive change is therefore far from simple. The best that can be said at present is that CYP 1A1 induction, disruption of steroid hormone metabolism and reproductive disorder are all related in some way to contaminant exposure, but whether the steroid hormone and reproductive effects are mediated partly or completely through CYP 1A1 induction is certainly not clear yet.

4.   Correlations between monooxygenase induction and whole-organism or community responses

Monooxygenase induction has been used extensively to indicate the presence (and effects) of environmental organic contaminants. In view of its apparent success, the induction response has been compared to other indices of environmental effect, particularly during a series of collaborative workshops organised by the Intergovernmental Oceanographic Commission's Group of Experts on the Effects of Pollutants (IOC/GEEP). These workshops have usually involved the application of various "biological effects monitoring" techniques to a marine site known from chemical analysis to contain a gradient of contamination which is usually sediment-associated. The object of the workshops has been to demonstrate the suitability of approaches to assessing the significance of pollutants.

The first of the IOC/GEEP Workshops was held in Oslo in 1986 (Bayne et al., 1988) and involved, among other studies, an examination of an industrialised fjord (Langesundfjord) southwest of Oslo. The sediment of the fjord were contaminated with a range of materials including PAH, industrial chlorinated hydrocarbons and metals, and concentrations of these in sediments and biota declined from the head of the fjord towards its mouth. European flounder (Platichthys flesus) caught at sites ascending this contaminant gradient showed increasing hepatic mono-oxygenase (EROD) activity (Addison and Edwards, 1988) and also increasing amounts of P450E, the analogue of CYP 1A1 (Stegeman et al., 1988) --- in other words, evidence of monooxygenase induction. Several other biological effects were also measured in organisms sampled along this contaminant gradient, including "Scope for Growth" (SFG) in the mussel, Mytilus edulis. SFG is a measure of energy partitioning, and in the mussel it has been shown repeatedly to be sensitive to various kinds of environmental stress, including contamination (Widdows, 1985). SFG is therefore a whole organism sub-lethal bioassay of environmental quality. In Langesundfjord, SFG steadily declined in mussels sampled at points ascending the contaminant gradient, paralleling exactly monooxygenase induction in the flounder (Widdows and Johnson, 1988).

There is, of course, no mechanistic connection between the biochemical response of monooxygenase induction in flounder liver with the whole-organism physiological response of reduced SFG in the mussel. However, both biological effects appear to respond to the same cause, the declining environmental quality associated with the increasing contaminant gradient in sediments. The correlation between the biochemical and whole-organism responses suggest that monooxygenase induction may be an early warning of some unconnected, but more general, response.

A second IOC/GEEP Workshop held in Bermuda (Addison and Clarke, 1990) led to similar conclusions. A slight gradient of varied contaminants, including PAH in sediments and tributyl tin (a component of marine anti-fouling paints) in surface waters resulted in some spatial variation in hepatic mono-oxygenase induction in several species of coral reef fish (Stegeman et al., 1990); this was manifested as changes in EROD activity and also in the concentrations of P450E (the analogue of CYP 1A1). Measurements of SFG in a local bivalve mollusc, Arca zebra, were usually inversely correlated with those of monooxygenase induction in the fish (Widdows et al., 1990), as had been the case in Langesundfjord. Finally, similar trends were found in benthic macrofaunal community structure, as indicated by species abundance and diversity measurements at the same sampling sites (Warwick et al., 1990). In other words, the biochemical response in fish, the whole-organism physiological response in the mollusc, and the structure of the benthic macrofaunal community all appeared to respond to trends in environmental quality. There is (obviously) no cause-effect relationship linking these responses, but all of them appear to reflect the same general trend in environmental quality as measured by chemical analysis. This correlation does not "prove" that monooxygenase induction may be an indication of more complex environmental effects, but it provides some support for monooxygenase measurements as an early warning of more serious impacts.

Conclusions

There is now good evidence that monooxygenase induction (particularly of CYP 1A1 by some organic contaminants) is not just a biochemical curiosity, but that it has some important implications for the organism.   First, it may play a significant role in the degradation and eventual clearance of contaminants --- particularly PAH --- from the organism.   This may not be altogether beneficial, however, as some of the intermediates in the degradation process may be metabolised to derivatives which may bind covalently to DNA; this, in turn may lead to carcinogenesis.   However, although both these general processes (clearance and excretion, and DNA-adduct formation) can occur, it is not clear yet whether both always do occur, or whether other factors can affect their relative importance.

The relationship between monooxygenase induction and steroid hormone metabolism (and eventual interference with reproduction) is equally unclear. There seems no doubt that a complex relationship exists, and there are theoretical reasons for believing that some steroid derivatives might be directly affected by monooxygenase induction by xenobiotics; there is also field evidence which shows that reproduction is affected by contamination. Whether this is mediated through monooxygenase induction is not clear. The details of this connection remain to be worked out, and with them some understanding of the significance of monooxygenase induction by xenobiotics to the whole reproductive process.

Finally, there is good evidence, even if based on correlations between responses, rather than on direct experimental evidence, that monooxygenase induction in fish by xenobiotics may be a useful early warning of effects at the whole organism or community level. The present approach used in some international monitoring programmes of measuring monooxygenase induction to complement chemical analyses therefore seems reasonable, provided it is recognised that monooxygenase induction does not indicate general environmental quality, but instead, the probable impact of a rather limited suite of organic compounds which interact with the Ah receptor.

References

Bayne BL, Clarke KR, Gray JS (Eds.) (1988) Biological effects of pollutants: results of a practical workshop. Mar Ecol Prog Ser 46: nos. 1 - 3

Collier TK, Gruger EH Jr., Varanasi U (1985) Effect of Aroclor 1254 on the biological fate of 2,6-dimethylnaphthalene in Coho salmon (Oncorhynchus kisutch). Bull Env Contam Toxicol 34: 114 - 120

Dunn BP (1991) Carcinogen adducts as an indicator for the public health risks of consuming carcinogen-exposed fish and shellfish. Env Health Pers 90: 111 - 116

Edwards AJ, Addison RF, Willis DE, Renton KW (1988) Seasonal variation of hepatic mixed function oxidases in winter flounder (Pseudopleuronectes americanus). Mar Env Res 26: 299 - 309

Förlin L (1980) Effects of Clophen A50, 3-methyl-cholanthrene, pregnenolone-16α-carbonitrile, and phenobarbital on the hepatic microsomal cytochrome P-450-dependent monooxygenase system in rainbow trout, Salmo gairdneri, of different age and sex. Toxicol Appl Pharmacol 54:420 - 430

Förlin L, Haux C (1985) Increased excretion in the bile of 17ß-[^{3}H]estradiol-derived radioactivity in rainbow trout treated with ß-naphthoflavone. Aquat Toxicol 6: 197 - 208

Förlin L, Andersson T, Koivusaari U, Hansson T (1984) Influence of biological and environmental factors on hepatic steroid and xenobiotic metabolism in fish: interaction with PCB and ß-naphthoflavone. Mar Env Res 14: 47 - 58

Graham MJ, Lucier GW, Linko P, Maronpot R, Goldstein JA (1988) Increases in cytochrome P-450 mediated 17ß-estradiol 2-hydroxylase activity in rat liver microsomes after both acute administration and subchronic administration of 2,3,7,8-tetrachlorodibenzo-p-dioxin in a two-stage hepatocarcinogenesis model. Carcinogenesis 9: 1935 - 1941

Gray JS (1992) Biological and ecological effects of marine pollutants and their detection. Mar Poll Bull 25: 48 - 50

Hansson T, Rafter J, Gustafsson J-A (1980) Effects of some common inducers on the hepatic microsomal metabolism of androstenedione in rainbow trout with special reference to cytochrome P-450-dependent enzymes. Biochem Pharmacol. 29: 583 - 587

Kleinow KM, Melancon MJ, Lech JJ (1987) Biotransformations and induction: implications for toxicity, bioaccumulation and monitoring of environmental xenobiotics in fish. Env Health Pers 71: 105 - 119

McMaster ME, van der Kraak GJ, Portt CB, Munkittrick KR, Sibley PK, Smith IR, Dixon DG (1991) Changes in hepatic mixed- function oxygenase (MFO) activity, plasma steroid levels and age at maturity of a white sucker (Catostomus commersoni) population exposed to bleached kraft pulp mill effluent. Aquat Toxicol 21: 199 - 218

Melancon MJ, Lech JJ (1979) Uptake, biotransformation, disposition, and elimination of 2-methylnaphthalene and naphthalene in several fish species. In: Aquatic Toxicology, Eds. L.L. Marking and R.A. Kimerle, ASTM STP 667, pp 5 - 22

Melancon MJ, Saybolt J, Lech JJ (1977) Effect of piperonyl butoxide on disposition of di-2-ethylhexyl phthalate by rainbow trout. Xenobiotica 7: 633 - 640

Munkittrick KR, Portt CB, van der Kraak GJ, Smith IR and Rokosh DA (1991) Impact of bleached kraft mill effluent on population characteristics, liver MFO activity, and serum steroid levels of a Lake Superior white sucker (Catostomus commersoni) population. Can. J. Fish. Aquat. Sci. 48: 1371 - 1380

Okey AB (1990) Enzyme induction in the cytochrome P-450 system. Pharmacol Ther 45: 241 - 298

Schiewe MH, Weber DW, Myers MS, Jacques FJ, Reichert WL, Krone CA, Malins DC, MacCain BB, Chan S-L, Varanasi U (1991) Induction of foci of cellular alteration and other hepatic lesions in English sole (Parophrys vetulus) exposed to an extract of an urban marine sediment. Can. J. Fish. Aquat. Sci. 48: 1750 - 1760

Spies RB, Rice DW Jr, Felton J (1988a) Effects of organic contaminants on reproduction of the starry flounder Platichthys stellatus in San Francisco Bay. 1. Hepatic contamination and mixed-function oxidase (MFO) activity during the reproductive season. Mar Biol 98: 181 - 189

Spies RB, Rice DW Jr, Felton J (1988b) Effects of organic contaminants on reproduction of the starry flounder Platichthys stellatus in San Francisco Bay. 2. Reproductive success of fish captured in San Francisco Bay and spawned in the laboratory. Mar Biol 98: 191 - 200

Statham CN, Elcombe CR, Szyjka SP, Lech JJ (1978) Effect of polycyclic aromatic hydrocarbons on hepatic microsomal enzymes and disposition of methylnaphthalene in rainbow trout in vivo. Xenobiotica 8: 65 - 71

Stegeman JJ, Chevion M (1980) Sex differences in cytochrome P-450 and mixed-function oxygenase activity in gonadally mature trout. Biochem Pharmacol 29: 553 - 558

Stegeman JJ, Lech JJ (1991) Cytochrome P-450 monooxygenase systems in aquatic species: carcinogen metabolism and biomarkers for carcinogen and pollutant exposure. Env Health Persp 90: 101 - 109

Stegeman JJ, Renton KW, Woodin BR, Zhang Y-S, Addison RF (1990) Experimental and environmental induction of cytochrome P-450E in fish from Bermuda waters. J Exp Mar Biol Ecol 138: 49 - 67

Stegeman JJ, Woodin BR, Goksøyr A (1988) Apparent cytochrome P-450 induction as an indication of exposure to environmental chemicals in the flounder Platichthys flesus. Mar Ecol Prog Ser 46: 55 - 60

Stein J.E., Hom T. and Varanasi U. (1984). Simultaneous exposure of English sole (Parophrys vetulus) to sediment-associated xenobiotics: Part 1 --- uptake and disposition of ^{14}C-polychlorinated biphenyls and ^{3}H-benzo(a)pyrene. Mar. Env. Res. 13: 97 - 119

Stein J.E., Hom T., Sanborn H.R. and Varanasi U. (1991). Effects of exposure to a contaminated-sediment extract on the metabolism and disposition of 17ß-estradiol in English sole (Parophrys vetulus). Comp. Biochem. Physiol., 99C: 231 - 240

Stein JE, Reichert WL, Nishimoto M, Varanasi U (1990) Overview of studies on liver carcinogenesis in English sole from Puget Sound; Evidence for a xenobiotic chemical etiology II: Biochemical studies. Sci Tot Environ 94: 51 - 69

Varanasi U, Nishimoto M, Reichert WL, Eberhart B-T L (1986) Comparative metabolism of benzo(a)pyrene and covalent binding to hepatic DNA in English sole, Starry flounder and rat. Cancer Res 46: 3817 - 3824

Varanasi U, Stein JE, Nishimoto M, Reichert WL, Collier TK (1987) Chemical carcinogenesis in feral fish: Uptake, activation, and detoxication of organic xenobiotics. Environ Health Pers 71: 155 - 170

Vodicnik MJ, Lech JJ (1983) The effect of sex steroids and pregnenolone-16α-carbonitrile on the hepatic microsomal monooxygenase system of rainbow trout (Salmo gairdneri). J. Steroid Biochem. 18: 323 - 328

Warwick RM, Platt HM, Clarke KR, Agard J, Gobin J (1990) Analysis of macrobenthic and meiobenthic community structure in relation to pollution and disturbance in Hamilton Harbour, Bermuda. J Exp Mar Biol Ecol 138: 119 - 142

Waterman MR, John ME, Simpson ER (1986) Regulation of synthesis and activity of cytochrome P-450 enzymes in physiological pathways. In: Cytochrome P-450 structure, mechanism and biochemistry, Ed. P.R. Ortiz de Montellano, pp 345 - 386, Plenum, NY

Widdows J (1985) Physiological procedures. In: The effects of stress and pollution in marine animals, Eds. B.L. Bayne, D.A.Brown, K. Burns, D.R. Dixon, A. Ivanovici, D.R. Livingstone, D.M. Lowe, A.R.D. Stebbing and J. Widdows. Praeger, N.Y., pp 161 - 178

Widdows J, Johnson D (1988) Physiological energetics of Mytilus edulis: Scope for growth. Mar Ecol Prog Ser 46: 113 - 121

Widdows J, Burns KA, Menon NR, Page DS, Soria S (1990) Measurement of physiological energetic (scope for growth) and chemical contaminants in mussels (Arca zebra) transplanted along a contamination gradient in Bermuda. J Exp Mar Biol Ecol 138: 99 - 117

List of Invited Contributors

Richard F. Addison, Ocean Chemistry Division, Department of Fisheries and Oceans, P. O. Box 6000, Sidney, B. C.., V8L 4B2, Canada

Diana Anderson, Department of Genetic Toxicology, British Ind. Biol. Res. Assoc., BIBRA, Carshalton, Surrey, SM5 4DS, England

Emel Arýnç, Joint Graduate Program in Biochemistry, Department of Biology, Middle East Technical University, 06531 Ankara, Turkey

Donald R. Buhler, Toxicology Program, Agriculture and Life Sciences, Oregon State University, Corvallis, OR 97331-7307, U. S. A.

Johannes Doehmer, Molecular Toxicology, Institute of Toxicology and Environmental Health, Munich Technical University, D-8000 Munich, Germany

Yoshiaki Fujii-Kuriyama, Department of Chemistry, Faculty of Science, Tohoku University, Aoba-Ku, Sendai, 980, Japan

Ernest Hodgson, Department of Toxicology, North Carolina State University, P. O. Box 7633, Raleigh, NC, 27695, U. S. A.

Eric F. Johnson, Department of Basic and Clinical Research, Division of Biochemistry, The Scripps Research Institute, LaJolla, California, 92037, U. S. A.

Wolfgang Klinger, Institute of Pharmacology and Toxicology, Faculty of Medicine, Friedrich Schiller University, D-07743, Jena, Germany

Sergey V. Kotelevtsev, Biology Department, Moscow State University, Moscow, 119899, Russia

David Kupfer, Worcester Foundation for Experimental Biology, Shrewsbury, MA, 01545, U. S. A.

Anthony Y. H. Lu, Merck Research Laboratories, P. O. Box 2000-RY80A2, Rahway, NJ, 07065, U. S. A.

Franz Oesch, Institute of Toxicology, University of Mainz, Obere Zahlbacher Strasse 67, D-6500, Mainz, Germany

Richard M. Philpot, Laboratory of Cellular and Molecular Pharmacology, NIEHS/NIH, P. O. Box 12233, Research Triangle Park, NC, 27709, U. S. A.

John B. Schenkman, Department of Pharmacology, University of Connecticut, Health Center, Farmington, CT, 06032, U. S. A.

John J. Stegeman, Biology Department, Woods Hole Oceanographic Institute, Woods Hole, MA, 02543, U. S. A.

Chung S. Yang, Laboratory for Cancer Research, College of Pharmacy, Rutgers University, Piscataway, NJ, 08855, U. S. A.

Index

NATO ASI Series H

NATO ASI Series H

NATO ASI Series H

NATO ASI Series H

NATO ASI Series H

NATO ASI Series H

Springer-Verlag
and the Environment

We at Springer-Verlag firmly believe that an international science publisher has a special obligation to the environment, and our corporate policies consistently reflect this conviction.

We also expect our business partners – paper mills, printers, packaging manufacturers, etc. – to commit themselves to using environmentally friendly materials and production processes.

The paper in this book is made from low- or no-chlorine pulp and is acid free, in conformance with international standards for paper permanency.

DATE DUE

MAY 2 7 1998	